COMPLETE ADVANCED LEVEL

Pure Mathematics

TEACHER RESOURCE FILE

Andy Martin • Kevin Brown • Paul Rigby • Simon Riley

STANLEY THORNES

First published in 2000 by
Stanley Thornes (Publishers) Ltd
Ellenborough House
Wellington Street
CHELTENHAM
GL50 1YW

A catalogue record for this book is available from the British Library.

ISBN 0 7487 3561 5

00 01 02 03 04 / 10 9 8 7 6 5 4 3 2 1

Typeset by The Alden Group, Oxford
Printed in Great Britain by Ashford Colour Press,
Gosport, Hampshire

Contents

This *Complete Advanced Level Mathematics* Teacher Resource File has been specially developed to support fully your teaching of Pure Mathematics. It should be used alongside the core book in the series. The file has been broken down into sections for flexibility and ease of use with your students and according to your needs.

Acknowledgements

We are grateful to the following Unitary Authorities and Examination Boards for permission to reproduce questions from their past examination papers:

- Assessment and Qualifications Alliance (AQA), including Northern Examinations and Assessment Board (NEAB), Associated Examining Board (AEB), Secondary Mathematics Project (SMP) and Joint Matriculation Board (JMB).
- The Edexcel Foundation, including University of London Examinations and Assessment Council (ULEAC).
- Oxford, Cambridge and RSA (OCR), including University of Cambridge Local Examinations Syndicate (UCLES), Oxford & Cambridge Schools Examination Board (OCSEB) and Mathematics in Education and Industry (MEI).
- Northern Ireland Council for the Curriculum, Examinations and Assessment (NICCEA).
- Welsh Joint Education Committee (WJEC)

All answers and worked solutions provided for examination questions are the responsibility of the authors. Every effort has been made to contact copyright holders. The authors and publishers apologise to anyone whose rights have been overlooked and will be happy to rectify any errors or omissions at the earliest opportunity.

CHAPTER OBJECTIVES

1 Algebra I

After completing this chapter, the student should be able to:

- state and apply the **properties of surds**, and rationalise denominators
- use the **properties of indices**
- explain what is meant by the term 'polynomial'
- extract **common factors** from algebraic expressions
- **factorise quadratics**, using the **PAFF method**
- factorise quadratics by **completing the square**
- factorise quadratics involving the **difference of two squares**
- solve quadratic equations by applying the **quadratic formula**
- explain that the **discriminant indicates the shape of a quadratic graph**
- solve **simultaneous equations**, including the case of one **linear** equation and one **quadratic** equation.

2 Coordinate Geometry

After completing this chapter, the student should be able to:

- **calculate the distance between two points** A (x_1, y_1) and B (x_2, y_2) using $AB = \sqrt{(x_2 - x_1)^2 + (y_2 - y_1)^2}$
- extend this formula to $AB = \sqrt{(x_2 - x_1)^2 + (y_2 - y_1)^2 + (z_2 - z_1)^2}$, where A (x_1, y_1, z_1) and B (x_2, y_2, z_2) are two points in 3-dimensional space
- recognise that $y = mx + c$ and $ax + by + c = 0$ are **two forms of the equation of a straight line**
- use the formula $y - y_1 = m(x - x_1)$ to find the **equation of a straight** line of gradient m passing through the point (x_1, y_1)
- use the formula $\dfrac{y - y_1}{y_2 - y_1} = \dfrac{x - x_1}{x_2 - x_1}$ to find the **equation of the straight line** passing through (x_1, y_1) and (x_2, y_2)
- solve **linear and quadratic inequalities**, and use graphs to check results
- write down **the equation of a circle** as either:

  $(x - a)^2 + (y - b)^2 = r^2$ or $x^2 + y^2 - 2ax - 2by + c = 0$
- find, by **completing the square,** the **radius and centre of a circle** from its equation
- use the result $m_1 \times m_2 = -1$ for **perpendicular lines** with gradients m_1 and m_2
- explain how to find the **equations of the tangent and the normal** to a circle at a given point.

3 Trigonometry I

After completing this chapter, the student should be able to:

- use the **trigonometric ratios** sine, cosine and tangent, with **angles of any size**
- state the exact values of the **trigonometric ratios of the special angles, 30°, 45° and 60°** and use them in calculations
- sketch the **graphs of sin θ, cos θ** and **tan θ, correctly stating the period** of each one
- define the **reciprocal trigonometric ratios** as $\operatorname{cosec}\theta = \dfrac{1}{\sin\theta}$, $\sec\theta = \dfrac{1}{\cos\theta}$ and $\cot\theta = \dfrac{1}{\tan\theta}$
- use the **Pythagorean identities** $\sin^2\theta + \cos^2\theta \equiv 1$, $\tan^2\theta + 1 \equiv \sec^2\theta$ and $1 + \cot^2\theta \equiv \operatorname{cosec}^2\theta$ in a variety of problem solving situations
- use the **sine** and **cosine rules** to solve problems in a variety of contexts
- find the **area of a triangle** using three distinct methods
- use **radian measure,** and find **arc lengths** and **sector areas** in a range of problems.

4 Functions

After completing this chapter, the student should be able to:

- distinguish between the different **types of mappings** that exist between two sets
- define the **domain, co-domain** and **range of a function,** and distinguish between the co-domain and the range
- calculate the **ranges of functions** for both **discrete** and **continuous** domains
- explain how to **find inverse functions algebraically**
- sketch the graph of the inverse function $y = f^{-1}(x)$ from the graph of the function $y = f(x)$ by reflecting it in the line $y = x$
- recognise **composite functions**, and derive algebraic expressions for a function of a function
- **translate** the graph of the function $y = f(x)$ to obtain the graphs of $y = f(x) \pm a$, $y = f(x \pm a)$ and $y = af(x)$
- distinguish between **odd, even and periodic functions.**

5 Differentiation I

After completing this chapter, the student should be able to:

- explain how the **gradient of a curve** at a particular point can be approximated by finding the **gradient of a chord** drawn to a nearby point
- **differentiate quadratic and cubic functions** from 'first principles'.
- calculate the gradient at any point on the graph of $y = x^n$, where n is a **rational power** of x
- locate and determine the nature of any **stationary points** on a curve
- distinguish between values of x for which a function is increasing or decreasing
- use differentiation to find the equation of the tangent and normal at any point on a curve
- apply these techniques to **optimisation** and other practical problems
- use the **chain rule** to differentiate composite functions
- **differentiate functions involving multiple angles** and/or rational powers of sine and cosine.

6 Algebra II

After completing this chapter, the student should be able to:

- **divide a polynomial by a linear factor** and check the result using either multiplication or substitution.
- state the **remainder** and **factor theorems,** and use the factor theorem to **factorise cubic expressions**
- add, subtract, multiply and divide **algebraic fractions**
- **classify partial fractions** into one of three categories: linear denominators; quadratic factors in the denominator; repeated factors in the denominator
- use the '**cover-up' rule** to calculate the numerators of partial fractions with linear denominators
- sketch the **graphs of continuous and discontinuous functions,** after checking:
  - where the graph crosses the axes
  - the location and nature of stationary points
  - the behaviour of the function as $x \to \pm\infty$
  - where the function is undefined
- **sketch the graph of the modulus of a function,** $|f(x)|$, by reflecting in the x-axis all parts of the graph for which the function is negative.

7 Exponentials and Logarithms

After completing this chapter, the student should be able to:

- define the word **logarithm** from the equivalent statements $y = a^x$ and $x = \log_a y$
- distinguish between **common** and **natural logarithms**
- state the four **basic properties of logarithms** as:
  - $\log ab = \log a + \log b$
  - $\log a^n = n\log a$
  - $\log\left(\frac{a}{b}\right) = \log a - \log b$
  - $\log 1 = 0$ and $\log_a a = 1$
- use the exponential function e^x to **model growth and decay** in a variety of contexts
- **transform exponential curves** of the form $y = ab^x$ into straight-line graphs by plotting $\log y$ against x
- **transform polynomial curves** of the form $y = ax^b$ into straight-line graphs by plotting $\log y$ against $\log x$.

8 Sequences and Series

After completing this chapter, the student should be able to:

- recognise a **sequence** and know that a **series is the sum of a sequence**
- find the next term of an **iteratively generated sequence**
- express a given series in **sigma notation,** $\sum_{r=1}^{n} f(r)$, and evaluate such a series
- write down the nth term of an **arithmetic progression** (AP) and find the sum of the first n terms of such a series
  - nth term $u_n = a + (n - 1)d$, where a is the first term, and d is the common difference
  - sum of n terms $S_n = \frac{1}{2}n[2a + (n - 1)d]$
  - sum of n terms $S_n = \frac{1}{2}(a + l)$, where l is the last term
- write down the nth term of a **geometric progression** (GP) and find the sum of the first n terms of such a series
  - nth term $u_n = ar^{n-1}$
  - sum of n terms $S_n = \dfrac{a(1 - r^n)}{1 - r}$
  - sum to infinity $S_\infty = \dfrac{a}{1 - r}$ for $|r| < 1$
- write down the **binomial theorem** as $(1 + x)^n = 1 + \binom{n}{1}x + \binom{n}{2}x^2 + \ldots + \binom{n}{r}x^r + \ldots$, and use $(a + x)^n = a^n + na^{n-1}x + \dfrac{n(n-1)}{2!}a^{n-2}x^2 + \ldots + x^n$, where n is a positive integer
- use the **binomial expansion** for negative or fractional values of n:
  $(1 + x)^n = 1 + nx + \dfrac{n(n-1)}{2!}x^2 + \dfrac{n(n-1)(n-2)}{3!}x^3 + \ldots$ provided $-1 < x < 1$
- find **power series expansions** in a variety of contexts and use them to make approximations.

9 Integration I

After completing this chapter, the student should be able to:

- state the **notation** used for integration
- use the formula $\int x^n \, dx = \frac{x^{n+1}}{n+1} + c$ and apply this to the integration of **simple functions**
- check the results of integration by using the reverse process of **differentiation**
- use integration to find the **area between a curve and the x- or y-axis, between two limits**
- appreciate that the area below the x-axis is 'negative'
- use integration to find the **area enclosed between two curves**
- use integration to find the **volume of a solid of revolution** about either the x- or y-axis
- **integrate trigonometric functions**, including $\sin^2\theta$ and $\cos^2\theta$, by using the double angle formulas.

10 Trigonometry II

After completing this chapter, the student should be able to:

- use the **factor formulas** to **convert sums and differences of sines and cosines into products**
- **transform the function $a\sin x + b\cos x$** into one of the following four forms:
  - $R\sin(x + \alpha)$
  - $R\sin(x - \alpha)$
  - $R\cos(x + \alpha)$
  - $R\cos(x - \alpha)$
- solve equations of the type $a \sin x + b \cos x = c$
- state the **general solution to a trigonometric equation.**

11 Differentiation II

After completing this chapter, the student should be able to:

- use the **product and quotient rules** to differentiate suitable functions
- state the **derivatives of the exponential function** $y = e^x$ and **natural logarithmic function** $y = \ln x$ and use the **chain**, **product** and **quotient rules** to differentiate related functions
- state the **derivatives of tan** θ, **cot** θ, **sec** θ and **cosec** θ and the **inverse trigonometric functions** $\sin^{-1}\theta$, $\cos^{-1}\theta$ and $\tan^{-1}\theta$
- use the **second derivative** to distinguish between the **maximum and minimum points on a curve**
- find the location of any **general points of inflexion** on a curve
- **differentiate implicitly** functions that cannot be expressed in the form $y = f(x)$
- find the derivative of **functions defined parametrically**
- write down the **Maclaurin series** of a given function.

12 Integration II

After completing this chapter, the student should be able to:

- use the **fundamental theorem of calculus** to form a list of **standard integrals**
- develop a clear method for **integrating a rational function**
- use **integration by parts** to **integrate more complex functions** that are expressed in the form of a product
- use the method of **substitution** to **transform integrals** into more manageable forms
- use integration to find the **area under a parametrically defined curve.**
- select a correct method to integrate a particular function and have a clear overview of all the different techniques used to integrate different types of functions
- apply the methods of integration to solve **simple differential equations**, including the case where the variables are **separable**
- use differential equations to **model** certain situations.

13 Numerical Methods

After completing this chapter, the student should be able to:

- find an approximate solution to an equation using **graphical methods**, **interval bisection** and **linear interpolation**
- recognise that the **interval** within which a **root** of the equation $f(x) = 0$ is located can be found by looking for a **change in sign** in the value of $f(x)$
- by first rearranging equations into a suitable form of $x = g(x)$, solve them using the **fixed point iteration formula** $x_{n+1} = g(x_n)$.
- use the **Newton–Raphson method** to find a solution to the equation $f(x) = 0$ to a required level of accuracy
- use the **mid-ordinate** rule, the **trapezium rule** and **Simpson's rule** to calculate numerically the **area under a curve.**

14 Vectors

After completing this chapter, the student should be able to:

- use **simple vector algebra** to **manipulate vectors** in 2-dimensional and 3-dimensional space
- use the **scalar product** to find the **angle between two vectors**
- find the **vector equation of a straight line**
- find the **equation of a plane** in various situations
- deal with a range of problems involving **line and planes**, including calculating the **perpendicular distance from a point to a given line or plane.**

15 Proof and Mathematical Argument

After completing this chapter, the student should be able to:

- state that a **mathematical argument** is one in which each step can be deduced from a previous proven result.
- explain that exhausting every possibility is a recognised form of proof known as **proof by exhaustion**
- **disprove a statement** by finding an appropriate counter example and realise that the statement cannot then be proved
- use the **proof by contradiction method** in appropriate situations such as proving the irrationality of $\sqrt{2}$
- explain that some results can be **proved by induction.**

TEACHING NOTES

1 Algebra I

General Notes

Ensure students:

- are totally familiar with the algebraic techniques included in all intermediate GCSE syllabuses; in particular how to **collect like terms**, **expand brackets** and **solve linear equations**
- can **factorise simple algebraic expressions**, including algebraic terms raised to powers as common factors
- are familiar with a **division algorithm** that they can perform accurately without a calculator – using the **numerical long division algorithm** will help with **polynomial division**, which is covered in Chapter 6 (Algebra II).

Specific Notes

1.1 The Number System and Surds

- Make sure students understand that **surds have no sign** and that they are assumed to be positive.
- Make sure students are aware that **rationalising a denominator** means having **no square root sign in the bottom of the fraction**. Students will need practice identifying the **conjugate** before this technique can be completed quickly.
- Very able students find the concept of the **countability of integers and rationals** fascinating. The real numbers being uncountable can introduce discussion on the infinite and infinity. Encourage students not to think of infinity as a single number.

1.2 Indices

- Ensure students learn the **rules of indices** and can apply them in a variety of contexts. These will be needed in work on **exponentials** and **logarithms** in Chapter 7 (Exponentials and Logarithms).

1.3 Polynomials

- It is important that students can spot the **degree or order of polynomials** properly. Make sure this can be done when the terms appear in any order.
- This section introduces new key words and technical phrases. Students need to hear each new term in context a number of times, with a definition, until they become familiar with the term, e.g. the coefficient, the number or multiplier of the unknown quantity x. Then the definition can be dropped.

1.4 Factorisation

- Check that students are familiar with the **three basic techniques of factorisation**. Encourage them always to start by extracting any common factors. The **PAFF method** of factorising quadratics is one that allows weaker students access to more demanding algebra.

1.5 Solving Quadratic Equations

- Encourage students to solve quadratic equations by the four main techniques developed in the text. Remember that a graphical calculator can be used to check their algebraic solutions graphically.
- Make sure students learn the **quadratic formula**. This can often be quoted and used in other parts of Advanced Level mathematics.
- Check that students can **use the discriminant, $b^2 - 4ac$, to get a mental picture of the quadratic graph**. This helps students focus on whether there are no solutions, one repeated solution or two distinct solutions.

1.6 Simultaneous Equations

- Ensure students are confident in solving **simultaneous equations**. Both the elimination and substitution methods should become 'second nature' to students. Students should be encouraged to check their solutions by substituting back into the original equations.

2 Coordinate Geometry

General Notes

This chapter introduces students to a context for many of the algebra skills and techniques from GCSE. Ensure students:

- can cope with the **subscripts** to letters so that (x_1, y_1) and (x_2, y_2) are viewed as distinct coordinates
- are able to use **Pythagoras' theorem** and **expand bracketed terms raised to the power of 2**.

Specific Notes

2.1 Coordinate Geometry

- Check that students learn the **distance formula for two points** in both 2-D and 3-D.
- Students need to know the **definition of gradient**. Make sure all students can quote the conditions for lines to be parallel or perpendicular.
- Use the context of tangents and normals to apply the **condition for two straight lines to be perpendicular to each other:** $m_1 \times m_2 = -1$.

2.2 The Equation of a Straight Line

- Students need to recognise the following equivalent forms for the **equation of a straight line**:

$$y = mx + c$$
$$y = a_1x + a_0$$
$$ax + by + c = 0$$

- Check that students can deduce the **gradient and intercept of a straight line** when given its equation in any form.

2.3 More on the Straight Line

- The results for finding the **equation of a straight line** in various situations should be learnt thoroughly by students. In particular, $y - y_1 = m(x - x_1)$ and $\dfrac{y - y_1}{y_2 - y_1} = \dfrac{x - x_1}{x_2 - x_1}$ should be formulas that students can quote and use accurately.

2.4 Inequalities

- Check that students can **manipulate inequalities properly**. Discourage students from substituting 'equals' into an inequality and then reintroducing the inequality in the final line of work. Use the method of critical values to check the sign of expressions. This can then be used by students to check the location of the graph above or below the x-axis.

2.5 The Equation of a Circle

- Make sure students learn both forms for the general equation of a circle.

3 Trigonometry I

General Notes

The material in this chapter includes the higher tier GCSE syllabus content on trigonometry. Ensure students are:

- familiar with the **trigonometric ratios of acute angles**, measured in degrees
- able to use a scientific calculator efficiently to find ratios and angles
- familiar with **obtuse and reflex angles** in the context of bearings.

Specific Notes

3.1 Trigonometric Functions

- Ensure students know that arcsin, arccos and arctan notation is the same as $\sin^{-1}$, $\cos^{-1}$ and $\tan^{-1}$ notation.
- Encourage students to learn **CAST mnemonics** and the graphs of the trigonometric ratios so that angles other than acute angles can be identified as solutions.
- Ensure students become familiar with the surd form of the ratios of the special angles 30°, 45° and 60°.
- Emphasise that more than one angle can have a specific sine, cosine or tangent value; a calculator only gives the **principal value**.
- Link the **reciprocal ratios** to other application areas. Projectile formulas contain the sec ratio and may be familiar to mechanics/physics students.

3.2 Equations and Identities

- Check that students are clear in the distinction between an **identity** and an **equation**. This is particularly important when students first meet trigonometric proofs.

3.3 Compound and Double Angle Formulas

- **Compound angle formulas** are also known as **sum** and **difference rules**. They have applications in many areas of the mathematics syllabus. Encourage students to learn them.
- **Double angle formulas** appear in many integration topics to simplify integrals involving $\sin^2\theta$ or $\cos^2\theta$.
- Students need to be familiar with all three versions of the expression for $\cos 2\theta$.

3.4 The Cosine Rule and the Sine Rule

- The cosine rule is easier for students to learn if first compared to Pythagoras' theorem. Stress the convention of lower-case letters for sides and capital letters for angles.
- Show students both forms of the sine rule. Make it clear that having the unknown side or angle as the numerator does simplify the algebra.

3.5 The Area of a Triangle

- Students should remember that there are several ways to find the area of a triangle. The symmetry of the sine version and Hero's formula can be emphasised to aid student learning.

3.6 Radian Measure

- Radians can be introduced experimentally by using the **RAD mode** on a calculator. Remind students that multiples of π are not easy to spot when expressed in decimal form on a calculator. How many recognise $\frac{\pi}{2}$ or $\frac{\pi}{3}$?
- The results for **arc length** and **sector area** are frequently tested at Advanced Level. Make it clear to students that they are expected to use radians.

4 Functions

General Notes

Ensure students:

- have competence in algebraic techniques, particularly solving simultaneous equations, changing the subject of an equation and methods for solving quadratic equations, including completing the square
- are able to recognise the graphs of **standard mathematical functions**: **linear**, **quadratic**, **cubic**, **reciprocal** and **trigonometric functions**.

- Discuss the graphs before starting this work and draw students attention to characteristics like **symmetry**, **periodicity**, the **number of turning points** and **asymptotes**. Encourage students to plot graphs of functions as they work, using the calculator, but ensure that, by the end of the chapter, they can sketch graphs when one of the standard mathematical functions is transformed by a **reflection**, **translation** and/or **stretch**.
- Use the application and activities section to reinforce the distinction between the **co-domain and range of a function**.
- It is desirable for students to have access to a graphical calculator.

Specific Notes

4.1 Mappings and Functions

- Encourage students to develop their own examples for **one-to-one**, **many-to-one**, **one-to-many** and **many-to-many** mappings for a context other than football.
- Ensure students understand that a **function** is defined as a mapping under which **each object in the domain maps to only one image in the co-domain**. Students also need to recognise mappings from the graph that represents the mapping and they need to be able to distinguish between **functions and non-functions**. This is particularly important when deciding which functions are one-to-one and therefore will have **inverse functions**.

4.2 Inverse Functions

- Ask students to plot the graphs of $y = f(x)$, its inverse $y = f^{-1}(x)$, and the line $y = x$ on their graphical calculators. This should illustrate the **reflective symmetry that exists between a function and its inverse**. Ensure that students use a 'square view'; that is, one in which the scale is the same on both axes. Most graphical calculators use a display that consists of a rectangular grid of pixels.
- Students need to realise that the line of symmetry $y = x$ passes through the point(s) of intersection of the graphs of $y = f(x)$ and $y = f^{-1}(x)$. The equation $f^{-1}(x) = f(x)$ can at times be more easily solved using $x = f(x)$ or $x = f^{-1}(x)$, especially if the function or its inverse involves a root. **This point is commonly examined.**

4.3 Composite Functions

- Ensure that students are familiar with the different styles of notation for composite functions: $fg(x)$ and $f \circ g(x)$.

4.4 Transformations of Graphs and Functions

- With careful explanation, most students should quickly understand how the graphs of $y = f(x) + a$ and $y = af(x)$ are obtained by, respectively, translating the graph of $y = f(x)$ by $+a$ units parallel to the y-axis, or stretching it by a factor a parallel to the y-axis. However, emphasise that the graph of $y = f(x + a)$ is obtained by translating the graph of $y = f(x)$ by $-a$ units parallel to the x-axis and *not* by $+a$ units, as most students might initially suggest.

4.5 Even, Odd and Periodic Functions

- Students should become familiar with the modulus function and how to sketch the graph of the modulus of a simple function. Encourage students to check their sketches using a graphical calculator. This work is revisited in Chapter 6 (Algebra II).

5 Differentiation I

General Notes

Ensure students:

- are familiar with the **equation of a straight line** and know how to calculate the **gradient of a line segment**
- are competent at manipulating **index notation**, in order to be able to **differentiate roots** and other functions involving **rational powers**
- have experience of solving **simple trigonometric equations**, including those involving double angles, before attempting to locate the maximum and minimum points on the graphs of trigonometric functions.

- It is important that students recognise what is required in a particular problem and interpret different phrases correctly. For example, the phrase 'rate of change of v' would usually refer to changes in v with respect to time t, or $\frac{dv}{dt}$.
- Students should be encouraged to leave any fractions, roots or multiples of π in their workings and answers, and to give a decimal approximation in the final answer only if required, usually to three significant figures.

Specific Notes

5.1 Finding the Gradient of a Curve

- As an introduction to 'differentiation from first principles' in the text, the **gradient of the curve** $y = x^2$ at different points is found by **drawing chords** to nearby points. This work could be extended by carrying out similar calculations for the curves $y = x^3$, $y = x^4$. Many students will incorrectly predict that the corresponding gradient functions will be $2x$, $3x$, $4x$, ...

5.2 Stationary Points

- The nature of any stationary point on a curve is determined by looking at the gradient of the curve on either side of the point. Discussion of the second derivative is left until Chapter 11, although some teachers may wish to introduce both techniques at the same time. For those following modular courses, the syllabus and accompanying guidance notes should help teachers decide which method is expected.
- When finding the **maximum or minimum value of a given function**, $f(x)$, students need to appreciate that it is often not enough just to solve the equation $f'(x) = 0$. The corresponding maximum or minimum value may have to be calculated and, by considering the gradient on either side, students may have to show that this is either a maximum or minimum value, as required.
- When solving **optimisation problems**, discuss the limitations of the particular mathematical model and any assumptions upon which it is based. It is particularly important that students are given optimisation problems where the function is dependent on two variables, one of which can be eliminated using additional information about the model. This type of question often appears in examinations.

5.3 Further Applications of Differentiation

- When finding the equation of a tangent or normal to a curve, students could be encouraged to **check** their calculations by plotting the curve and the line on a graphical calculator. However, they will need to ensure that they have a 'square view' on their display.
- It is important that students appreciate that displacement, velocity and acceleration are **vector quantities** and therefore have both **magnitude** and **direction**. You will need to explain that a **positive direction has to be defined**, and to interpret the meaning of **negative displacements**, **velocities** and **accelerations**.

5.4 The Chain Rule and Related Rates of Change

- Although it is important for students to be able to **differentiate composite functions** or 'functions of functions' by introducing a **third variable** and using the **chain rule**, they should also be encouraged to understand how to differentiate functions such as $y = (2x + 9)^5$ and $y = \sin 3\theta$ 'on sight'.

5.5 Differentiation of Trigonometric Functions

- It is very important that students remember to work in radians when differentiating trigonometric functions. Emphasise how radian measure creates the useful result for sector area used in the inequalities that establish the small angle approximations.

6 Algebra II

General Notes

Ensure students:

- are familiar with the terms numerator, denominator, factors, linear, quadratic, roots and polynomial
- recognise that functions can be continuous or discontinuous
- recognise the graphs of $y = mx + c$, $y = ax^2 + bx + c$ and $y = \frac{1}{x}$ and are familiar with the work on graph transformations from Chapter 4 (Functions).

Specific Notes

6.1 Polynomial Division

- If students struggle with the division algorithm, an alternative is to rearrange the numerator so it can be written in terms of the denominator. This technique is useful for improper algebraic fractions and integration (see Chapter 12, Integration II).
- It is essential that students can substitute a number into a polynomial before they use the remainder and factor theorems.

6.2 Algebraic Fractions

- A revision of the four rules with fractions may be necessary before the work on algebraic fractions is attempted.

6.3 Partial Fractions

- Stress that the 'cover-up' rule can only be used with linear denominators.
- Ensure students learn the different types of partial fractions. Equating coefficients is a technique frequently needed at Advanced Level.

6.4 Curve Sketching

- Students should have plenty of practice at curve sketching and be able to check their work using a graphical calculator.
- Check that students use the four-step approach to curve sketching, especially with rational functions:
  1 find the points of intersection with the axes
  2 locate and determine the nature of any stationary points
  3 examine the behaviour of the function as $x \to \pm\infty$
  4 identify any asymptotic values of x for which the function is undefined, and at which the graph is discontinuous.
- For graphs of modulus functions, make it clear that only the first and second quadrants are needed. The function is always positive, and the graph is therefore above the x-axis.

7 Exponentials and Logarithms

General Notes

Ensure students:

- are familiar with index notation and can apply the rules of indices introduced in Chapter 1 (Algebra I)
- have experience of solving equations and changing the subject of the formula
- can use a graphical calculator; in particular, they should be able to calibrate the axes and recognise graphical translations
- know the definition of an irrational number; this should have been covered on the higher tier GCSE, but will be unfamiliar to those from the intermediate tier.

Specific Notes

7.1 The Exponential Function

- Emphasise the application of exponential functions to growth and decay. You could link to syllabus content in Advanced Level geography, biology, chemistry and economics, for example.
- Ensure students realise that **e is an irrational number**. It is the limit of $\left(1+\frac{1}{n}\right)^n$ and the sum of an infinite series. Both can be used as calculator exercises to establish good rational approximations on a calculator.

7.2 Logarithms

- Check that students learn the **definition of logarithm** and are familiar with both **log and ln notation**. The location of the keys on the calculator needs to be quickly established.

7.3 Laws of Logarithms

- Encourage students to learn a **summary of logarithm properties**. These can be verified regularly using the calculator functions.

7.4 Solving $a^x = b$

- Check that students can sketch both $y = e^x$ and $y = \ln x$. Link these graphs to the work on functions in Chapter 4 (Functions), particularly the reflection in the line $y = x$ of the graphs of a function and its inverse.
- When using reduction to linear form, one method of distinguishing between an exponential or power form is to check if the curve passes through the origin. If it does, assume the power form $y = ax^b$.

8 Sequences and Series

General Notes

Ensure students:

- are familiar with **basic sequences**: odds, evens, triangular numbers, prime numbers, and so on.

- Much of this work has historical references of interest to students. Encourage students to research work of Gauss, Pascal and Newton to provide another context for these techniques.

Specific Notes

8.1 Sequences and Series

- Ensure students are aware of the distinction between a **sequence** and a **series**.

8.2 Arithmetic Progression

- The **formulas for APs** should be learnt, particularly for the nth term and the sum to n terms.

8.3 Geometric Progression

- The **formulas for GPs** should be learnt, particularly for the nth term and the sum to n terms.

8.4 Convergence, Divergence and Oscillation

- The condition $|r| < 1$, or $-1 < r < 1$ for a GP to converge to a sum to infinity is often tested at Advanced Level. Stress that if $|r| > 1$ the GP will diverge.

8.5 The Binomial Theorem and Power Series

- Encourage students to **evaluate binomial coefficients** $\binom{n}{r}$ without a calculator.
- Distinguish between the **binomial theorem** for positive integral values and **fractional** or **negative values**. For the latter, an **infinite power series** is created and often used to approximate calculations.
- Students need to develop familiarity with **sigma notation** in a variety of contexts. This is particularly important for work on **Proof** (Chapter 15, Proof and Mathematical Argument).

9 Integration I

General Notes

Ensure students:

- can **differentiate powers of x**, **polynomials** and **basic trigonometric functions**
- realise that **integration and differentiation are reverse processes** (this fact is called the **fundamental theorem of calculus**, and it can be used as a tool to check that integration or differentiation has been done correctly)
- appreciate that integration can also be seen as a summation process, particularly with regard to calculating areas and volumes of revolution.

Specific Notes

9.1 Indefinite Integration

- Ensure students understand the words **integrand**, **integral**, **definite** and **indefinite**.
- Stress that the operator $\mathrm{d}x$, $\mathrm{d}n$, $\mathrm{d}t$, $\mathrm{d}\theta$, etc. should always appear after the integrand.
- Encourage students to remember the **constant of integration with indefinite integrals**.
- Encourage students to use differentiation to check their integration.

9.2 The Area Under a Graph

- Check that students know that **integration can be considered to be a summation process**. It is used to find the area under a curve; this area is 'negative' if it appears below the x-axis.

9.3 The Area between a Vertical Axis and a Curve

- Make sure that students use the correct form for **area** when calculations are done using the y-axis instead of the x-axis.

9.4 The Area between Two Curves

- Encourage students to check whether $f(x) > g(x)$ before attempting to integrate $f(x) - g(x)$ to calculate the area between two curves.

9.5 Volumes of Revolution

- Check that students use the correct form for volume integrals, in particular when rotations are about the y-axis.

9.6 Integration of Sine and Cosine

- Do not accept **integration of trigonometric functions** with angles measured in degrees. **Radians must be used.**
- When integrating $\sin^2 x$ and $\cos^2 x$ insist on the student showing the correct rearrangement of the formula containing $\cos 2x$. This is frequently tested at Advanced Level.

10 Trigonometry II

General Notes

Ensure students:

- have covered Chapter 3 (Trigonometry I)
- are experienced with **radian measure, including scaling axes for graphs**
- can distinguish clearly between **compounding trigonometric ratios** and **compounding angles**
- fully understand the **compound angle formulas** before commencing this chapter.

- Encourage students to leave angles as multiples of π, where appropriate. **Decimal approximations** are best left to the final numeric answer.

Specific Notes

10.1 The Factor Formulas

- Encourage students to learn the **factor formulas**. (Make sure they are aware that they are also known as the **sum and product formulas**.) These can be remembered in words as well as mathematical symbols.
- Make sure that students follow the **structure of a proof of an identity**. Always begin with one side and deduce the other using known results. This often requires practice at **factorisation with trigonometric terms**.
- When **checking solutions of equations graphically**, ensure that students have an appropriate range for each axis; otherwise solutions disappear from view. This makes it more difficult for students to understand why there are so many acceptable, yet different, values.

10.2 Functions of the Form $f(x) = a\sin x + b\cos x$

- Solving $a\sin x + b\cos x = c$ using $R\sin(x \pm \alpha)$ or $R\cos(x \pm \alpha)$ is a type of question that often appears in examinations. Spend time checking that students **formulate the correct simultaneous equations** before they attempt to solve them. A **graphical calculator** can be used effectively here to check their work by superimposing the graphs.
- Encourage students to think carefully before **maximising** or **minimising rational trigonometric functions with calculus**. Frequently all that is required is the identification of the angles that maximise $R\sin(x \pm \alpha)$ or $R\cos(x \pm \alpha)$.

10.3 General Solutions of Trigonometric Equations

- Check the syllabus carefully before working through the material on general solutions. This topic is not part of the inter-board subject criteria and consequently students may not need to cover this part of the material. Consolidation questions on this material appear at the end of each set of consolidation exercises.

11 Differentiation II

General Notes

Ensure students:

- have become competent in **differentiating polynomial and rational powers of** x, and sine and cosine functions, including those requiring the use of the chain rule
- are able to locate and determine the nature of any **stationary points** and find the **equation of the tangent or normal** to a curve at a given point
- know how to **solve exponential and logarithmic equations** before attempting problems that deal with some of the applications of differentiation for exponential and logarithmic functions, such as the location of stationary points.

- Students should be encouraged to leave any terms involving e or ln a, where a is a positive real number, in their workings and answers, and to give a **decimal approximation** in the final answer only if required.

Specific Notes

11.1 Differentiating Products and Quotients

- Able students should be challenged to establish the quotient rule from first principles using a similar approach to that used to derive the product rule in the textbook.

11.2 Differentiation of Exponentials and Logarithms

- Link the various methods of establishing e at this point. Use the power series derivation from Chapter 7 (Exponentials and Logarithms) and the calculus version here, and highlight the irrationality of e.
- Differentiation of logarithmic functions is used again in Chapter 12 (Integration II) to integrate rational functions. Ensure that students learn the result for differentiation here.

11.3 Further Trigonometric Differentiation

- Check that students are familiar with the reciprocal ratios sec θ, cosec θ and cot θ before beginning this work.
- Emphasise the use of the chain rule to differentiate trigonometric functions, and ensure that students learn the results for the inverse trigonometric functions.

11.4 Using the Second Derivative

- It is important for students to understand that once a stationary point has been located, the sign of the second derivative distinguishes between maxima and minima.
- $\frac{d^2y}{dx^2} = 0$ does not imply a stationary point of inflexion. Use $y = x^4$ at the point (0, 0) as an example.
- Students should learn how to classify the four types of points of inflexion. Note clearly the signs of $\frac{dy}{dx}$ and $\frac{d^3y}{dx^3}$.

11.5 Implicit Differentiation

- Emphasise the use of the chain rule to differentiate implicit terms such as y, y^2 and xy with respect to x.
- Show students how to avoid tedious algebra by finding the gradient at a given point using the equation involving x, y and $\frac{dy}{dx}$. Substitution into this equation is often the quickest method.

11.6 Parametric Differentiation

- Stress the importance of using the chain rule to eliminate the parameter when finding the gradient of these curves.
- Advanced Level questions on this topic require a **high level of algebraic manipulative skill**, particularly in factorisation and solving simultaneous equations.

11.7 Maclaurin series

- Encourage students to plot the graphs of $y = f(x)$ and its Maclaurin series expansion to investigate how good this polynomial approximation is at the values of x of interest.

12 Integration II

General Notes

Ensure students:

- can **differentiate** and **integrate a variety of standard mathematical functions**; in particular, they should have covered the work in Chapter 5 (Differentiation I), Chapter 9 (Integration I) and Chapter 11 (Differentiation II)
- can use **partial fractions** and **rewrite improper algebraic fractions** by expressing the numerator in terms of the denominator, or by using long division as appropriate
- are aware that differentiation of their answer can always be used as a check of their integration.

Specific Notes

12.1 Standard Integrals

- Make sure students check whether the integral can be written as a standard integral first.

12.2 Rational Functions

- Check that algebraic fractions are **proper** before integrating. **Improper fractions** can be expressed as the **sum of a polynomial and a proper fraction** by using polynomial division or by writing the numerator in terms of the denominator.

12.3 Integration by Parts

- Check that the student can recognise when **integration by parts** is required and that it *does* matter which part is labelled u and which part is labelled $\frac{dv}{dx}$.

12.4 Integration by Substitution

- Ensure students remember to express the operator 'dx' in terms of the new variable when using **integration by substitution**.

12.5 The Area under a Parametrically Defined Curve

- Ensure that students have experience of integrating by substitution before beginning this topic.
- Emphasise that the lower and upper limits on the integral in terms of the parameter correspond to the lower and upper values of x respectively for the area being calculated.

12.6 Differential Equations

- Encourage students to **solve differential equations** either directly using **integration** or by **separating the variables**. Examples in the text show how this type of equation can model real-life situations.
- Remind students to include a constant of integration when solving differential equations. Ensure that they are aware of the different ways in which this constant can be manipulated.

13 Numerical Methods

General Notes

Ensure students:

- are competent at **rearranging equations** and **differentiating polynomial**, **trigonometric**, **exponential** and **logarithmic functions** using the **chain**, **product** and **quotient rules**, if necessary
- have used a variety of **integration techniques** to find the area under a curve
- have access to a graphical calculator; they should be encouraged to use it as opportunities arise in the text.

Specific Notes

13.1 Approximate Solutions of Equations

- Make sure students understand that the root(s) of the equation $f(x) = 0$ are the value(s) of x where the graph of $y = f(x)$ crosses the x-axis, and that the root(s) of the equation $f(x) = g(x)$ correspond to the points at which the graphs of $y = f(x)$ and $y = g(x)$ intersect.
- Throughout the discussion of the **Newton–Raphson method** and **fixed point convergence** in the text, the first numerical approximation to the solution of an equation is denoted x_0. Be aware, however, that **some examination questions might denote the initial approximation as x_1.**
- As discussed in the text, there may be several ways of expressing a given equation in the form $x = g(x)$. Students need to appreciate that some of these arrangements result in an **iteration formula $x_{n+1} = g(x_n)$** that, for a given initial value, will diverge or converge to a different root. If they have access to graph plotting software, get them to print out copies of the graphs of $y = x$ and $y = g(x)$ on the same axes for each of the different possible arrangements they have identified. Using a pencil and ruler, they can then **draw cobweb or staircase diagrams** from their chosen initial approximation x_0 and see whether this particular rearrangement leads to convergence or divergence.

13.2 Numerical Integration

- It is important to emphasise that the **mid-ordinate**, **trapezium** and **Simpson's rules of numerical integration** only work if the area to be found is divided into vertical strips of equal width.
- In addition, students must remember that **Simpson's rule** will only work if the area is divided into an even number of strips. They should appreciate that the reason for this is that the curve at the top of pairs of adjacent strips is approximated by a **quadratic curve**.
- When using the **trapezium rule** with your students, ask them to predict whether the numerical value obtained will be an underestimate or an overestimate of the exact value of an integral. This point is commonly examined and depends upon whether the tops of the trapeziums approximating the area are mainly above or below the curve itself.

14 Vectors

General Notes

Ensure students:

- are competent in algebraic techniques; in particular they should be able to solve **linear simultaneous equations**, and **trigonometric equations with angles measured in radians**
- are familiar with the **geometry of straight lines and planes**; in particular, they should have covered **parallel and perpendicular lines**.

Specific Notes

14.1 Vector Geometry

- Check that students can **write vectors in the many equivalent forms**: lower-case letters underlined, capital letters ordered with an arrow above, using base vectors **i**, **j** and **k** or as column vectors.

14.2 Vectors in Two and Three Dimensions

- The **modulus** of a vector $|\mathbf{a}|$ can be related to the modulus function already met in Chapter 6 (Algebra II). Encourage students to make this link.
- Emphasise that **unit vectors exist in all directions** and ensure that students know how to find them.

14.3 The Scalar Product

- Ensure students realise that **a.b** is a scalar quantity and is used to find angles between vectors. It is often tested at Advanced Level in the context of showing that two vectors are perpendicular.
- Encourage students to write the equivalence statement when starting questions on the scalar product:

$$\mathbf{a}.\mathbf{b} = ab\cos\theta = a_1b_1 + a_2b_2 + a_3b_3 \quad \begin{cases} > 0 & \Rightarrow \quad \theta \text{ is acute} \\ = 0 & \Rightarrow \quad \theta = 90^\circ, \text{ so } \mathbf{a} \text{ and } \mathbf{b} \text{ are perpendicular} \\ < 0 & \Rightarrow \quad \theta \text{ is obtuse} \end{cases}$$

14.4 The Vector Equation of a Straight Line

- Make sure students can identify a vector **parallel** to a line when presented with the vector equation of a line.
- Encourage students to work with both **column vectors** and **base vectors** when finding equations of lines.
- Check that students learn how to distinguish between lines that are parallel, lines that intersect and lines that are skew.
- Emphasise the use of the **scalar product** to find the shortest distance of a point from a line.

14.5 The Vector Equation of a Plane

- Check that students can write the **vector equation of a plane** in both forms and can interchange between vectors and Cartesian representation.
- Ensure students can identify a **vector perpendicular** to a plane when its equation is given as $\mathbf{r}.\mathbf{n} = d$ or $ax + by + cz = d$.
- Emphasise the use of the **scalar product** to find the shortest distance between a point and a plane.

15 Proof and Mathematical Argument

General Notes

Ensure students:

- are competent in algebraic manipulation
- are experienced in transforming the subject of an equation and in factorisation.

- This chapter is designed to introduce greater rigour in the work of the student. Many students will be unfamiliar with the process of deduction and will often use results without considering their validity. Consequently, a section on **absurd results** has been included to generate discussion, particularly on algebraic techniques.
- The chapter draws on many topics from the book. Consequently trigonometry and calculus need to be developed if the student is to benefit from many of the ideas.

Specific Notes

15.1 Mathematical Argument

- Discuss carefully with students what is meant by **mathematical argument**. The **use of paradoxes** will create debate on exactly where the error has occurred. This then provides the basis for a notation that is both consistent and easily understood.
- Many nonsense results are created using division by zero. Encourage students to spot this trick, especially when algebraic expressions are being used.

15.2 Proof by Exhaustion

- Many students believe that things can be proved by checking every case. Use the **divisibility rule for 3** to show the problems with this viewpoint. Even with fast processing time on a computer, this style of proof can be impractical.
- Emphasise that the '**disproof by counter example**' is used frequently in number theory. The search for patterns in prime numbers, and particularly algebraic formulas that generate primes, has historical references. Encourage students to research some of these references.
- Make it clear to students that there are still many statements that cannot be proved or disproved. These are often called **conjectures** and are named after the mathematician who proposed them.

15.3 Proof by Contradiction

- It is important for students to realise that **proof by contradiction is acceptable as a mathematical argument**. This can be illustrated using the irrationality of $\sqrt{2}$. Encourage students to look for other applications, as illustrated in Exercise 15.3 Contextual.

15.4 Proof by Induction

- Many students find **proof by induction** daunting because of the higher level of algebraic skill required. Emphasise that the result to be proved is known and that this result can be checked carefully as they progress from $n = k$ to $n = k + 1$.

QUESTION BANK

MIRROR EXERCISES

Review

1 Write the following in index form:

a $y \times y \times y$
b $5 \times 5 \times 5$
c $y \times y \times y \times y \times y$
d $a \times a \times b \times b \times a$
e $3x \times 2x$
f $5a \times a \times 2a$

2 Find the exact value of the following:

a $\sqrt{7} \times \sqrt{7}$
b $\sqrt{5} \times \sqrt{5}$
c $\sqrt{12} \times \sqrt{12}$
d $\sqrt{6} \times \sqrt{6}$
e $\sqrt{y} \times \sqrt{y}$
f $\sqrt{b} \times \sqrt{b}$

3 By collecting like terms, simplify the following expressions:

a $5x + 2x$
b $3x^2 + 7x^2 - 5x^2$
c $7y - 3y - 2y$
d $5xy + 2yx + 3x$
e $2x^2 - 3x + 5x$
f $7x + 9x - 2y$

4 Multiply out, or expand, the following terms in brackets, simplifying the answer where possible:

a $3(x + 4)$
b $5(x - 2)$
c $-3(x - 1)$
d $-4(x - 3)$
e $-2(x + 2y - 3)$
f $5(3x - y - 3)$
g $(x + 2)(x - 1)$
h $(x + 2)(x + 3)$
i $(x - 4)(x - 1)$
j $(x + 6)(x - 1)$
k $(x + 2)(x - 3)$
l $(x - 3)^2$

5 Solve the following equations:

a $3x = 21$
b $5x - 3 = 7$
c $7x - 1 = 9x - 5$
d $5(x + 3) = 35$
e $-2(x - 1) = -6$
f $-6(x - 6) = 18$

6 Factorise these expressions:

a $5x + 5y$
b $8x - 8y$
c $3x - 12y$
d $10x + 15y$
e $21x^2 + 7xy$
f $15xy^3 + 35x^3y$

7 Without using a calculator evaluate the following (leaving remainders in fraction form where necessary):

a $10\,944 \div 12$
b $3762 \div 11$
c $7945 \div 13$
d $10\,251 \div 14$
e $6757 \div 16$
f $9267 \div 17$

1.1 Technique

1 Simplify the following:

a $\sqrt{12}$
b $\sqrt{50}$
c $\sqrt{99}$
d $\sqrt{363}$
e $\sqrt{28}$
f $\sqrt{32}$
g $\sqrt{80}$
h $\sqrt{98}$

2 Simplify the following and express each as a single surd:

a $9\sqrt{3} - 5\sqrt{3}$
b $8\sqrt{2} + \sqrt{2}$
c $\sqrt{80} - 2\sqrt{5}$
d $\sqrt{80} + \sqrt{45}$
e $\sqrt{112} + \sqrt{63} - \sqrt{28}$
f $\sqrt{80} + \sqrt{45} - \sqrt{20}$

3 Expand and simplify the following:

a $(\sqrt{5} - \sqrt{2})(\sqrt{5} + \sqrt{2})$
b $(\sqrt{5} - \sqrt{3})(\sqrt{5} + \sqrt{3})$
c $(7\sqrt{2} - \sqrt{3})(7\sqrt{2} + \sqrt{3})$
d $(\sqrt{7} - \sqrt{2})(\sqrt{7} + \sqrt{2})$
e $(2\sqrt{7} + 1)(2\sqrt{7} - 1)$
f $(3\sqrt{5} - 1)(3\sqrt{5} + 1)$
g $(\sqrt{11} - 2)(\sqrt{11} - 2)$
h $(2\sqrt{11} - 3)(2\sqrt{11} - 3)$

4 Rationalise the denominator of the following fractions:

a $\dfrac{1}{\sqrt{7}}$
b $\dfrac{2}{\sqrt{11}}$
c $\dfrac{6}{\sqrt{2} + \sqrt{5}}$
d $\dfrac{6}{\sqrt{7} + 1}$
e $\dfrac{\sqrt{3} + 1}{\sqrt{3} - 1}$
f $\dfrac{\sqrt{a} - 1}{\sqrt{a} + 1}$
g $\dfrac{\sqrt{b} + 1}{\sqrt{b} - 1}$
h $\dfrac{\sqrt{7} - \sqrt{5}}{\sqrt{7} + \sqrt{5}}$

1.2 Technique

1 Simplify the following expressions:

a $\dfrac{5^3 \times 5^4}{5^6}$
b $\dfrac{3^3 \times 3^7}{3^4}$
c $\dfrac{4^{11}}{4^3 \times 4^6}$
d $\dfrac{8^{18}}{8^5 \times 8^6}$
e $\dfrac{10x^4y}{2x^3}$
f $\dfrac{27x^3y^4}{9x^3y^2}$

2 Simplify the following expressions:

a $(x^4)^3$
b $(x^5)^2$
c $4(x^5)^2$
d $5(x^{10})^2$
e $(2x^2)^2$
f $(3x^3)^2$

3 Write the following expressions in index form:

a $\dfrac{5}{x^3}$
b $\dfrac{3}{5x^2}$
c $\dfrac{7^3 \times 7^2}{7^8}$
d $\dfrac{9^2 \times 9^2}{9^9}$
e $\dfrac{5x^3y^2}{15xy^7}$
f $\dfrac{2x^6y^3}{8x^7y^3}$

4 Write the following expressions in their simplest form:

a $\dfrac{\left(x^3\right)^{\frac{1}{3}}}{(x^6)^{\frac{1}{2}}}$
b $\dfrac{\left(x^5\right)^{\frac{2}{5}}}{x^{-3} \times x^2}$
c $\sqrt{\dfrac{16x^6}{y^2}}$
d $\sqrt{\dfrac{2x^8}{18x^2}}$
e $\sqrt[3]{\dfrac{54x^6}{2x \times x^{-1}}}$
f $\sqrt[5]{\dfrac{x^4 \times x^2}{(x^{-2})^2}}$

5 Evaluate the following expressions without using a calculator:

a $(16^{\frac{1}{4}})^{-2}$
b $25^{\frac{3}{2}}$
c $(9^{\frac{1}{2}})^{-2}$
d $(-7^{\frac{3}{2}})^0$
e $(2\frac{1}{4})^{\frac{1}{2}}$
f $(5\frac{1}{16})^{\frac{1}{4}}$
g $(625)^{-\frac{3}{4}}$
h $(7\frac{1}{9})^{-\frac{1}{2}}$

6 Using trial and improvement, or otherwise, solve (find the value of x for) these equations:

a $2^x = 64$
b $3^x = 27$
c $4^x = \dfrac{1}{16}$
d $11^x = \dfrac{1}{11}$

e $15^x = 1$ **f** $x^{-3} = \frac{8}{27}$

g $5^{2x} = \frac{1}{25}$ **h** $\left(\frac{1}{2}\right)^x = 32$

1.3 Technique

1 For each of the following polynomials, write down
i the order (degree) **ii** the coefficient of x^2.

a $5x^3 + 2x^2 - 7$
b $6x^5 + 3x^2 + x - 7$
c $12x^2 + x + 1$
d $3x^8 + 6x$
e $5x^4 - 2x^2 + 2$
f $3x^2 + x^{-1} - 7x^{-3}$

2 Add:
a $2x^2 + 7x - 4$ and $5x^2 - 3x + 4$
b $3x^2 + 7x - 2$ and $2x^2 + 3x - 2$
c $6x^3 - 5x^2 - 2x + 13$ and $-3x^3 + 2x^2 + 1$
d $2x^4 + 3x^2 - 7x - 11$ and $4x^4 - 3x^3 + 7x + 12$

3 Find $y_1 - y_2$ when:
a $y_1 = 6x^2 + 4x + 8$ and $y_2 = 3x^2 + 6x + 1$
b $y_1 = 9x^2 - 3x - 8$ and $y_2 = 6x^2 - 4x - 1$
c $y_1 = -3x^3 + 7x^2 - 2x + 13$ and $y_2 = 6x^3 - 4x^2 + 6x + 6$
d $y_1 = 7x^4 - 2x^3 - 2x^2 + 6x$ and $y_2 = -2x^4 + 3x^2 + 6x + 2$

4 Multiply and simplify the following:
a $(2x + 1)$ and $(2x + 1)$
b $(3x - 2)$ and $(3x - 2)$
c $(3x + 1)$ and $(x^2 - 5)$
d $(x + 1)$ and $(x^2 - 2x + 3)$
e $(2 - x)$ and $(x^2 + 3x + 7)$
f $(4 + x)$ and $(2x^2 - 2x + 1)$

5 Expand and simplify the following:
a $(x + 1)(2x^2 - 3x + 1) + (x - 1)(x^2 + x + 5)$
b $(2x - 1)(2x^2 + 2x - 7) + (x + 1)(x^2 + 4x + 2)$
c $(3x - 2)(x^2 + x - 1) + (2x + 1)(2x^2 - x + 1)$
d $(5x + 1)(2x^2 - 3x + 1) + (x + 1)(2x^2 - x + 2)$
e $(4x - 3)(2x^2 - x + 1) - (x - 1)(2x^2 - 4x - 1)$
f $(2x + 3)(3x^3 - x + 2) - (x - 2)(x^3 + x + 1)$

6 Expand and simplify the following:
a $(3x + 1)(3x + 1)$
b $(2x + 3)(x - 2)^2$
c $(x - 2)^4$
d $(1 + 3x)^4$
e $(x + 2)^4$
f $(2x + 1)^4$

7 Write the following as proper fractions:
a $\frac{x+3}{x+4}$
b $\frac{2x+3}{x+3}$
c $\frac{4x+7}{x-1}$
d $\frac{6x+7}{2x+1}$
e $\frac{3x-1}{x+3}$
f $\frac{3x+2}{x-2}$

1.4 Technique

1 Factorise the following quadratic expressions:
a $x^2 + x - 6$
b $x^2 - 3x - 10$
c $x^2 + x - 2$
d $x^2 - 7x + 6$
e $x^2 - 2x - 15$
f $x^2 - 8x + 12$
g $x^2 - x - 2$
h $x^2 - 14x + 13$

2 Factorise the following quadratic expressions using the difference of two squares:
a $x^2 - 25$
b $y^2 - 100$
c $81x^2 - 4$
d $9y^2 - 1$
e $\tan^2\theta - 1$
f $4\sin^2\theta - 1$
g $9x^2 - 16y^2$
h $121x^2 - 144y^2$

3 Factorise the following expressions completely:
a $2x^2 - 72$
b $3y^2 - 243$
c $10x^2 - 250$
d $7y^2 - 28$
e $5t^3 - 5t$
f $3\cos^2\theta - 3$

4 Factorise the following expressions:
a $2x^2 - 11x + 15$
b $2x^2 - 5x - 3$
c $6x^2 + 19x + 10$
d $3x^2 + 5x + 2$
e $4x^2 + 23x + 15$
f $11x^2 + 21x - 2$
g $3x^2 + 10x - 8$
h $12x^2 - 7x - 12$

5 Factorise the following expressions:
a $3x^2 - 18x - 21$
b $2x^2 - 18$
c $2x^2 - 6x - 20$
d $5x^2 + 10x - 40$
e $4x^2 - 8x + 4$
f $7x^2 + 35x + 42$
g $9x^2y + 69xy + 42y$
h $10x^2y - 25xy + 10y$

1.5 Technique

1 Solve the following:
a $(x + 5)(x - 2) = 0$
b $(x + 1)(x + 8) = 0$
c $(x - 3)(x + 4) = 0$
d $(x - 2)(x + 6) = 0$
e $(2x - 1)(2x + 7) = 0$
f $3(x + 1)(x - 1) = 0$
g $x(x - 3) = 0$
h $5(x + 2)(3x - 4) = 0$

2 Solve, by factorising, the following equations:
a $x^2 + x - 6 = 0$
b $x^2 - 12x - 13 = 0$
c $x^2 - 9x + 14 = 0$
d $x^2 - 7x + 10 = 0$
e $9x^2 + 6x - 8 = 0$
f $9x^2 - 12x - 12 = 0$

3 Solve, by completing the square, the following equations:
a $x^2 - 2x - 5 = 0$
b $x^2 + 3x + 1 = 0$
c $x^2 + 7x + 5 = 0$
d $x^2 - 6x - 9 = 0$
e $2x^2 + 2x - 3 = 0$
f $2x^2 + 3x - 4 = 0$

4 Solve, by using the quadratic formula, the following equations:
a $2x^2 + 2x - 3 = 0$
b $2x^2 + 9x + 8 = 0$
c $3x^2 + 5x - 3 = 0$
d $x^2 - 5x + 1 = 0$
e $x^2 - 2x - 5 = 0$
f $5x^2 + 2x - 1 = 0$

5 Solve the following equations. In each case check the solutions by using a graphical calculator to find the points of intersection between the quadratic and the x-axis.
a $8x^2 - 8x - 6 = 0$
b $3x(x - 6) + 15 = 5$
c $2(x^2 + 2) = 3(x + 4)$
d $x^2 = 3(x + 1) + x$
e $18x^2 = 12x - 2$
f $\frac{x^2 - 15}{x} = -2$

6 Write the following expressions in the form $a(x+p)^2+q$

a $x^2 - 2x - 5$
b $x^2 + 3x + 1$
c $-x^2 + 3x + 3$
d $x^2 + 3x - 8$
e $-2x^2 + 2x + 8$
f $3x^2 + 2x - 1$

1.5 Contextual

1 The formula $h = ut - \frac{1}{2}gt^2$ gives the height h of a body, after time t, when thrown vertically upwards with velocity u, where g is constant. Calculate t when $g = 9.8$, $u = 17$ and $h = 7$. Why are there two answers?

2 The flat washer illustrated here has a diameter of 4.4 cm, and the hole has a diameter of 1.6 cm. Without using a calculator, find the shaded area.

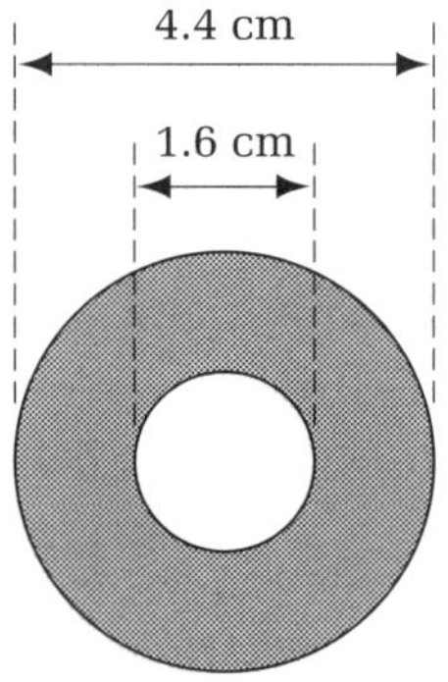

3 The sum of the first n natural numbers $(1 + 2 + \ldots + n)$ is given by the formula $S = \frac{1}{2}n(n+1)$. If the sum of the numbers is 120, how many numbers have been added?

4 The work done by a racing car travelling along a track is given by $mv^2 - mu^2$, where m is the mass of the car and u and v are its initial and final speeds, respectively. Without using a calculator, find the work done by the car in terms of m, when the initial speed $u = 39\,\text{m s}^{-1}$ and the final speed $v = 61\,\text{m s}^{-1}$.

5 The sum of the first n even numbers is given by the formula $s = n(n+1)$. How many numbers must be added to give a sum of 462?

6 A basketball playing mathematician devises the following formula for scoring a basket: $3\tan^2\alpha - 16\tan\alpha + 5 = 0$. At what angle must the ball be thrown? (Hint: Let $x = \tan\alpha$, solve the quadratic equation for x, and then find α.)

1.6 Technique

1 Solve the following pairs of linear simultaneous equations:

a $2y - x = 1$ and $x + 2y = 7$
b $3y + 2x = 12$ and $3x - 2y = 5$
c $x + y = -1$ and $9x + 8y = -7$
d $2x + 5y = 16$ and $x - y = 1$
e $5x - 8y = 0$ and $2x - 4y = 4$
f $x + y = 2$ and $2x - y = 2$

2 Solve the following pairs of simultaneous equations:

a $y = x + 2$ and $y = x^2 - 4$
b $y = 1 - x$ and $y = 1 - x^2$
c $y = 5(x+1)$ and $y = x^2 + 5x - 4$
d $y = 10 - x$ and $y = x^2 - 3x + 2$
e $2x + y = 1$ and $x^2 - xy - y^2 = -11$
f $x - y = -1$ and $x^2 + y^2 = 25$

3 Solve the following pairs of simultaneous equations:

a $60 - T = 6a$ and $T - 30 = 3a$
b $\frac{1}{3}(2x + y) = 4$ and $\frac{1}{11}(3x - 2y) = 1$
c $y = 2x - 1$ and $2x^2 + 3xy - 2y^2 = 8$
d $2x = 2y - 13$ and $y = \dfrac{12}{x}$

1.6 Contextual

1 The graph $y = x^2 - 3x - 10$ crosses the line $y = 6x$ at the points A and B. Find the coordinates of the points.

2 For a football match the attendance was 49 000 people: x people paid £40, y people paid £25 and the total receipts for the game came to £1.51 million. How many people paid for the lower price tickets?

3 Five CDs and three tapes cost £148. Two CDs and seven tapes cost £123. Find the cost of a CD and a tape.

4 A straight line passes through the points (1, 9) and (5, 13). Using the general equation for a straight line, $y = mx + c$, find m and c.

5 In 13 years time Sîan will be three times as old as Helen was four years ago. At the moment twice their combined ages is 62. Find their ages.

Consolidation C

1 Lesley throws a ball vertically upwards. Given the formula $h = ut + \frac{1}{2}gt^2$, where h is the height, u is the initial velocity of the ball and g is the gravitational acceleration, calculate the time t for the ball to reach a height $h = 11$ m, when $u = 18\,\text{m s}^{-1}$ and $g = -9.8\,\text{m s}^{-2}$.

2 The diagram show graphs $y = 5 - 2x$ and $x^2 + 2y = 15$, which cross at A and B. Find the coordinates of A and B.

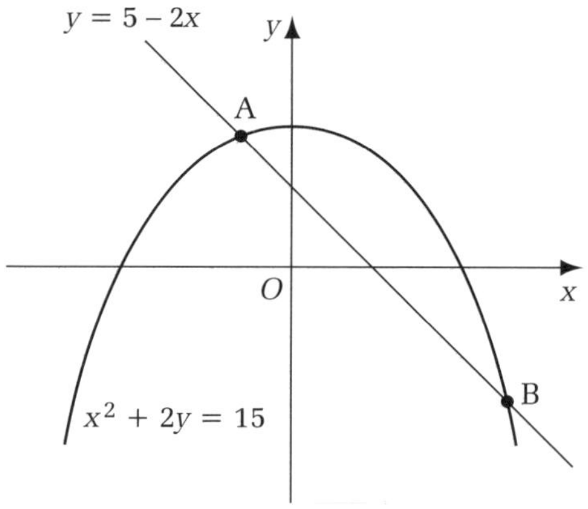

3 An exhibition is attended by 900 people: x people pay £24 at the door and y people pay £19 for tickets in advance. The organiser took a total of £17 500.

a Form two equations using this information.

b Solve them to find out how many paid at the door and how many bought tickets.

4 A pen company requires 700 m of plastic tubing in order to create felt-tip pens. Show that the area of the cross-section is $\pi(R - r)(R + r)$. If $R = 6.5$ mm and $r = 5.5$ mm, find the volume of plastic used to make the felt-tip pens.

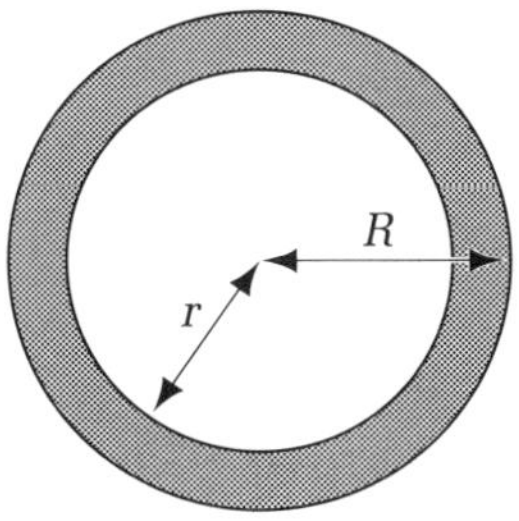

5 Tom hits a golf ball over a tree. His colleague Bob uses the formula $3 = 19\tan\theta - 3(1 + \tan^2\theta)$ in order to find the angle at which the ball was initially hit. Solve the equation and find what the angle might be.

6 Here is a sketch of the curve $y = 4 - 12x - 3x^2$ and its axis of symmetry $x = -2$.

a Write down the coordinates of the maximum point, A, and the y-intercept, B.

b Find the values of the constants p and q such that $4 - 12x - 3x^2 = p(x + 2)^2 + q$

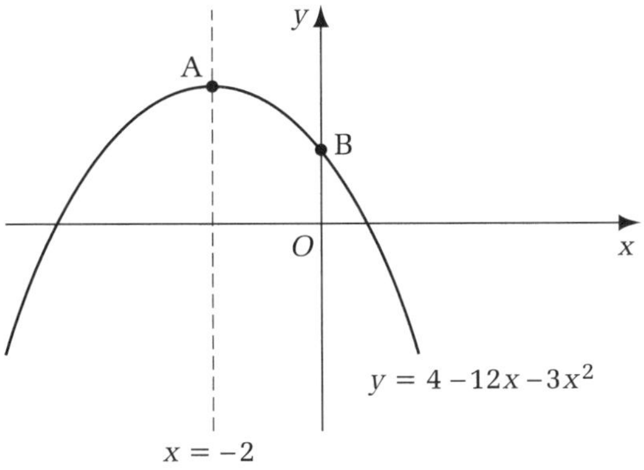

Review

1 Make y the subject of the following equations:

a $y - x = 9$ **d** $y - 3 = 2(x + 2)$

b $y + x + 7 = 13$ **e** $\dfrac{y - 1}{x + 2} = 3$

c $y + 3 = 3(x - 1)$ **f** $\dfrac{y + 2}{x - 6} = \dfrac{1}{3}$

2 For each triangle **a**–**c**, find the length of the lettered side without using a calculator.

a

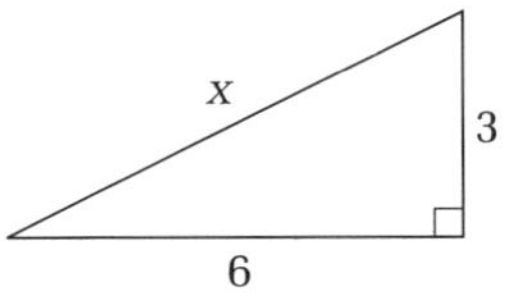

b

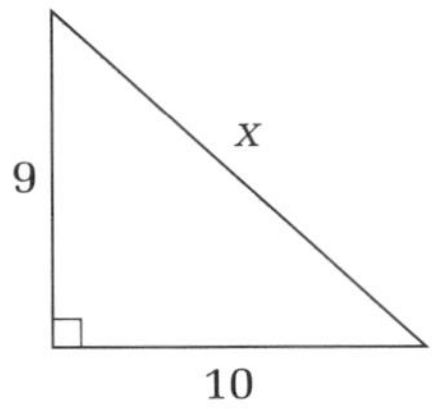

c

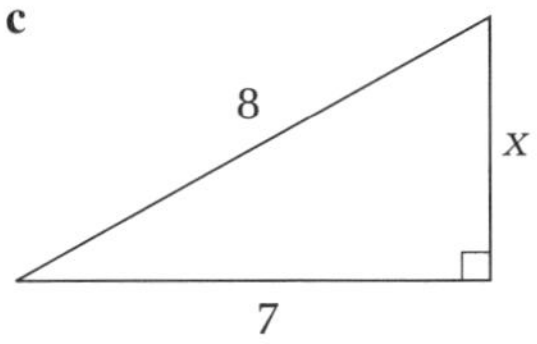

3 Expand the following expressions:

a $(x + 4)^2$ **b** $(x + 1)^2$ **c** $(x + 5)^2$
d $(3x + 1)^2$ **e** $(5x + 1)^2$ **f** $(2x + 3)^2$

4 Expand the following expressions:

a $(x - 4)^2$ **b** $(x - 5)^2$ **c** $(c - d)^2$
d $(3x - 1)^2$ **e** $(4x - 1)^2$ **f** $(5x - 2)^2$

2.1 Technique

1 Find the shortest distance between the following pairs of points:

a (11, 9) and (10, 10) **e** (−6, −2) and (−4, 0)
b (2, 8) and (4, 7) **f** (3, −3) and (−1, −4)
c (1, 9) and (−2, 3) **g** (−2, 3, 7) and (0, 9, 3)
d (7, 1) and (−2, 2) **h** (3, 4, 5) and (1, 0, 10)

2 Find the length of the straight line joining the following pairs of points:

a A (6, 4) and B (8, −1)
b A (−3, −1) and B (−5, −5)
c A (7, 1) and B (−3, −2)
d A (−3, −5) and B (0, 2)
e A (3, −7) and B (0, 2)
f A (3, 3) and B (−2, 1)
g A (9, −2, −7) and B (3, 0, −4)
h A (1, 3, 5) and B (2, −2, 3)

3 Find the gradient of the straight line formed by joining the following pairs of points:

a (1, 2) and (6, 22) **d** (0, 5) and (−3, −2)
b (7, 2) and (10, 14) **e** (−8, −2) and (−4, −1)
c (3, 9) and (1, 12) **f** (7, 1) and (7, −3)

4 For the following pairs of points, A and B, find:

i the gradient of the line AB
ii the gradient of a line perpendicular to AB.

a A (0, 5) and B (2, 11)
b A (2, 1) and B (−2, 3)
c A (−5, 0) and B (2, −5)
d A (−1, 6) and B (−2, −3)
e A (−2, −1) and B (4, −3)
f A (−3, 1) and B (1, 7)

5 Rearrange the following equations to make y the subject. State whether the pairs of lines are parallel or perpendicular to each other:

a $y = 5x + 3$ and $y = -\frac{1}{5}x + 3$
b $y = -2x - 7$ and $y = \frac{1}{2}x + 7$
c $y = 3x - 1$ and $y = 5 + 3x$
d $y - 2x + 3 = 0$ and $2y = 4x + 7$
e $9x + 6y = 4$ and $4y = 2 - 6x$
f $2y = 4(x + 2)$ and $8y + 4x = -1$

2.1 Contextual

1 Martina moves her position from the point (1, 3) to the point (3, 8) on a park map. The map is drawn to a scale of 1:10 000 cm. Find:

a the shortest distance on the map that Martina covers
b her actual distance moved in kilometres.

2 Twins John and James radio their respective coordinates to each other. John is in position (2, 7) whilst James is situated at (–3, –2). How far apart are the twins?

3 The night before military manoeuvres an army troop are given starting coordinates (878, 382) and finishing coordinates (841, 371). By considering only the final two digits of each coordinate, find the distance on the Ordnance Survey Map moved by the troop. If the scale of their map is 1:25 000 cm find the actual distance covered in kilometres.

4 According to the garden plan, the cottage (7, 5) is the same distance from the ash tree (3, 3) and the beech tree (5, 1). Investigate this statement.

5 A parallelogram is formed by lines joining the points P, Q, R and S. Given the coordinates of the points P (4, –2), Q (2, 4) and R (8, 6):

a find the coordinates of S
b show that PQRS is a square.

6 The diagram shows a sketch of a cuboid. Given that the coordinates of A are (1, 4, 2) and the coordinates of F are (5, 9, 8), find the shortest distance between A and F.

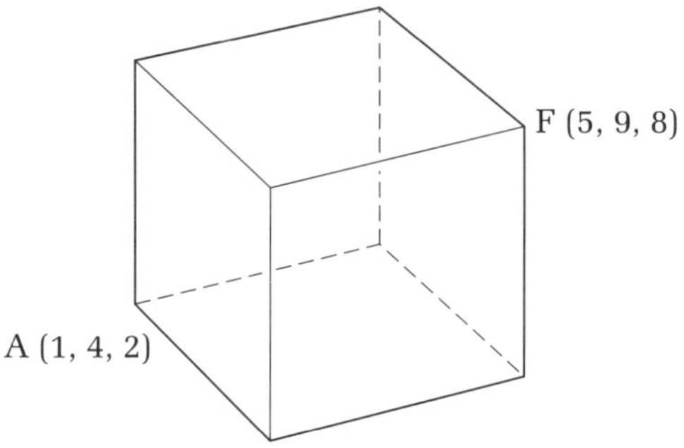

2.2 Technique

1 Rearrange the following equations into the form $y = mx + c$:

a $2x + y + 6 = 0$
b $-5x + y - 2 = 0$
c $x - y + 8 = 0$
d $4x - y - 5 = 0$
e $\dfrac{y-3}{x-2} = 3$
f $\dfrac{y+1}{x-4} = -2$
g $\dfrac{y-6}{5} = 2x$
h $\dfrac{4-y}{3} = -x$

2 State the gradient and the y-intercept of the straight line graphs produced by the following equations:

a $y = 4x + 3$
b $y = -5x + 1$
c $y = 7 - 4x$
d $y = -\frac{1}{3}x$

3 For the straight lines produced by the following equations, find the gradient, and coordinates of the y-intercept:

a $x + y + 3 = 0$
b $2x - y - 5 = 0$
c $3x + y + 7 = 0$
d $4x + 2y + 3 = 0$
e $x + 5y - 1 = 0$
f $2x - 3y = 8$

4 The equation of the line shown is given by $2x - 3y + 18 = 0$. Find the gradient and the coordinates of A and B.

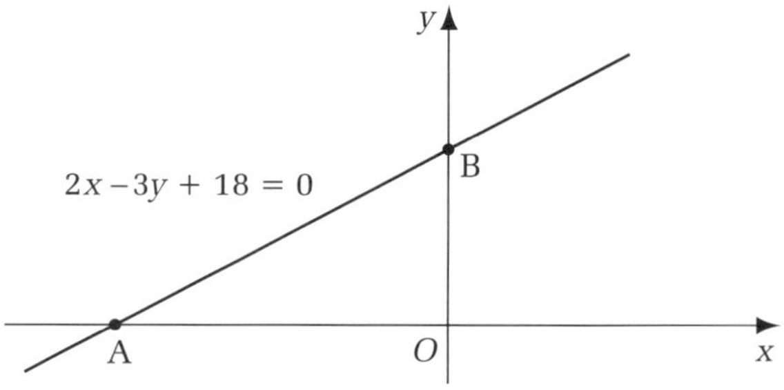

5 Rearrange the equation $y - 4 = -\frac{1}{3}(x - 2)$ into the form $ax + by + c = 0$. Sketch the graph of this equation.

6 Solve the simultaneous equations $x + y = 4$ and $2x - y + 1 = 0$ using a graphical method.

2.3 Technique

1 Find the equation of the straight line with the given gradient passing through the stated point in the following:

a gradient 2, point (1, 3)
b gradient 4, point (3, 1)
c gradient 1, point (–1, 4)
d gradient –2, point (3, 0)
e gradient $\frac{1}{3}$, point (6, –1)
f gradient $-\frac{1}{4}$, point (–1, 4)

2 Find the equation of the straight line joining the following pairs of points:

a A (4, 5) and B (5, 7)
b P (–1, –4) and Q (0, 1)
c R (–3, 5) and S (3, –7)
d A (2, –7) and B (4, –13)
e F (–2, 2) and G (–4, 1)
f (–6, 0) and (–3, –1)

3 Find the equation of the line perpendicular to the line AB, that also passes through the midpoint of AB, when:

a A (2, 0) and B (4, 8)
b A (1, 2) and B (−1, 3)
c A (2, −1) and B (4, −3)
d A (−2, −3) and B (−4, −5)

2.3 Contextual

1 A straight line that passes through the points A (−2, −3) and B (p, 7), has gradient 2. Determine the value of p and find the equation of the straight line.

2 Consider the two points, P (−1, 6) and Q (5, 10).

a Find the midpoint of PQ.
b Find the gradient of PQ.
c Write down the gradient of a line perpendicular to the line PQ.
d Find the equation of the perpendicular bisector of PQ. Write it in the form $ax + by + c = 0$.

3 Sketch a diagram to show the points A (−2, −2), B (8, 4) and C (5, −3). Let M be the midpoint of AB. Find the coordinates of M and write down the equation of the straight line that passes through M and C.

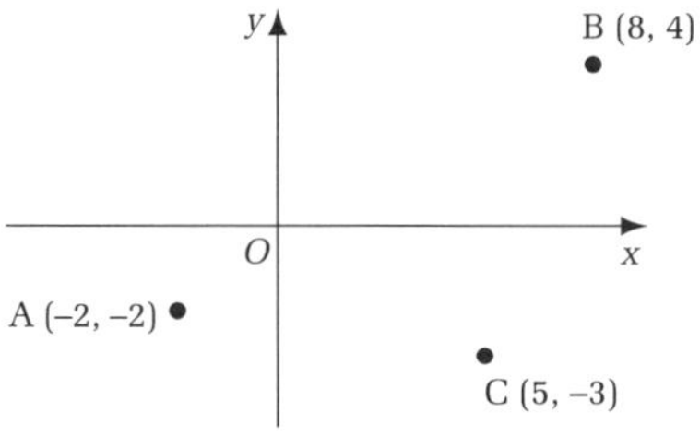

4 A is the point (3, 5), and B (6, 2) lies on the straight line $x - y - 4 = 0$.

a Find the equation of the straight line parallel to $x - y - 4 = 0$ that passes through A. Write it in the form $ax + by + c = 0$.
b Show that the straight line joining A and B is perpendicular to the line $x - y - 4 = 0$.
c Find the perpendicular distance between the two parallel lines.

2.4 Technique

1 Solve the following linear inequalities:

a $x + 7 > 9$
b $2x + 5 \leq 3$
c $3x + 4 \geq x + 6$
d $\frac{3+x}{4} < x - 6$

2 Solve the following quadratic inequalities:

a $x^2 > 25$
b $x^2 \leq 100$
c $x^2 - 121 > 0$
d $x^2 - 225 < 0$

3 Solve the following quadratic inequalities:

a $x^2 - 3x - 10 > 0$
b $x^2 - x - 12 \geq 0$
c $x^2 + x - 6 < 0$
d $x^2 - 2x - 35 < 0$
e $x^2 - 5x - 24 \leq 0$
f $x^2 - 6x + 8 > 0$

4 Transform the following statements into quadratic inequalities involving zero and solve the inequality in each case:

a $x^2 - x \geq 20$
b $x(x + 3) > 108$
c $x^2 - 3x > 28$
d $x^2 + 2x > -1$
e $x^2 \geq 8x - 16$
f $x^2 \leq 4x + 21$

5 Solve the following quadratic inequalities:

a $2x^2 + x - 21 > 0$
b $6x^2 - 13x + 6 \leq 0$
c $2x^2 < 3x + 2$
d $2x^2 + 7x > 15$
e $6x^2 \geq x + 1$
f $4x^2 > 3x + 10$

2.4 Contextual

1 Solve the inequality $x^2 \geq 2x + 5$, leaving your answer in the form $a + b\sqrt{c}$, where a, b and c are integers.

2 Find an inequality represented by the dotted section of this graph.

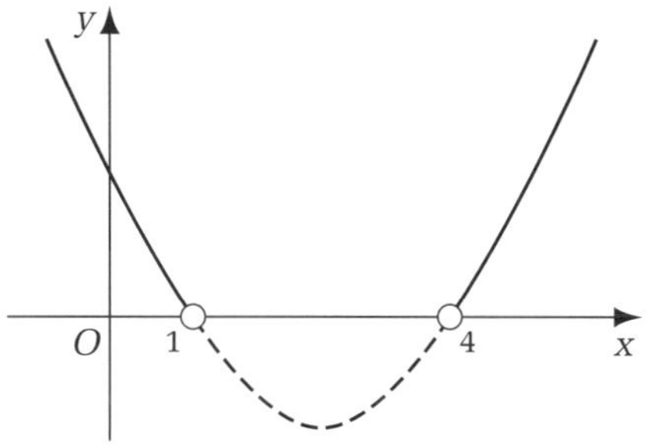

3 Find the set of values of x for which $(2x - 1)^2 - 4 < x(3x - 2)$.

4 Solve the equation $x^2 + 3\sqrt{3}x - 30 = 0$, writing your answer using surds. Hence, or otherwise, solve $x^2 + 3\sqrt{3}x - 30 \geq 0$.

2.5 Technique

1 Find the centre and radius of each of the following circles:

a $(x - 3)^2 + (y - 2)^2 = 10^2$
b $(x + 5)^2 + (y + 2)^2 = 81$
c $(x - 7)^2 + y^2 = 1$
d $(x + 3)^2 + (y - 6)^2 = 10$

2 Write the equations of the circles with the given centres and radii in the form $x^2 + y^2 - 2ax - 2by + c = 0$:

a centre (−2, 4), radius 3
b centre (3, 0), radius 1
c centre (5, −1), radius $\sqrt{2}$
d centre (2, −2), radius $3\sqrt{2}$

3 Find the centre and radius of the following circles:

a $x^2 + y^2 - 14x - 2y + 46 = 0$
b $x^2 + y^2 - 6y + 6 = 0$
c $x^2 + y^2 - 8x + 4y + 18 = 0$
d $x^2 + y^2 + 6x + 4y + 8 = 0$

2.5 Contextual

1 The point (2, k) lies on the circle $x^2 + y^2 + 4x - 6y = 28$. Find the values of k.

2 The point (k, 0) lies on the circle with centre (5, −2) and radius $2\sqrt{10}$. Find two possible values of k.

3 The point A lies on the circle with centre (2, −2) and radius $\sqrt{53}$. Given that A lies on the y-axis, find the possible coordinates of A.

4 Find the equation of the tangent to the circle $x^2 + y^2 - 14y + 44 = 0$ at the point (1, 5). Find also the equation of the normal at this point.

5 A straight line touches the circle $x^2 + y^2 + 2x - 16 = 0$ at the point (3,−1). Find the equation of this line and the equation of the perpendicular line passing through the point (2, 2).

6 The points A (3, −2) and B (1, 0) form the diameter of a circle. Find the equation of the circle.

7 Find the distance of a tangent from the point (3, 5) to the circle $x^2 + y^2 + 4x + 2y + 4 = 0$.

Consolidation C

1 A (1, 3) and B (5, −1) have a midpoint at C. Find the coordinates of C and the gradient of AB. Hence, find the equation of the perpendicular bisector of AB, giving your answer in the form $y = mx + c$, where m and c are integers.

2 Given that the straight line AB passes through (4, −2) and is perpendicular to the line PQ, whose equation is $2x + y - 3 = 0$, find the equation of AB. Find the point where AB and PQ cross and the perpendicular distance from the line PQ to the point (5, 8).

3 Find the length of the tangent from the point (−6, −7) to the circle $x^2 + y^2 + 10y = 0$.

4 Solve the inequality $(x + 3)(x - 4) \leq -2(2x + 1)$.

5 The straight line passing through the points A (4, −2) and B (k, 7) has gradient $-\frac{1}{2}$.

a Find the value of k.

b Find the equation of line AB.

c Calculate the length AB in the form $a\sqrt{b}$ where a and b are integers.

6 **a** State the radius and coordinates of the centre of the circle whose equation is $(x + 3)^2 + (y - 4)^2 = 25$.

b Given that (k, 6) lies on this circle, find the two values of k in the form $a + \sqrt{b}$, where a and b are integers.

7 A circle, centred at P, passes through A (9, 18), B (4, 19) and C (−1, −6).

a Find the equation of the perpendicular bisector of AB.

b Find the equation of the perpendicular bisector of BC.

c Using your answers to **a** and **b** solve the equations simultaneously to find the coordinates of the centre P.

d Find the distance AP.

e Hence write down the equation of the circle.

Review

1 Use Pythagoras' theorem to find the length of side AB in the following triangles:

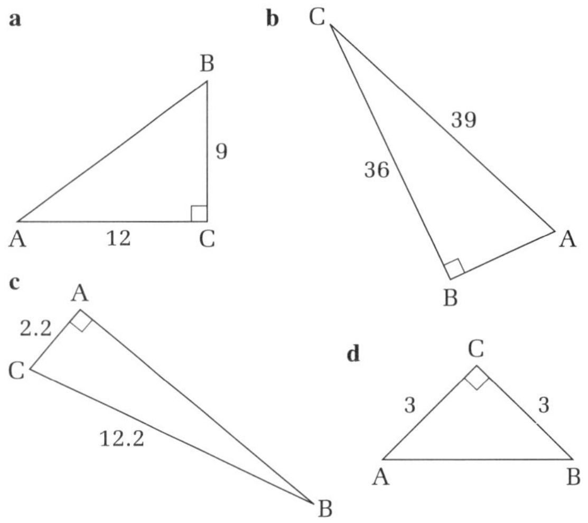

2 Factorise each of the following expressions and solve the equations, where appropriate:

a $p^2 - 4q^2$

b $r^2 + 5r + 4$

c $y^2 + 3y - 18$

d $m^2 - 11m + 30$

e $k^2 - 3k - 28 = 0$

f $2p^2 - 11p + 12 = 0$

3 Write down the values of $\sin x$, $\cos x$ and $\tan x$ for the following triangles:

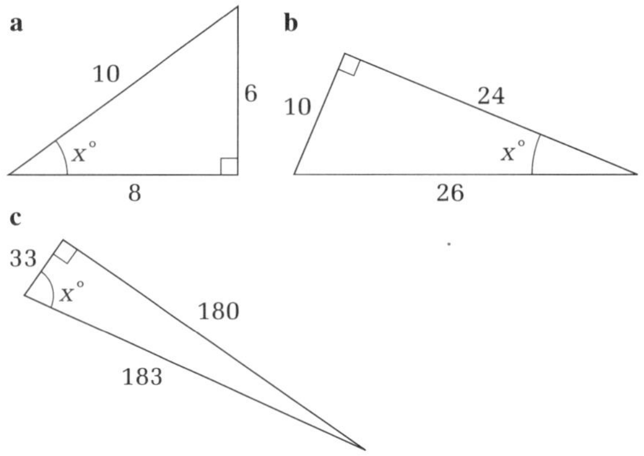

4 Calculate, to three significant figures, the areas of the following triangles:

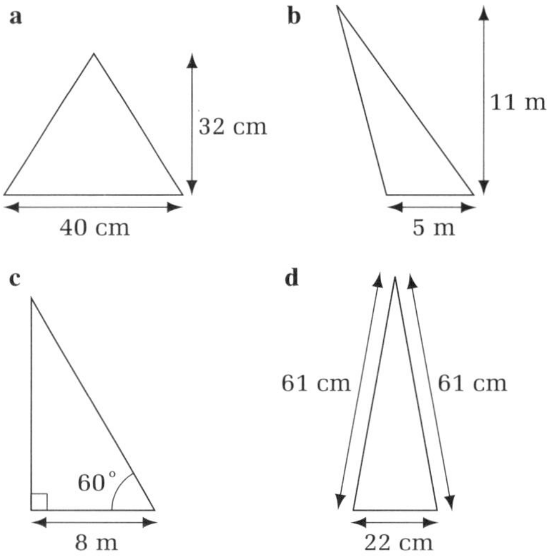

5 Solve the following triangles:

a

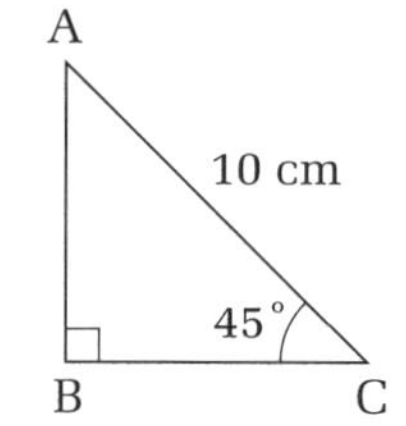

b

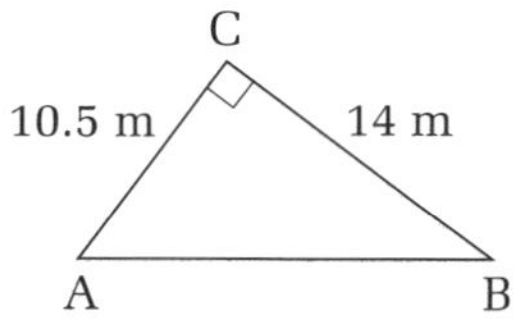

c

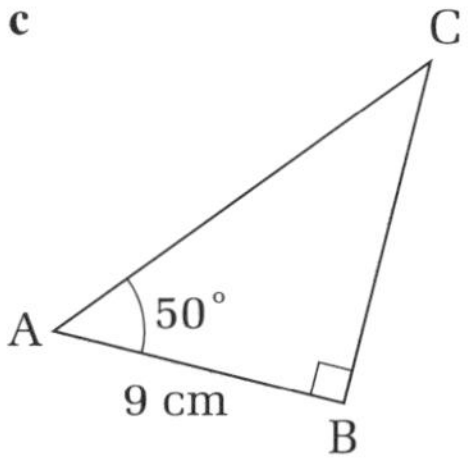

d

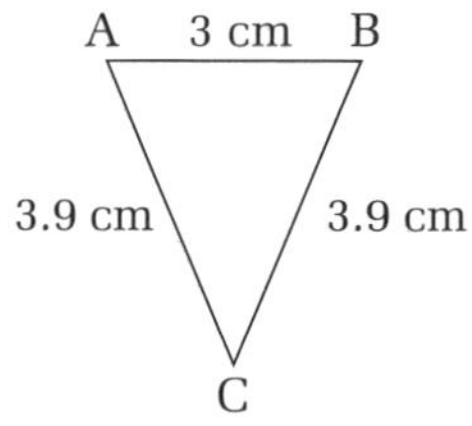

6 For each of the following diagrams write down the bearing of B from A:

a

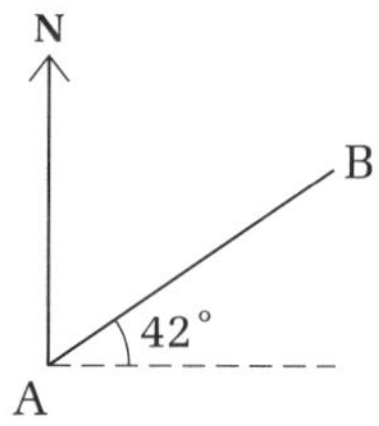

b

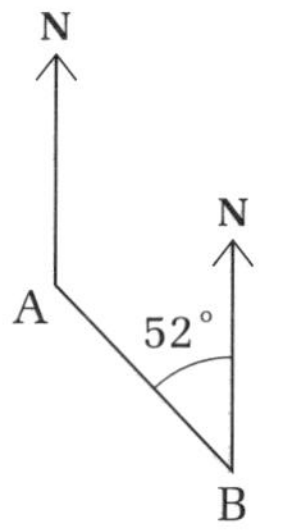

c

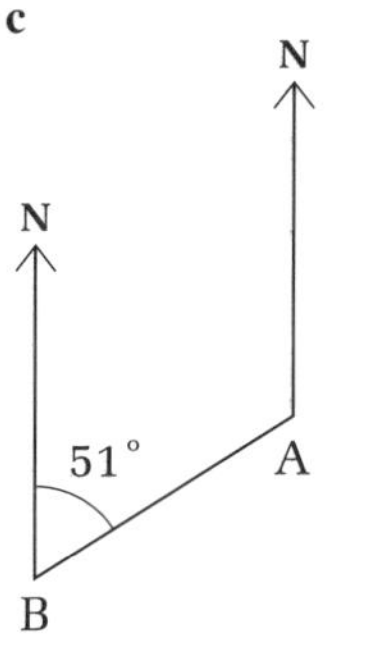

d

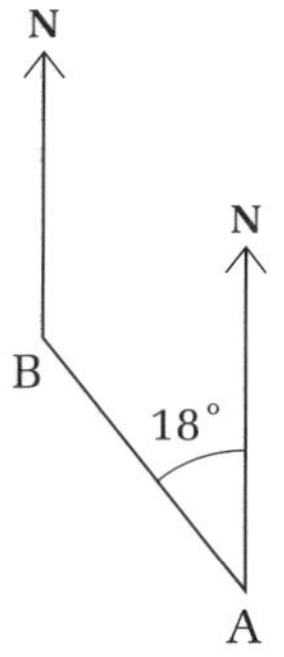

3.1 Technique

1 Write down the following as trigonometric ratios of positive acute angles:

a $\cos 129°$ **e** $\tan 388°$
b $\tan 215°$ **f** $\cos(-125°)$
c $\cos 310°$ **g** $\sin(-305°)$
d $\sin 232°$ **h** $\tan(-420°)$

2 Write down the exact values of the following, leaving answers in terms of surds if appropriate:

a $\tan 150°$ **e** $\sin 210°$
b $\cos 225°$ **f** $\cos 420°$
c $\sin 300°$ **g** $\sin(-300°)$
d $\cos 330°$ **h** $\tan(-420°)$

3 Solve the following trigonometric equations for $0° \le \theta \le 360°$. Give your answers correct to one decimal place.

a $\sin\theta = 0.619$ **d** $4\tan\theta = 7$
b $\cos\theta = -0.46$ **e** $\sin\theta = 2\cos\theta$
c $\tan\theta = 1.782$ **f** $5\sin\theta = 11\cos\theta$

4 Find, correct to four significant figures, the values of:

a cosec 59° **e** cosec 299°
b sec 72° **f** cot 451°
c cot 114° **g** cosec (−17°)
d sec 156° **h** sec (−999°)

5 Find the principal value solutions to the following:

a $10\sin\theta = -7$ **d** $3\cot\theta - 2 = 0$
b $\tan\theta + 2 = -5$ **e** $5 - 2\tan\theta = 13$
c $\sec\theta = 5$ **f** $2\operatorname{cosec}\theta = 5$

6 Construct a table giving values of $\sin\theta$, $\cos\theta$ and $\tan\theta$ for appropriate values of θ in the range $-90° \le \theta \le 450°$. On separate pieces of paper, draw the graphs of $\sin\theta$, $\cos\theta$ and $\tan\theta$ for this range of angles. Use the graphs to solve the following equations:

a $\sin\theta = -0.2$ **b** $\cos\theta = 0.9$ **c** $\tan\theta = 4.5$

7 Draw the graph of $y = \cos\theta$ for values of θ in the range $-90° \le \theta \le 90°$. Use your graph to find:

a $\cos^{-1}(0.3)$ **b** $\cos^{-1}(-0.2)$

3.1 Contextual

1 The depth of water in a harbour, y metres, can be modelled by an equation $y = 7\sin(30t)° + 13$, where t is the time in hours from midnight on a particular day.

a Draw the graph of this function over a period of 24 hours.
b From the graph find the times of the high and low tides.
c Use the graph to find the period of time for which the harbour has water deeper than 18 metres.

2 The height of a tide can be modelled by a function of the form $h = a\cos(bt°) + c$, where h is the height of the water in metres, and t is the time in hours after midnight on Sunday. Find the values of a, b and c for the following tide table:

Tide	Time	Height (m)
High	0000	28
Low	1200	2
High	2400	28
Low	3600	2

3 The depth of water at the entrance to a harbour is y metres at time t hours after low tide. The value of y is given by $y = 10 - 3\cos kt$, where k is a positive constant. Write down, or obtain, the depth of water in the harbour:

a at low tide **b** at high tide.

Show by means of a sketch graph how y varies with t between successive low tides. Given that the time interval between a low tide and the next high tide is 6.2 hours, calculate, correct to two decimal places, the value of k.

(NEAB)

3.2 Technique

1 Solve the following trigonometric equations for $0° \le \theta \le 360°$, giving angles correct to one decimal place:

a $2\sin\theta - 5\cos\theta = 0$
b $8\sin\theta + \cos\theta = 0$
c $5\sin\theta = \cos\theta$
d $2\cot\theta = \tan\theta$
e $4\cos\theta = 3\sec\theta$
f $2\sin\theta = \text{cosec}\,\theta$

2 Solve the following equations completely for $-180° \le \theta \le 180°$:

a $\tan^2\theta - 8\tan\theta + 15 = 0$
b $10\sin^2\theta + 3\sin\theta - 1 = 0$
c $16\cos^2\theta - 1 = 0$
d $2\cos^2\theta = 1$
e $3\sin^2\theta + 13\sin\theta = 10$
f $\tan^2\theta + 3\tan\theta = 0$

3 Solve the following equations completely in the range $0° \le \theta \le 360°$, giving answers correct to one decimal place where necessary:

a $\sin 2\theta = \frac{1}{3}$
b $\tan(\theta - 53°) = 0.528$
c $\cos(2\theta - 40°) = 0.75$
d $\tan(3\theta + 10°) = 4.1$
e $\sin(\theta - 90°) = 0.5$
f $\tan 2\theta = \frac{1}{2}$

4 Use Pythagorean identities to solve the following equations for the range $-180° \le \theta \le 180°$:

a $3\cos^2\theta + 4\sin\theta - 4 = 0$
b $4\sin^2\theta + 4\cos\theta - 5 = 0$
c $6\sec^2\theta + 7\tan\theta - 11 = 0$
d $\cot^2\theta + \text{cosec}\,\theta - 5 = 0$
e $\tan^2\theta + \sec^2\theta = 51$

5 Prove the following identities:

a $\text{cosec}\,\theta - \sin\theta \equiv \cos\theta\cot\theta$
b $\sec^2\theta - \sin^2\theta \equiv \tan^2\theta + \cos^2\theta$
c $\sec\theta + \tan\theta \equiv \dfrac{1}{\sec\theta - \tan\theta}$
d $(\cot\theta + \text{cosec}\,\theta)^2 \equiv \dfrac{1+\cos\theta}{1-\cos\theta}$
e $\dfrac{1}{1+\cos\theta} + \dfrac{1}{1-\cos\theta} \equiv 2\text{cosec}^2\theta$

3.2 Contextual

1 The trajectory of a golf ball struck from a tee can be modelled by the equation $y = x\tan\theta - \dfrac{x^2\sec^2\theta}{60}$, where θ is the angle of projection (θ is acute), x is the horizontal distance of the ball from the tee in metres and y is the vertical height of the ball above the fairway, also measured in metres. If the ball just clears a bush 0.77 metres high, whose trunk is 29 metres from the tee, find, to one decimal place, the angle of projection.

2 The equation of the path of a projectile referred to horizontal and upward vertical axes, Ox, Oy for a golf ball is $y = x\tan\theta - \dfrac{x^2\sec^2\theta}{30}$. Show how to reduce this equation to a quadratic equation in $\tan\theta$. Show that there are two distinct values of θ for which the ball can pass through a given point (X, Y), where $X > 0$, provided $30Y < 225 - X^2$. Interpret this result in terms the golfer would understand.

3.3 Technique

1 Without using a calculator, find the exact value of the following:

a tan 75°
b sin 165°
c cos 165°
d tan 165°
e sin 195°
f tan 195°

2 Write down the exact value of the following expressions:

a $\cos 25°\cos 65° - \sin 25°\sin 65°$
b $\sin 30°\cos 15° + \cos 30°\sin 15°$
c $\dfrac{\tan 63° - \tan 18°}{1 + \tan 63°\tan 18°}$
d $\sin 73°\cos 43° - \cos 73°\sin 43°$
e $\cos 83°\cos 23° + \sin 83°\sin 23°$
f $\dfrac{\tan 120° + \tan 15°}{1 - \tan 120°\tan 15°}$

3 Solve the following equations for $0° \le \theta \le 360°$, giving angles correct to one decimal place:

a $\sin(\theta + 60°) = 2\cos\theta$
b $\sin(\theta + 30°) = 3\cos(\theta - 30°)$
c $\cos(\theta - 30°) = \frac{1}{2}\sin\theta$
d $\sin(\theta + 60°) = -3\cos\theta$

4 Prove the following identities:

a $\sin 3\theta \equiv 4\sin\theta - 3\sin^3\theta$
[Hint: First write $\sin 3\theta$ as $\sin(2\theta + \theta)$]
b $\cos 3\theta \equiv 4\cos^3\theta - 3\cos\theta$

5 If $\sin A = \frac{11}{61}$ and $\cos B = -\frac{3}{5}$, find the exact values of the following:

a $\sin(A + B)$
b $\sin(A - B)$
c $\cos(A + B)$
d $\cos(A - B)$
e $\tan(A + B)$
f $\tan(A - B)$
g $\sin 2A$
h $\sin 2B$
i $\cos 2A$
j $\cos 2B$
k $\tan 2A$
l $\tan 2B$

6 Prove the following identities:

a $\cot\theta - \tan\theta \equiv 2\cot 2\theta$
b $\dfrac{2\tan\theta}{1+\tan^2\theta} \equiv \sin 2\theta$
c $\dfrac{1}{(1-\tan\theta)} - \dfrac{1}{(1+\tan\theta)} \equiv \tan 2\theta$
d $\text{cosec}\,2\theta - \cot 2\theta \equiv \tan\theta$

3.3 Contextual

1 **a** A wheelchair ramp, as illustrated in the diagram, can be moved by being lifted at B and wheeled on a castor at A.

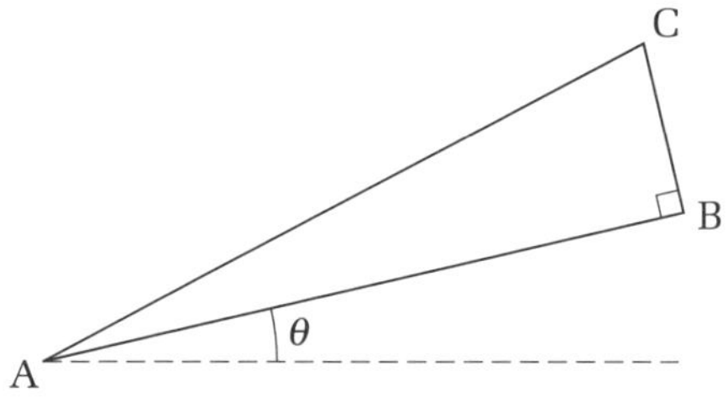

The dimensions of the ramp are $CB = 40$ cm, $AC = 2.3$ m. The base of the ramp is inclined at an angle θ, as shown in the diagram, as it is moved through a 2 m high doorway.

i Explain why $\sin(\theta + 10°) \leq 0.8696$.

ii Use this inequality to determine θ_{max}, the maximum possible value of θ.

iii Comment on the practicality of θ achieving this maximum value.

b If this ramp was replaced by a more user friendly type, in which $BC = 40$ cm and $AC = 3$ m, how would this affect your calculations?

3.4 Technique

1 Use the cosine rule to find the length of b in the following triangles:

a $B = 65°$, $a = 10$ cm, $c = 14$ cm

b $B = 115°$, $a = 9$ m, $c = 17$ m

c $B = 43°$, $a = 3.6$ cm, $c = 8.1$ cm

2 Use the cosine rule to find the angles in the following triangles:

a $a = 11$ cm, $b = 6$ cm, $c = 14$ cm

b $a = 9$ cm, $b = 4$ cm, $c = 12$ cm

c $a = 7$ cm, $b = 9$ cm, $c = 3$ cm

3 Calculate the length of the lettered side in the following triangles:

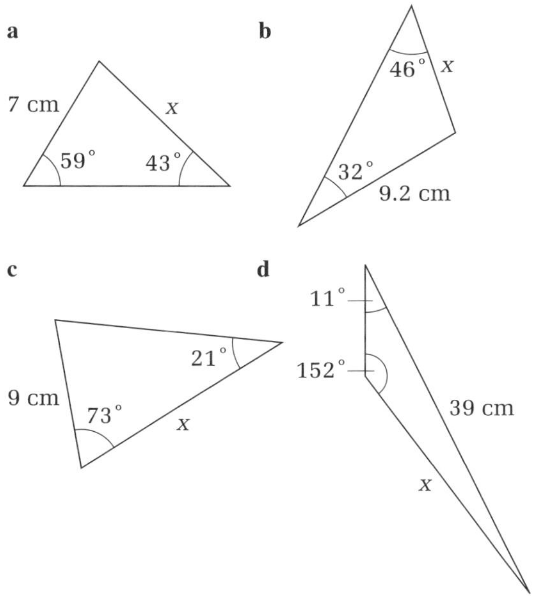

4 Solve the following triangles:

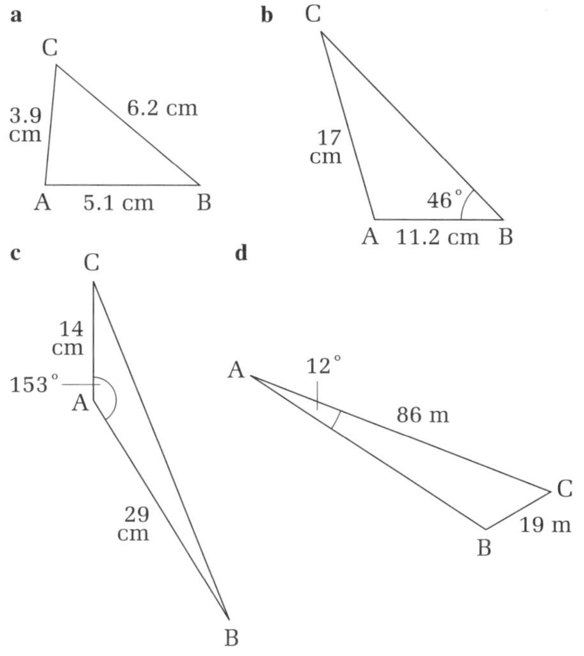

3.4 Contextual

1 A ship sails 5 nautical miles from A to B on a bearing of 086°. It then changes to a bearing of 153° and sails 11 nautical miles to C. Calculate the distance AC and the bearing of C from A.

2 A ship A is 15 miles from a lighthouse B on a bearing of 070°. A second ship C is 17 miles from the same lighthouse on a bearing of 315°. Find the distance between A and C and the bearing of C from A.

3 A, B and C are three triangulation points used on an Ordnance Survey Map. B is 11 km due south of A. C is 8 km from B on a bearing of 289°. Find the distance from A to C and the bearing of C from A.

4 From a point A on the same level as the base of a church tower, the angle of elevation of the top of the tower is 21°. From a point B, 30 m closer to the church, and on the same level, the angle of elevation is 34°. Find the height of the church tower.

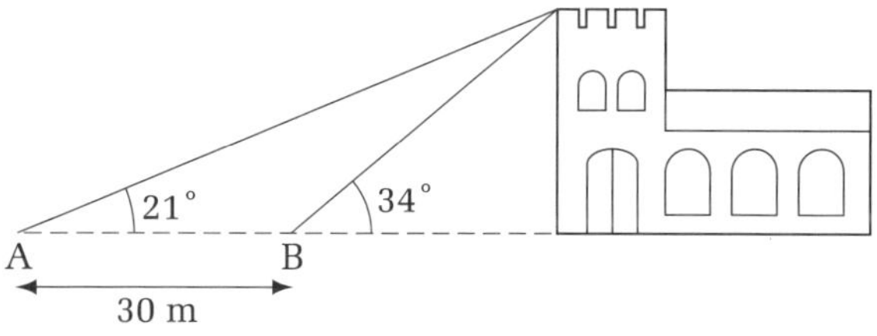

3.5 Technique

1 For the right-angled triangle with sides 5 cm, 12 cm and 13 cm find the area using:

a area $= \frac{1}{2}$(base × perpendicular height)

b area $= \frac{1}{2}ab\sin C$

c Hero's formula.

2 Find the area of the following triangles:

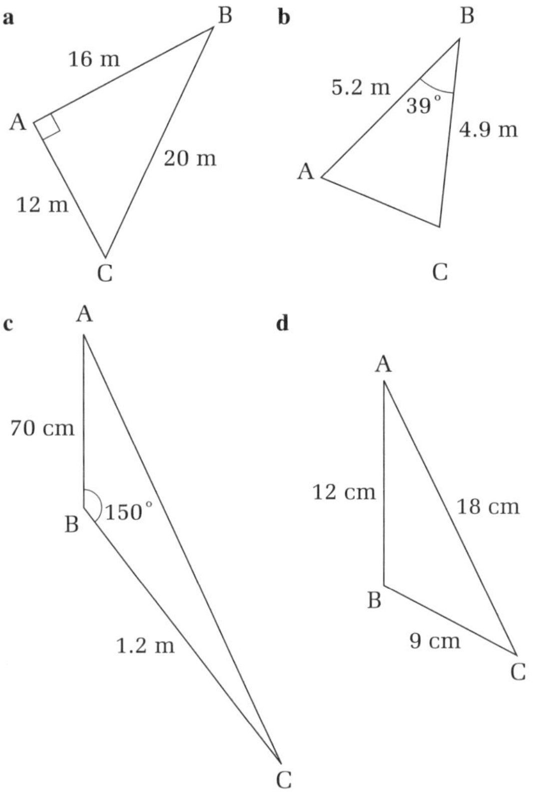

3.5 Contextual

1 The area of an acute angled triangle ABC is 60 cm^2. Given that $a = b = 13$ cm, find c.

2 A surveyor wishes to estimate the area of a small woodland patch. A sketch shows an area that is roughly triangular. Estimate the area of the woodland.

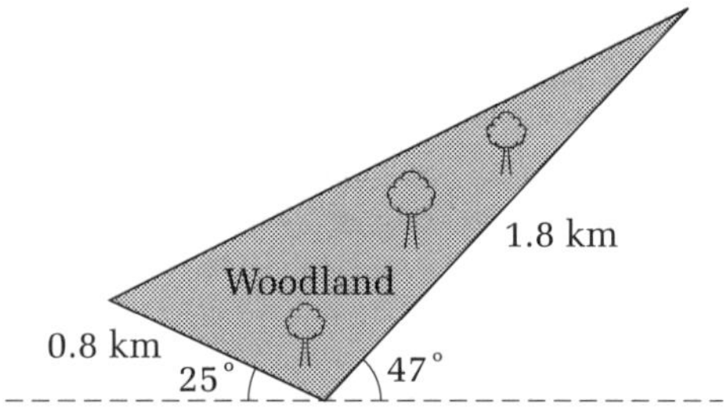

3 Beth, a mathematician's daughter, wishes to share her square birthday cake equally amongst her seven friends. She wants everyone to have the same amount of cake and icing. Siobhan, her friend, suggests dividing the perimeter by seven and cutting from each marked point to the centre of the cake. Assume that the cake is iced on the top and around the sides. If the perimeter of the cake is 112 cm and it is 10 cm high:

a will Siobhan's suggestion work?

b what area of icing does each person get?

c approximately what volume of cake will each receive?

3.6 Technique

1 Without using a calculator convert the following angles into radians, leaving your answers in terms of π:

a 10° **b** 140° **c** 170°
d 220° **e** 310° **f** 350°

2 Convert the following into radians, writing answers correct to three significant figures:

a 5° **b** 57° **c** 93°
d 161° **e** 229° **f** 301°

3 Without using a calculator convert the following angles to degrees.

a $\frac{\pi}{9}$ **b** $\frac{5\pi}{12}$ **c** $\frac{7\pi}{18}$
d $\frac{7\pi}{15}$ **e** $\frac{11\pi}{6}$ **f** $\frac{19\pi}{9}$

4 Find the arc lengths and sector areas of the following sectors, correct to one decimal place:

a $\theta = 0.4$ rad and $r = 9$ cm

b $\theta = 1.2$ rad and $r = 6$ cm

c $\theta = 2.6$ rad and $r = 11.3$ cm

d $\theta = 4.1$ rad and $r = 6.2$ cm

5 A sector OPQ has arc length 20 cm and sector area 40 cm^2 (as shown). Write down two equations involving r and θ (the radius and sector angle). Solve these equations to find the values of r and θ.

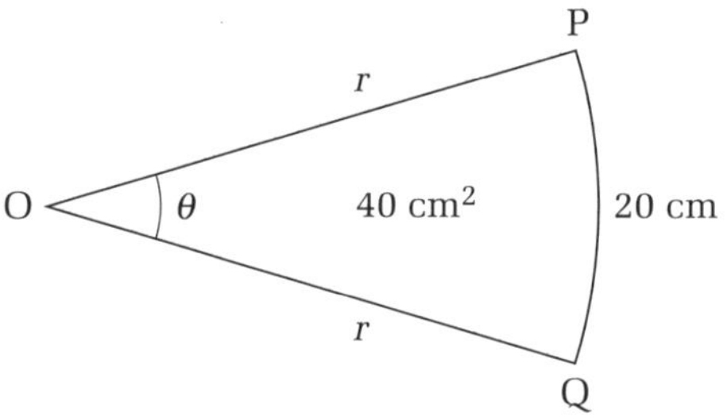

6 Given that the area of a sector OPQ is 80 cm^2 and that $\theta = 0.6$ rad, find the radius of the circle.

7 Now go back to Exercise 3.2 Technique and do questions 1, 2 and 4 using radians.

3.6 Contextual

1 A wedge of cheese, 1.5 cm thick, has a cross-section which is a sector of a circle. If the radius of the circle is 8 cm and the arc length is 6 cm, find the volume of the cheese.

2 A paper hat is made by joining, edge to edge, the two radii bounding a sector of a circle of radius 30 cm, as shown in the following diagram. If the circumference of the circular base AB, of the hat is 50 cm, find:

a the angle θ, in radians, of the sector of the circle

b the surface area of the outside of the hat.

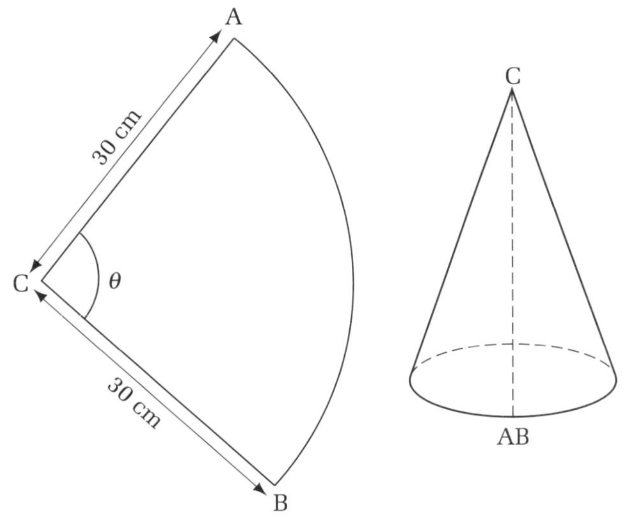

(*NICCEA*)

3 A partly filled cylindrical oil drum of diameter 80 cm lies with its curved surface in contact with a horizontal floor. The diagram shows a circular cross-section of the drum with centre O. The chord AB denotes the horizontal surface of the oil.

a Given that the oil surface is 20 cm above the floor level, find the value of angle AOB in radians.

b The length of the drum is 173 cm. Calculate the volume of oil in the drum, giving your answer in cm^3 to three significant figures.

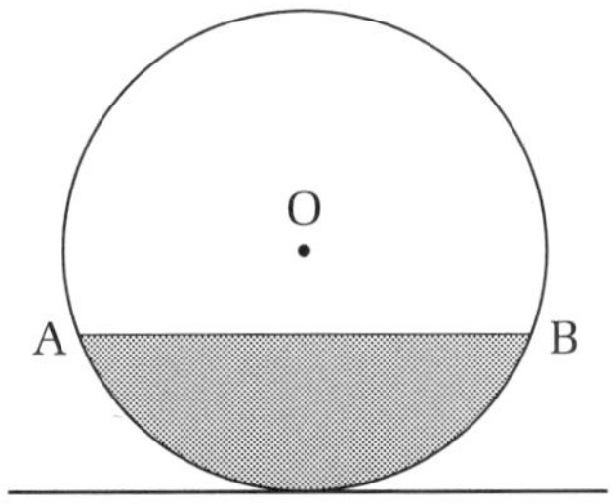

(*AEB*)

4 A windscreen wiper clears a region ABCD of a car windscreen (assumed to be flat), where AB and DC are circular arcs centred at O. Given that $OD = 10$ cm, $DA = 40$ cm and $\angle$ AOB $= 2.1$ rad, calculate the area of region ABCD.

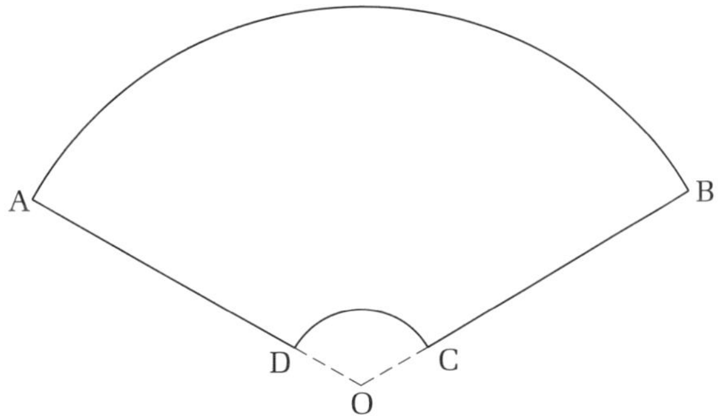

Consolidation C

1 It is given that $0° < \alpha < 360°$. State the two values of α for which $\tan \alpha$ has the same value as $\tan(-30°)$.

(*UCLES*)

2 Find to the nearest 0.1° the angles x between 0° and 180° for which:

a $\sin 3x = \frac{1}{2}$ **b** $\sin^2 x = \frac{1}{2} \cos x$

(*ULEAC*)

3 A student models the evening lighting-up time by the equation $L = 6.125 - 2.25\cos\left(\frac{\pi t}{6}\right)$, where the time, L hours pm, is always in GMT (Greenwich Mean Time) and t is in months, starting in mid-December. The model assumes that all months are equally long.

a Calculate the value of L for mid-January and for mid-May.

b Find, by solving an appropriate equation, the two months in the year when the lighting up time will be 5 pm (GMT).

c Write down the equation for L if t were to be in months starting in mid-March.

(*NEAB*)

4 Two parts A and B of a wooden jigsaw fit together as shown in the diagram. Part A is a circle, centre O of radius 6 cm. The chord PQ has length 10 cm. Part B is bounded by the tangents to the circle at P and Q, which meet at R, and the minor arc PQ of the circle.

a Find the area of the minor sector OPQ.

b Find the area of part B.

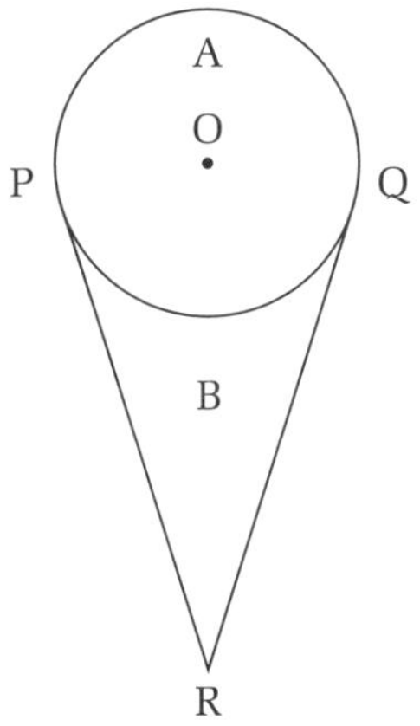

(*NICCEA*)

5 In the figure $AB = x$ cm, $BC = 3x$ cm, $CD = 20$ cm, $\sin \alpha = \frac{4}{5}$ and angle ABC $= 120°$.

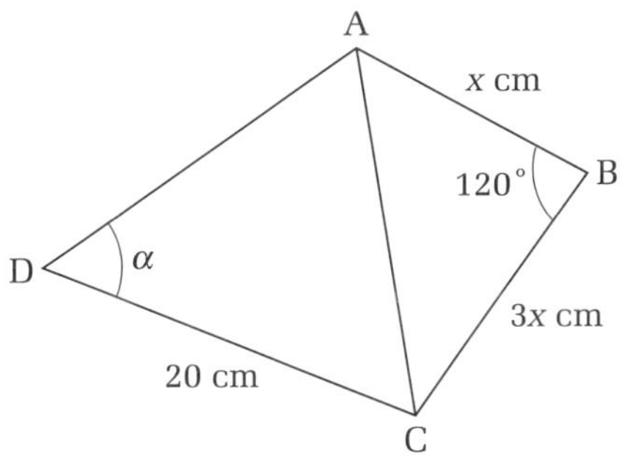

a If the area of triangle ADC is 112 cm^2, prove that $AD = 14$ cm.

b Given that α is an acute angle, state the value of $\cos \alpha$ and hence calculate AC^2.

c Prove that $x^2 = 20$.

d Calculate the angle BCA.

(*OCSEB*)

6 The figure shows the graphs:

[1] $y = 5\cos 2x^\circ + 2$

[2] $y = \cos x^\circ$

for $0 \le x \le 180$.

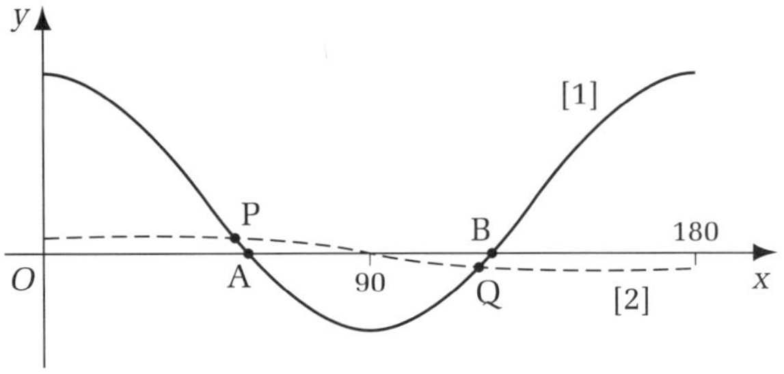

a Find the coordinates of the points A, B where graph [1] meets the x-axis.

b Use a formula for $\cos 2x^\circ$ to write the equation of graph [1] in terms of $\cos x^\circ$. Hence show that the y-coordinates of the points P and Q, where graphs [1] and [2] intersect, satisfy the equation $10y^2 - y - 3 = 0$. Solve this equation, and hence find the coordinates (both x and y) of P and Q. Give your answers correct to one decimal place.

(*OCSEB*)

7 The diagram shows a sector OAB of a circle, with centre O, in which AOB = 120°.

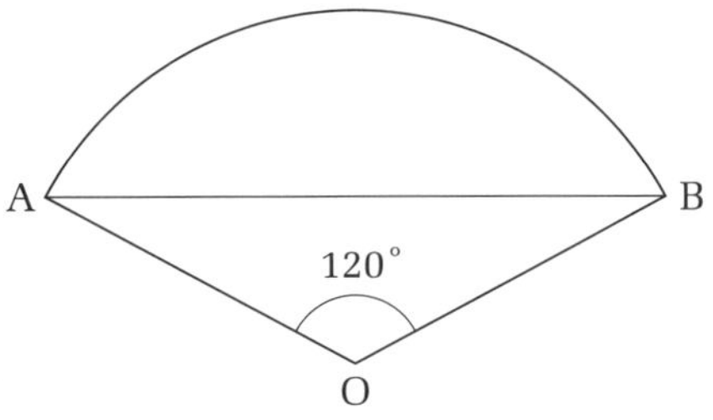

a Show that $\dfrac{\text{area of sector OAB}}{\text{area of triangle OAB}} = \dfrac{4\pi}{3\sqrt{3}}$.

b Find the exact value of $\dfrac{\text{length of arc AB}}{\text{length of chord AB}}$.

(*Cambridge*)

8 **a** Starting from the identity for $\cos(A + B)$ prove that $\cos 2x \equiv 1 - 2\sin^2 x$.

Find, in radians to two decimal places, the values of x in the interval $0 \le x < 2\pi$ for which

b $2\cos 2x + 1 = \sin x$

c $2\cos x + 1 = \sin \frac{1}{2}x$

(*London*)

9 The diagram shows the triangle OCD with $OC = OD = 17$ cm and $CD = 30$ cm. The midpoint of CD is M. With centre M, a semicircular arc A_1 is drawn on CD as diameter. With centre O and radius 17 cm, a circular arc A_2 is drawn from C to D. The shaded region R is bounded by the arcs A_1 and A_2.

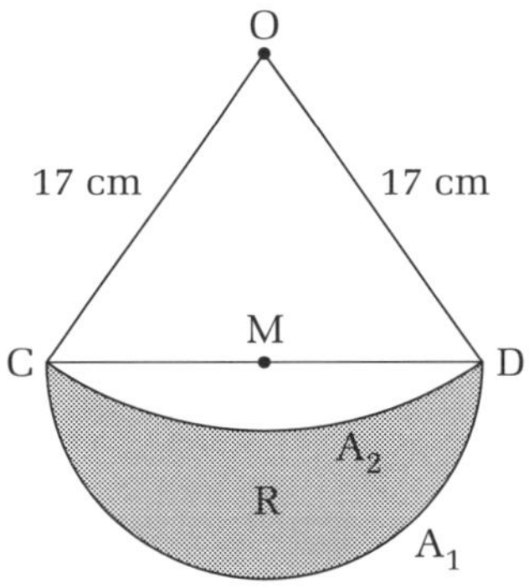

Calculate, giving answers to two decimal places:

a the area of triangle OCD

b the angle COD in radians

c the area of the shaded region R.

(*UCLES*)

10 Given that $-90 < x < 90$ find the values of x for which:

a $4\sin^2 x^\circ = 3$ **b** $\sec(2x - 15)^\circ = 2$

(*UCLES*)

Review

1 Make x the subject of each of the following equations:

a $3y - 4x = 1$

b $y = 5 - 2x^2$

c $y + 1 = \sqrt{x + 5}$

d $y = \sqrt[4]{x} - 3$

e $2y = \dfrac{3}{x - 1}$

f $y = \dfrac{6x - 1}{x}$

2 Express each of the following expressions in the form $(x + a)^2 + b$ or $c(x + a)^2 + b$, where a, b and c are constants:

a $x^2 - 6x + 16$

b $x^2 + 8x + 11$

c $x^2 - x + 2$

d $x^2 + 5x + 7$

e $2x^2 - 20x + 36$

f $3x^2 + 12x + 11$

3 Find the exact solutions of the following quadratic equations:

a $x^2 - 2x - 35 = 0$

b $4x^2 - 7 = 0$

c $2x^2 + 9x + 4 = 0$

d $x^2 - 4x - 1 = 0$

e $x^2 + 8x + 5 = 0$

f $9x^2 + 30x + 23 = 0$

4 Rearrange each of the following into expressions that eliminate improper algebraic fractions:

a $\dfrac{x - 4}{x - 3}$

b $\dfrac{x - 3}{x - 4}$

c $\dfrac{3x + 2}{x + 2}$

d $\dfrac{4x - 7}{x - 5}$

e $\dfrac{6x + 7}{3x + 2}$

f $\dfrac{5x}{x + 2}$

4.1 Technique

1 The functions below are defined for the given discrete domains. In each case:

i draw a mapping diagram

ii state the range of the function

iii state the type of mapping.

a $f : x \to 3 - x$ for $\{x: -4, -2, 0, 2, 4\}$

b $f : x \to \sqrt[3]{x}$ for $\{x: -8, -1, 0, 1, 8\}$

c $f : x \to 1 + x^2$ for $\{x: 0, 5, 10, 15, 20\}$

d $f : x \to \dfrac{1}{x+2}$ for $\{x: -1, 0, 1, 2, 3\}$

2 The functions below are defined for the given continuous domains. In each case:

i sketch the graph of the function

ii state the range of the function

iii state the type of mapping.

a $f{:}x \to -3x + 2$ for $\{x \in \mathbb{R}: -1 \le x \le 3\}$

b $f{:}x \to (x-2)^2$ for $\{x \in \mathbb{R}: -1 \le x \le 5\}$

c $f{:}x \to x^3 - 1$ for $\{x \in \mathbb{R}: -2 \le x \le 2\}$

d $f{:}x \to \dfrac{2}{x+1}$ for $\{x \in \mathbb{R}: -1 < x \le 2\}$

3 State whether or not each of the following graphs represents a function, giving reasons for your answers:

a

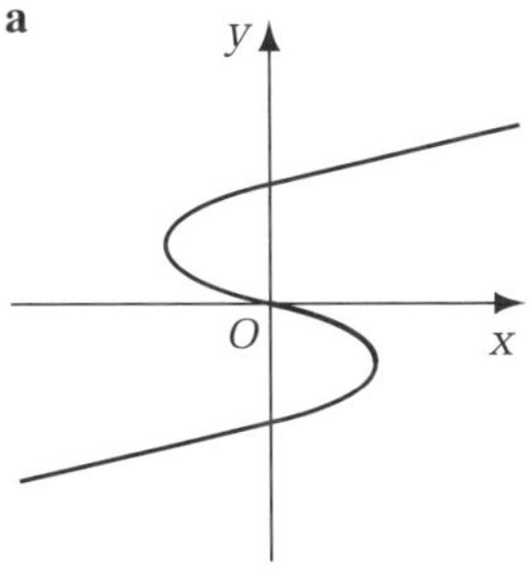

b

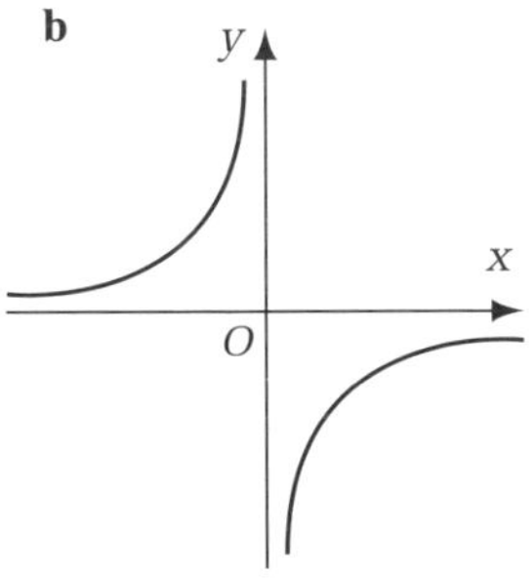

c

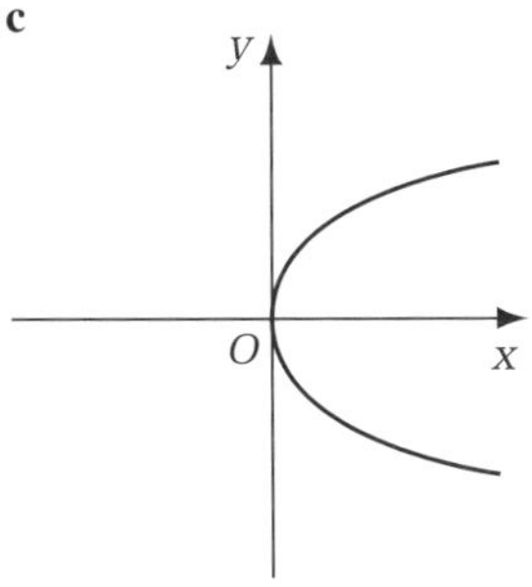

d

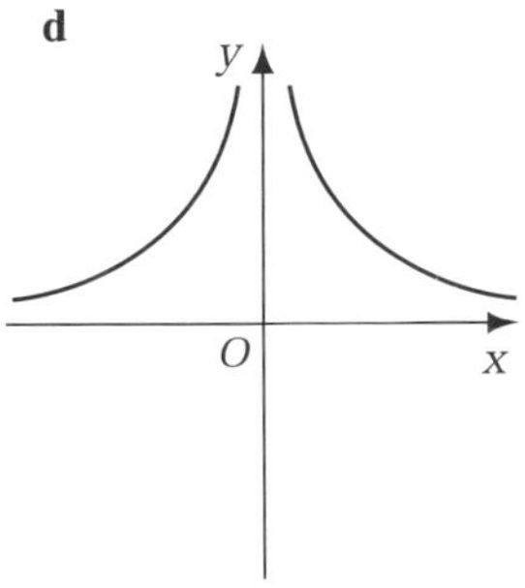

4 The mappings f and g are defined by:

$$f : x \to \begin{cases} \frac{1}{3}x + 1, & x \le 3 \\ x^2 - 7, & x \ge 3 \end{cases}$$

$$g : x \to \begin{cases} x^3 + 1, & x \le 1 \\ \dfrac{2}{x+1}, & x \ge 1 \end{cases}$$

Explain why f is a function, but g is not a function.

5 The functions f and g are such that $f(x) = x^2 - 3x + 2$ and $g(x) = x^2 - 10x - 6$.

a Find the values of a for which $f(a) = 0$.

b Find the values of b for which $g(b) = 5$.

c Find the values of c for which $2f(c) + g(c) = 10$.

6 The linear function f is defined by $f{:}\, x \to ax + b$, where a and b are constants. Given that $f(5) = -2.5$ and $f(-1) = -5.5$, find a and b. Hence calculate the value of c for which $f(c) = 9$.

7 The quadratic function g is defined by $g{:}\, x \to ax^2 + bx + c$, where a, b and c are constants. Given that $g(0) = 5$, $g(2) = 5$ and $g(-2) = -1$, find a, b and c. Hence calculate the values of d for which $g(d) = -1$.

4.2 Technique

1 For each of the following functions:

i state whether an inverse function exists, giving a reason for your answer

ii sketch the graph of the inverse function if it exists.

a

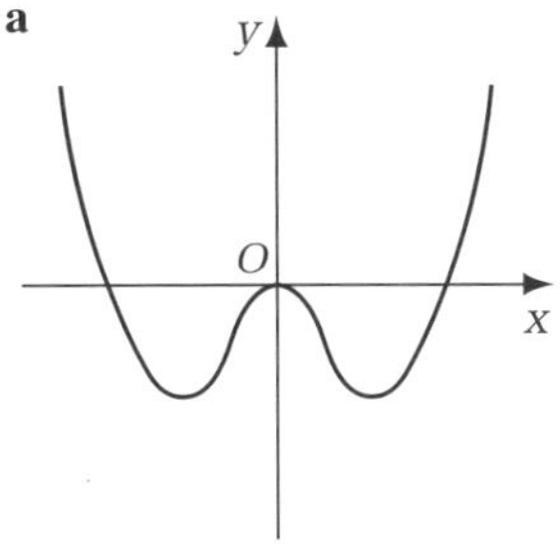

b

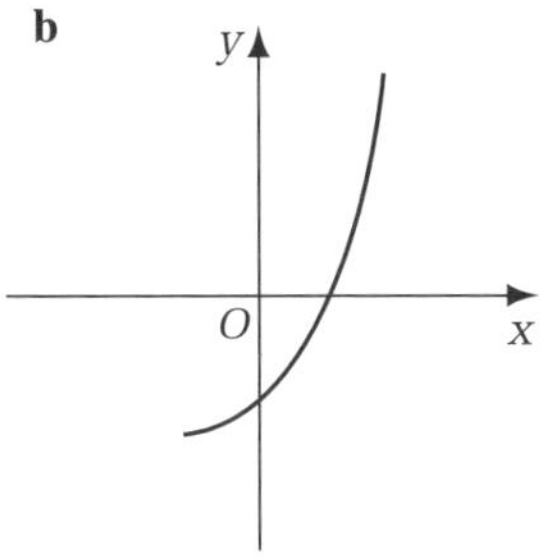

c

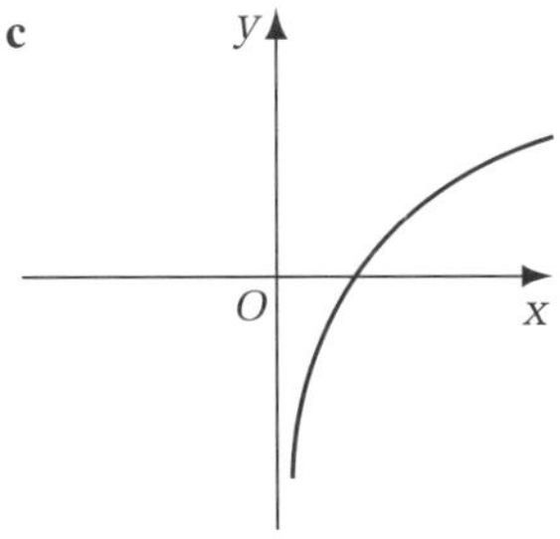

d

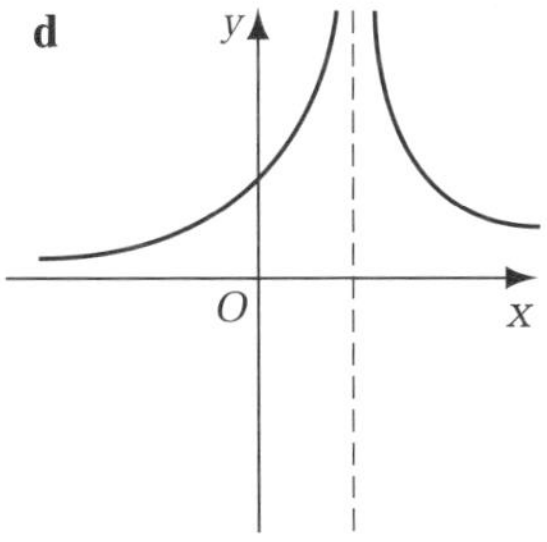

2 Find an expression for f^{-1} for each of the following functions:

a $f(x) = 5x + 11$

b $f(x) = \dfrac{10}{3x+1},\ x \ne -\dfrac{1}{3}$

c $f(x) = 4x^3 + 3$

d $f(x) = \sqrt[3]{x+2},\ x \le -2$

e $f(x) = \dfrac{3}{\sqrt{x-4}},\ x > 4$

f $f(x) = \dfrac{5}{(x-1)^2},\ x > 1$

3 The rational functions f, g and h are defined for the set of real numbers such that:

$$f(x) = \frac{5}{3-x},\ g(x) = \frac{3x}{2x-1} \text{ and } h(x) = \frac{x-4}{3x+2}$$

a State the values of x for which each of these functions is undefined.

b Find expressions for each of the inverse functions, and state their domains.

4 The functions f and g are defined such that $f(x) = \dfrac{x-15}{4x+a}$ and $g(x) = \dfrac{bx}{x+10}$, where a and b are constants.

a Find expressions for $f^{-1}(x)$ and $g^{-1}(x)$.

b State the values of a and b for which functions f and g are self-inverses.

5 **a** Express $x^2 + 6x + 4$ in the form $(x + a)^2 + b$, where a and b are integers.

b Function $f(x) = x^2 + 6x + 4$ is defined for the domain $x \in \mathbb{R}, x \geq q$. Find the least value of q for which function f is a one-to-one mapping.

c State the range of f.

d Find an expression for the inverse function f^{-1}.

e Solve the equation $f(x) = f^{-1}(x)$.

6 The function f is defined for the domain $-\pi \leq x \leq \pi$, by $f(x) = 4\sin\left(\frac{x}{2}\right)$.

a Find an expression for f^{-1}. State, or obtain, the domain of f^{-1}.

b Calculate the value of $f^{-1}(2\sqrt{3})$ in terms of π.

7 The diagram shows the graph of $y = f(x)$ where $f(x) = \cos\left(\frac{x}{2}\right)$.

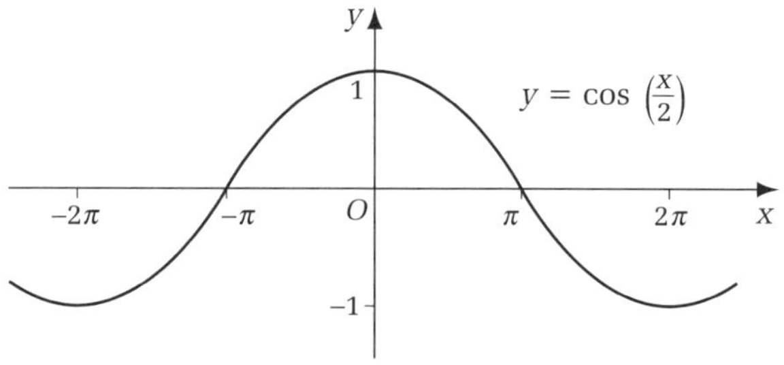

a State the restricted domain for which function f is a one-to-one mapping.

b For this restricted domain:

i find an expression for f^{-1}

ii state the domain of f^{-1}

iii sketch the graphs of $y = f(x)$ and $y = f^{-1}(x)$ on one set of axes, using the same scale on both axes.

c State the domain of the function $g(x) = \cos(ax)$, where a is a positive constant, if the inverse function g^{-1} exists.

4.3 Technique

1 The functions f, g and h are defined by $f(x) = 2x - 3$, $g(x) = \sqrt{x}$ and $h(x) = x^2 + 1$. Find an expression for each of the following composite functions in terms of x.

a fg **e** gh **h** g^2

b gf **f** hg **i** h^2

c fh **g** f^2 **j** fgh

d hf

2 For each of the following, express function p as a composite of the functions

$f: x \to 2x$ $g: x \to \tan 3x$ $h: x \to 5 - x$

a $p: x \to 5 - \tan 3x$ **d** $p: x \to 5 - 2x$

b $p: x \to \tan(15 - 3x)$ **e** $p: x \to 10 - 2x$

c $p: x \to 4x$ **f** $p: x \to 5 - 2\tan 3x$

3 Given that $f(x) = x - 2$, $g(x) = 4x$ and $h(x) = \dfrac{1}{x+1}$:

a find an expression for $fgh(x)$

b solve the equation $fgh(x) = 3x$.

4 Functions f and g are defined by $f(x) = x^2 + 3$ and $g(x) = \sqrt{x - 3}$.

a Show that $fg(x) = x$ and $gf(x) = x$.

b Comment on the significance of your result in **a**.

5 The functions f and g are defined by:

$f: x \to \dfrac{1}{x+1}, x \in \mathbb{R}, x \geq 0$

$g: x \to 2x - 3, x \in \mathbb{R}, x \geq 0$

a State the ranges of f and g for the given domains.

b Find an expression for fg, and determine its domain.

c Find an expression for gf, and determine its domain.

6 The functions f and g are defined by

$f: x \to 2 - x^2, x \in \mathbb{R}$

$g: x \to \dfrac{1}{2x+4}, x \in \mathbb{R}, x \neq -2$

Find expressions for the following, in each case stating any values of x for which these composite functions are undefined:

a $fg(x)$ **b** $gf(x)$

c $f^2(x)$ **d** $g^2(x)$

7 Given that $f(x) = 3x + 1$, $g(x) = 2 - x$ and $h(x) = \dfrac{1}{x^3}$ $(x \neq 0)$, find expressions for:

a f^{-1}, g^{-1} and h^{-1} **b** fg, $(fg)^{-1}$ and $g^{-1}f^{-1}$

c hg, $(hg)^{-1}$ and $g^{-1}h^{-1}$ **d** fh, $(fh)^{-1}$ and $h^{-1}f^{-1}$.

4.4 Technique

1 The diagram shows the graph of $y = f(x)$, where $f(x) = 3 - 2x - x^2$.

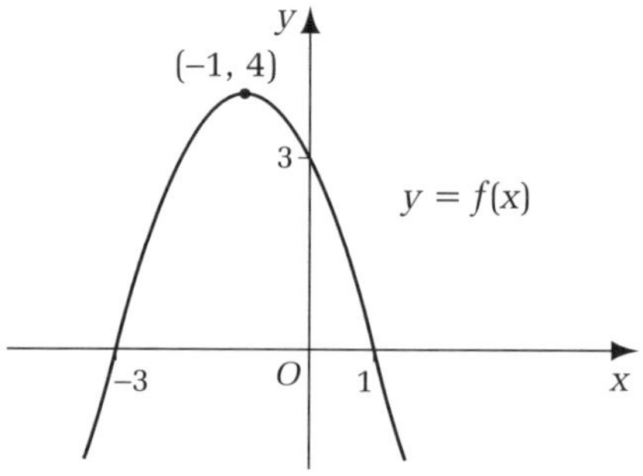

For each of the following transformations of this function:

i describe the tranformation(s) geometrically

ii write down an expression for the new function

iii sketch the graph of the transformed function, indicating the coordinates of its intersection with the axes and the turning point.

a $y = f(x) - 3$ **d** $y = f(x - 2) - 3$

b $y = f(x - 1)$ **e** $y = f(-x)$

c $y = -\frac{1}{2}f(x)$ **f** $y = 2f(x - 3)$

2 The function f is defined by $f(x) = x^2 - 2x + 3$.

a Express f in the form $f(x) = (x + a)^2 + b$, where a and b are integers.

b Describe the transformation by which the graph of $y = x^2$ is mapped to the graph of $y = f(x)$.

3 The function g is defined by $g(x) = 3x^2 + 6x + 1$.

a Express g in the form $g(x) = 3(x + a)^2 + b$, where a and b are integers.

b Describe the transformations by which the graph of $y = x^2$ is mapped to the graph of $y = g(x)$. Clearly state the order in which they must be applied.

4 Each of the following cubic curves is a transformation of the $y = x^3$ curve. In each case, state the transformation(s) that have occurred. State also the order in which they must be applied.

a $y = x^3 - 2$ **d** $y = -x^3 + 1$

b $y = \frac{1}{2}x^3$ **e** $y = (3x - 1)^3$

c $y = (x - 7)^3$ **f** $y = (x + 5)^3 - 3$

5 The diagram shows the graph of $y = x + \dfrac{1}{x}$.

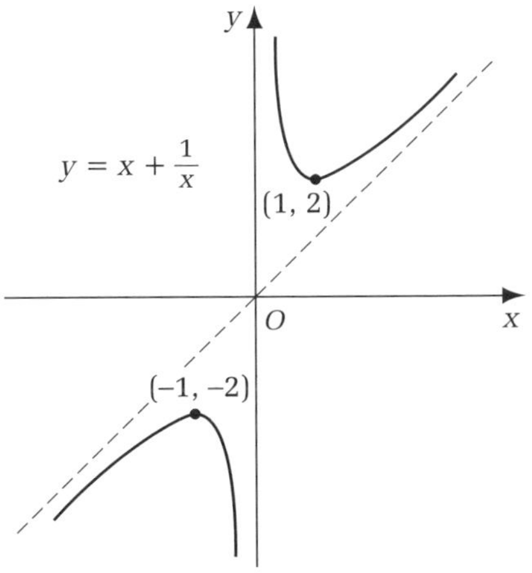

i Describe the transformations that map this graph to each of the graphs below. State the order in which they must be applied.

ii Use these transformations to write down the equation of each graph.

a

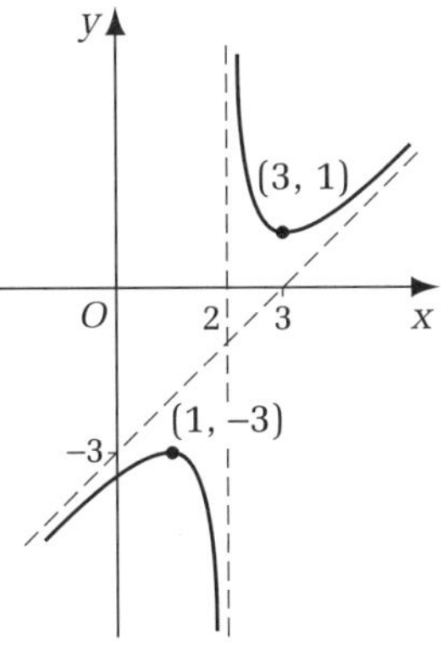

b

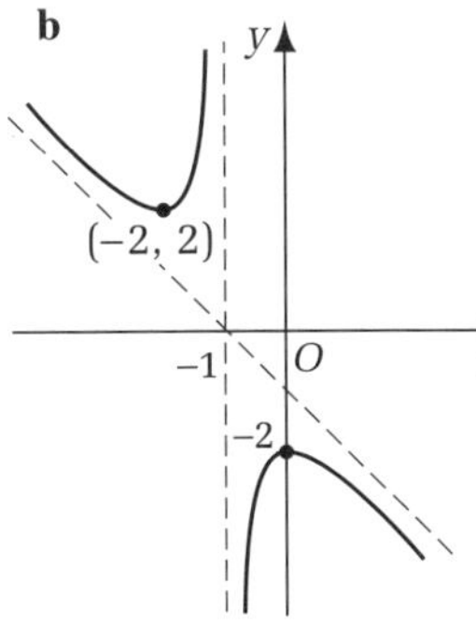

c

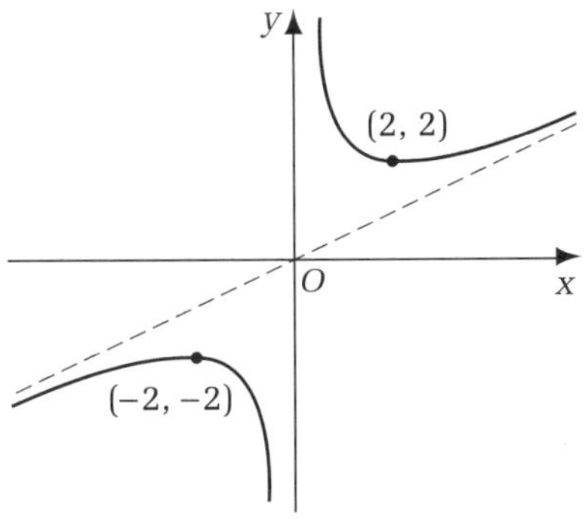

6 **i** Describe the transformations by which the graph of $y = \cos x$ is mapped to each of the curves below.

ii Make a sketch of the graph of each of the transformed functions for $-360° \le x \le 360°$.

a $y = \cos(x + 180°)$ **b** $y = \cos(-x)$

c $y = 3\cos x$ **d** $y = 2 - \cos x$

7 The function $f : x \rightarrow \dfrac{x}{x+1}$, $x \le 1$, is transformed to give the function g. For each of the following transformations:

i find an expression for g

ii state the value of x for which function g is undefined:

a a stretch of factor $\frac{1}{2}$ parallel to the x-axis

b a translation of -5 units parallel to the x-axis

c a stretch of factor 2 parallel to the y-axis, followed by a reflection in the x-axis

d a translation of $+2$ units parallel to the y-axis.

4.5 Technique

1 Which of the following graphs represent the graph of an even or odd function?

a

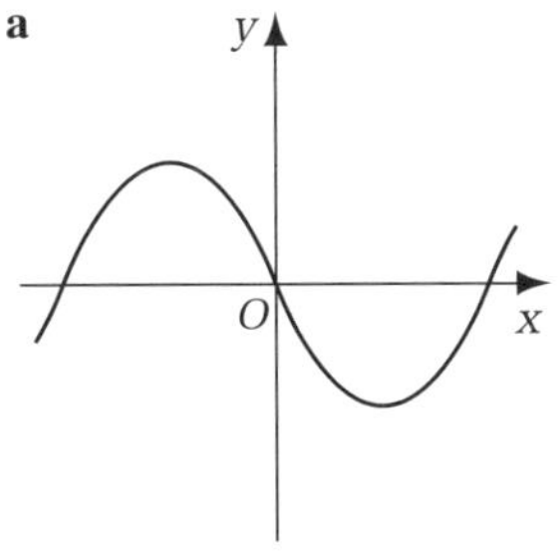

b

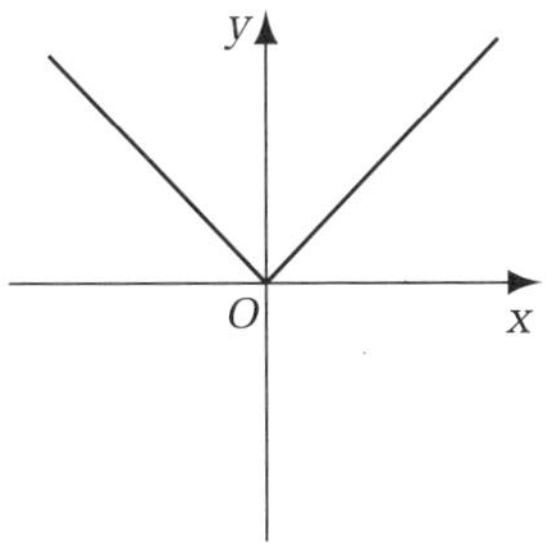

c

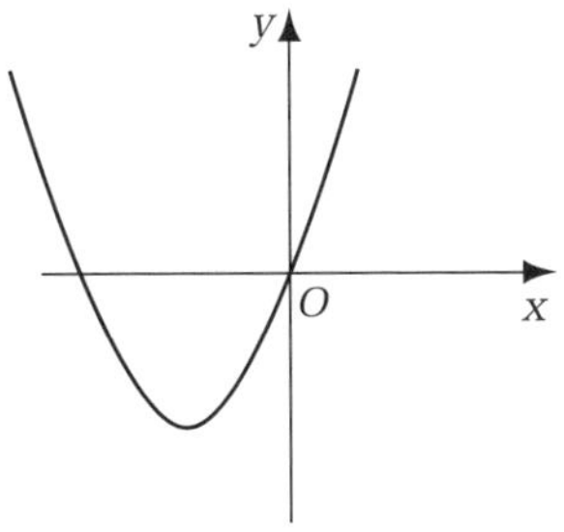

d

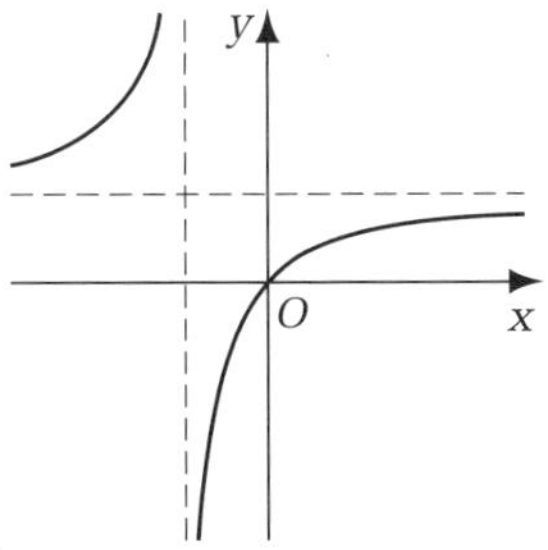

e

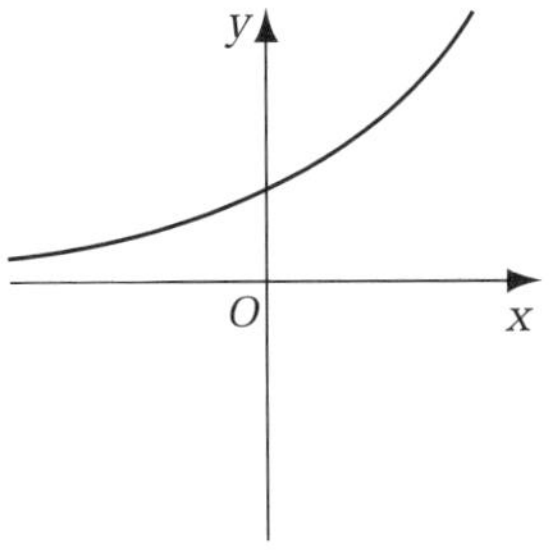

f

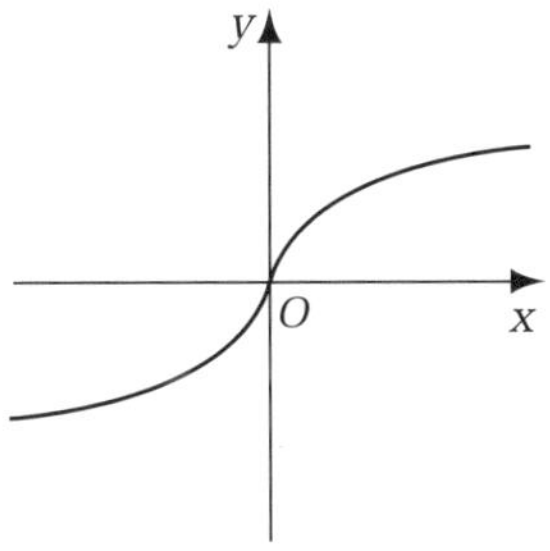

2 Which of the following functions are even, odd, or neither even nor odd?

a $f(x) = x^2 + \dfrac{1}{x}$ **e** $f(x) = |x + 4|$

b $f(x) = x^2 + 2|x|$ **f** $f(x) = |x| + 4$

c $f(x) = -\cos x$ **g** $f(x) = x^3 + 4x$

d $f(x) = (x - 2)^2$ **h** $f(x) = -\tan x$

3 The diagram shows the graph of $y = |f(x)|$ for some quadratic function $f(x)$.

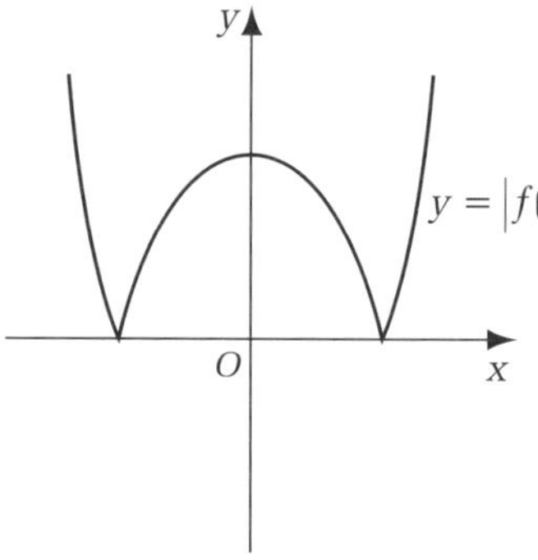

a On separate diagrams, sketch the two possible graphs of $y = f(x)$.

b Briefly explain how to distinguish between these two possibilities.

4 The function $f(x)$ is periodic with a period of 2, and is defined by

$f(x) = \sqrt[3]{x}, \qquad 0 \le x < 1$

$f(x) = (x - 2)^2, \qquad 1 \le x < 2$

a Sketch the graph of $y = f(x)$ for $-2 \le x \le 4$.

b Determine the value of $f(-1.875)$ and $f(5.5)$.

Consolidation C

1 The graph of $y = f(x)$ is shown in the diagram. On separate diagrams sketch the graphs of

a $y = f(x + 1)$ **b** $y = |f(x + 1)|$

showing the coordinates of the points where the graphs meet the x-axis.

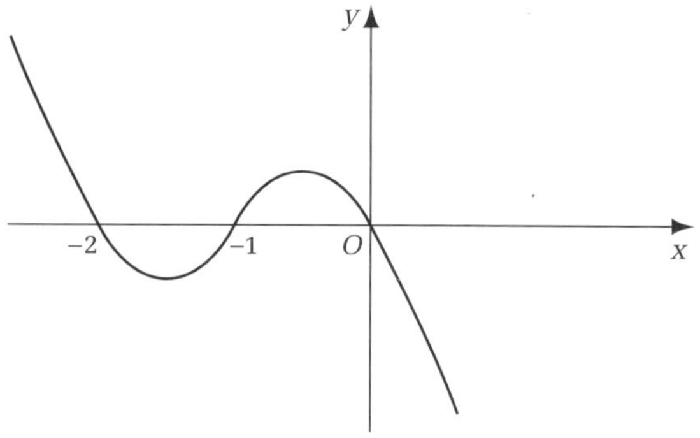

(*UCLES*)

2 The functions f and g are defined by

$f: x \to x^2 + 3, \qquad x \in \mathbb{R}$

$g: x \to 2x + 1, \qquad x \in \mathbb{R}$

a Find, in a similar form, the function fg.

b Find the range of the function fg.

c Solve the equation $f(x) = 12g^{-1}(x)$.

(*ULEAC*)

3 The graph of $y = f(x)$ is sketched below and is such that $f(x) = 1$ for $x \le 1$ and for $x \ge 4$.

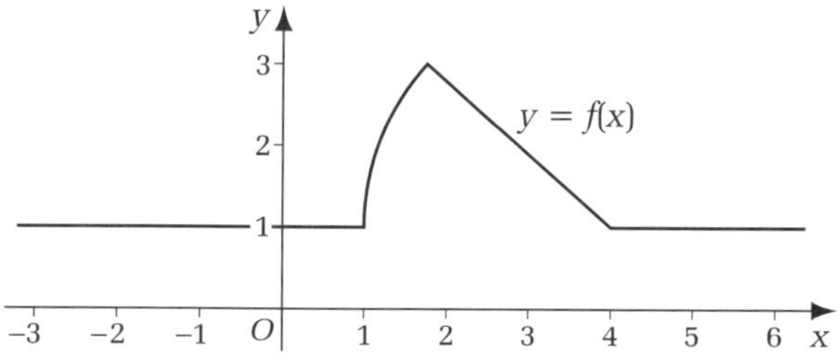

Sketch the graph of each of the following on separate axes, giving a clear indication of the scale on the axes.

a $y = f(x - 2)$ **b** $y = f(-x)$ **c** $y = f(\frac{1}{2}x) - 2$

(*AEB*)

4 The function f with domain $x \in \mathbb{R}, x \ge 0$ is defined by $f(x) = 4x^2 - 1$.

a State the range of f and sketch the graph of f.

b Explain why the inverse function f^{-1} exists and find $f^{-1}(x)$.

c Given that g has domain $x \in \mathbb{R}, x \ge 0$ and is defined by $g(x) = \sqrt{x + 6}$, solve the inequality $fg(x) \ge f(x)$.

(*OCSEB*)

5 The functions f and g are defined for $x \in \mathbb{R}$ by $f: x \to x^3$ and $g: x \to 2 - 3x$. Find:

a $fg(x)$ **b** $(fg)^{-1}(x)$

(*UCLES*)

6 Functions f and g, each with domain $\mathbb{R}$, are defined by $f: x \to x + 1$ and $g: x \to |x|$.

a Write down an expression for $f^{-1}(x)$, and state, giving a reason, whether g has an inverse.

b Sketch the graph of $y = gf(x)$.

c Find the solution set of the equation $gf(x) = fg(x)$.

(*UCLES*)

7 The function f is defined on the domain $x > 0$ by $f(x) = 1 + \dfrac{2}{x}$.

a i Find an expression for $f^{-1}(x)$.

ii State the domain of f^{-1}.

b The composite function g is defined by $g = ff$.

i Find an expression for $g(x)$.

ii State the range of g.

(*WJEC*)

8 Functions f and g are defined by

$f: x \to 4 - x, \qquad x \in \mathbb{R}$

$g: x \to 3x^2, \qquad x \in \mathbb{R}$

a Find the range of g.

b Solve $gf(x) = 48$.

c Sketch the graph of $y = |f(x)|$ and hence find the values of x for which $|f(x)| = 2$.

(*ULEAC*)

9 The graph of the continuous function $f(x)$ is sketched below.

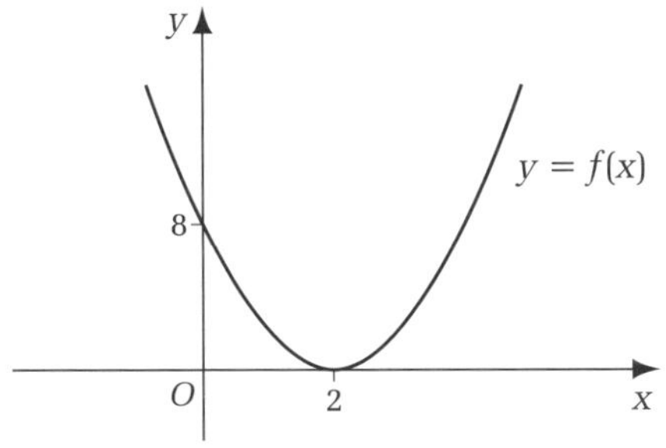

Draw similar sketches for the functions:

a $y = 2f(x)$ **b** $y = f(x - 2)$ **c** $y = f(2x)$

(NICCEA)

10 The functions f and g are defined by

$f: x \to x^2 - 1, \quad x \in \mathbb{R}, x \geq 0$

$g: x \to (x - 1)^2, \quad x \in \mathbb{R}, x \geq 0$

a State the range of f.

b Find the inverse function f^{-1} and state its domain.

c Explain, briefly, why the function g does not have an inverse for the given domain.

d Write down a domain for which g has an inverse.

(NICCEA)

Review

1 Find the gradient of the straight line joining the following pairs of points:

a (1, 3) and (4, 15) **b** (−4, 16) and (1, −9)

c (2, −5) and (5, 2) **d** (−3, 0) and (0, −4)

2 Find the value of the following functions at the given value of x:

a $f(x) = 2x^3 - 7x + 10$, when $x = -1$

b $f(x) = \frac{2}{x} + \frac{x}{2}$, when $x = 3$

c $f(x) = \sin x + \sin 2x$, when $x = \frac{\pi}{3}$

3 Find the equation of the straight line that passes through the given point with the gradient indicated, giving your answer in the form $y = mx + c$:

a (2, 3), gradient = 4 **b** (−3, 1), gradient = $\frac{1}{3}$

c (5, −6), gradient = −5 **d** (−5, 5), gradient = $-\frac{4}{3}$

4 Solve the following trigonometric equations for $-\pi \leq \theta \leq \pi$:

a $\tan\left(\theta + \frac{\pi}{4}\right) = 1$ **b** $2\sin\theta - 3\cos\theta = 0$

c $\sin 2\theta + \sqrt{3}\cos\theta = 0$ **d** $\cos 2\theta = 3\cos\theta + 1$

5 Write each of the following in index notation:

a $\frac{4}{x^2}$ **d** $\frac{9}{\sqrt[3]{x}}$

b $\sqrt[3]{x^2}$ **e** $(\sqrt[4]{x})^3$

c $\frac{2}{(x-5)^2}$

5.1 Technique

1 Differentiate each of the following from first principles to find $\frac{dy}{dx}$:

a $y = x^4$ **b** $y = 2x^2 + 7x - 3$

c $y = \frac{1}{x^2}$

2 Differentiate each of the following with respect to x:

a $y = x^9$ **e** $y = 4x - \frac{1}{x^2}$

b $y = \frac{1}{4}x^4 + \frac{1}{3}x^3$ **f** $y = \frac{3x^2 - 5}{x^3}$

c $y = (4x + 3)(x - 5)$ **g** $y = \frac{6}{\sqrt[3]{x}}$

d $y = \sqrt{x}^3$ **h** $y = \frac{2x - x^3}{\sqrt{x}}$

3 Find the first derivative of each of the following functions:

a $f(x) = \frac{1}{5}x^2 + \frac{1}{4}x^3$ **d** $h(p) = \frac{3p - 4p^3}{p^2}$

b $f(t) = (t + 2)(3 - t)$ **e** $\theta(t) = 6t^{\frac{3}{2}}$

c $f(s) = s^4 + 2s^3 - 4s$

4 Find the gradient of each of the following curves at the point indicated:

a $y = x^2 - 7x + 2$, at (4, −10)

b $y = x^4 - 3x^2 + 1$, at (−2, 5)

c $y = \frac{3}{x} + \frac{x}{3}$, at (3, 2)

d $y = (x^2 - 2)(x^2 + 3)$, at (−1, −4)

5 Find the coordinates of the points on each of the following curves at which the gradient has the value indicated:

a $y = x^3 - 3x^2 - 18x + 10$, where $\frac{dy}{dx} = 6$

b $y = x^3 - 2x + 3$, where $\frac{dy}{dx} = 25$

c $y = 3x + 1 - \frac{2}{x}$, where $\frac{dy}{dx} = 5$

d $y = (x + 9)(x - 1)$, where $\frac{dy}{dx} = 0$

6 Find the gradient of each of the following curves at their points of intersection with the x- and y-axes:

a $y = x^2 - 6x + 8$ **b** $y = (3x + 1)(x - 2)$

7 Find the first, second and third derivatives of each of the following:

a $y = 1 + 5x - 9x^2 + 3x^3 - 2x^4$

b $f(x) = 2x - \frac{4}{x}$

5.2 Technique

1 For each of the following quadratic functions:

i find the coordinates of the points where the graph of $y = f(x)$ crosses the axes

ii find the coordinates of the stationary point on the graph, and determine its nature

iii sketch the graph of $y = f(x)$, and check your graph using a graphical calculator.

a $f(x) = x^2 - 8x + 15$ **b** $f(x) = 7 + 6x - x^2$

c $f(x) = x^2 - 4x + 5$ **d** $f(x) = 2x^2 - 14x + 20$

2 For each of the following cubic functions, find the coordinates of any stationary points, and determine their nature:

a $y = x^3 - 9x^2 + 27x - 22$

b $y = x^3 + 2x^2 + 2x + 5$

c $y = 8x^3 + 12x^2 + 6x - 1$

d $y = x^3 - 3x^2 - 24x + 20$

3 **a** Find the coordinates of the points where the quartic curve $y = x^4 + 4x^3$ crosses the axes.

b Find $\frac{dy}{dx}$. Hence find the location, and determine the nature, of the stationary points on the curve.

c Sketch $y = x^4 + 4x^3$. Check your graph using a graphical calculator.

4 For each of the following functions, find an expression for $f'(x)$, and hence locate and determine the nature of any stationary points on the graph of $y = f(x)$.

a $f(x) = \frac{x}{2} + \frac{8}{x}$ **b** $f(x) = x + \frac{32}{x^2}$

c $f(x) = x^2 + \frac{16}{x^2}$ **d** $f(x) = \frac{2}{x} - \frac{3}{x^2}$

5 Given that $A = 4r^2 + 4r - 3$, find the minimum value of A and the value of r for which this occurs.

6 Given that $W = 20 - t - t^2$, find the maximum value of W and the value of t for which this occurs.

7 Find the values of x for which $f(x) = x^2 - 13x + 42$ is an increasing function of x.

8 Find the values of u for which $P(u) = 2u^3 - 11u^2 - 8u + 5$ is a decreasing function of u.

5.2 Contextual

1 A window box is made by cutting squares of side length y cm from each corner of a 240 cm × 65 cm rectangular sheet of aluminium. The remaining flaps are then folded up to make the vertical sides

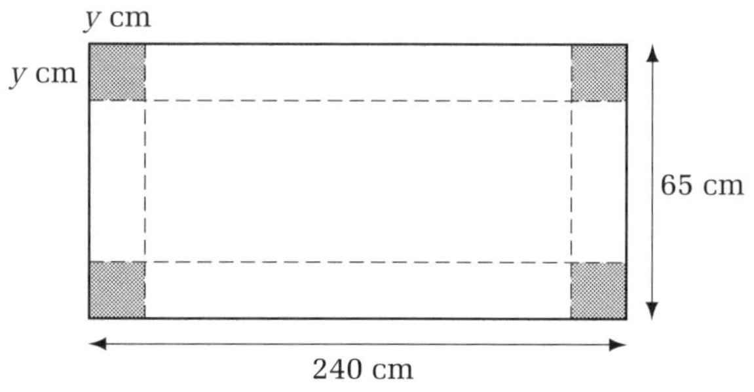

a Find an expression in terms of y for the volume V (in cm^3) of the box.

b Find $\frac{dV}{dy}$, and then solve the equation $\frac{dV}{dy} = 0$.

c Find the maximum possible volume of the box. Give its dimensions and justify your answer.

2 The owner of a fish farm wishes to construct 12 salt-water interlinked ponds, as shown in the diagram. Each pond is to be 3 metres deep, x metres long and y metres wide. The sides and bottom of the ponds are to be made of netting, but the top of each pond is to be left open. Adjacent ponds share the netting for their adjoining walls. The owner calculates that the total volume required is 13 500 m^3.

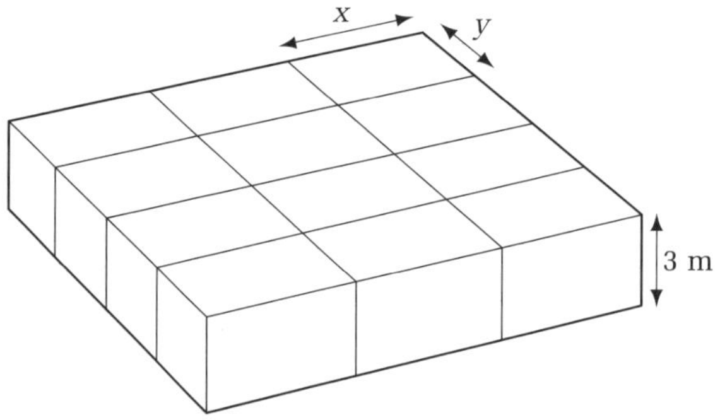

a Find an expression for the total volume V (in m^3), and hence express y in terms of x.

b Find an expression for the total area of netting required, A (in m^2), in terms of x only.

c Hence find the dimensions of each rectangular pond if the total area of netting is to be kept to a minimum. Justify your answer.

d What is the minimum area of netting required?

3 The price of copper on the London Commodity Market during the course of a particular day's trading can be modelled by $C = 380 - 9t^2 + 5t^3 - \frac{1}{2}t^4$, where C is the price per unit of copper (in pence) and t is the time in hours after the 8 am start of trading.

a Write down the opening price, and calculate the price per unit at noon.

b Calculate the highest price that copper reaches during the day. At what time does this happen?

c If trading closes at 4 pm, by how much has the price of copper risen or fallen during the course of the day's trading?

4 A vase in the shape of a hexagonal prism has a capacity of $1\frac{1}{2}$ litres (1 litre ≡ 1000 cm^3). Its height is h cm, and each of the sides of its hexagonal base is x cm.

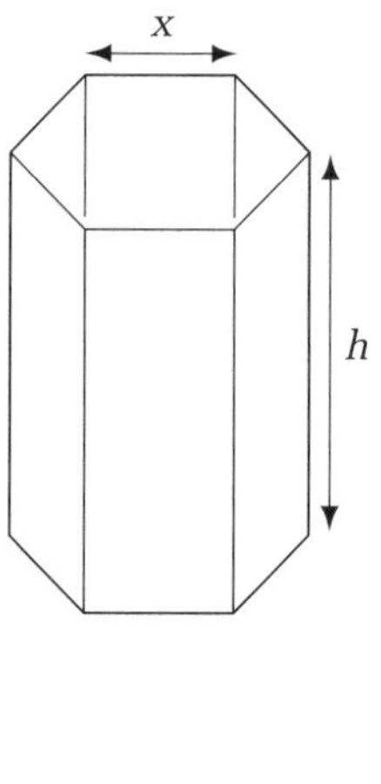

a Show that the area of its base is $\dfrac{3\sqrt{3}x^2}{2}$ cm^2.

b Show that $h = \dfrac{1000}{\sqrt{3}x^2}$.

c Express the total external surface area of the vase, S, in term of x only.

d Find the values of x and h that correspond to the minimum value of S, giving your answers to the nearest millimetre. What is this minimum surface area?

5 A department store wants to sell as much of its leftover Christmas stock as possible in its January sale. It estimates that it could sell 200 musical Santas if it reduced the price of each one to £3, but that only 100 could be sold at £5 each. The store originally bought the Santas for £2 each.

a The number of musical Santas sold can be modelled by $n = a + bs$, where £s is the selling price, and a and b are constants. Find a and b.

b Show that the cost £c to the store of those Santas it manages to sell is given by $700 - 100s$.

c Show that, in terms of the selling price s, the profit generated by the musical Santas is given by $p = 450s - 50s^2 - 700$, and find the selling price that would maximise the profit. Calculate the maximum profit and the number of musical Santas that would have to be sold to achieve it.

5.3 Technique

1 Find the equations of the tangent and normal to each of the following curves at the points indicated:

a $y = x^2 + 5x + 2$, at $(-1, -2)$

b $y = (3x - 1)(x - 2)$, at $(3, 8)$

c $y = x^4 - 7x^2$, at $(-2, -12)$

d $y = x - \dfrac{4}{x}$, at $(4, 3)$

e $y = \sqrt[3]{x}$, at $(-8, -2)$

f $y = x + x^3$, at $\left(\frac{1}{2}, \frac{5}{8}\right)$.

2 Find the equations of the tangents to the curve $y = 15 - 2x - x^2$ at the points where it crosses the x-axis. Show that these tangents intersect at the point $(-1, 32)$.

3 Show that the equation of the tangent to the curve $y = x^3 + 9x^2 - 2x + 5$ at the point where it crosses the y-axis is $y = -2x + 5$. Find the coordinates of the other point on the curve whose tangent is parallel to $y = -2x + 5$. Find the equation of this second tangent.

4 Find the equation of the normal to the curve $y = x - \dfrac{4}{x}$ at the point where $x = 2$. Find the coordinates of the other point where this normal intersects $y = x - \dfrac{4}{x}$.

5.3 Contextual

1 The height h metres of an object thrown vertically upwards at time t seconds after its release is given by $h = 25t - 5t^2$.

a Calculate how long it takes for the object to return to its point of projection.

b Find the value of t at which the object is momentarily stationary, and hence calculate the maximum height reached by the object.

2 The displacement, s metres, of an object from some fixed point O at time t seconds is given by $s = t^3 - 2t^2 + 3t - 4$.

a Find expressions for the velocity, v, and acceleration, a, at time t.

b Show that the object always moves in the same direction.

c Calculate the average velocity of the object during the first five seconds.

3 A cylinder has a base radius of 9 cm and a height of 8 cm. Calculate the small changes:

a in its volume **b** in its surface area

that result from a small change of 0.2 cm in its height. Leave your answers in terms of π, and assume that the radius remains constant.

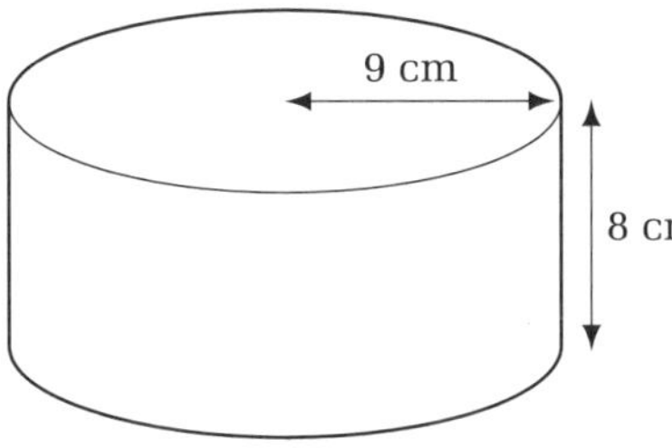

4 A cuboidal box has a height $2x$, a length $8x$ and a width $3x$. Calculate the percentage change in:

a its surface area **b** its volume

if the value of x decreases by 3%.

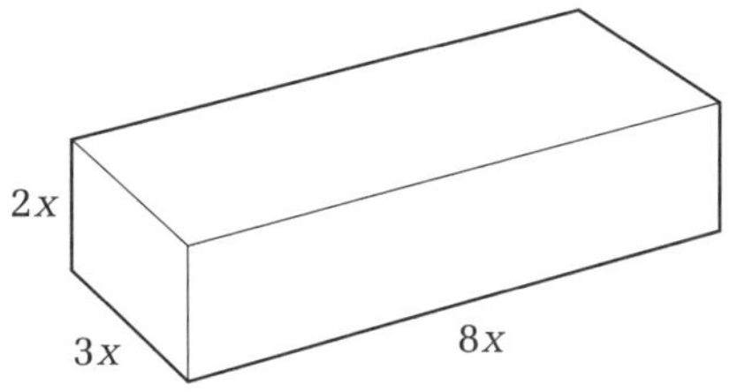

5.4 Technique

1 Use the chain rule to find $\dfrac{dy}{dx}$ for each of the following:

a $y = (x + 7)^5$

b $y = (3x - 4)^6$

c $y = (5 - 2x)^4$

d $y = (x^3 - 8)^3$

e $y = (x^2 + 5x - 2)^2$

f $y = \dfrac{2}{3x - 11}$

g $y = \sqrt{4x - 5}$

h $y = (x^5 - 1)^{\frac{6}{5}}$

i $y = \dfrac{1}{(9 - 4x)^2}$

j $y = \dfrac{1}{\sqrt{16 + x^2}}$

2 Use the chain rule to find the derivative of each of the following functions.

a $f(t) = (6t + 1)^5$

b $p(c) = (c - 6)^{-3}$

c $h(x) = \sqrt{x^4 + 1}$

d $g(a) = \sqrt[3]{12a - 6}$

e $r(\theta) = \dfrac{6}{\sqrt{\theta + 1}}$

f $s(t) = \dfrac{10}{t^4 + 1}$

g $I(w) = \left(w - \dfrac{1}{w}\right)^3$

h $A(\theta) = (6 + \sqrt{\theta})^4$

3 Find the gradient of each of the following curves at the point indicated:

a $y = (x + 4)^3$, at $(-2, 8)$
b $y = (3x - 10)^5$, at $(3, -1)$
c $y = (13 - 4x)^4$, at $(4, 81)$
d $y = \dfrac{3}{2x - 7}$, at $(5, 1)$
e $y = \sqrt{100 - x^2}$, at $(8, 6)$.

4 Find the coordinates of the point(s) on each of the following curves at which the gradient has the value indicated:

a $y = (x - 7)^5$, where $\dfrac{dy}{dx} = 80$
b $y = (7x + 6)^4$, where $\dfrac{dy}{dx} = -28$
c $y = \sqrt{x + 10}$, where $\dfrac{dy}{dx} = \dfrac{1}{8}$

5 Given that $h = y^2 - 5y - 6$ and $y = 6 - x^2$ find $\dfrac{dh}{dy}$ and $\dfrac{dy}{dx}$. Hence, use the chain rule to find an expression for $\dfrac{dh}{dx}$ in terms of x only.

6 Given that $P = (4s + 7)^3$ and $t = 2s - 3$ use the chain rule to find an expression for $\dfrac{dP}{dt}$ in terms of t only.

5.4 Contextual

1 The radius of a circular wine stain on a white tablecloth is increasing at the rate of $\dfrac{30}{10 + r}$ centimetres per second. Find the rates at which the stain's perimeter and area are changing when the radius of the stain is 2 centimetres (leave your answers in terms of π).

2 As air is pumped into a spherical balloon, the rate at which its volume increases remains a constant 30π cm^3 s^{-1}.

a Find the rate at which the radius of the balloon is changing when $r = 5$ cm.
b What is the rate at which the surface area of the balloon is increasing when $r = 4$ cm?

3 Find the coordinates of the stationary points on the graph of $y = 2x + \dfrac{1}{2x - 5}$, and determine their nature.

4 A cuboidal metal block, of length $15x$ cm, width $3x$ cm and height x cm, is being uniformly heated, and is expanding such that its volume increases at the constant rate of 432 cm^3 s^{-1}.

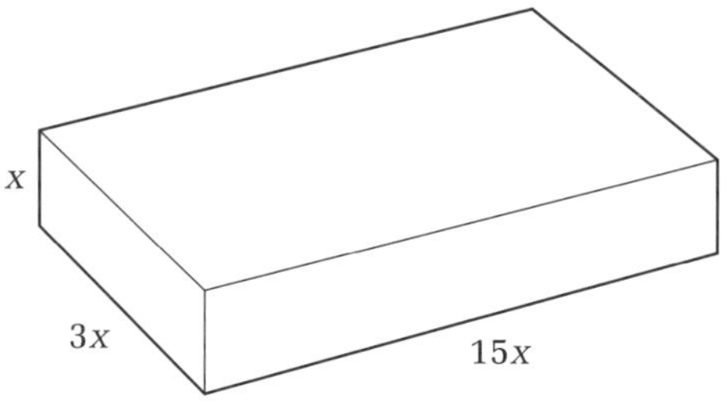

a Express V in terms of x, and hence find $\dfrac{dV}{dx}$.
b Use the chain rule to calculate the rate at which x is changing when the height is 8 cm.
c Find an expression for the surface area S of the block. Hence calculate the rate at which it is changing when the height is 8 cm.

5 Sand flows into the bottom flask of a conical egg timer at the rate of 35 cm^3 s^{-1}. The height of the flask is 20 cm, and the radius of its base is 10 cm. At any given time, t seconds, the height of the sand in the bottom of the flask is h cm.

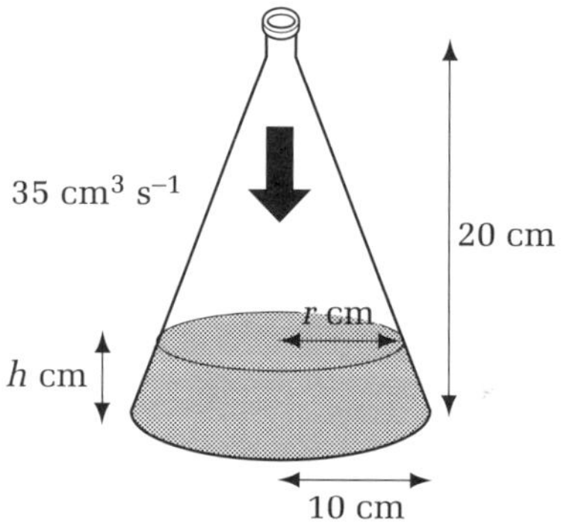

a Explain why the radius r cm of the top of the sand at any given time is $\frac{1}{2}(20 - h)$ cm.
b Find an expression for the volume V of sand in the flask in terms of h only.
c Show that $\dfrac{dV}{dh} = \dfrac{\pi}{4}(20 - h)^2$.
d Calculate the rate at which the height of the sand is rising when $h = 10$ cm.

5.5 Technique

1 Differentiate each of the following with respect to x:

a $y = \sin 4x$
b $y = \cos 9x - \cos 7x$
c $y = \sin \frac{5}{2}x$
d $y = -3\cos 3x$
e $y = \sin (x + \pi)$
f $y = \cos^4 x$
g $y = \sin^3 5x$
h $y = \sin 8x - \sin 4x$
i $y = 6 \sin^2\left(\dfrac{x}{3}\right)$
j $y = a\sin^2 (px) + b\cos^2 (qx)$, where a, b, p and q are constants.

2 Find the gradient of each of the following curves at the point indicated:

a $y = 4 \sin x$, where $x = \dfrac{\pi}{3}$
b $y = \cos^2 x$, where $x = \dfrac{\pi}{4}$
c $y = x^3 - \sin 2x$, where $x = \pi$
d $y = \sin 5x - \sin 3x$, where $x = 0$

3 Find the coordinates of the points on each of the following curves at which the gradient has the value indicated.

a $y = 5\cos x$, for $0 \leq x \leq 2\pi$, where $\dfrac{dy}{dx} = -5$

b $y = \sin^2 x$, for $0 \le x \le \pi$, where $\frac{dy}{dx} = 0$

c $y = 2x - \sin x$, for $0 \le x \le 2\pi$, where $\frac{dy}{dx} = 3$

d $y = \cos 2x$, for $0 \le x \le \pi$, where $\frac{dy}{dx} = -1$

4 For each of the following trigonometric functions, find an expression for $f'(x)$. Hence locate and determine the nature of any stationary points on the graph of $y = f(x)$ in the interval $0 \le x \le 2\pi$.

a $f(x) = 2\cos\frac{3x}{4}$ **b** $f(x) = \frac{1}{2}x - \sin x$

c $f(x) = x + \sin^2 x$

5 For each of the following trigonometric functions, in the interval $0 \le x \le 2\pi$:

i find the coordinates of the points where the graph of $y = f(x)$ crosses the axes

ii find the coordinates of the stationary points on the graph, and determine their nature

iii sketch the graph of $y = f(x)$. Check your results using a graphical calculator.

a $f(x) = \cos x - \sqrt{3}\sin x$ **b** $f(x) = \sin x + 4\cos x$

c $f(x) = 2\sin x - 5\cos x$

6 Find the equations of the tangent and normal to the curve $y = \sin 2x - 3\cos x$ at the point where:

a $x = \frac{\pi}{2}$ **b** $x = \pi$

5.5 Contextual

1 The depth D metres of water in a harbour during a 24 hour period is given by $D = 8 + \sin\left(\frac{2\pi t}{13}\right) - 3\cos\left(\frac{2\pi t}{13}\right)$, where t is the time in hours after midnight.

a Write down the initial depth of the water at midnight.

b Find an expression for $\frac{dD}{dt}$.

c Calculate the rate at which the depth of water is changing at midnight.

d Calculate the rate at which the depth of water is changing at 8.00 am.

e At what time does the first high tide occur?

2 The displacement, x metres, of an object from some fixed point 0, at time t seconds after it is released, is given by $x = 4\sin 2t - 3\cos t$.

a Find expressions for the object's velocity and acceleration at time t.

b Calculate the values of its initial displacement, velocity and acceleration.

c Find the times at which the object is momentarily stationary during the first 2π seconds of its motion, and calculate its displacement at these times.

Consolidation C

1 You are given that $y = 2x^3 + 3x^2 - 72x + 130$.

a Find $\frac{dy}{dx}$.

P is the point on the curve where $x = 4$.

b Calculate the y-coordinate of P.

c Calculate the gradient of P, and hence find the equation of the tangent to the curve at P.

d There are two stationary points on the curve. Find their coordinates.

(OCSEB)

2 The volume, V, of a sphere of radius r is given by $V = \frac{4}{3}\pi r^3$.

a Obtain an expression for $\frac{dV}{dr}$.

b A balloon when almost inflated can be modelled as a sphere. When the radius of the balloon is 15 cm, it is observed that the rate of increase of the radius is 0.1 cm s^{-1}. Find, to two significant figures, the rate of increase of the volume at this time.

(NEAB)

3 A curve has the equation $y = \frac{5x^2 + 8}{x}$. Find an expression for $\frac{dy}{dx}$ and hence find, in terms of p, where p is small, the approximate increase in y as x increases from 2 to $2 + p$.

(UCLES)

4 A light aircraft starts from rest and travels along a straight level runway until it has sufficient speed for take-off. During its motion along the runway the aircraft travels s metres in t seconds, where $s = 0.4t^3 - 0.02t^4$. Given that at the instant of take-off the horizontal acceleration of the aircraft is zero, calculate:

a the time taken from rest until take-off

b the distance it travels along the runway.

(NEAB)

5 An open rectangular tank has a horizontal square base of side x metres. The height of the tank is y metres. The tank is made of thin sheet metal. The external surface area of the tank is S square metres, and the volume is V cubic metres.

a Show that $S = \frac{4V}{x} + x^2$.

Given that $V = 62.5$ and that x varies,

b Find $\frac{dS}{dx}$.

c Determine the value of x for which $\frac{dS}{dx} = 0$ and find the corresponding value of S.

d Show that the value of S is the minimum possible value.

(ULEAC)

6 A new make of personal organiser costs £100 each to manufacture and market. Market research indicates that the number, N thousand, of organisers that will be sold in the first year is given by $N = aS + b$, where a and b are constants and £S is the selling price. The numbers which will be sold at two particular selling prices are given in the table.

S	N
120	30
140	10

a Find the values of a and b.

b Express the profit per organiser in terms of S and hence show that the total profit in the first year is £T thousand, where $T = -15000 + 250S - S^2$.

c Find the selling price that will produce the maximum possible total profit in the first year and find this total profit. (You should show that this profit is a maximum and not a minimum.)

(*UCLES*)

7 The diagram shows the flat surface of a tray consisting of a rectangular region WXYZ and a semi-circular region at each end with WX and YZ as diameters. The rectangle WXYZ has area 200 cm^2.

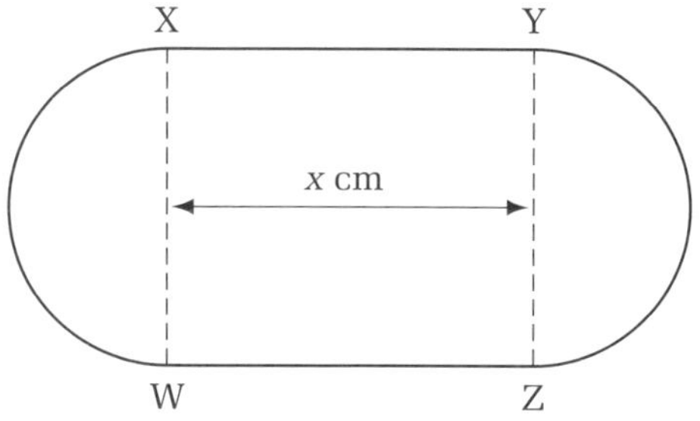

a Given that $XY = x$ cm, show that the perimeter P cm of the tray is given by the formula $P = 2x + \dfrac{200\pi}{x}$.

b Find the minimum value of P as x varies.

(*ULEAC*)

8 A sheet of paper is folded to make a box of width w cm and volume V cm^3, where $V = w(15 - w)(20 - w)$. Use differentiation to find the value of w for which the maximum volume occurs. Find this maximum volume, giving your answer to two decimal places.

(*NEAB*)

9 **a**

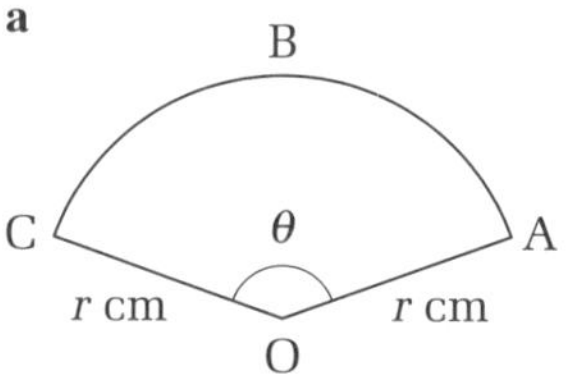

The diagram shows a piece of wire bent to form the perimeter OABCO of a sector of a circle, centre O, radius r cm, where angle AOC is θ radians. The wire is of length 100 cm and r and θ may vary. Find:

i the value of r for which the area enclosed by the wire is a maximum,

ii the corresponding value of the angle AOC in degrees.

b Given that $y = -\dfrac{2}{3x^3} + \dfrac{5}{2x^2} - \dfrac{2}{x}$, find the two values of x for which y is stationary. Show that the larger of these values of x corresponds to a minimum value of y.

(*UCLES*)

10 A particle P moves along a straight line. At time t seconds after the motion begins, the displacement of P from a fixed point O on the line is s metres, where $s = 2\sin t + \cos 2t$. Find:

a the initial velocity and acceleration of P.

b the times in the interval $0 \leq t \leq 2\pi$ seconds at which particle P is at rest.

c the greatest distance of P from O during this time interval, and the time at which this occurs.

Review

1 Use the long division algorithm (method) to find the following, leaving any remainders as fractions:

a $1253 \div 13$ **b** $3121 \div 7$

c $9258 \div 8$ **d** $71261 \div 11$

2 Write each of the following in its simplest form:

a $\frac{1}{3} + \frac{1}{5}$ **b** $\frac{2}{7} + \frac{1}{3}$

c $\frac{1}{4} + \frac{3}{10}$ **d** $\frac{2}{5} - \frac{1}{7}$

3 Express the following as a mixed fraction (that is, a number and a proper algebraic fraction):

a $\dfrac{x+3}{x+1}$ **b** $\dfrac{x-1}{x+4}$

c $\dfrac{3x+2}{x-2}$ **d** $\dfrac{5x}{1-x}$

4 Match each of the following equations to the graphs sketched below:

a $y = x^3$ **c** $y = (x-4)^2 - 3$

b $y = \dfrac{1}{x}$ **d** $y = |x^2 - 9|$

e $y = 2x + 1$

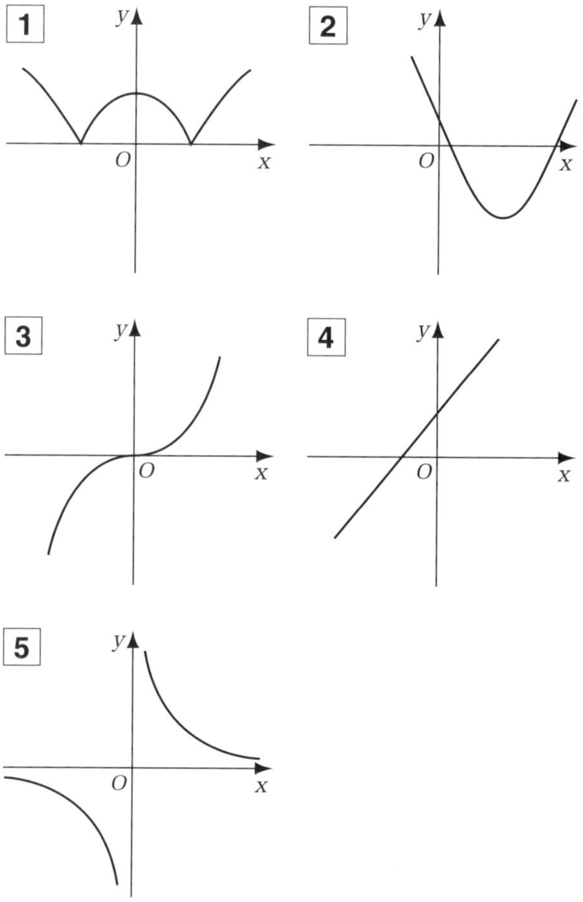

6.1 Technique

1 Use the long division algorithm to find:

a $\dfrac{x^3 + 4x^2 + 5x + 2}{x+1}$ **b** $\dfrac{x^3 + 4x^2 - 2x - 3}{x-1}$

c $\dfrac{2x^3 - 5x^2 + 13x + 8}{2x+1}$ **d** $\dfrac{2x^3 - x^2 + 9}{2x+3}$

2 Use the remainder theorem to find the remainder in the following:

a $\dfrac{x^2+2x-5}{x+4}$ **b** $\dfrac{2x^2+3x-3}{x-2}$

c $\dfrac{4x^2+2x-1}{2x-1}$ **d** $\dfrac{x^3+2x^2+2x-1}{x+1}$

3 Factorise completely the following polynomials:

a x^3-4x^2+x+6 **d** $x^3-3x^2-16x-12$

b x^3-7x-6 **e** $3x^3+7x^2-4$

c x^3+3x^2-x-3 **f** $3x^3+13x^2+8x-12$

4 Solve the following inequalities:

a $x^3+2x^2-29x+42>0$

b $x^3-10x^2+19x+30<0$

c $x^3-5x^2-12x+36\leq 0$

d $x^3+x^2-12x\geq 0$

e $x^3-3x^2-10x>0$

f $2x^3-9x^2-53x-24<0$

6.1 Contextual

1 $f(x)=x^4-5x^3-7x^2+29x+30$

a Factorise $f(x)$ completely.

b When is $f(x)<0$?

2 $(x-3)$ and $(x+4)$ are both factors of $2x^3+ax^2+bx-12$. Find a, b and the third linear factor.

3 Find the coordinates of the points where $y=2x^4-9x^3-21x^2+65x+75$ crosses the x-axis.

4 When $x^3+ax^2+bx-12$ is divided by $(x-2)$ and $(x+3)$ the remainders are -18 and 12, respectively. Find the values of a and b.

5 Solve the equation $4x^3-8x^2-x+2=0$.

6 If $f(x)=8x^3-10x^2+x+1$, find the range of values for x for which $f(x)\geq 0$.

6.2 Technique

1 Find:

a $\dfrac{2}{x+1}+\dfrac{3}{x-1}$ **d** $\dfrac{6}{x+3}+\dfrac{6}{x+2}$

b $\dfrac{5}{x+2}-\dfrac{3}{x-1}$ **e** $\dfrac{7}{x+1}+\dfrac{3}{x-1}$

c $\dfrac{4}{x+2}-\dfrac{1}{x+1}$ **f** $\dfrac{6}{x-2}-\dfrac{4}{x+2}$

2 Express the following as single fractions:

a $\dfrac{4}{2x+1}+\dfrac{2}{x-2}$ **d** $\dfrac{3}{x-2}-\dfrac{2}{7x+1}$

b $\dfrac{2}{x-1}-\dfrac{5}{4x+3}$ **e** $\dfrac{2}{3x-2}-\dfrac{1}{x-1}$

c $\dfrac{2}{3-2x}+\dfrac{3}{2x+1}$ **f** $\dfrac{3}{3x-2}+\dfrac{2}{1-2x}$

3 Identify the LCM of these denominators and use it to write each of the following as single algebraic fractions:

a $\dfrac{3}{x^2-9}+\dfrac{2}{x-3}$

b $\dfrac{3}{(x+2)(x-1)}+\dfrac{2}{3x-1}$

c $\dfrac{x}{(x+2)(2x-1)}+\dfrac{3}{x+1}$

d $\dfrac{2x+3}{x^2+2x-8}-\dfrac{2x+1}{(x+4)(x-3)}$

e $\dfrac{2}{3(x+2)}-\dfrac{2}{(x+2)^2}$

f $\dfrac{3}{(x+1)^2}-\dfrac{1}{x+1}$

4 Express each of the following as a single fraction:

a $\dfrac{2}{3x}\times\dfrac{x}{6}$ **d** $\dfrac{2x}{5x^2-500}\times\dfrac{x+10}{4}$

b $\dfrac{4(x+1)}{9}\times\dfrac{3}{16x}$ **e** $\dfrac{x+2}{x}\times\dfrac{2x}{x+5x+6}$

c $\dfrac{2(x-2)}{5}\times\dfrac{x+6}{x^2-4}$ **f** $\dfrac{8}{x-3}\times\dfrac{x^2-8x+15}{16x}$

5 Simplify:

a $\dfrac{5x}{6}\div\dfrac{12x}{3}$ **d** $\dfrac{x}{x^2-25}\div\dfrac{3}{4x+20}$

b $\dfrac{3x}{15}\div\dfrac{15}{10(x+3)}$ **e** $\dfrac{4}{x^2+3x-10}\div\dfrac{8}{x+5}$

c $\dfrac{x+1}{4}\div\dfrac{x^2-1}{2}$ **f** $\dfrac{3}{x+5}\div\dfrac{21x}{x^2+9x+20}$

6.3 Technique

1 Express the following in partial fractions (denominators with quadratic factors):

a $\dfrac{2x}{(x+1)(x+3)}$ **d** $\dfrac{56x}{(x+2)(x-5)}$

b $\dfrac{x}{(x+2)(x+3)}$ **e** $\dfrac{1}{x^2-x-6}$

c $\dfrac{7x-29}{(x-7)(x-3)}$ **f** $\dfrac{5}{x^2+2x-8}$

2 Express the following in partial fractions (denominators with quadratic factors):

a $\dfrac{x(x+5)}{(x-1)(x^2+2x+3)}$ **d** $\dfrac{4x^2-x+15}{(x-1)(x^2+x+4)}$

b $\dfrac{2x^2-3x+15}{(x-3)(x^2+3)}$ **e** $\dfrac{4x^2+15x+3}{(x-2)(x^2+x+1)}$

c $\dfrac{4x^2-3x+5}{(x-2)(x^2+1)}$ **f** $\dfrac{2(8x-1)}{x^3-5x^2+x-5}$

3 Express the following in partial fractions (denominators with a repeated factor):

a $\dfrac{x-5}{(x+3)^2}$ **d** $\dfrac{3x-5}{(x-1)^2}$

b $\dfrac{x+1}{(x-3)^2}$ **e** $\dfrac{2x-1}{(x+1)(x-3)^2}$

c $\dfrac{x+2}{(x+5)^2}$ **f** $\dfrac{x+2}{(x+1)(x-3)^2}$

4 Express these improper algebraic fractions as partial fractions:

a $\dfrac{3(x^3+x+2)}{(x-1)(x+2)}$ **d** $\dfrac{4x^3-3x^2+5x-12}{(x-2)(x^2+2)}$

b $\dfrac{x^3-6x^2+11x-15}{(x-5)(x-2)}$ **e** $\dfrac{x^2+8x+14}{(x+3)^2}$

c $\dfrac{3x^4-x^3+11x^2-x+9}{(x-1)(x^2+2)}$ **f** $\dfrac{x^3-5x}{(x-1)(x+3)^2}$

5 Express the following in partial fractions:

a $\dfrac{11x^2-29x+10}{(x-5)(x^2+x+5)}$ **d** $\dfrac{5x^2+32x+34}{(x-3)(x+2)^2}$

b $\dfrac{x^3+3x^2-4x-8}{(x+1)(x+3)}$ **e** $\dfrac{x^2-26x-29}{(7-x)(x+2)^2}$

c $\dfrac{x^3-4x^2+2x-6}{(x-5)(x^2+4)}$ **f** $\dfrac{5x^2-9x-12}{(x-5)(x^2+x+4)}$

6.4 Technique

1 Sketch the graph of the following continuous functions:

a $g(x) = x^3 + 4x^2 - x - 4$

b $g(x) = |x^2 - 4x - 5|$

c $h(x) = x^3 + 3x^2 - 13x - 15$

d $f(x) = |x^3 + 5x^2 - 4x - 20|$

2 Sketch the graph of the following rational functions:

a $g(x) = \dfrac{2}{x-3}$ **b** $h(x) = \dfrac{12}{x-5}$

c $g(x) = \dfrac{1}{5-2x}$ **d** $h(x) = \left|\dfrac{1}{3x+2}\right|$

3 Sketch the graph of the following rational functions:

a $f(x) = \dfrac{x+2}{x-3}$ **b** $g(x) = \dfrac{x+3}{x+2}$

c $f(x) = \dfrac{2x+3}{x+1}$ **d** $h(x) = \left|\dfrac{2x+5}{3x-2}\right|$

4 Solve the following:

a $|x| = |4x + 15|$ **d** $|x| \le |8 - 3x|$

b $|x| = |18 - 2x|$ **e** $|x| \ge |9x - 20|$

c $|x + 3| = |5x - 13|$ **f** $|3x - 2| \le |2x + 7|$

6.4 Contextual

1 Given that $f(x) = 6x^3 - x^2 - 11x + 6$:

a factorise $f(x)$ and then sketch the graph of $y = f(x)$

b on a separate diagram sketch the graph of $y = |f(x)|$.

2 Given that $f(x) = 3x - 1$ and $g(x) = x + 2$. Find the range of values of x for which:

a $f(x) > 0$ **b** $g(x) > 0$

c $\dfrac{f(x)}{g(x)} > 0$

3 Sketch the graph of $y = \dfrac{3x+1}{3x-1}$. Hence or otherwise find the range of values of x for which $y < 0$.

4 On separate diagrams sketch the graphs of:

a $y = \dfrac{6x+8}{3x-1}$ **b** $y = \left|\dfrac{6x+8}{3x-1}\right|$

Consolidation C

1 Find the values of a and b if $2x^3 + 3x^2 + ax + b$ is exactly divisible by $(2x + 1)$ and $(x - 2)$. Find also the third linear factor.

2 Given that $p(x) = a(x + 3)^4 + b$, find a and b if the remainder when $p(x)$ is divided by $x + 2$ is 7 and the remainder when $p(x)$ is divided by $x + 1$ is 37.

3 The function $f(x) = (5x + 1)(x - 1)(x - 2)$.

a Express $f(x)$ in the form $Ax^3 + Bx^2 + Cx + D$.

b Express $\dfrac{2x^2-5x-9}{f(x)}$ in partial fractions.

c Solve $f(x) > 0$.

4 With the aid of a sketch, solve the inequality $g(x) < 0$, given that $g(x) = -4x^3 + 19x^2 - 24x + 9$.

5 A curve has the equation $y = \dfrac{3}{2(x-1)(x+2)}$.

a Express $\dfrac{3}{2(x-1)(x+2)}$ in partial fractions.

b Find $\dfrac{dy}{dx}$.

c Lynn thinks the answer is $\dfrac{1}{2(x+2)^2} - \dfrac{1}{2(x-1)^2}$.

Andy thinks the answer is $\dfrac{-3(2x+1)}{2(x-1)^2(x+2)^2}$.

Who is correct? Explain your answer.

6 Sketch the graph of the curve $y = \dfrac{x+4}{x-2}$. State the values of x for which $y > 0$.

7 Sketch the graph of the curve $y = \left|\dfrac{5x-1}{x-1}\right|$.

8 Given that $y = \dfrac{3x^2+2x+5}{(x-2)(x-1)(2x+3)}$, express y in partial fractions.

Review

1 Write the following numbers in standard (index) form:

a 72 000 000 **b** 0.000 073 5

c five million **d** six thousandths

2 Simplify the following:

a $a^5 \times a^3$ **d** $a^x \div a^y$
b $\left(\dfrac{a^{12}}{a^3}\right)$ **e** y^0
c $(x^5)^3$ **f** $a^{2x} \times a^{2y}$

3 Without using a calculator, work out:

a $9^{\frac{1}{2}}$ **d** 12^0
b $\left(\frac{5}{6}\right)^{-1}$ **e** $81^{-\frac{3}{4}}$
c $64^{\frac{1}{3}}$ **f** $\left(1\frac{9}{16}\right)^{\frac{1}{2}}$

4 Use your calculator to find the values of the following, correct to three significant figures:

a $12^{\frac{1}{2}}$ **b** $(3.8)^{-2}$
c $(100)^{-\frac{3}{4}}$ **d** $(-7)^{-4}$

7.1 Technique

1 Using a calculator, find correct to three significant figures the value of:

a e^3 **b** e^{-2}
c $e^{\frac{1}{2}} + e^{\frac{3}{2}}$ **d** $e^{\frac{1}{3}}$

2 Use a graphical calculator to sketch the graphs of the following. In each case describe the transformation that maps $y = e^x$ onto each of the functions:

a $y = e^{(x+3)}$ **b** $y = e^x + 3$
c $y = 2e^x$ **d** $y = e^{2x}$

3 Draw the graph of $y = 3^x$. From your graph find, as accurately as you can, the value of:

a $3^{1.5}$ **b** $3^{0.6}$
c $3^{-0.5}$ **d** 3^{-2}

4 Using a calculator, sketch the graph of $y = e^x + 1$. Use the trace and zoom functions to find the values of x correct to three significant figures for which:

a $e^x = 5$ **b** $e^x = 7.12$
c $e^x = 1.5$ **d** $e^x = \pi$

7.1 Contextual

1 The mass of bacteria, measured in grammes, on a stale burger, is given by the formula $M(t) = 9.2 \times 2^t$, where t is the time in days since the burger was declared stale. Find the initial mass of the bacteria and the number of days needed for the total mass to exceed 100 g.

2 A professor discovers that the number of bacteria in her experiment at time t days was given by $n = n_0 e^{5t}$ where n_0 was the original count when 'clean'. Find the value of t when the colony was treble its initial size.

3 A brand of engine oil cools according to the model $T = 68e^{-kt} + 12$, where T is temperature in °C and t is time in minutes from the moment the engine is switched off.

a What is the initial temperature of the oil when the engine is switched off?

b If the oil cools to 37°C after 5 minutes, find the value of k.

4 An oil painting increases in value by 6% per year. By how much does the value of the painting increase in:

a 1 month **b** 6 months **c** 6 years?

How long will it take for the picture to double its present value?

5 A classic car bought for £9000 depreciates at 15% per year. After t years it is worth £x.

a Sketch a graph of x against t.

b Use the graph to calculate:

i when it is worth half its original value
ii when its value first falls below £1000.

7.2 Technique

1 Without using a calculator, find the values of:

a log 100 **b** log 0.001
c log 1 000 000 **d** log 0.000 000 01

2 Find the value of the following, to three significant figures:

a log 28 **b** log 3
c $\log\left(\frac{1}{5}\right)$ **d** log (0.7)

3 Without using a calculator find the exact value of x in each of the following:

a $x = \log_4 64$ **b** $x = \log_3 9$
c $x = \log_5\left(\frac{1}{25}\right)$ **d** $x = \log_8\left(\frac{1}{64}\right)$

4 Find the base of the following logarithms:

a $2 = \log_a 81$ **b** $3 = \log_a 216$
c $\frac{1}{2} = \log_a 16$ **d** $-3 = \log_a\left(\frac{1}{8}\right)$

5 Find the value of the following, to three significant figures:

a ln 4 **b** ln 120
c ln 8 **d** ln 20

6 Evaluate the following, without using a calculator:

a ln e **b** $\ln \sqrt[3]{e}$
c $\ln \sqrt[5]{e^4}$ **d** $\ln\left(\dfrac{1}{\sqrt[4]{e^5}}\right)$

7 Copy and complete the following table:

Exponent form	Logarithmic form
$2^3 = 8$	$\log_2 8 = 3$
$3^{-1} = \frac{1}{3}$	$\log_3\left(\frac{1}{3}\right) = -1$
$3^4 = 81$	
	$\log_4 64 = 3$
$8^2 = 64$	
$5^{-1} = \frac{1}{5}$	
$25^{\frac{1}{2}} = 5$	

7.2 Contextual

1 The Richter scale measures the magnitude of earthquakes using logarithms. The Richter value is only part of the number describing an earthquake's strength. If it has a Richter value of 6, the magnitude of the earthquake is 10^6. Look at the table below.

Country	Year	Magnitude	Deaths
China	1976	8.0	655 000
China	1976	7.8	240 000
Alaska	1964	8.5	178
Italy	1980	7.2	3000

a How many times greater was the first earthquake in China than the second? Give your answer in the form $K:1$.

b How many times greater was the earthquake in Alaska than the one in Italy? Give your answer in the form $K:1$.

2 Change in sound level in terms of power is given by

$$\text{change in dB} = 10\log_{10}\left(\frac{\text{new power}}{\text{old power}}\right)$$

a Find the change in dB when power output increases from 100W to 700W.

b How many times approximately will the power level emitted by a source of sound have increased if the sound level increases by 60dB?

c Police order the organisers of a rave to turn down their loudspeaker power output from 600 MW to 250 MW. What is the corresponding fall in sound level in dB?

d During a pop concert the power output from the sound system changes from 600 MW to 300 MW. What is the corresponding fall in sound level in dB?

3 In chemistry the acidity or alkalinity of a solution is measured by its pH factor defined by $\text{pH} = -\log_{10}[\text{H}^+]$ where $[\text{H}^+]$ is the quantity of hydrogen ions present in the solution. The pH value of 7 indicates a neutral solution for which $[\text{H}^+] = 10^{-7}$ moles per litre.

a Calculate the pH value of *Braindead* shampoo of strength 1.9×10^{-9} moles per litre.

b *Tengola's* isotonic drink has a pH value of 5.7. Calculate the concentration of hydrogen ions.

c Sulphuric acid has a pH of 2, and ethanoic acid has a pH of 5. How many more times concentrated is the sulphuric acid?

7.3 Technique

1 Write the following as single logarithms:

a $\ln 7 + \ln 4$ **d** $2\ln 3 - \ln 15 - 3\ln 6$

b $\log 4 + \log 3 - \log 2$ **e** $\ln 4 + 2\ln 10 - 2\ln 5$

c $\log 14 - \log 9$ **f** $\log 3 - \log 4$

2 Given $\log 2 = 0.3$ and $\log 5 = 0.7$ find, without using a calculator:

a $\log 16$ **b** $\log 25$ **c** $\log 32$

d $\log 2.5$ **e** $\log 0.5$ **f** $\log 100$

3 Simplify:

a $\log x + \log y + 2\log z$ **d** $\log x + \log y - \log z$

b $3\ln x - \ln y$ **e** $-\log a - \log b$

c $3\ln x + \frac{1}{2}\ln y$ **f** $1 + 2\log x$

4 Express in terms of $\ln x$, $\ln y$ and $\ln z$.

a $\ln(xyz^2)$ **d** $\ln(x\sqrt[3]{y})$

b $\ln\left(\frac{xy^2}{z^2}\right)$ **e** $\ln\left(\frac{1}{xy}\right)$

c $\ln(x\sqrt{y})$ **f** $\ln\left(x\sqrt{\frac{y^2}{z}}\right)$

5 Solve the following logarithmic equations.

a $\log_x 16 = 0.5$ **b** $\log_x 8 = -3$

c $\ln x^2 - \ln x = 4$ **d** $\log_x 16 + 2\log_x 5 - 2 = 0$

7.3 Contextual

1 Explain why $xy = 1\,000\,000$ and $\log_{10}x + \log_{10}y = 6$ are equivalent interchangeable statements.

2 On the same axes sketch the graphs of $y = \log x^2$ and $y = \log\left(\frac{1}{x^2}\right)$. Using laws of logarithms suggest a possible transformation for these graphs. Describe an inverse function for $y = \log\left(\frac{1}{x^2}\right)$.

3 **a** Show that $\log_2 8 = \log_4 64$.

b Evaluate $\log_2 8 - \log_4 64$ without a calculator.

4 By taking natural logarithms, or otherwise find $g(x)$ such that $x^{2x} = e^{xg(x)}$ for all values of x.

5 Given that $\log_8 n = \log_2 3 + 2\log_4 3$, explain why $n = 3^6$.

7.4 Technique

1 Solve the following equations:

a $8^x = 100$ **b** $6^x = 156$

c $12^x = 2001$ **d** $9^x = 8$

2 Solve the equations:

a $11^{(2x-3)} = 100$ **b** $13^{(x+2)} = 33$

c $4^{(x-2)} = 25$ **d** $2^{2(x+3)} = 212$

3 Find the value of x in the following:

a $6^{2x} = 7^{x+1}$ **b** $4^{2x+1} = 3^x$

c $8^{x+2} = 2^{x-3}$ **d** $7^{2x+3} = 8^{3x-2}$

4 Solve the following exponential equations.

a $e^{2x} + 2e^x - 3 = 0$ **b** $4e^{2x} - 4e^x - 3 = 0$

c $e^{4x} - 2e^{2x} - 8 = 0$ **d** $6e^{4x} - 17e^{2x} + 7 = 0$

5 Find the smallest integer p such that

a $2^p > 500$ **b** $4^p > 5000$

c $0.7^p < 0.002$ **d** $(6.125)^p > 499$

6 Find equations for the following sets of data:

a

x	0	1	2	3	4
y	2	6	18	54	162

b

x	0	1	2	3	4
y	0	2	16	54	128

c

x	0	1	2	3	4
y	0	13	416	3159	13 312

d

x	0	1	2	3	4
y	2	1.6	1.28	1.024	0.8192

7 Solve the following equations:

a $100^x - 10^{(x+1)} + 9 = 0$

b $10^{2x} - 10^{(x+1)} - 11 = 0$

c $4 \times 10^{2x} - 4 \times 10^{(x+1)} + 19 = 0$

d $100^x = \dfrac{16}{9}$

7.4 Contextual

1 In a new town, the population P thousands is calculated every t years from 1997 onwards. The results are summarised below.

t	1	2	3	4	5
P	9.6	15.4	24.6	39.3	62.9

It is believed that P and t are connected by an exponential relationship of the form $P = ab^t$, where a and b are constants. When will the population be expected to exceed 250 thousand?

2 Show that the equation $e^{4x} - 2e^{2x} = 8$ has one real solution. Why does the second solution of the related quadratic not produce a solution to the exponential equation?

3 A technician for the local water authority measured N cubic metres (m^3) of water flowing per second into a canal lock when the difference in water levels was x metres (m). From her results she was able to form an equation of the form $N = ax^b$. Using the table below, find approximate values for a and b.

x	0.8	1.0	1.2	1.4	1.6
N	2.175	3.8	5.994	8.813	12.305

4 Julie, the mad professor, believes her pre-heated beer is the elixir of perpetual life. She heats the beer and allows it to cool. The decrease in temperature is recorded as shown below.

t (min)	1	2	3	4
T (°C)	42.5	25.75	15.6	9.5

It is believed that T and t are connected by the exponential relationship $T = Ae^{\lambda t}$ where A and λ are constants.

a Draw a graph of t against $\ln T$. Use your graph to estimate the values of the constants A and λ.

b How long does it take, to the nearest second, for the beer to reach 30°C?

Consolidation C

1 Express as single logarithms:

a $\log uv^2 - \log u^2v$

b $\log (wuv) + \log (uv) - \log \left(\dfrac{1}{w}\right)$

2 Solve the following inequalities:

a $\ln (x+1)^3 > 5$ **b** $(0.3)^{2x+1} > 7$

3 The value £V of a sports car can be modelled by the equation $V = Ae^{-kt}$, where t is the age of the car in years. The car cost £32 000 when new, and after two years its value was £24 000.

a State the value of the constant A.

b Calculate the value of the constant k, giving your answer to three significant figures.

(AEB)

4 The value of a car depreciates by 10% each year. Initially the car costs £8500.

a Explain why the value of the car after t years can be modelled by the equation $V = 8500 \times (0.9)^t$.

b Express t in terms of V.

c After how many years will the car be valued at $\frac{2}{3}$ of its original price?

5 Show that $x + 1$ is a factor of $x^3 - 3x^2 + 4$. Hence solve the equation $x^3 - 3x^2 + 4 = 0$. Given that $x = 3^y$, show that $x^3 - 3x^2 + 4 = 3^{3y} - 3^{2y+1} + 4$. Hence, solve the equation $3^{3y} - 3^{2y+1} + 4 = 0$, giving your answer to three significant figures.

(AEB)

6 The number of nuclei (N) of a radioactive isotope present at time t seconds is given by the formula $N = N_0e^{-0.0133t}$, where N_0 is the number of nuclei present when $t = 0$. Find the half-life of this radioactive isotope (that is the time required for the number of nuclei to decay to half of the initial value).

7 **a** Simplify $2\log_5 10 + \log_5 50 - \log_5 40$.

b Solve $3^{(2x-1)} = 17$.

c Using the result $\log_a b = \dfrac{1}{\log_b a}$, solve the equation $\log_2 x - 6\log_x 2 = 1$.

8 The following table gives the population, P thousand, of a new town t years after it is founded.

t	5	10	15	20	25
P	14.6	35.9	60.8	88.4	118.2

The town planning officer suggests that the relationship between P and t can be modelled by the equation $P = \alpha t^{\beta}$, where α and β are constants.

a Explain why, under this model, the relationship between $\ln P$ and $\ln t$ is a linear one.

b Use the data to plot $\ln P$ against $\ln t$ on graph paper. Explain why your results confirm that the suggested model does fit the data.

c By substituting the values of P when $t = 5$ and $t = 25$ into the above equation, or otherwise, find the values of α and β correct to two significant figures.

d Use your results to predict the value of P when $t = 50$. Comment on the reliability of your prediction.
(*WJEC*)

9 Write $\ln x^3 + \ln xy - \ln y^2$ as a single term. Hence obtain an expression for y in terms of x if $\ln x^3 + \ln xy - \ln y^2 = 0$.
(*NEAB*)

10 **a** Simplify the following expressions:
i $\log x^3 - \log xy^2$
ii $\log x + \log\left(\frac{1}{x}\right)$
iii $\dfrac{\log x^3y - \log xy}{\log x^2 + \log x}$
b Solve the equation $(0.57)^{3x-1} = 2.7^x$.

Review

1 By collecting together like terms, simplify the following expressions:
a $a + 2a + 2d$ **b** $3a + a + nd$
c $a + ar + 2ar$ **d** $1 + x^2 + x^3 - 2(x^2 + x^3)$

2 Expand the following;
a $(n + 2)(n + 3)$ **b** $(r + 3)(r - 2)$
c $r(r + 1)(r + 2)$ **d** $(1 - x)^2$

3 Evaluate:
a $2n + 3$, when $n = 4$
b $3n - 2$, when $n = 5$
c $n^2 + 1$, when $n = 3$
d $\dfrac{n(n-1)(n-2)}{6}$, when $n = 5$
e ar^3, when $a = 3$ and $r = 2$
f $\dfrac{n(n+1)}{2}$, when $n = 4$

4 Express each of the following in terms of their partial fractions:
a $\dfrac{6}{r(r+1)}$ **b** $\dfrac{7}{(r+1)(r+2)}$
c $\dfrac{8}{(r+4)(r+5)}$ **d** $\dfrac{11}{(n+1)(n+2)}$

8.1 Technique

1 Find the first four terms of the sequences with the following nth terms:
a $u_n = 7n - 2$ **d** $u_n = (n - 1)(n + 2)$
b $u_n = -n + 3$ **e** $u_n = 2^{n+1}$
c $u_n = n^2 + 1$ **f** $u_n = n^3 + n$

2 Find u_n, the nth term, for each of the following sequences:
a 1, 5, 9, 13, . . . **e** 1, 4, 9, 16, . . .
b 4, 9, 14, 19, . . . **f** 3, 8, 15, 24, . . .
c 23, 19, 15, 11, . . . **g** $\frac{3}{5}, \frac{5}{7}, \frac{7}{9}, \frac{9}{11}, \ldots$
d −1, −6, −11, −16, . . . **h** 0, 7, 26, 63, 124, . . .

3 Use the following recurrence relations to find the first four terms of each sequence:
a $u_n = u_{n-1} + 2, u_1 = 3$
b $u_n = 2u_{n-1} + 3, u_1 = 1$
c $u_n = 3u_{n-1} - 1, u_1 = 3$
d $u_n = 2u_{n-1} - u_{n-2}, u_2 = 3, u_1 = 1$
e $u_n = 3u_{n-1} + u_{n-2}, u_2 = 2, u_1 = 1$
f $u_n = \frac{1}{2}(u_{n-1}) + 3, u_1 = 8$

4 Express the following series in sigma notation:
a $2 + 5 + 10 + 17$
b $4 + 7 + 12 + 19$
c $5 + 6 + 7 + 8 + 9 + 10 + 11$
d $4 + 7 + 10 + 13 + 16 + 19$
e $3 + 8 + 15 + 24 + 35$
f $1 + 8 + 27 + 64 + 125$

5 Evaluate:
a $\sum_{r=1}^{3} r^2$ **d** $\sum_{r=2}^{3} (3r - 2)$
b $\sum_{r=2}^{4} r^2$ **e** $\sum_{r=3}^{6} r(r - 1)$
c $\sum_{r=3}^{5} \left(r^3 + 1\right)$ **f** $\sum_{r=1}^{10} \left(r^2 - 3\right)$

6 Express the following series in the given form of sigma notation:
a $3 + 5 + 7 + 9, \sum_{r=3}^{n} f(r)$
b $3 + 5 + 7 + 9 + 11 + 13, \sum_{r=3}^{n} f(r)$
c $8 + 10 + 12 + 14, \sum_{r=4}^{n} f(r)$
d $11 + 16 + 21 + 26 + 31 + 36, \sum_{r=3}^{n} f(r)$
e $3 + 10 + 17 + 24, \sum_{r=2}^{n} f(r)$
f $8 + 12 + 16 + 20 + 24, \sum_{r=5}^{n} f(r)$

8.2 Technique

1 Find:
i the next two terms
ii the nth term, u_n, of each of the following:
a 3, 5, 7, 9, . . . **b** 7, 11, 15, 19, . . .
c 29, 27, 25, 23, . . . **d** −1, −3, −5, −7, . . .

2 Given the arithmetic progressions

2, 5, 8, 11, . . . u_n
6, 10, 14, 18, . . . v_n
67, 65, 63, 61, . . . w_n

find:

a u_7 **b** u_{20} **c** v_{10}
d v_{30} **e** w_9 **f** w_{20}

3 Find the sum of the following arithmetic series:

a $3 + 9 + 15 + \ldots + 45$
b $12 + 15 + 18 + \ldots + 36$
c $1 + 8 + 15 + \ldots + 43$
d $15 + 20 + 25 + \ldots + 45$

4 Use the method of Gauss to calculate:

a $81 + 79 + 77 + \ldots + 35$
b $48 + 42 + 36 + \ldots + 12$
c $21 + 17 + 13 + \ldots + -11$
d $30 + 28 + 26 + \ldots + 18$
e $30 + 25 + 20 + \ldots - 15$
f $-2 - 5 - 8 - 11 - \ldots - 20$

5 Evaluate:

a $\sum_{r=1}^{8} (3r + 2)$ **b** $\sum_{r=3}^{10} (2r + 3)$
c $\sum_{r=4}^{10} (6r - 7)$ **d** $\sum_{r=6}^{21} (3r - 8)$

8.2 Contextual

1 Girders used to construct a bridge form the sequence 10, 16, 22, 28, ... What is the nth term of this sequence?

2 Calculate the 30th term of the series $-7 - 3 + 1 + 5 + \ldots$.

3 Find the sum of the arithmetic series $3 + 18 + 33 + 48 + \ldots + 288$.

4 Find the sum of the first 21 terms of the arithmetic series $6 + 15 + 24 + 33 + \ldots$

5 The eleventh term of an arithmetic series is five times the second term, and the third term is 13. Find:

a the first term
b the common difference of the AP
c the common factor of the 8th and 19th terms.

6 The sum of the first six terms of an AP is 72. The sum of the first twelve terms is 288. Find the sum of the first 18 terms.

7 If 4, x, y, 25 are in arithmetic progression, find x and y. Find the sum of the first 100 terms of this sequence.

8 Emily saves £120 of her first year's salary. She then saves £20 more the following year. Every successive year her savings increase by £20. How much will she have saved after 25 years?

9 An AP has three consecutive terms with a sum of 39 and a product of 1365. Find the terms.

10 An AP has first term 8 and common difference 0.6. If the sum of the first n terms is to exceed 270, find the value of n.

8.3 Technique

1 Find:

i the next two terms
ii the nth term of the following geometric progressions:

a 1, 10, 100, . . . **b** 3, 9, 27, 81, . . .
c 6, 9, 13.5, . . . **d** 12, −6, 3, −1.5, . . .

2 Given the sequences

3, 6, 12, 24, . . . u_n
5, 15, 45, 135, . . . v_n
100, 50, 25, 12.5, . . . w_n

find:

a u_7 **b** v_8 **c** v_6
d w_7 **e** w_6 **f** u_6

3 Given the first term and common ratio of a geometric series, find the sum of the terms indicated:

a $a = 16, r = 1.1 : S_{10}$ **b** $a = 14, r = -\frac{1}{2} : S_8$
c $a = 30, r = 0.9 : S_{12}$ **d** $a = 12, r = \frac{4}{5} : S_{10}$

4 Find the sum of the following geometric series:

a $5 + 10 + 20 + \ldots + 5120$
b $1 + 3 + 9 + 27 + \ldots + 6561$
c $8 - 20 + 50 - \ldots + 1953.125$
d $8 - 20 + 50 - \ldots - 781.25$
e $2 + 10 + 50 + \ldots + 781\,250$
f $\frac{3}{4} + 1\frac{1}{2} + 3 + 6 + \ldots + 768$

8.3 Contextual

1 Find the twelfth term of the geometric series $16 + 12.8 + 10.24 + \ldots$ correct to three significant figures.

2 Find the common ratio of a geometric series given that the second term is 12 and the sixth term is 60.75.

3 Find the sum of the first ten terms of a geometric series with common ratio 3 and first term 0.75.

4 Write down the first three terms of $\sum_{n=1}^{12} 18\left(\frac{2}{9}\right)^n$. State clearly:

a the first term of the GP
b the common ratio
c the sum of this series.

5 Evaluate:

a $\sum_{n=1}^{6} 400(1.2)^n$ **b** $\sum_{n=2}^{8} 100\left(\frac{7}{10}\right)^n$

8.4 Technique

1 Find the sum to infinity for a geometric progression with:

a $a = 200, r = \frac{2}{3}$ **d** $a = 70, r = 0.9$
b $a = 80, r = \frac{1}{8}$ **e** $a = 600, r = 1.1$
c $a = 400, r = 0.6$ **f** $a = 2000, r = \frac{1}{5}$

2 Describe the behaviour of the following sequences:

a $u_{n+1} = |u_n - 7|$, when $u_1 = 13$
b $\sin(30n)^\circ$
c $\ln 1.4, (\ln 1.4)^2, (\ln 1.4)^3, \ldots$
d $\ln 4, (\ln 4)^2, (\ln 4)^3, \ldots$
e $u_n = \dfrac{12}{(n+1)(n+2)}$
f $u_n = \dfrac{8}{n(n+1)}$

3 Find the sum to infinity of the following geometric series:

a $270 + 90 + 30 + \ldots$ **d** $500 + 300 + 180 + \ldots$
b $270 - 90 + 30 - \ldots$ **e** $500 - 200 + 80 - \ldots$
c $256 + 64 + 16 + \ldots$ **f** $400 - 133\frac{1}{3} + 44\frac{4}{9} - \ldots$

4 For the following geometric progressions find the value of the unknown terms:

a $r = \frac{3}{7}, S_\infty = 70, a = ?$ **d** $r = \frac{19}{20}, S_\infty = 80, a = ?$
b $a = 36, S_\infty = 44, r = ?$ **e** $a = 3, S_\infty = 15, r = ?$
c $a = 8, S_\infty = 80, r = ?$ **f** $r = \frac{13}{20}, S_\infty = 120, a = ?$

8.4 Contextual

1 Find the sum to infinity of a geometric series with first term 28 and common ratio $\frac{3}{7}$.

2 The first term of a geometric series is 68 and the common ratio is $-\frac{1}{3}$. Find the sum to infinity.

3 The sum to infinity of a GP is 75 and the first term is 25. Find the common ratio.

4 Given that the common ratio of a geometric series is $\frac{5}{9}$ and the sum to infinity is 819, find the first term.

5 The sum to infinity of a geometric progression is 16 times the first term. Find the common ratio.

6 The sum to infinity of a GP is 125 and the sum of the first three terms is 98. Find the first term and the common ratio.

7 Jenny says, "The sum to infinity of a GP with first term 42 and common ratio $\frac{5}{4}$ is 294". Anne immediately says that she must be wrong.

a Explain why Anne is correct.
b Given that the sum to infinity and the first term are correct, find the common ratio.

8.5 Technique

1 Find the expansion of the following binomials using Pascal's triangle:

a $(1+x)^5$ **b** $(3-x)^3$
c $(4+x)^4$ **d** $(2-x)^6$

2 Find the value of the following binomial coefficients:

a $\binom{5}{2}$ **b** $\binom{9}{2}$
c $\binom{7}{3}$ **d** $\binom{6}{4}$

3 Expand the following using the binomial theorem:

a $(1+x)^4$ **b** $(5-x)^4$
c $(1-2x)^6$ **d** $\left(x+\dfrac{3}{x}\right)^4$

4 Write the following as a series of ascending powers of x, up to and including the term x^3:

a $(1+x)^{11}$ **b** $(3+x)^4$
c $(5-2x)^4$ **d** $(3+2x)^5$

5 Find the range of values of x for which series expansions of the following are valid:

a $(5-3x)^{\frac{3}{5}}$ **b** $(1-3x)^{-7}$
c $\sqrt[5]{1+\frac{2}{5}x}$ **d** $(7-5x)^{\frac{1}{2}}$

6 Find the first four terms of the binomial series for the following functions:

a $(1+2x)^{-5}$ **b** $\sqrt{2-3x}$
c $(1+5x)^{\frac{3}{5}}$ **d** $\dfrac{1-3x}{(1+x)^5}$

8.5 Contextual

1 Find the binomial expansion of $(3+x)^5$. When is this expansion valid?

2 Find the term in x^8 in the expansion of $(1+3x)^{10}$.

3 Find the coefficient of the x^{11} term in the expansion of $(1-2x)^{14}$.

4 The expansion of $\left(x-\dfrac{2}{x}\right)^8$ has a term that does not contain x. State the value of this term.

5 Find the value of $(0.99)^7$ to four decimal places using the binomial expansion of $(1-x)^7$.

6 Expand $\sqrt[3]{1-2x}$ as a series of ascending powers of x, up to and including the terms in x^3. State the range of values of x for which the expansion is valid.

Consolidation C

1 Find the sum of the arithmetic series $12 + 16 + 20 + \ldots 128$.

2 **a** The tenth term of an AP is six times the second term. Given that the first term is 3, find:

i the common difference
ii the sum of the first 12 terms.

b The fifth term of a GP is $156\frac{1}{4}$ and the third term is 25. Find:

i the common ratio
ii the first term
iii the sum of the first ten terms, correct to three significant figures.

3 The sum of the first two terms of an arithmetic series is 47. The 30th term of this series is −62. Find:

a the first term of the series and the common difference

b the sum of the first 60 terms of the series.

(ULEAC)

4 A savings scheme pays 5% per annum compound interest. A deposit of £100 is invested in the scheme at the start of each year.

a Show that at the start of the third year after the annual deposit has been made, the amount in the scheme is £315.25.

b Find the amount in the scheme at the start of the 40th year, after the annual deposit has been made.

(ULEAC)

5 The sequence $u_1, u_2, u_3, \ldots$, where u_1 is a given real number, is defined by $u_{n+1} = |4 - u_n|$.

a Given that $u_1 = 1$, evaluate u_2, u_3 and u_4, and describe the behaviour of the sequence.

b Given that $u_1 = 6$, describe the behaviour of the sequence.

c For what value of u_1 will all terms of the sequence be equal to each other?

(UCLES)

6 Given that $u_n = 50\left(\frac{3}{5}\right)^n$, where $n = 1, 2, 3, \ldots$, find:

a the first three terms of the sequence

b u_8, to three significant figures

c $\sum_{n=1}^{10} u_n$, to three significant figures

d the sum to infinity of the sequence.

7 **a** **i** Obtain the first four terms in the expansion, in ascending powers of x, of $(1 - 2x)^{-1}$.

ii State the range of values of x for which the expansion is valid.

b Obtain the first four terms in the expansion, in ascending powers of x, of $\dfrac{1+3x}{1-2x}$.

(NEAB)

8 **a** Express $\dfrac{7-4x}{(2x-1)(x+2)}$ in partial fractions as $\dfrac{A}{(2x-1)} + \dfrac{B}{(x+2)}$, where A and B are to be found.

b Find the expansion of $\dfrac{1}{1-2x}$ in the form $a + bx + cx^2 + \ldots$, where a, b and c are to be found. Give the range of values of x for which this expansion is valid.

c Find the expansion of $\dfrac{1}{2+x}$ as far as the term containing x^2. Give the range of values of x for which this expansion is valid.

(OCSEB)

Review

1 Find:

a $\dfrac{d}{dx}(2x^3 - 6x^2 - 1)$

b $\dfrac{d}{dx}\left(\frac{1}{3}x^3 - x^2 + x - 6\right)$

c $\dfrac{d}{dx}\left(\dfrac{10}{x^2}\right)$

d $\dfrac{d}{dx}(5 \sin x)$

e $\dfrac{d}{dx}(\cos 5x)$

f $\dfrac{d}{dx}(2 \sin 3x + 5 \cos 2x)$

2 Write each of the following expressions as Ax^n or $(Ax^m + Bx^n)$ where A and B are constants:

a $\dfrac{3}{5x^2}$

b $\dfrac{4\sqrt{x}}{x}$

c $\dfrac{10 - 3x^2}{x^3}$

d $\dfrac{7}{\sqrt{x}}$

e $\sqrt{x}(x - \sqrt{x})$

f $\dfrac{x^3}{2/3}$

3 Find the value of each of the following expressions for the given value of x:

a $2x^3 - 5x^2 - x$, when $x = 3$

b $-\frac{1}{3}x^3 - x^2 + x$, when $x = -2$

c $\frac{1}{5}x^5 - x^3 + x$, when $x = 1$

4 Sketch the graphs of:

a $y = -3$

b $y = -x$

c $y = x^2 + 1$

d $y = x^3 + 1$

e $y = 1 + \cos x$

f $y = 2 \sin x$

5 Simplify the following using sigma notation:

a $1 + 3 + 5 + 7 + 9$

b $1 + \frac{1}{2} + \frac{1}{3} + \frac{1}{4}$

c $1 + 2 + 4 + 8 + 16 + 32 + 64$

d $3 + 5 + 7$

9.1 Technique

1 Use integration notation to describe each of the following statements:

a integrate $t^2 - 2t + 1$ with respect to t

b integrate e^{-2x} with respect to x

c integrate $\cos 2\theta$ with respect to θ.

2 Integrate each of the following with respect to x:

a 5

b $10 - 3x$

c $7 - x + x^2$

d $x(5 - 2x)$

e $(x + 2)(x - 7)$

f $(2x - 3)^2$

3 Find:

a $\int dt$

b $\int 32\, dx$

c $\int (t^2 - t)\, dt$

d $\int \sqrt[3]{x}\, dx$

e $\int \dfrac{1}{x^3}\, dx$

f $\int x^{1.2}\, dx$

g $\int 16t^3\, dt$

h $\int \dfrac{10}{t^6}\, dt$

i $\int 50t^{1.5}\, dt$

j $\int 12x^{-1/2}\, dx$

k $\int \dfrac{1}{\sqrt[3]{x}}\, dx$

4 Find:

a $\int \left(1 - 2x - 3x^2\right)\, dx$ **d** $\int t(t+1)(t-1)\, dt$

b $\int 2x(1-x)\, dx$ **e** $\int (r+3)(r-1)\, dr$

c $\int (2-t)^2\, dt$ **f** $\int \sqrt{r}(\sqrt{r} - 2)\, dr$

5 Find:

a $\int \left(\frac{2-x}{x^3}\right) dx$ **d** $\int \left(1 - \sqrt{t}\right)^2 dt$

b $\int \left(\frac{5-x}{\sqrt{x}}\right) dx$ **e** $\int \frac{x-1}{\sqrt{x}}\, dx$

c $\int \left(\frac{12}{t^2} - t\right) dt$ **f** $\int \sqrt{x}(x+1)\, dx$

6 If $g'(x) = \frac{8}{x^2}$ and $g(2) = -5$, find $g(x)$.

7 The velocity of a particle, at time t, is given by $v = t^2 + 4t$. Find an expression for the distance s travelled by the particle at time t, given that $s = 27$ when $t = 3$.

9.1 Contextual

1 **a** Differentiate x^4 with respect to x, and explain how to use the result to find $\int x^3 dx$.

b Differentiate $x^{10} + x$, $x^{10} + x - 1$, $x^{10} + x + 1$ and $x^{10} + x + 2$ with respect to x. Hence explain why $\int (10x^9 + 1)dx = x^{10} + x + c$.

2 Copy and complete the following, where each '•' represents a missing term:

a $\int \left(x^4 - 6x^3 + \bullet\right) \bullet = \bullet - \bullet + 4x + c$

b $\int \left(6 - 2t^2\right) dt = \bullet \int \left(3 - t^2\right) dt$

$= \bullet(3t - \bullet) + c$

$= \bullet - \bullet + c$

c $\int dr = \bullet + c$

d $\int \bullet = 2y + c$

3 The gradient of a curve is given by $\frac{dy}{dx} = 3x^2 - 1$. Given that the curve passes through (0, 3), find the equation of the curve.

4 If $g'(x) = 1 - \frac{1}{x^2}$, and (1, 3) is a point on the curve $y = g(x)$, find an expression for $g(x)$.

5 The gradient at any point (x, y) on a curve is given by $\frac{dy}{dx} = -6\sqrt{x}$. If the curve passes through (4, 5), find the equation of the curve.

6 A particle travels in a straight line in such a way that, t seconds after passing through a fixed point O, its velocity in m s^{-1} is given by $v = t + 1$.

a Find an expression for the displacement s, after t seconds.

b How far has the particle travelled after one second?

7 A car starts from rest at a set of traffic lights, and 1 minute later stops at the next set of lights. Its velocity, t seconds after starting, is given by $v = 1.8t - 0.03t^2$ m s^{-1}.

a Find an expression for the distance travelled from the first set of lights.

b Find the distance between the two sets of lights.

8 A particle leaves a fixed point O with a velocity of 13 m s^{-1}. It travels in a straight line and its acceleration after t seconds is given by $a = 7 - 6t$ m s^{-2}.

a Find an expression for the velocity of the particle in terms of t.

b Find an expression for the displacement of the particle after leaving O.

9.2 Technique

1 Copy and complete the following:

a $\int_a^b (7 - 4x)\, dx = [\bullet]_a^b$

b $\int_a^b \left(x^2 - x - 1\right) dx = [\bullet]_a^b$

c $\int_a^b d\theta = [\bullet]_a^b$

d $\int_a^b (\bullet)\, dt = \left[5t^3 - 7t^2 + t\right]_a^b$

2 Use the notation for integration to describe the areas of the shaded region in the following:

a

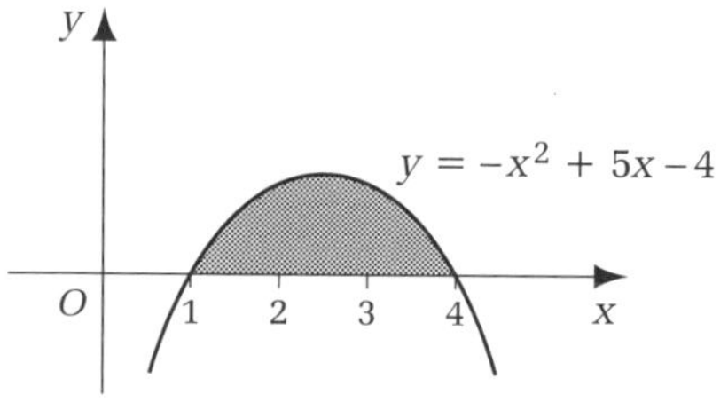

b

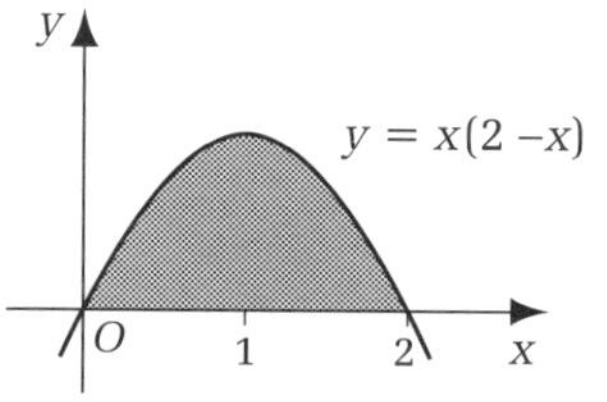

3 Copy and complete this working, to find the area of the region under the curve $y = (x - 2)^2$ between $x = 0$ and $x = 2$.

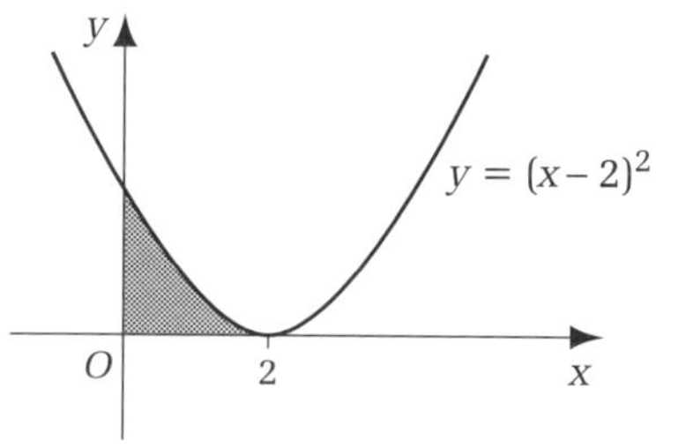

$$\text{area} = \int_0^2 (x-2)^2 \, dx$$

$$= \int_0^2 \left(x^2 - 4x + 4\right) dx$$

$$= \left[\frac{x^3}{3} - \frac{4x^2}{2} + 4x\right]_0^2$$

$$=$$

4 By sketching a graph and shading, show the areas described by:

a $\int_{-1}^{1} x^2 \, dx$ **b** $\int_0^3 (x+1)\, dx$ **c** $\int_1^3 2 \, dx$

5 Use integration to find the areas of the shaded regions in the following:

a

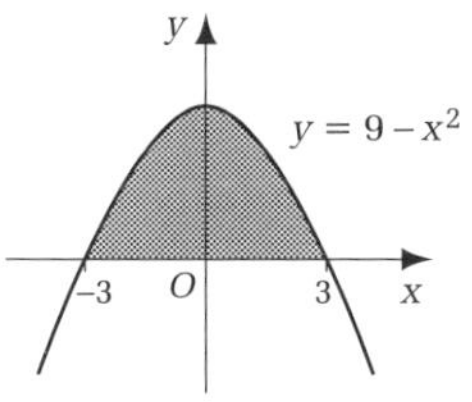

b

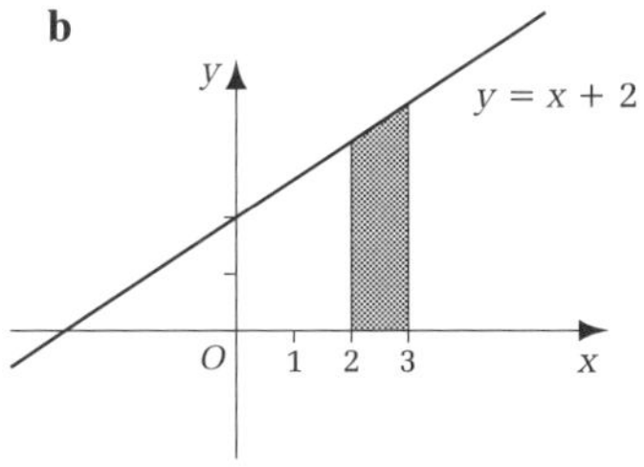

6 Use integration to find the area of the shaded region in each of the following. In each case first work out where the curve cuts the x-axis.

a

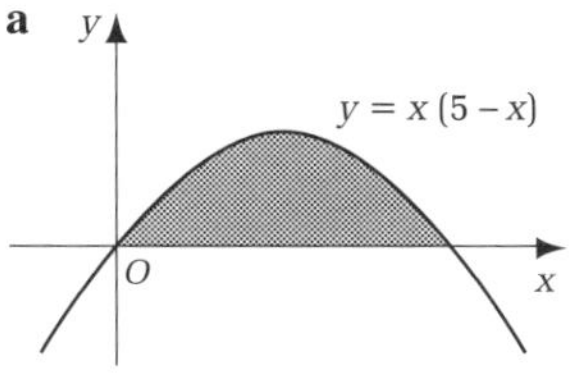

b

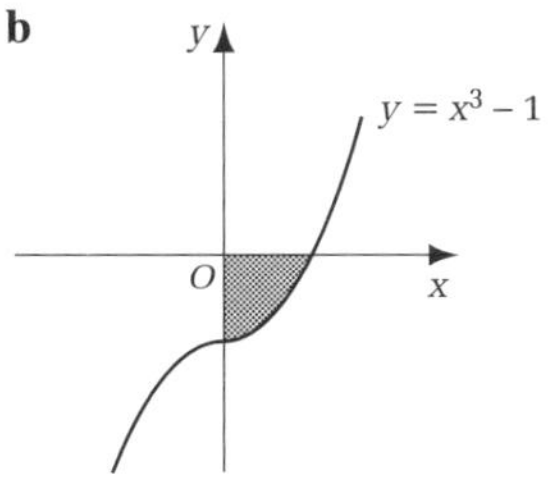

7 The graph shows the line $y = x$ and the curve $y = \frac{1}{x^2}$.

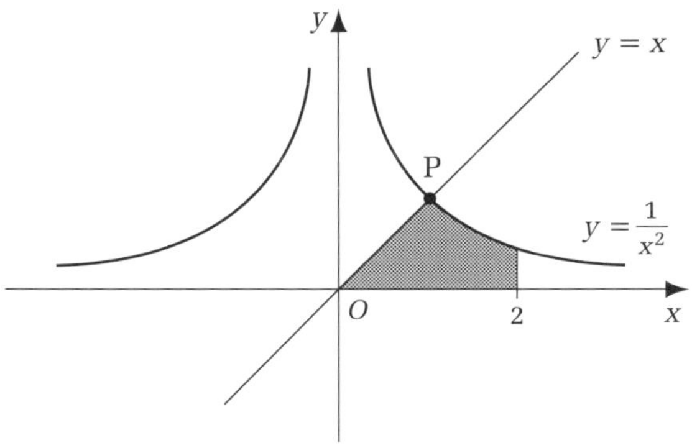

a Write down the coordinates of the point P.

b Calculate the shaded area.

8 **a** Find the area of the shaded region.

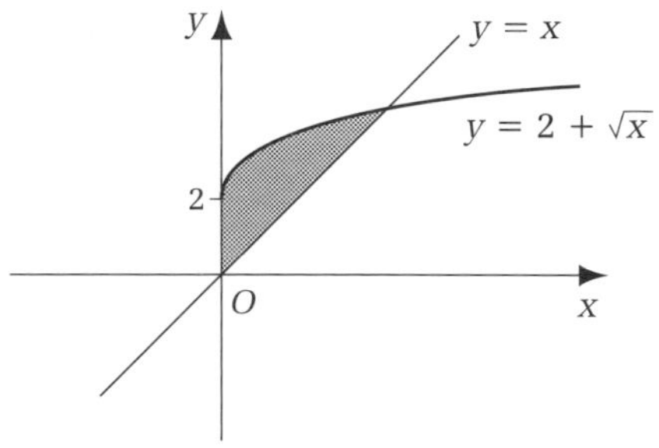

b Now find the area of this pattern.

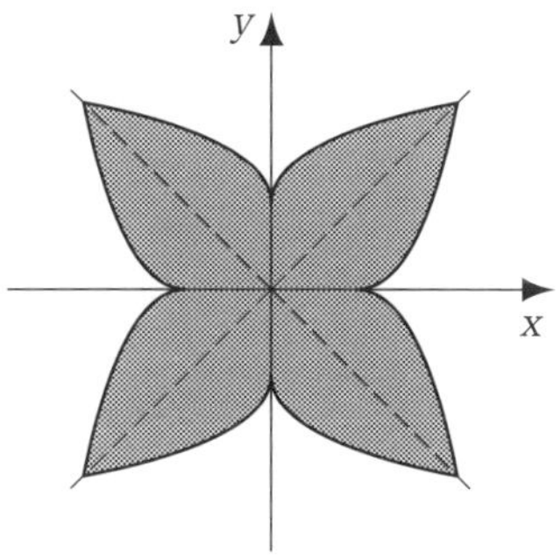

9 Evaluate the following integrals:

a $\int_0^1 \left(x^5 - x^6\right) dx$ **b** $\int_1^2 \left(4x^3 - 2x\right) dx$

9.2 Contextual

1

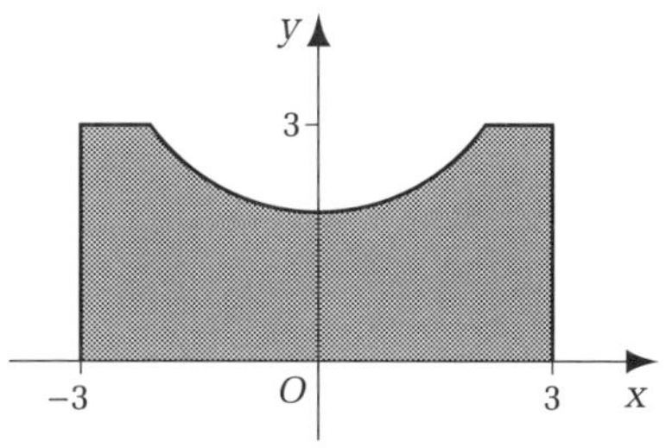

If the curved part of a metal plate can be modelled by the curve $y = \frac{1}{4}x^2 + 2$, find the area of the plate.

2 The graph shows the section of an archway. The equation of the arch is $y = 5 - \frac{1}{10}x^2 - \frac{1}{1000}x^4$. Use integration to calculate the area of the cross-section.

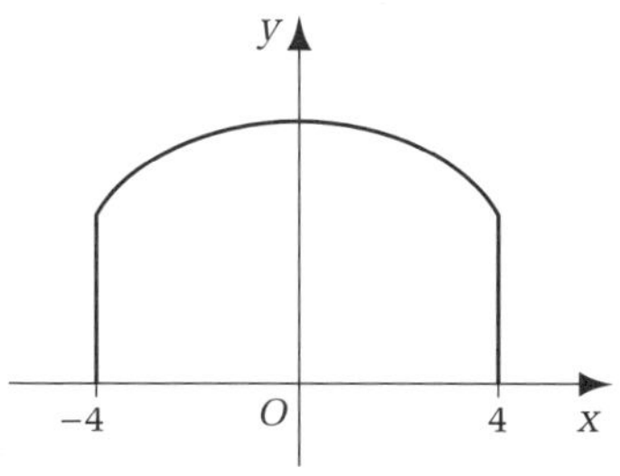

3 The diagram shows the velocity–time graph of an object. The object accelerates for the first four seconds then travels with constant velocity.

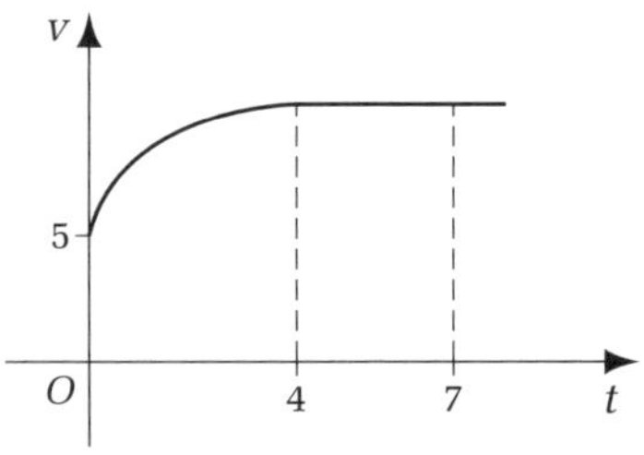

In the first four seconds the velocity is given by $v = 3t^{\frac{1}{2}} + 5$, where t is the time. Find the distance travelled in the first seven seconds.

4 The graph shows the relationship between the current I and potential difference V of a filament lamp. The result $\frac{1}{10}\int_0^{10} I \, \mathrm{d}V$ gives the mean current as V increases from 0 to 10. If $I = 0.5V^{\frac{1}{2}}$, calculate the mean current as V increases from 0 to 10.

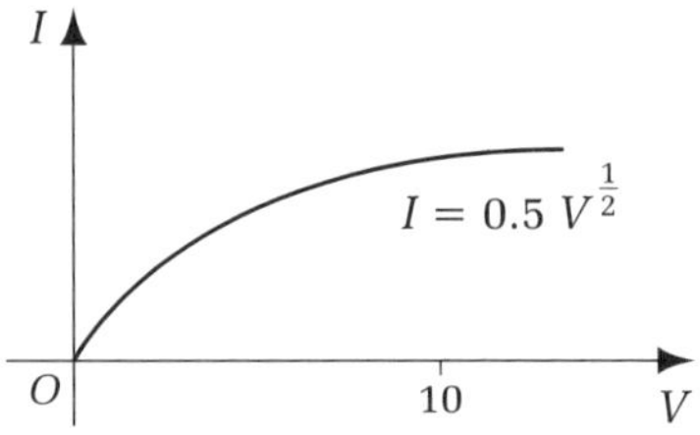

5 The diagram shows the curve $y = x^3$. Find a if the shaded area is 4 square units.

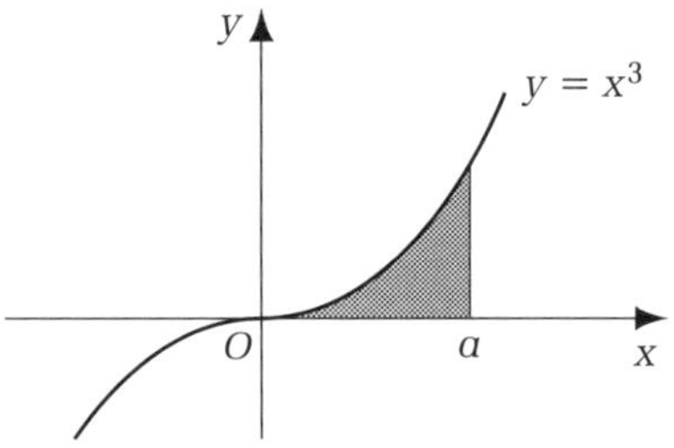

9.3 Technique

1 The graph shows the line $y = x - 2$.

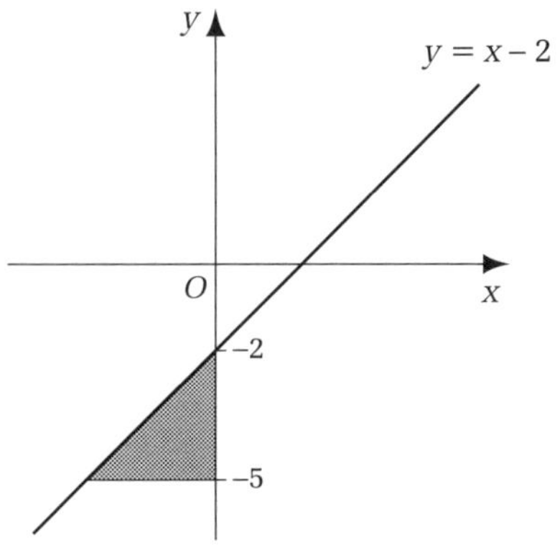

a Find x in terms of y.

b Find the area of the shaded region using integration. (Check this by using the formula for the area of a triangle.)

2 The graph shows part of the curve $y = \frac{1}{4}x^2$. Find x in terms of y and then find the area of the shaded region.

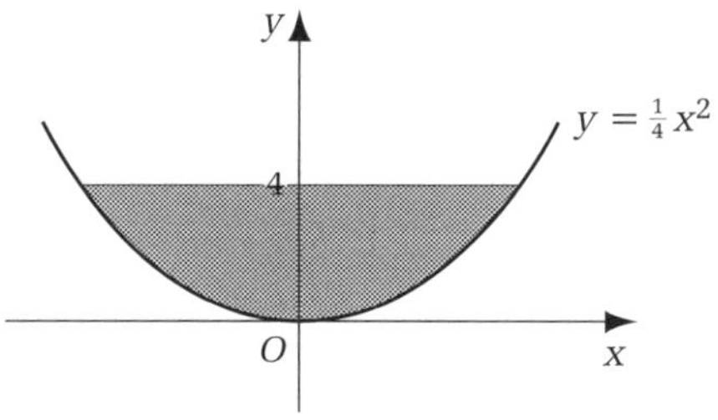

3 The graph shows the curve $y = \frac{1}{8}x^3$. Find x in terms of y. Then find the area of the shaded region, giving your answer correct to two decimal places.

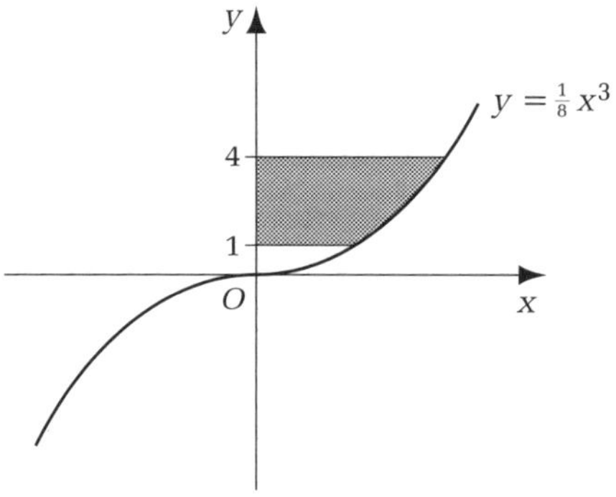

4 The graph shows the curve $y = 1 + \sqrt{x - 1}$. Find x in terms of y. Then find the exact area of the shaded region.

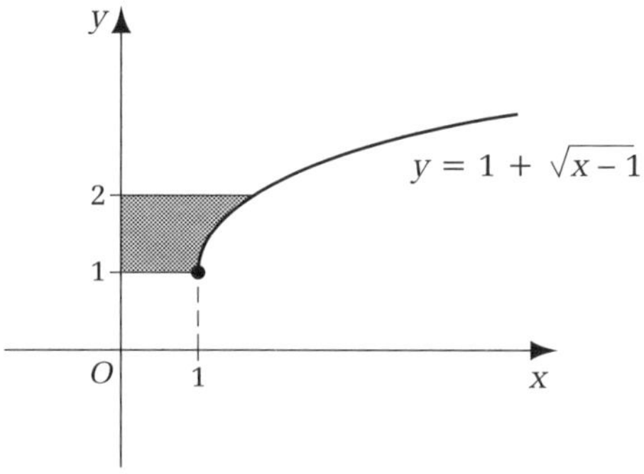

9.4 Technique

1 Find the area between the line $y = x + 1$ and the curve $y = x^2 - 1$.

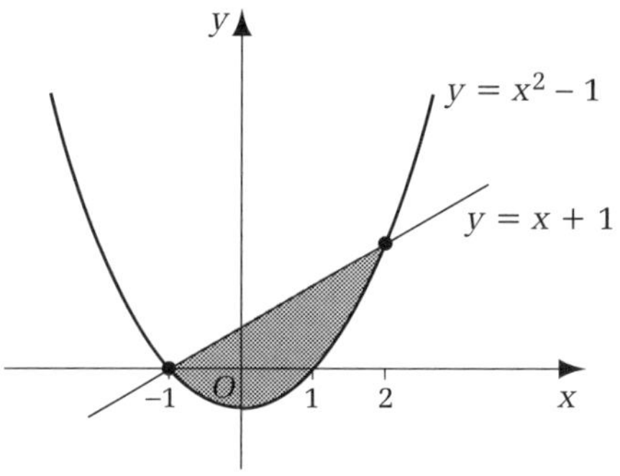

2 Find the area between the curves $y = 1 - x^2$ and $y = x^2 + 2x - 3$.

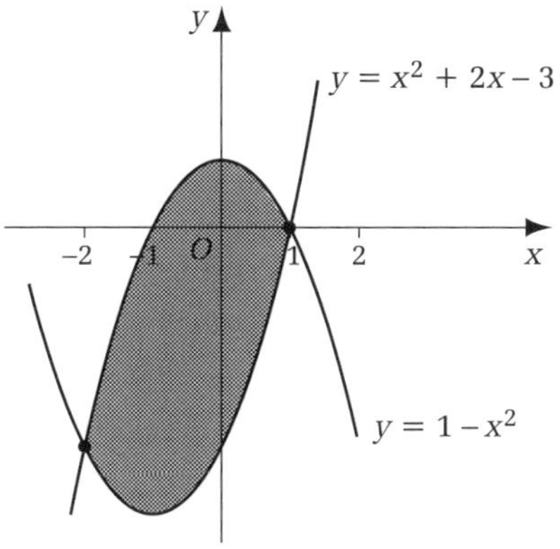

3 Find the area between the curve $y = x^2 - 2x + 2.25$, the line $y = x$ and the y-axis.

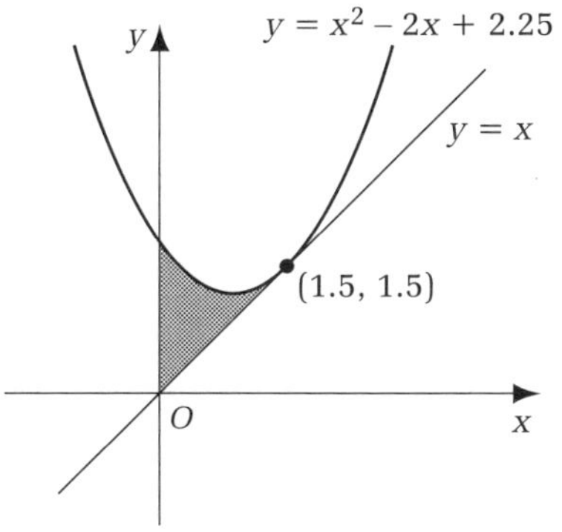

9.4 Contextual

1 The graph shows tangent $y = 4x - 4$ to the curve $y = x^2$.

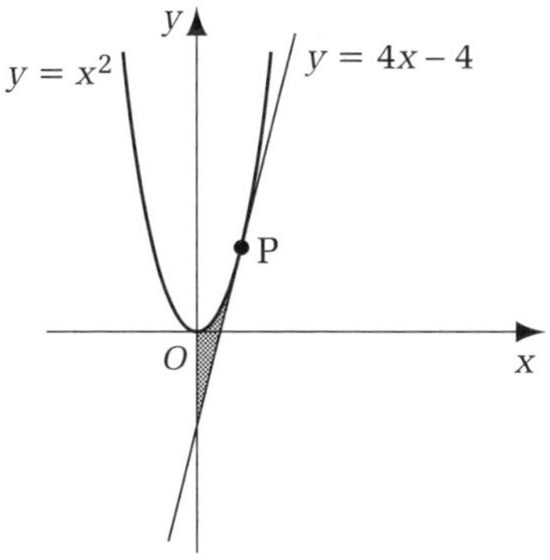

a Find the coordinates of P.

b Find the area of the shaded region.

2 The graph shows the line $y = 3 - x$ and the curve $y = x(x - 3)$.

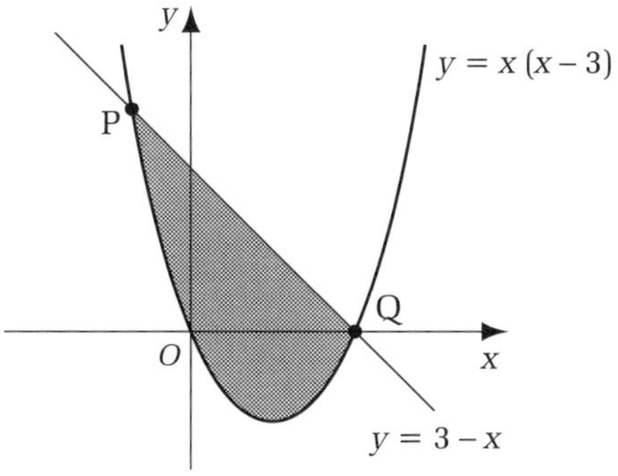

a Find the coordinates of P and Q.

b Calculate the area of the shaded region.

3 The graph shows the line $y = 4 - x$ and the curve $y = x^5(4 - x)$.

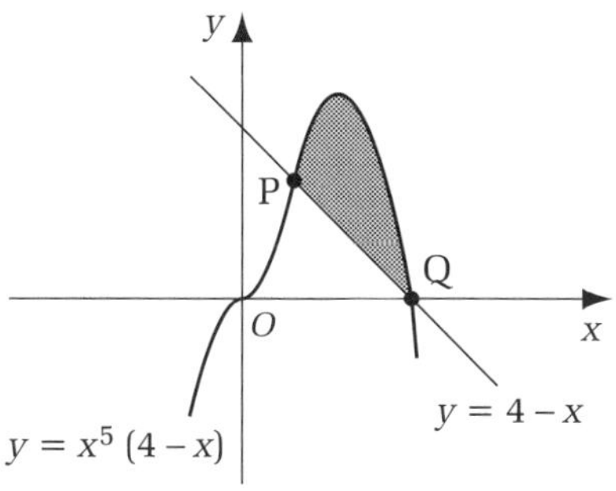

a Verify that the coordinates of P are (1, 3).

b Write down the coordinates of Q.

c Find the area of the shaded region, giving your answer to two decimal places.

4 Find the area between the curves $y = x(x - 2)$ and $y = 2x(5 - x)$.

5 Find the area between the following curves:

a $y = 8x$ and $y = \sqrt{x^5}$ **b** $y = 3x$ and $y^2 = 9x$

9.5 Technique

1 Sketch the solids formed when each of the shaded areas is rotated through 360° about the x-axis. In each case calculate the volume of the solid.

a

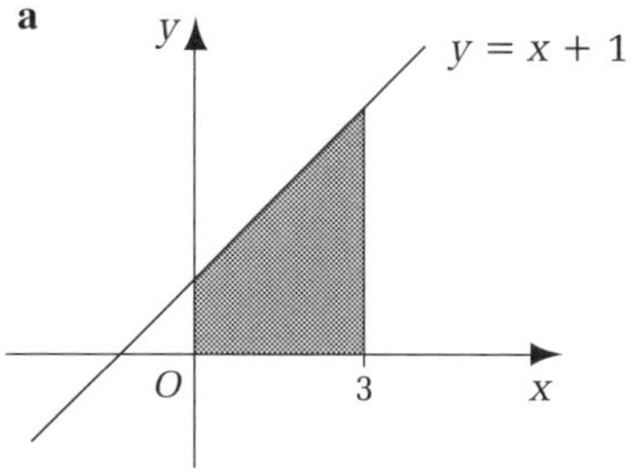

b

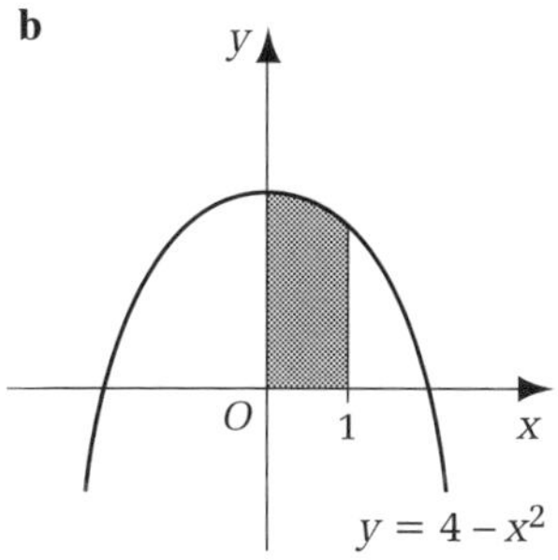

c

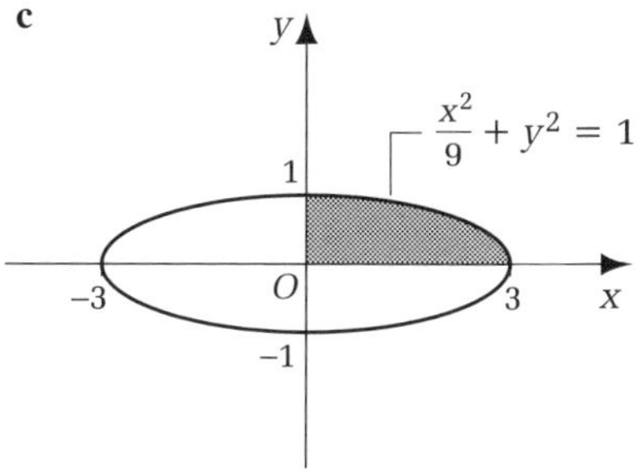

d

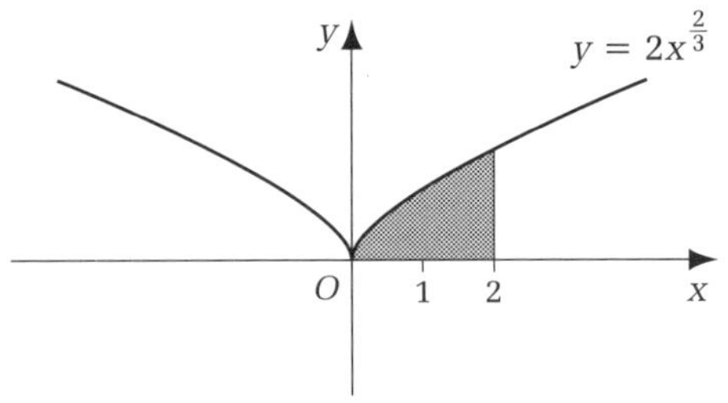

2 Sketch the solids formed when each of the shaded areas is rotated through 360° about the y-axis. In each case, calculate the volume of the solid.

a

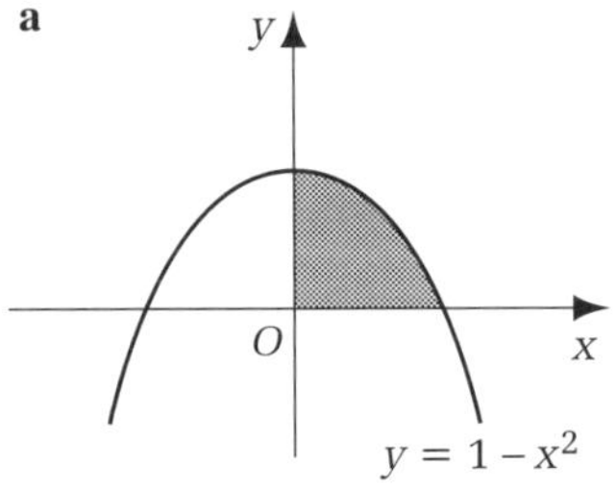

b

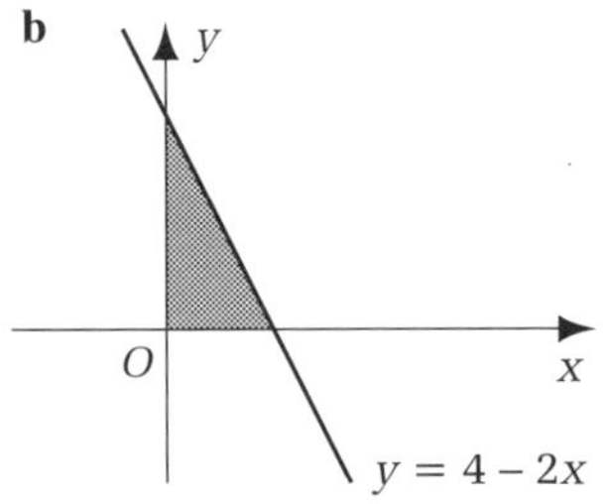

3 Sketch the solid formed when the area under the curve $y = \frac{1}{2}x^2 + 1$, from $x = -1$ to $x = 1$, is rotated through one complete turn about the x-axis. Calculate the volume of this solid.

2 A cone is formed when the shaded region is rotated through one complete turn about the x-axis.

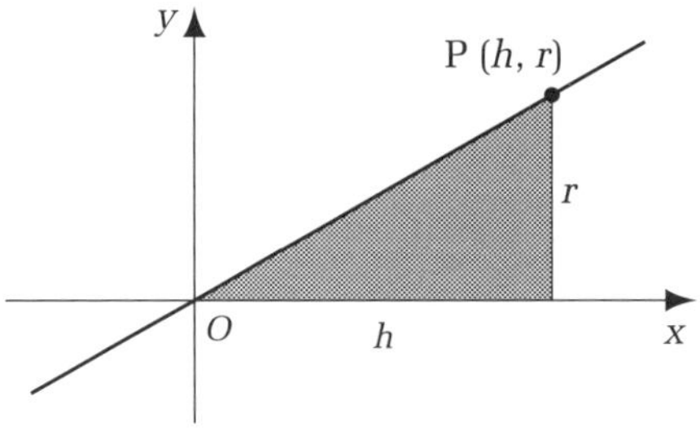

a Explain why the line OP has equation $y = \frac{r}{h}x$.

b Use integration to find the volume of a cone with height h and base radius r.

3 A paperweight is in the form of the solid generated when the shaded region is rotated about the y-axis.

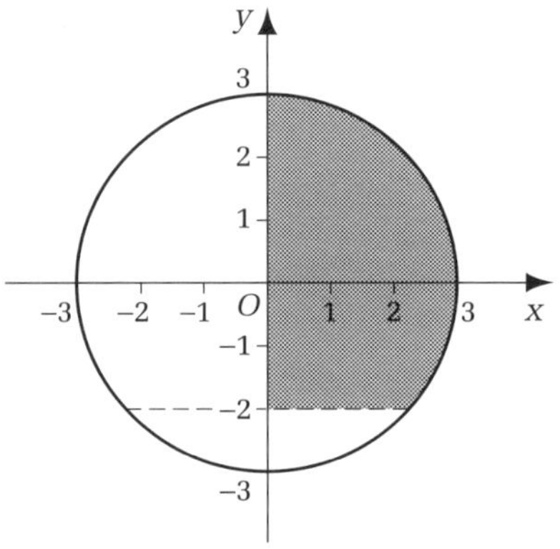

The equation of the circle is $x^2 + y^2 = 9$. Find the volume of the paperweight if this shaded region is rotated through one complete turn about the y-axis.

9.5 Contextual

1 A rugby ball style solid is formed when part of the curve $\frac{x^2}{100} + \frac{y^2}{25} = 1$ is rotated about the x-axis. Find the volume generated when the shaded region is rotated through one complete turn about the x-axis. Hence find the volume of the ball.

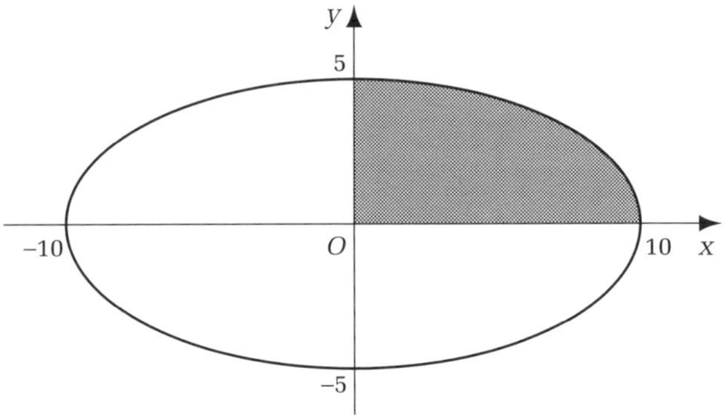

9.6 Technique

1 Integrate the following:

a $2 \cos x$ **b** $-2 \sin x$

c $2 + \cos x$ **d** $1 + \cos x + \sin x$

2 Find:

a $\int \cos 2x \; dx$ **d** $\int (x - 2 \sin 2x) \; dx$

b $\int 4 \cos 2x \; dx$ **e** $\int (3 + 10 \sin 5x) \; dx$

c $\int (1 - 4 \cos 2x) \; dx$ **f** $\int (\cos 2x - \sin 2x) \; dx$

3 Evaluate:

a $\int_0^{\frac{\pi}{2}} (\cos x - \sin x) \; dx$ **b** $\int_{\frac{\pi}{2}}^{\pi} \sin \tfrac{1}{3}x \; dx$

4 Find the area of the shaded region in each of the following, correct to two decimal places:

a

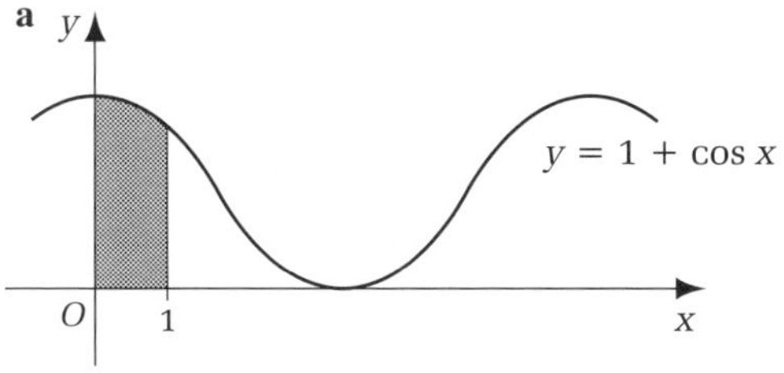

b

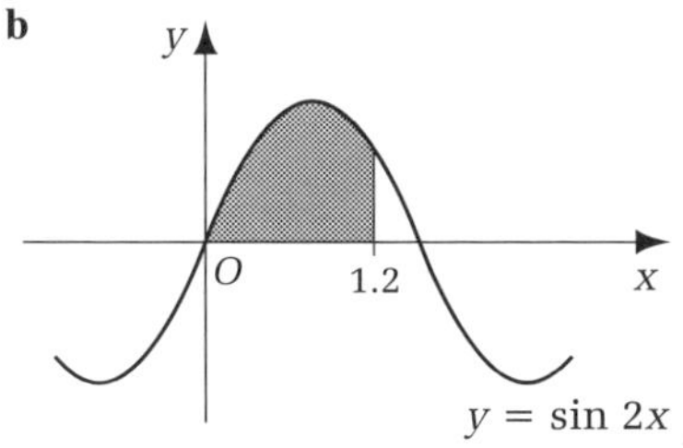

5 Find the exact area of the shaded region in each of the following:

a

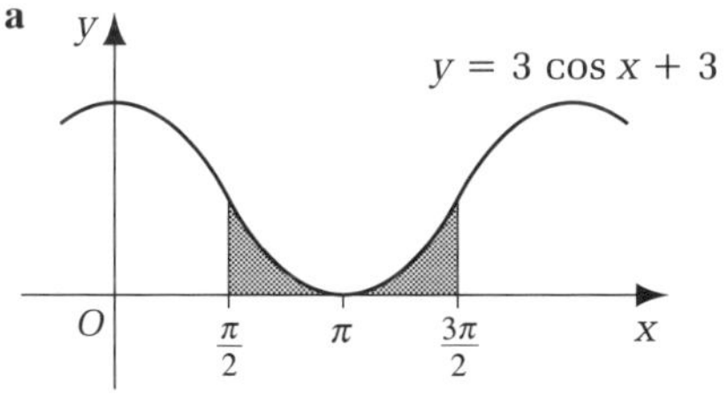

b

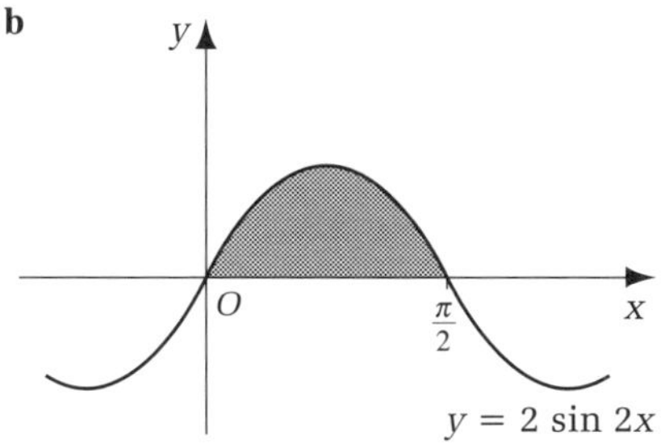

c

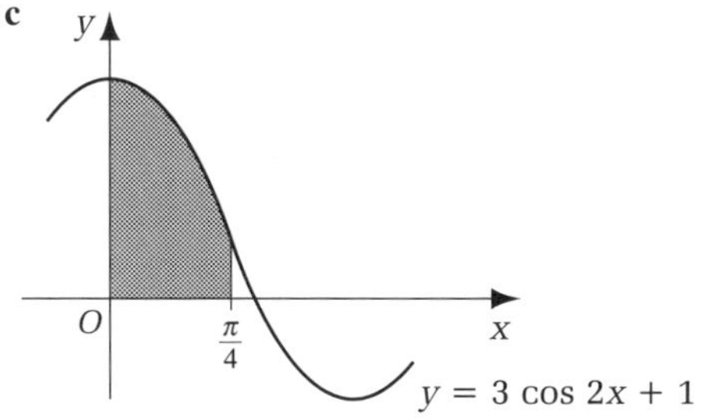

6 Find the area of the shaded region correct to two decimal places.

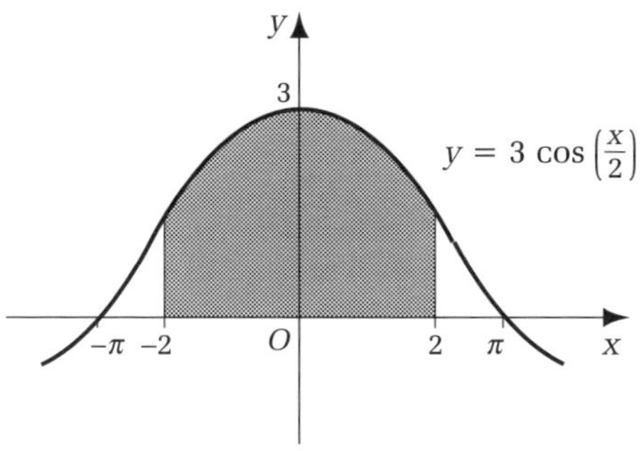

7 Find the exact areas of the following shaded regions:

a

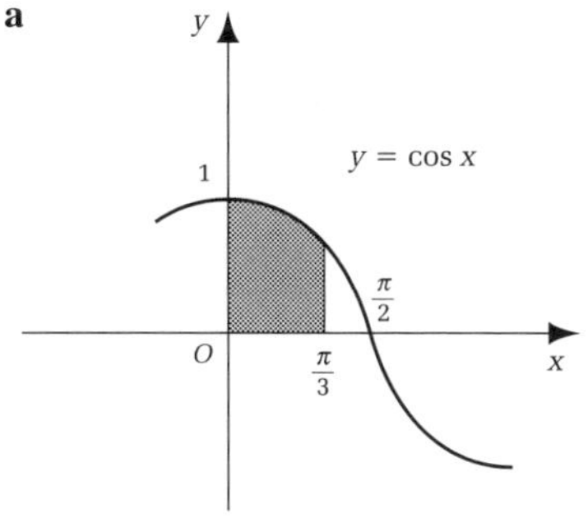

b

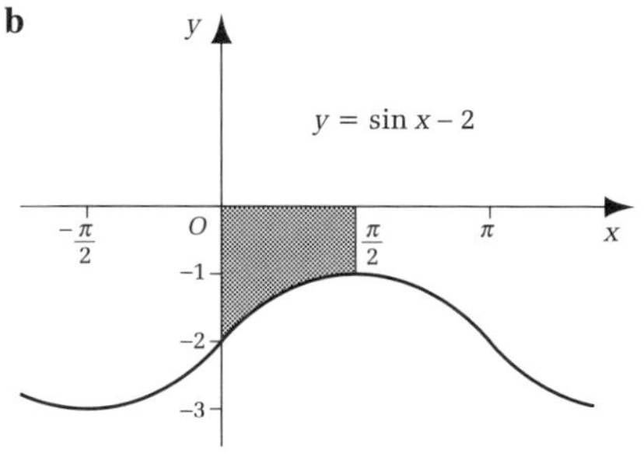

c

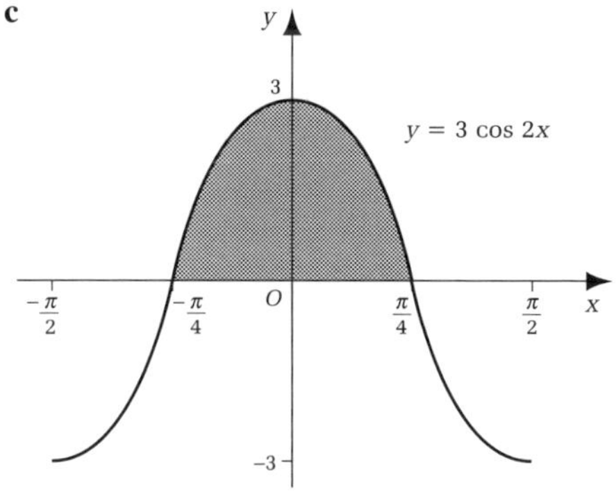

d

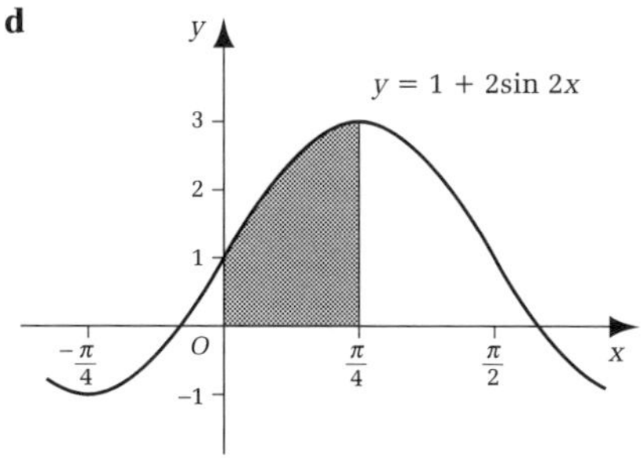

8 Find:

a $\int \sin^2 2x \, dx$ **b** $\int \cos^2\left(\frac{x}{2}\right) dx$

9.6 Contextual

1 The graph shows $y = \cos x$.

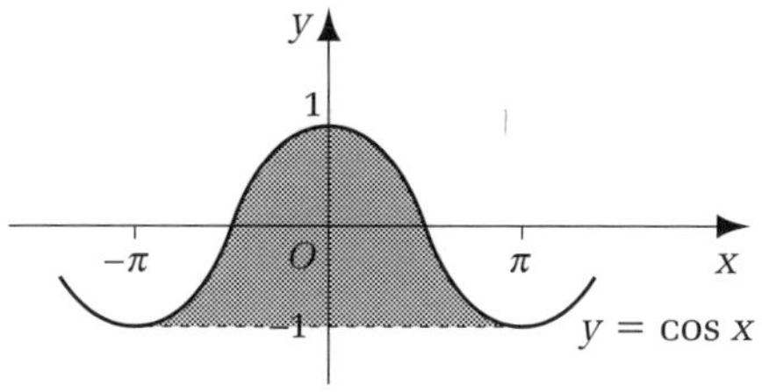

a Explain why $\int_{-\frac{\pi}{2}}^{0} \cos x \, dx = \int_{0}^{\frac{\pi}{2}} \cos x \, dx$.

b Find the area of the shaded region.

2 Evaluate $\int_{-\frac{\pi}{2}}^{0} \sin x \, dx$. Sketch the graph of $y = \sin x$ and illustrate what this integral represents. Comment on your answer.

3 Identify the transformation of the sine or cosine function in these graphs. Use your results to find the areas of the shaded regions.

a

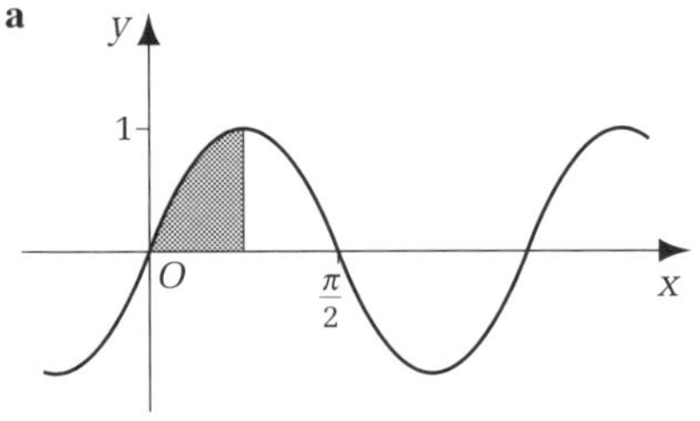

b

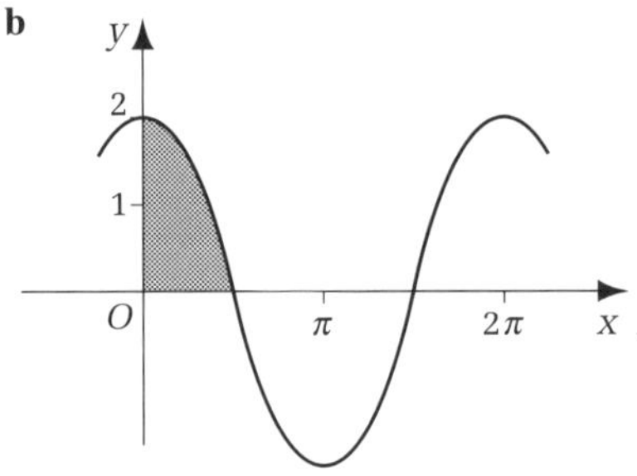

4 The graph shows parts of the curves $y = \sin x$ and $y = \sin 2x$.

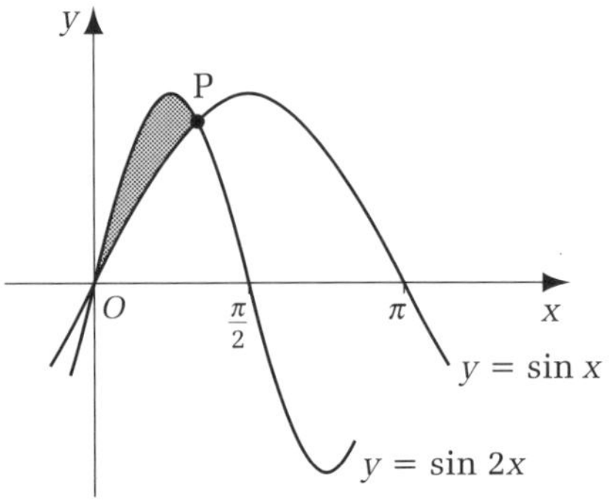

a Verify that the coordinates of P are $\left(\frac{\pi}{3}, \frac{\sqrt{3}}{2}\right)$.

b Find the area of the shaded region.

5 The graph shows the curve of $y = \sin^2 x$. Calculate the area of the shaded region.

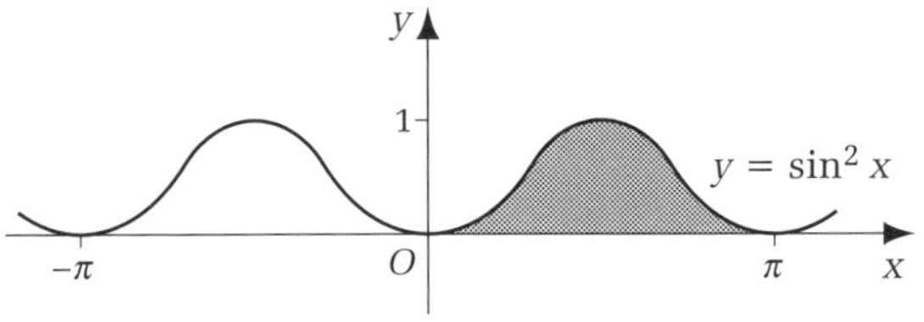

6 The graph shows part of the curve $y = 1 + \sin x$. Calculate the volume generated when the shaded area is rotated through one complete turn about the x-axis.

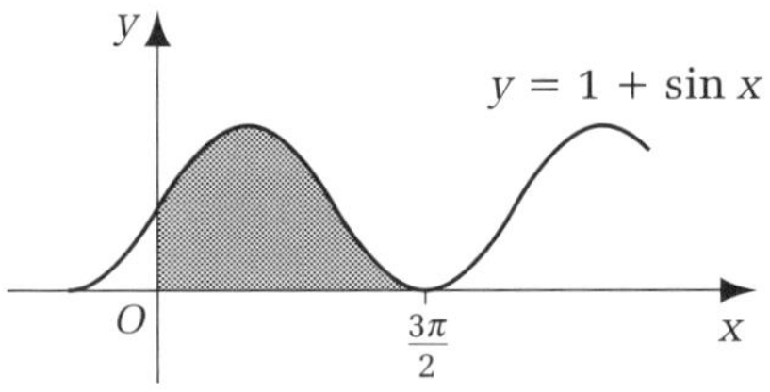

7 The graph shows part of the curve $y = 2\cos 2x$. The shaded region is rotated through 360° about the x-axis. Calculate the volume of the solid created.

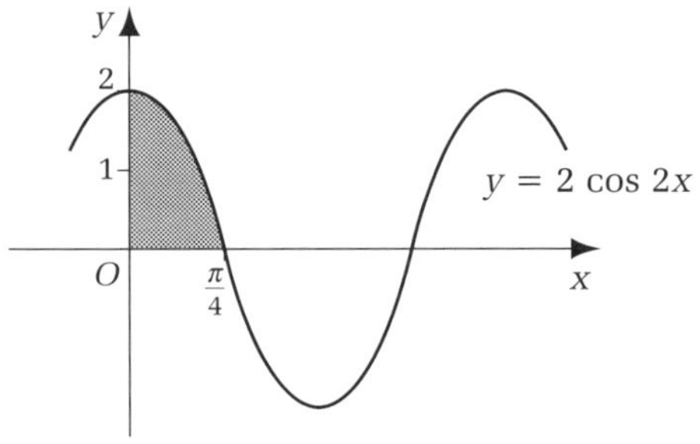

Consolidation C

1 Find:

a $\int (x-2)(x+5) \, dx$ **b** $\int x(2-3x) \, dx$

2 **a** Differentiate the function $f: x \to x^3 + 3\sin x$.

b Find $g(x)$ given that $g'(x) = 6x - \frac{4}{x^2}$ and $g(1) = 9$.

c Find $\int \left(x^3 - x + 2\right) dx$ and hence evaluate $\int_1^3 \left(x^3 - x + 2\right) dx$.

3 The particle P moves along the x-axis so that, after t seconds, its acceleration is $(6t - 24)$ m s^{-2}. When $t = 0$ the velocity of P is 21 m s^{-1} and $x = -1$.

a Find an expression for the velocity of P after t seconds and deduce the two values of t for which the velocity is zero.

b Find an expression for the displacement of P from the origin after t seconds and deduce the distance between the two points where P is stationary.

c Find the velocity and displacement of P when the acceleration is -12 m s^{-2}.

(OCSEB)

4

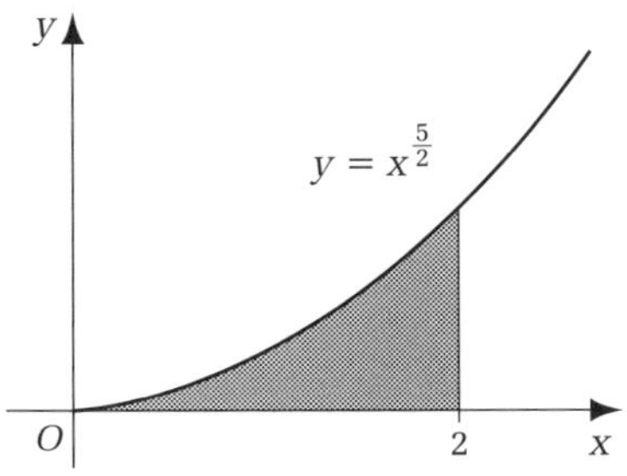

The shaded region is bounded by the graph of $y = x^{\frac{5}{2}}$, the x-axis for $0 \le x \le 2$ and the line $x = 2$. Calculate the shaded area giving your answer in the form $P\sqrt{2}$, where P is a rational number.

(NEAB)

5 In this question $f(x) = 1 + 16x - 10x^2 + 2x^3$ and $g(x) = 1 + 12x - 4x^2$.

a Verify that $f(1) = g(1)$ and $f(2) = g(2)$.

b Find $\int_0^2 [f(x) - g(x)]\,dx$.

(SMP)

6

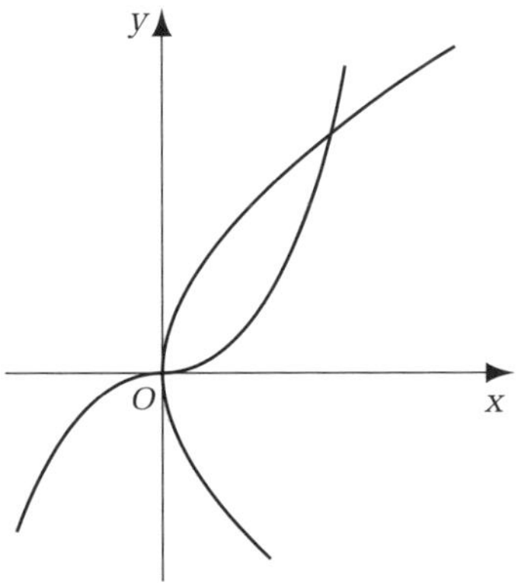

The figure shows parts of the graphs of $y = \frac{1}{8}x^3$ and $x = \frac{1}{16}y^2$.

a Find the coordinates of the points where the curves meet.

b Find the area of the region bounded by the two curves.

(SMP)

7

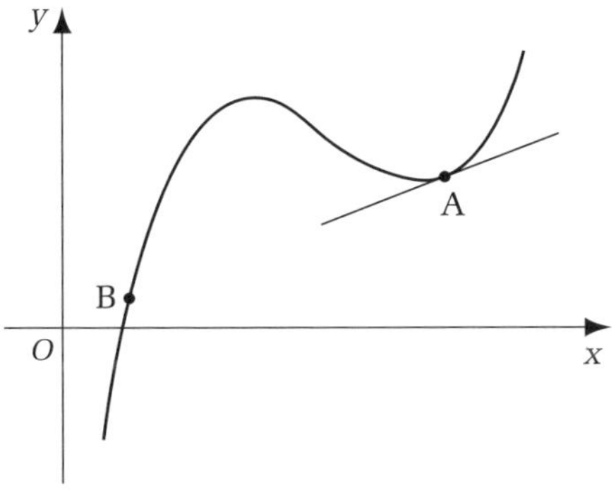

The graph shows the curve $y = x^3 - 13x^2 + 51x - 37$.

a Verify that the point where $x = 3$ is a maximum point.

b Find the gradient, the y-coordinate and the equation of the tangent at the point A, where $x = 6$.

c Verify that the point B on the curve, where $x = 1$, lies on the tangent at A.

d Find the area enclosed between the curve and the tangent line AB.

(MEI)

8 The diagram shows part of the curve $y = \dfrac{1}{\sqrt{x}}$.

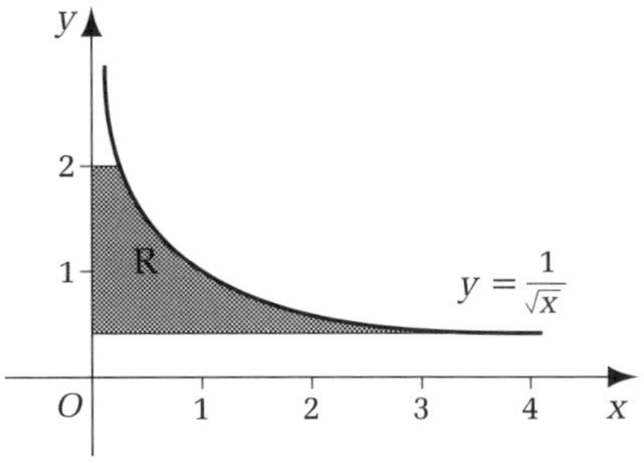

a Express x in terms of y and hence find the area of the region R.

b The region R is rotated about the y-axis through 360°. Calculate the volume generated.

9

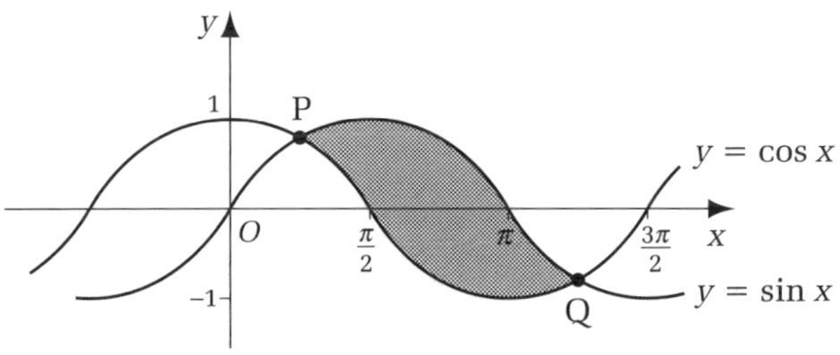

The diagram shows the curves $f(x) = \sin x$ and $g(x) = \cos x$.

a Verify that the x-coordinate of Q is $\dfrac{5\pi}{4}$ and find the x-coordinate of P.

b Explain why $\int_a^b (\sin x - \cos x)\,dx$ would evaluate the area of the shaded region for suitable values of a and b.

c Calculate the area of the shaded region.

10 **a** Find $\int(10 - \sin 2x + 6\cos 3x)dx$.

b

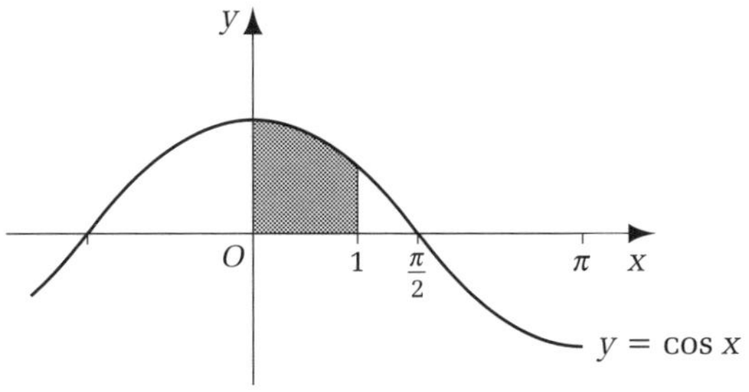

The graph shows part of the curve $y = \cos x$. Calculate the volume generated when the shaded region is rotated about the x-axis through four right angles. Give your answer correct to two decimal places.

Review

1 Write down the compound angle expansion for the following:

a $\sin(A - B)$

b $\cos(A + B)$

c $\tan(A - B)$

d $\cos(x - 30°)$

e $\sin(x + 45°)$

f $\tan\left(x + \dfrac{\pi}{4}\right)$

2 a Convert the following angles into radians, leaving the answer in terms of π:

i $60°$ ii $150°$ iii $300°$
iv $225°$ v $410°$ vi $-70°$

b Convert the following angles into degrees:

i $\frac{\pi}{3}$ ii $\frac{5\pi}{8}$ iii $\frac{\pi}{18}$
iv $\frac{11\pi}{6}$ v $\frac{7\pi}{2}$ vi $-\frac{7\pi}{18}$

3 Without using a calculator, write down the exact value of the following:

a $\sin\left(\frac{\pi}{4}\right)$ b $\tan 60°$ c $\cos 150°$
d $\sin\left(\frac{3\pi}{2}\right)$ e $\tan 240°$ f $\cos\left(\frac{\pi}{6}\right)$

10.1 Technique

1 Solve the following equations for $0° \le x \le 180°$:

a $\sin 5x - \sin x = \cos 3x$
b $\cos 2x + \cos 4x = \cos x$
c $\sin 3x - \sin x = \cos 2x$
d $\cos 5x + \cos x = \sin 5x + \sin x$

2 Solve the following equations for $0 \le x \le \pi$:

a $\cos 4x + \cos 2x = 0$
b $\cos 5x - \cos x = \sin 3x$
c $\sin x + \sin 2x + \sin 6x = \sin 3x$
d $\sin 3x + \sin x = \cos 3x + \cos x$

3 Prove the following identities:

a $\cos 3\theta + \cos\theta \equiv 4\cos^3\theta - 2\cos\theta$
b $\dfrac{\sin 2A + \sin 2B}{\sin 2A - \sin 2B} \equiv \dfrac{\tan(A+B)}{\tan(A-B)}$
c $\dfrac{\sin A + \sin B}{\cos A + \cos B} \equiv \tan\left(\dfrac{A+B}{2}\right)$
d $\dfrac{\cos\theta - \cos 3\theta}{\sin 3\theta - \sin\theta} \equiv \tan 2\theta$
e $\dfrac{\cos(\theta+30°)+\cos(\theta+60°)}{\sin(\theta+30°)+\sin(\theta+60°)} \equiv \dfrac{1-\tan\theta}{1+\tan\theta}$

4 Write each of the following as the sum or difference of two trigonometric functions:

a $2\cos 4\theta \sin\theta$ d $2\cos 5\theta\cos\theta$
b $2\sin 5\theta\cos 2\theta$ e $-2\sin 5\theta\sin 2\theta$
c $2\cos 5\theta\sin 2\theta$ f $2\sin 2\theta\sin 7\theta$

5 Without using a calculator, find the exact values of the following:

a $\sin 105° + \sin 15°$ b $\cos 75° + \cos 15°$
c $2\cos 52\frac{1}{2}°\cos 7\frac{1}{2}°$ d $-2\sin 37\frac{1}{2}°\sin 7\frac{1}{2}°$

10.2 Technique

1 Express each of the following in the form $R\sin(x+\alpha)$, with $0° \le \alpha \le 90°$:

a $6\sin x + 8\cos x$ b $12\sin x + 5\cos x$
c $60\sin x + 11\cos x$ d $3\sin x + 5\cos x$

2 Express each of the following in the form $R\sin(x-\alpha)$ with $0 \le \alpha \le \frac{\pi}{2}$:

a $\sin x - \sqrt{3}\cos x$ b $\sin x - \cos x$
c $4\sin x - 3\cos x$ d $5\sin x - 7\cos x$

3 Express each of the following in the form $R\cos(x+\alpha)$ with $0° \le \alpha \le 90°$:

a $12\cos x - 5\sin x$ b $24\cos x - 7\sin x$
c $2\cos x - 3\sin x$ d $\cos x - \sin x$

4 Express each of the following in the form $R\cos(x-\alpha)$ with $0 \le \alpha \le \frac{\pi}{2}$:

a $3\cos x + 4\sin x$ b $11\cos x + 60\sin x$
c $\sqrt{3}\cos x + \sin x$ d $4\cos x + 7\sin x$

5 Find the maximum and minimum values of the following expressions, in each case giving the corresponding values of x for $0° \le x \le 360°$:

a $3\cos x - 4\sin x$ b $5\sin x + 2\cos x$
c $3\sin x - 5\cos x$ d $12\cos x + 5\sin x$

6 Solve the following equations for $0° \le x \le 360°$:

a $6\sin x + 8\cos x = 5$ d $\sqrt{3}\cos x + \sin x = 1$
b $4\sin x - 3\cos x = 5$ e $5\sin x + 2\cos x = 0$
c $\cos x - \sin x = \sqrt{2}$ f $\sin x + 2\cos x = -1$

7 Solve the following equations for $-\pi \le x \le \pi$:

a $3\sin x + \cos x = 2$ b $\sin x - 2\cos x = 1$
c $5\cos x - \sin x = 3$ d $\sqrt{2}\cos x + 3\sin x = -1$

8 Find the maximum and minimum values of the following functions:

a $\dfrac{1}{3\sin x + 4\cos x + 6}$ b $\dfrac{2}{8\cos x - 6\sin x + 12}$
c $\dfrac{1}{\sin x - \sqrt{3}\cos x + 3}$ d $\dfrac{3}{12\cos x + 5\sin x + 20}$

10.2 Contextual

1 A mass is suspended from the end of a spring as shown in the diagram.

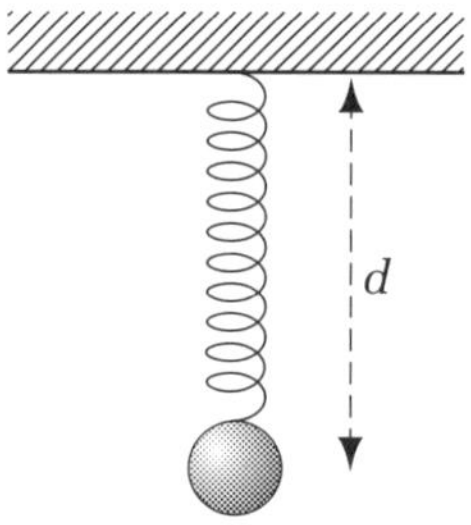

The mass is oscillating. The distance d cm between the fixture point and the mass is given by

$d = 16 + 12\sin(8t)^\circ - 9\cos(8t)^\circ$, where t seconds is the time after release. Find:

a the maximum and minimum distances from the fixture point

b the time at which the mass is first at its lowest point.

2 A fuel storage tank at a busy 24-hour service station has a variable depth. The depth of fuel can be modelled by the equation $d = 100 - [63\cos(2t)^\circ + 16\sin(2t)^\circ]$, where d centimetres is the depth at time t hours after midnight on Friday. Find:

a the least and greatest depths of fuel for this storage tank

b the time at which the depth first reaches these values.

10.3 Technique

1 Find the general solutions, in degrees, of the following equations:

a $\cos\theta = 0.2$
b $\cos\theta = -0.75$
c $\sin\theta = 0.9$
d $\sin\theta = -0.4$
e $\tan\theta = 0.1$
f $\tan\theta = -2.9$

2 Find the general solution of the following equations:

a $\cos\theta = \frac{1}{\sqrt{2}}$
b $\cos\theta = -\frac{1}{\sqrt{2}}$
c $\sin\theta = \frac{1}{\sqrt{2}}$
d $\sin\theta = 0$
e $\tan\theta = -1$
f $\tan\theta = \sqrt{3}$

3 Find the general solution of the following equations:

a $\cos 2\theta = \frac{1}{2}$
b $\sin 2\theta = \sin\theta$
c $\tan\theta = \tan 4\theta$
d $\sin 2\theta - \cos 2\theta = 0$

4 Find the general solution, in degrees, of the equation $16\cos\theta^\circ + 30\sin\theta^\circ = 17$.

5 Find the general solution of the equation $6\cos\theta - 4\sin\theta = 5$.

6 Find the general solution of the equation $\sin\theta + \sin 3\theta + \sin 5\theta = 0$.

Consolidation C

1 Solve the equation $\sin 6\theta + \sin 2\theta = 2\sin 4\theta$ for $0 \leq \theta \leq \pi$.

2 **a** Use the double angle formulae to show that

$$\frac{\sqrt{3}\sin 2\theta + \cos 2\theta + 1}{\cos\theta} \equiv 2\sqrt{3}\sin\theta + 2\cos\theta,$$

where $0^\circ \leq \theta \leq 90^\circ$.

b By writing $2\sqrt{3}\sin\theta + 2\cos\theta$ in the form $r\sin(\theta + \alpha)$, where r is real and α is acute, find the maximum value of this expression and the corresponding value of θ where $0^\circ \leq \theta \leq 90^\circ$.

(*NICCEA*)

3 **a** Express $63\cos 2\theta - 16\sin 2\theta$ in the form $R\cos(2\theta + \alpha)^\circ$ where $R > 0$ and α is an acute angle.

b Find all the solutions of the equation $63\cos 2\theta - 16\sin 2\theta = 26$ in the range $0^\circ \leq \theta \leq 180^\circ$.

(*WJEC*)

4 Express the function $2\sin x^\circ + \cos x^\circ$ in the form $R\sin(x + \alpha)^\circ$, stating the values of R and α. Using these values, write down the coordinates of the maximum turning point on the graph of $2\sin x^\circ + \cos x^\circ$ for $0 \leq x \leq 90$.

(*NEAB*)

5 $f(x) \equiv 5\cos x - 12\sin x, x \in \mathbb{R}$

Given that $f(x) = R\cos(x + \alpha)$, where $R > 0$ and $0 < \alpha < \frac{\pi}{2}$:

a find the values of R and α

b find the range of values of $f(x)$.

(*ULEAC*)

6 Given that $8\cos x - 15\sin x \equiv R\cos(x + A)$ find:

a the positive constant R

b the acute angle A, giving your answer in degrees to one decimal place.

c Hence, or otherwise, find the maximum value of $8\cos x - 15\sin x$.

d Find the value of x, in the range $0^\circ \leq x \leq 360^\circ$, which will give the maximum value, giving your answer in degrees to one decimal place.

(*ULEAC*)

7 Express $3\cos\theta - \sin\theta$ in the form $R\cos(\theta + \alpha)$, where $R > 0$ and $0 < \alpha < \frac{\pi}{2}$. Hence show that one of the values of θ, in the interval $0 < \theta < 2\pi$, for which $3\cos\theta - \sin\theta = 2$ may be expressed as $\cos^{-1}\left(\sqrt{\frac{2}{5}}\right) - \tan^{-1}\left(\frac{1}{3}\right)$ and find an expression for the other value of θ.

(*UCLES*)

8 Find the general solution of the equation $\cos 2x = \cos 3x$, giving your answer in radians in terms of π.

(*ULEAC*)

9 Find the general solution in radians of the equation $3\sin\theta - 4\cos\theta = 3\sin 2\theta - 4\cos 2\theta$.

(*AEB*)

10 $f(x) \equiv 7\cos x - 24\sin x$.

Given that $f(x) \equiv R\cos(x + \alpha)$, where $R \geq 0$, $0 \leq \alpha \leq \frac{\pi}{2}$, and x and α are measured in radians:

a find R and show that $\alpha = 1.29$ to two decimal places.

Hence write down:

b the minimum value of $f(x)$

c the value of x in the interval $0 \leq x \leq 2\pi$ which gives this minimum value.

d Find the smallest two positive values of x for which $7\cos x - 24\sin x = 10$.

(*ULEAC*)

Review

1 Differentiate each of the following with respect to x:

a $y = 6 - 3x - x^2$
b $y = 6x^5 - 2x^3 + x$
c $y = 4(\sqrt{x})^3$
d $y = x^{-2} + 3x^{-4}$
e $y = 6x^{\frac{2}{3}}$
f $y = (x^3 - 2)(x + 2)$
g $y = \cos 5x$
h $y = 4\sin\left(\frac{x}{2}\right)$
i $y = \cos\left(x + \frac{3\pi}{2}\right)$
j $y = \sin(\pi - 3x)$
k $y = \cos 6x - \sin^2 x$
l $y = x^3 - \cos^3 x$

2 Use the chain rule to find $\frac{dy}{dx}$ for each of the following:

a $y = (x - 9)^5$
b $y = (5x + 6)^4$
c $y = \sqrt{3 - x}$
d $y = \sqrt{x^3 - 27}$
e $y = \frac{10}{x^2 - 9}$
f $y = \frac{-3}{\sqrt{2x - 8}}$

3 Find the coordinates of the stationary points on the graph of $y = x^3 + 9x^2 + 27x + 10$ and determine their nature.

4 Find the equations of the tangent and the normal to the curve $y = (2x - 1)^2$ at the point where $x = -2$.

5 Find the exact solutions of the following exponential and logarithmic equations:

a $e^{x+1} = 9$
b $\ln(x + 5) = 2$
c $e^{2x} = 5$
d $\ln\left(\frac{x}{3}\right) = -2$

6 Write down the Cartesian equation for each of the following circles:

a radius 10, centre (0, 0)
b radius 12, centre (11, 0)
c radius 9, centre (7, −8)
d radius $\sqrt{5}$, centre (−5, 5)

7 Use the binomial theorem to express each of the following as a series of ascending powers of x, up to and including the x^3 term:

a $(1 - x)^{-3}$
b $(1 + 3x)^{\frac{2}{3}}$
c $\frac{1}{1 + 5x}$
d $\sqrt{1 - 2x}$

11.1 Technique

1 Use the product and quotient rules to differentiate each of the following with respect to x:

a $y = 2x(x - 3)^5$
b $y = (x - 1)(2x + 5)^4$
c $y = (x + 2)\sqrt{2x + 1}$
d $y = x\sin^2 x$
e $y = \frac{3x - 2}{x + 1}$
f $y = \frac{x^3}{4x + 1}$
g $y = \sin 5x \cos x$
h $y = \frac{\sin 5x}{\cos x}$

2 Find the gradient of each of the following curves at the point indicated:

a $y = x(x - 5)^4$ at (3, 48)
b $y = (2 - x)(x + 3)^3$ at (−1, 24)
c $y = \frac{4x + 3}{x - 3}$ at (2, −11)
d $y = \frac{2x^2}{3x - 2}$ at $(4, \frac{16}{5})$
e $y = (x^2 - 3)\cos x$ at $(\pi, 3 - \pi^2)$
f $y = (x - 1)\sqrt{x + 1}$ at (3, 4)

11.1 Contextual

1 Find the equations of the tangent and normal to the following curves at the points indicated:

a $y = \frac{x - 4}{3x - 2}$ at (1, −3)
b $y = (x + 1)(x + 2)^3$ at (−3, 2)
c $y = 2x\cos^2 x$ at $(\pi, 2\pi)$
d $y = \frac{x^2}{x + 5}$ at (−4, 16)

11.2 Technique

1 Differentiate each of the following with respect to x:

a $y = e^{3x+5}$
b $y = e^{x^4}$
c $y = e^{2-x}$
d $y = e^{\frac{1}{x}}$
e $y = e^{\sin 2x}$
f $y = x^3 e^x$
g $y = e^{-x}\cos^2 x$
h $y = \frac{e^x}{x - 1}$
i $y = e^x \cos 2x$
j $y = \frac{\sin^2 x}{e^x}$

2 Differentiate each of the following with respect to x:

a $y = \ln(7x)$
b $y = \ln(\sqrt{x})$
c $y = \ln(\sqrt[3]{x})$
d $y = \ln(1 - 2x)$
e $y = \ln(1 - x^2)$
f $y = \ln(\sin^2 x)$
g $y = x^4 \ln x$
h $y = \frac{\ln x}{x^3}$
i $y = \ln(1 + x) + \ln(1 - x)$
j $y = \log_{10}(e^x)$

3 For each of the following curves:

i find expressions for $\frac{dy}{dx}$ and $\frac{d^2y}{dx^2}$
ii locate and determine the nature of any stationary points.

a $y = (3x - 4)e^x$
b $y = \ln(2x - 1) + \frac{5}{(2x - 1)}$
c $y = e^{-x}\cos 2x$ (for $-\frac{\pi}{2} \le x \le \frac{\pi}{2}$)
d $y = (x^2 - 3)e^{4x}$

11.2 Contextual

1 Find the equations of the tangent and normal to the following curves at the points indicated:

a $y = \ln(2x - 5)$ at $x = 4$ **b** $y = e^{x^2-3}$ at $x = -2$

c $y = x^3 \ln x$ at $x = e$ **d** $y = (x^2 + 1)e^{-x}$ at $x = -1$

2 The number of visitors per day to a new out-of-town shopping complex can be modelled by the expression $N = 5900 + 100e^{0.05t}$, where t is the number of days after the complex opens.

a Calculate the number of visitors to the complex after 14 and 28 days.

b How many days after opening is it before the number of visitors per day exceeds 7000?

c Calculate the rate at which the number of visitors per day is increasing 30 days after the complex opens.

d State any limitations of this model.

3 During the first year of its manufacture, the cost, £C, of producing a particular microwave oven can be modelled by $C = 170 \ln(0.1t + 3)$, where t is the number of months since this new model was launched. In order to boost sales and establish the model, the manufacturer is willing to initially produce them at a loss. The wholesale price, £S, that they are sold to shops at is given by $S = 0.15t^3 - 0.6t^2 - 8t + 180$.

a Find an expression for $\dfrac{dS}{dt}$ and use it to calculate the lowest wholesale price that the manufacturer sets during this first year.

b Find an expression for the profit, £P, that the manufacturer makes on each microwave oven. Calculate the profit per oven after 3, 6, 9 and 12 months.

c Find an expression for $\dfrac{dP}{dt}$ and calculate the rate at which the profit per machine is changing after 3, 6, 9 and 12 months.

4 The annual number of visitors V (in 1000s) to a stately home t years after it is opened to the public in 1985 can be modelled by $V = 34 + 18 \ln(t + 1) - 3t$.

a After how many years did the number of visitors reach its maximum value? What was this maximum value, to the nearest thousand?

b How many visitors, to the nearest thousand, did the stately home have after two years, and what was the rate at which V was changing at this time?

11.3 Technique

1 Differentiate each of the following with respect to x:

a $y = \operatorname{cosec} 2x$

b $y = \cot 6x$

c $y = \tan^2 5x$

d $y = \operatorname{cosec}(x^3)$

e $y = \sec^2 3x$

f $y = \sin^{-1} 7x$

g $y = \tan^{-1}\left(\dfrac{x}{2}\right)$

h $y = \sin^{-1}(2x - 1)$

i $y = \tan^{-1} x^{\frac{3}{2}}$

j $y = \tan^{-1}(5x - 2)$

2 Use the product and quotient rules to differentiate each of the following with respect to x:

a $y = x\cot 2x$

b $y = (1 + x)^2 \tan x$

c $y = \dfrac{\sec x}{2x}$

d $y = x^3 \sin^{-1} x$

e $y = 3x\tan^{-1}\left(\dfrac{x}{3}\right)$

11.3 Contextual

1 Find the gradient of each of the following curves at the point indicated and the equations of the tangent and the normal at that point:

a $y = \sec 2x$ at $(\frac{\pi}{8}, \sqrt{2})$ **b** $y = \tan^{-1}\left(\dfrac{x}{5}\right)$ at $(5, \frac{\pi}{4})$

c $y = x \tan x$ at $(2\pi, 0)$

11.4 Technique

1 For each of the following cubic functions:

i find expressions for $f'(x)$ and $f''(x)$

ii find the coordinates of any stationary points on the graph of $y = f(x)$, and use the second derivative to determine their nature

iii sketch the graph of $y = f(x)$:

a $f(x) = x^3 + 6x^2 - 36x - 10$

b $f(x) = \frac{4}{3}x^3 - 2x^2 + x + 6$

c $f(x) = 72 + 120x - 3x^2 - 2x^3$

2 **a** State the coordinates of the points where the curve $y = (x - 4)(x + 2)^3$ crosses the axes.

b Expand $(x - 4)(x + 2)^3$.

c Locate and determine the nature of the stationary points on the curve $y = (x - 4)(x + 2)^3$. (Hint: Use the factor theorem.)

d Find the coordinates of the non-stationary point of inflexion on this curve. Use the first and third derivatives to determine its nature.

e Sketch $y = (x - 4)(x + 2)^3$.

3 Locate and determine the nature of the stationary points on the graph of $y = x^4 - 4x^3$. Show that there is also a non-stationary point of inflexion at (a, b), where a and b are integers to be determined.

4 **a** Find the coordinates of the stationary points on the curve $y = 4 - x + \sin^2 x$ in the interval $0 \le x \le 2\pi$. Use the second derivative to determine their nature.

b Find the coordinates of the non-stationary points of inflexion on the curve in this interval.

c Sketch $y = 4 - x + \sin^2 x$ for $0 \le x \le 2\pi$.

5 For each of the following curves:

i write down the coordinates of the points where the curve crosses the axes

ii write down the equation of any vertical asymptotes

iii find an expression for $\dfrac{dy}{dx}$

iv locate and determine the nature of any stationary points

v sketch the curve:

a $y = x(x - 2)^3$

b $y = (x - 3)(x + 1)^3$

c $y = \dfrac{(x + 2)^2}{x - 1}$

6 For each of the following curves:

i find expressions for $\frac{dy}{dx}$ and $\frac{d^2y}{dx^2}$

ii locate and determine the nature of any stationary points:

a $y = (x + 2)e^{-x}$

b $y = \ln(x - 1) + \frac{3}{x - 1}$

c $y = e^x \cos x$ (for $-\pi \le x \le \pi$)

d $y = (3 - x)^2 e^{\frac{x}{2}}$

11.4 Contextual

1

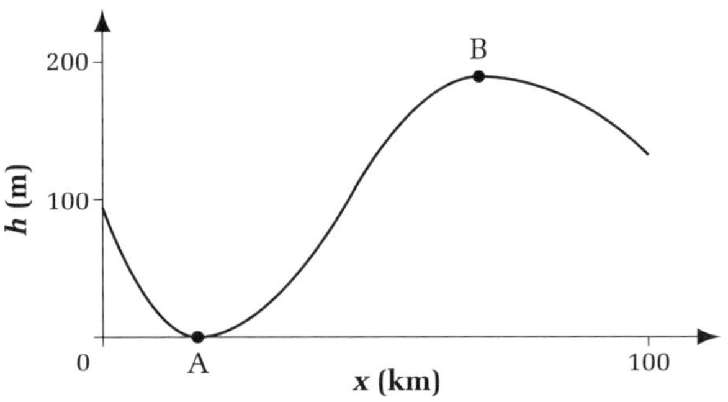

The height h metres above sea level that the course of a 100 km cycling road race takes can be modelled by $h = (10 - x)^2 e^{-(x + 1)/24}$, where x is the horizontal distance travelled by the cyclists from the start line.

a Calculate the heights above sea level at which the race starts and finishes.

b Find expressions for $\frac{dh}{dx}$ and $\frac{d^2h}{dx^2}$.

c After how many kilometres do the cyclists pass through the seaside town A?

d After how many kilometres does the course reach its highest point B? How high above sea level is this point?

e Calculate the value of x at which the climb from A to B is at its steepest, and find the height above sea level at this point.

11.5 Technique

1 For each of the following circles:

i write down its Cartesian equation

ii find an expression for $\frac{dy}{dx}$ in terms of x and y

iii calculate the gradient of the circle at the point P whose coordinates are indicated:

a radius 10, centre (5, 3); P (−1, 11)

b radius 17, centre (−9, −4); P (6, 4)

c radius $\sqrt{29}$, centre (−6, 7); P (−1, 5)

2 Find an expression for $\frac{dy}{dx}$ in terms of x and y for each of the following:

a $5x^2 - y^2 - 2xy = 0$

b $2x^3 - y + 4xy + 6 = 0$

c $3x^2 + 2y^2 + 5y = 0$

d $y^3 + 8x - 5xy - 5 = 0$

e $x - 5y + 4x^2y + 2 = 0$

f $4x + 3y^2 - xy^2 + 10 = 0$

3 Find the gradient of each of the following curves at points indicated:

a $3x + 2y^2 + x^2y = 0$ at (3, −3)

b $4y + 2x^2 - 7y^2 - 12 = 0$ at (4, 2)

c $x^3 + 8y^3 - 3xy - 28 = 0$ at (3, −1)

4 For each of the following curves:

i find an expression for $\frac{dy}{dx}$ in terms of x and y

ii find the coordinates of the stationary point(s) on the curve

iii find an expression for $\frac{d^2y}{dx^2}$ in terms of x, y and $\frac{dy}{dx}$

iv determine the nature of the stationary point(s):

a $2x^2 - y^2 + 4xy + 12 = 0$

b $5x^2 + 3y^2 + 6y - 9 = 0$

c $2x + 24y - x^2y - 10 = 0$

5 Find the equations of the tangent and the normal to each of the following curves at the points indicated:

a $3x + 2y + xy^2 + 6 = 0$ at (−1, 3)

b $y^2 + 3xy^2 - 8x - 16 = 0$ at (0, −4)

c $5y^3 + 10x + 2xy - 30 = 0$ at (3, 0)

11.6 Technique

1 Eliminate the parameter to find the Cartesian equations of the following curves:

a $x = 4t, y = t^3$

b $x = \frac{t}{2}, y = t^2 + t + 1$

c $x = 7\cos t, y = 7\sin t$

d $x = 2\cos\theta, y = 5\sin\theta$

e $x = -1 + 2\cos\theta, y = 4 + 3\sin\theta$

f $x = p + a\cos\theta, y = q + b\sin\theta$

2 Find an expression for $\frac{dy}{dx}$ for each of the following parametrically defined curves:

a $x = 3 - t, y = t^2 + 2t$

b $x = 1 - \cos\theta, y = \sin^3\theta$

c $x = \frac{t}{t + 4}, y = t^4 + 1$

3 Find the coordinates of the points on each of the following curves where the gradient has the value indicated:

a $x = 5t + 4, y = t^2 - 4t, \frac{dy}{dx} = -2$

b $x = \frac{1}{1 + t}, y = 2t, \frac{dy}{dx} = -8$

c $x = 2t, y = t + e^t, \frac{dy}{dx} = 3$

4 Find the equations of the tangent and the normal to the curve $x = \theta + \cos^2\theta, y = \sin 2\theta$, at the point where $\theta = \pi$.

5 The curve C has parametric equations $x = t^3$ and $y = 10 - t$. Find a general equation for the normal at some point $(t^3, 10 - t)$ on the curve. Hence, write down the equation of the normal to the curve at (1, 9).

6 Find the coordinates of the stationary points on each of the following parametrically defined curves. Find also an expression for the second derivative, and use it to determine the nature of these stationary points:

a $x = 2t + 5, y = 8t - t^2$
b $x = 2t - t^3, y = t^2$

11.7 Technique

1 Find the first three non-zero terms in the Maclaurin series expansions of:

a $\sin 5x$
b $\dfrac{2x}{e^{3x}}$
c $\ln\left(1 + \dfrac{x}{2}\right)$
d $\dfrac{e^x + e^{-x}}{2}$
e $\dfrac{e^x - e^{-x}}{2}$
f $e^x \sin x$

2 Use the quadratic approximations for $\cos 3x$ and $\ln\left(1 + \dfrac{x}{2}\right)$, obtained from their Maclaurin series, to solve the equation $\cos 3x = \ln\left(1 + \dfrac{x}{2}\right)$, correct to two decimal places, given that $0 < x < 1$.

3 Find the first three non-zero terms in the Maclaurin series expansion of $\ln(1 - 3x)$ and $\ln(1 + 4x)$. Then write down a cubic approximation for $\ln\left(\dfrac{1 - 3x}{1 + 4x}\right)$. Substitute $x = 0.1$ into this expression and find an approximate value of $\ln(0.5)$ to three decimal places.

Consolidation C

1 A curve has equation $y = (x + 3)e^{-x}$.

a Find $\dfrac{dy}{dx}$.
b Find the coordinates of the stationary point S on the curve.
c Calculate $\dfrac{d^2y}{dx^2}$ at the point S. What does its value indicate about the stationary point?

(MEI)

2 The parametric equations of a curve are $x = \cos t$, $y = 2 \sin t$, where the parameter t takes all values such that $0 \leq t \leq \pi$.

a Find the value of t at the point A where the line $y = 2x$ intersects the curve.
b Show that the tangent to the curve at A has gradient -2 and find the equation of this tangent in the form $ax + by = c$, where a and b are integers.

(UCLES)

3 The curve C has equation $x^2 + 4y^2 - 4x - 12y - 12 = 0$.

a By differentiation find the gradient of C at the point (6, 3).
b Find an equation of the tangent to the curve C at the point (6, 3).

(ULEAC)

4 Use the product rule to differentiate $x^2 \tan^2 x$ with respect to x.

(NICCEA)

5 A curve is defined parametrically by the equation

$x = t^3 - \dfrac{3}{t}, y = 2t^{\frac{3}{2}}$ $(t > 0)$.

a Show that $\dfrac{dy}{dx} = \dfrac{t^{\frac{5}{2}}}{t^4 + 1}$.
b Find the equation of the normal to the curve at the point where $t = 1$.

(AEB)

6 Obtain the first three terms in the Maclaurin series for $\ln(3 + x)$.

(UCLES)

7 Find the equation of the normal at the point (2, 1) on the curve $x^3 + xy + y^3 = 11$, giving your answer in the form $ax + by + c = 0$.

(UCLES)

8 Differentiate with respect to x:

a $e^{3x} \cos x$
b $(e^{3x} \cos x)^2$

(OCSEB)

9 A curve is defined by the parametric equations $x = t^2 + 3$, $y = t^3$, $t \in \mathbb{R}$. Show that the equation of the tangent at the point P at which $t = 2$ is $3x - y = 13$.

Find the equation of the normal at P.

The point Q on the curve has parameter $t = q$. Show that Q lies on the tangent at P if $q^3 - 3q^2 + 4 = 0$.

Express $q^3 - 3q^2 + 4$ as a product of three linear factors, not all distinct. Hence, find the coordinates of the point at which the tangent at P meets the curve again.

Show that the normal does not meet the curve again.

(NEAB)

10 Using differentiation, find the coordinates of the turning points of the curve $y = \dfrac{x^2}{1 + x}$. Determine the nature of the turning points.

(NICCEA)

Review

1 Differentiate $\sin 2x$. Explain how to find $\int \cos 2x \, dx$.

2 Integrate the following with respect to x.

a $x^5 + 5x^4 - x + 10$
b $(x + 1)^2$
c $\cos x$
d $\sin 5x$
e x^{-2}
f $\cos ax$

3 Sketch the curve $y = x^3 + 1$.

a Find the area under the curve between $x = -1$ and $x = 1$.
b Find the volume generated when this area is rotated about the x-axis through one complete turn.

4 Differentiate the following functions with respect to x:

a $\sin(8x+1)$ **e** e^{x^2} **h** $\ln(x^2+x-1)$
b $\cos(4x-1)$ **f** $e^{(2-5x)}$ **i** $\ln(\sin x)$
c e^{2x} **g** $\ln 2x$ **j** $\ln(x^4-1)$
d $e^{\frac{1}{2}x}$

5 **a** Express as a proper fraction:

i $\dfrac{x+3}{x-1}$ ii $\dfrac{x}{x+3}$

b Express $\dfrac{1}{x^2-4}$ in partial fractions.

c Express $\dfrac{5}{(x-3)(x^2+1)}$ in partial fractions.

d Express $\dfrac{x^2+4x-3}{(x+1)(x+4)}$ in partial fractions.

6 Find $\dfrac{dy}{dx}$ for each of the following:

a $y^2 = x^2 - x + 1$ **b** $y^2 = \cos x$

12.1 Technique

1 Find:

a $\int \sin 2x \, dx$ **d** $\int \sin(1-3x)\, dx$
b $\int \cos\frac{x}{2}\, dx$ **e** $\int \sec^2 2x \, dx$
c $\int \cos(2x+1)\, dx$ **f** $\int \tan^2 x \, dx$

2 Evaluate:

a $\int_0^{\pi} \sin\frac{\theta}{2}\, d\theta$ **b** $\int_0^{\frac{\pi}{4}} \cos 2\theta \, d\theta$

c $\int_0^{\frac{\pi}{4}} \tan^2 x \, dx$ **d** $\int_0^{\frac{\pi}{12}} \sec^2 4x \, dx$

3 Find:

a $\int \sin 3x \cos x \, dx$ **b** $\int \cos x \cos 3x \, dx$

c $\int_0^{\frac{\pi}{4}} \sin 5x \sin x \, dx$ **d** $\int_0^{\frac{\pi}{2}} \sin x \cos 3x \, dx$

4 Find:

a $\int e^{6x}\, dx$ **f** $\int e^{-\frac{1}{3}x}\, dx$
b $\int e^{-2x}\, dx$ **g** $\int 6e^{3x}\, dx$
c $\int \frac{1}{e^{4x}}\, dx$ **h** $\int 12e^{\frac{2}{3}x}\, dx$
d $\int e^{\frac{1}{2}x}\, dx$ **i** $\int 5e^{0.2x}\, dx$
e $\int e^{\frac{2}{5}x}\, dx$

5 Find:

a $\int (1+e^{-3x})\, dx$ **e** $\int_0^{10} \left(1+2e^{\frac{1}{10}x}\right) dx$

b $\int \left(e^{2x}+\frac{1}{e^{2x}}\right) dx$ **f** $\int_0^{1} \left(e^{x}-e^{-2x}\right) dx$

c $\int e^{1-\frac{1}{2}x}\, dx$ **g** $\int_1^{2} e^{1-x}\, dx$

d $\int (2-e^{-x})^2\, dx$

6 Find:

a $\int \frac{6}{x}\, dx$ **e** $\int \frac{1}{3x-5}\, dx$
b $\int \frac{1}{10x}\, dx$ **f** $\int \frac{2}{5-6x}\, dx$
c $\int \frac{1}{x-3}\, dx$ **g** $\int \frac{10}{5x+7}\, dx$
d $\int \frac{1}{3-x}\, dx$ **h** $\int \frac{1}{1-2x}\, dx$

7 Find:

a $\int \frac{3}{3x+5}\, dx$ **d** $\int \frac{2x-9}{x^2-9x}\, dx$
b $\int \frac{x}{1+x^2}\, dx$ **e** $\int \frac{12x-8}{3x^2-4x+1}\, dx$
c $\int \frac{x^3}{1+x^4}\, dx$ **f** $\int \frac{e^x}{1+e^x}\, dx$

8 Find:

a $\int \tan x \, dx$ **b** $\int \tan 2x \, dx$
c $\int \frac{\sec^2 x}{1+\tan x}\, dx$

9 Find:

a $\int \frac{1}{\sqrt{25-x^2}}\, dx$ **d** $\int \frac{1}{5+x^2}\, dx$
b $\int \frac{1}{\sqrt{100-x^2}}\, dx$ **e** $\int \frac{1}{\sqrt{3-x^2}}\, dx$
c $\int \frac{1}{4+x^2}\, dx$ **f** $\int \frac{4}{8+x^2}\, dx$

12.1 Contextual

1 Sketch the curve $y = 1 + \cos 2x$ for $0 \leq x \leq \pi$. Find the area bounded by the curve, the y-axis and the positive x-axis, leaving your answer in terms of π.

2 The graph shows the curve $y = \cos(1 + 2x)$.

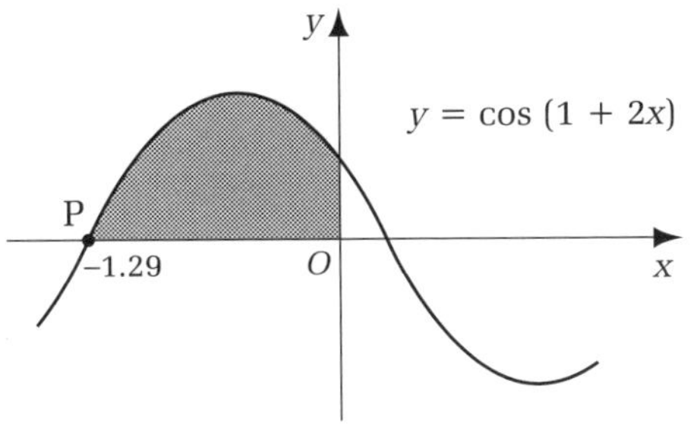

a Verify that the x-coordinate of the point P is -1.29.

b Calculate the area of the shaded region, giving your answer correct to two decimal places.

3 The diagram shows part of the curve $y = \sec x$. Find the volume generated when the shaded area is rotated about the x-axis through 360°.

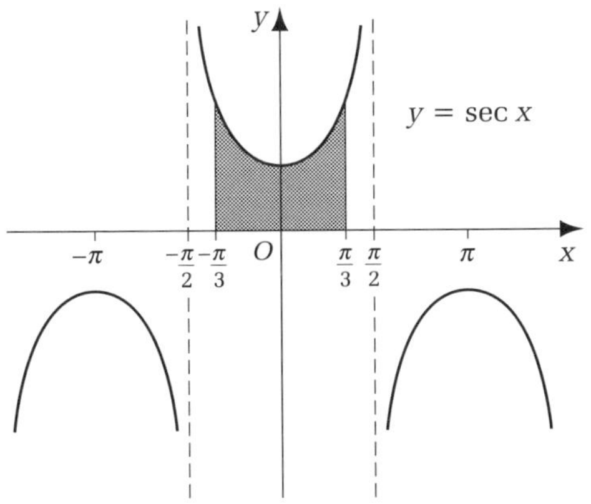

4 **a** Use $1 + \tan^2 A = \sec^2 A$ to express $\tan^2 \frac{x}{2}$ in terms of $\sec^2 \frac{x}{2}$.

b The graph shows part of the curve $y = \tan \frac{x}{2}$. Calculate the volume of the solid formed when the shaded area is rotated about the x-axis through 360°.

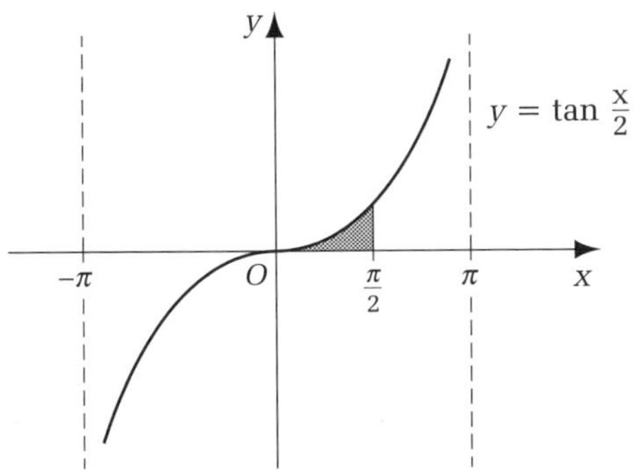

5 The graph shows part of the curve $y = \cos 2x \cos x$. Calculate the area of the shaded region.

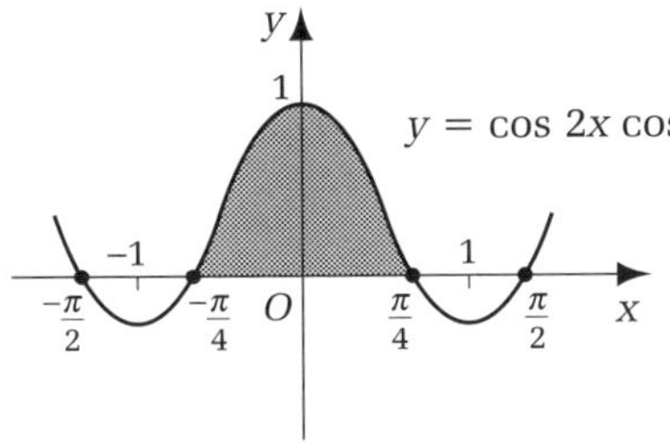

6 The graph shows the curve $y = 1 + e^{-x}$.

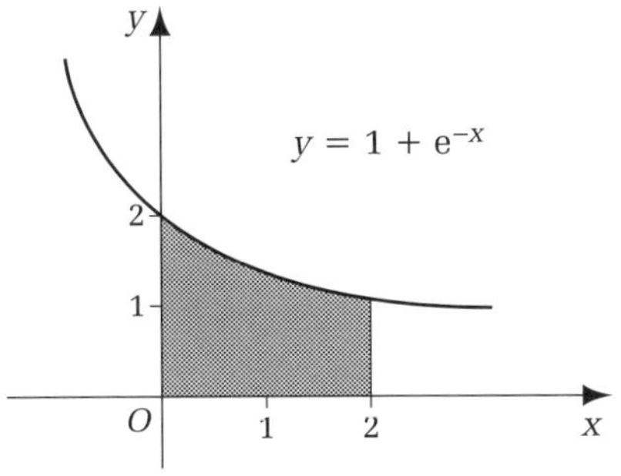

a Calculate the area of the shaded region.

b Calculate the volume of the solid formed if the shaded region is rotated about the x-axis through 360°, leaving your answer in terms of e.

7 **a** Find an expression for P in terms of x if $a^x = e^P$, where $a \in \mathbb{R}$.

b Use your answer to **a** to find $\int a^x \, dx$

8 The graph shows the curve $y = \dfrac{x}{1 + x^2}$. Calculate the area of the shaded region.

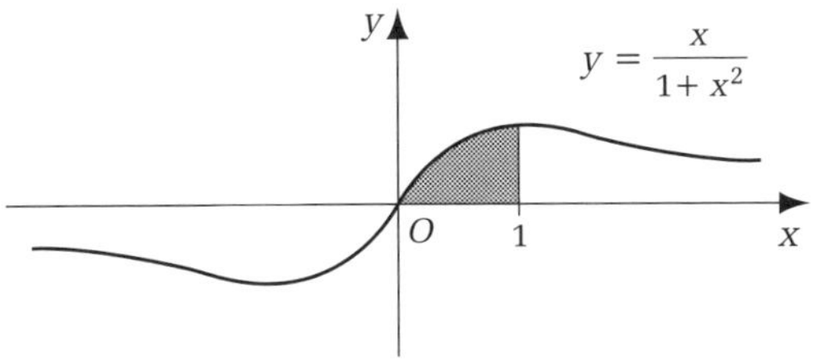

9 The graph shows the curve $y = \dfrac{1}{\sqrt{4 - x^2}}$. Calculate the area of the shaded region.

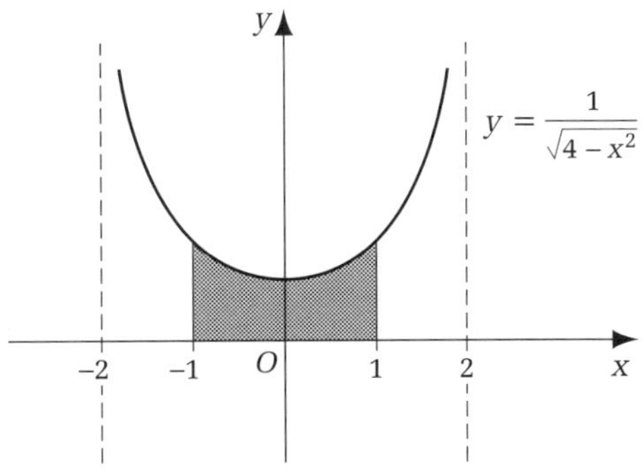

10 The graph shows the curve $y = \dfrac{1}{\frac{1}{4} + x^2}$. Find the area of the shaded region.

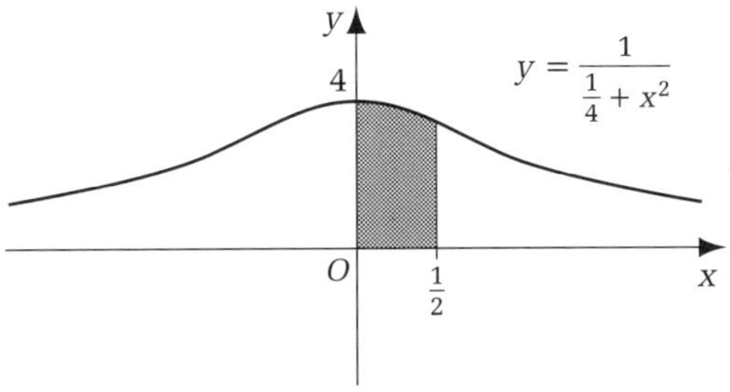

11 Calculate the areas of the shaded regions:

a

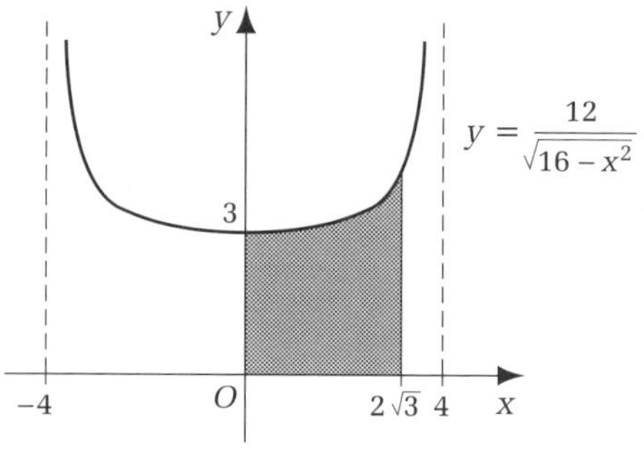

b

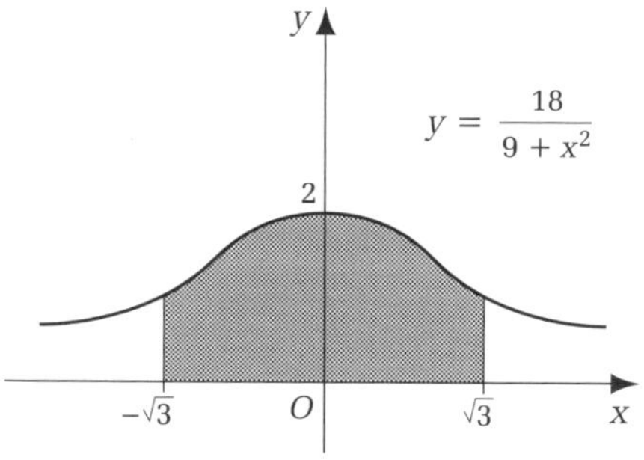

12.2 Technique

1 Integrate the following, first expressing each as a proper fraction:

a $\frac{2x+1}{x+1}$ **b** $\frac{x+7}{x+5}$

c $\frac{x}{3x+1}$ **d** $\frac{x}{1-x}$

2 Find the exact value of each of the following:

a $\int_1^4 \frac{8x+6}{4x+5}\,dx$ **b** $\int_0^2 \frac{x^2}{x^2+4}\,dx$

3 Evaluate these integrals, writing your answers correct to two decimal places:

a $\int_2^3 \frac{6x}{1+3x}\,dx$ **b** $\int_0^1 \frac{x}{7-x}\,dx$

4 Use partial fractions to find each of the following:

a $\int \frac{27-2x}{x^2+9x}\,dx$ **b** $\int \frac{7-4x}{x^2+x-2}\,dx$

c $\int \frac{1}{(x+2)(x+3)}\,dx$ **d** $\int \frac{1}{16-x^2}\,dx$

5 Find the exact value of each of the following:

a $\int_2^4 \frac{1}{x+2x^2}\,dx$ **d** $\int_0^2 \frac{1}{(3-x)(4-x)}\,dx$

b $\int_{-1}^1 \frac{3}{x^2-9}\,dx$ **e** $\int_0^4 \frac{6x}{(x+1)(x-5)}\,dx$

c $\int_{-2}^1 \frac{1}{x^2-6x+8}\,dx$ **f** $\int_3^4 \frac{x-1}{x(x-2)}\,dx$

6 Express $f(x) = \frac{2}{x^2(x-2)}$ in partial fractions.

Use your result to find $\int_3^4 f(x)\,dx$.

7 Write $\frac{3x^2-2x-47}{(x+3)(x-4)}$ as a proper fraction.

Hence find $\int \frac{3x^2-2x-47}{(x+3)(x-4)}\,dx$.

8 Express the integrand as a proper fraction before finding $\int \frac{x^2+1}{x^2-x-6}\,dx$.

9 $f(x) = \frac{1}{(x+1)(x^2+4)}, x \neq -1$

a Express $f(x)$ in the form $\frac{A}{x+1} + \frac{Bx+C}{x^2+4}$.

b Find the exact value of $\int_0^2 f(x)\,dx$.

12.2 Contextual

1 The graph shows the curve $f(x) = \frac{x}{x-5}$.

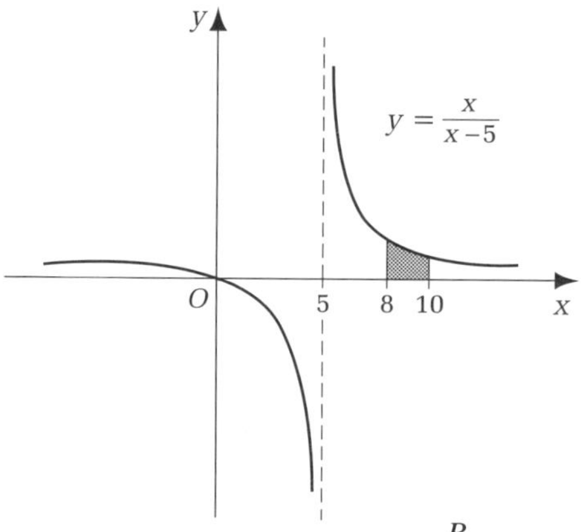

a Express $f(x)$ in the form $A + \frac{B}{x-5}$.

b Calculate the area of the shaded region.

2 **a** Express $\frac{x-3}{x+3}$ in the form $A + \frac{B}{x+3}$.

b The graph shows part of the curve $y = \frac{x-3}{x+3}$. Calculate the area of the shaded region.

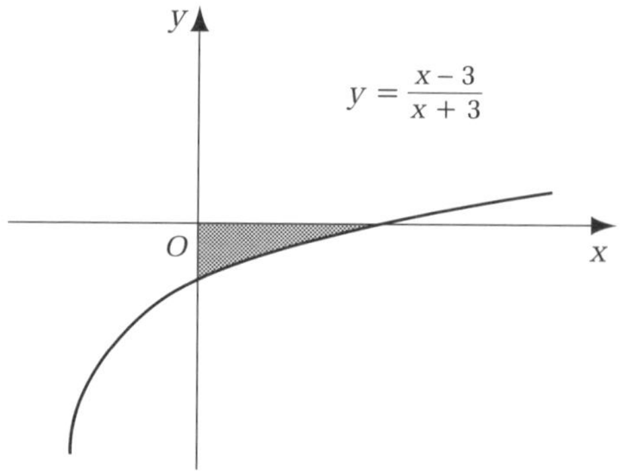

3 The graph shows the curve $f(x) = \dfrac{2}{(x+2)(4-x)}$. Find the area of the shaded region.

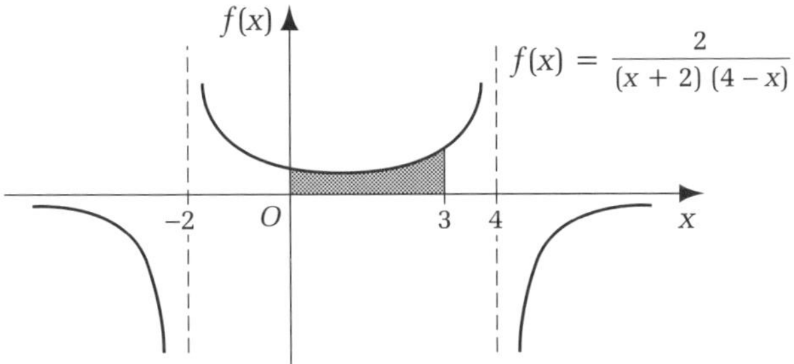

4 Given that $g(x) = \dfrac{5x+8}{x(x-8)}$, express $g(x)$ in the form $\dfrac{A}{x} + \dfrac{B}{(x-8)}$. Hence evaluate $\displaystyle\int_2^6 g(x)\,\mathrm{d}x$.

5 Find the value of a if $\displaystyle\int_1^3 \frac{4}{(3x-1)(x+1)}\,\mathrm{d}x = \ln(a)$.

6 Express $\dfrac{4}{x^2(2-3x)}$ in the form $\dfrac{A}{x^2} + \dfrac{B}{x} + \dfrac{C}{(2-3x)}$.

Hence find the exact value of $\displaystyle\int_1^2 \frac{4}{x^2(2-3x)}\,\mathrm{d}x$.

7 Express $g(x) = \dfrac{2x^2-5x+5}{(x-1)(x-3)}$ in the form

$A + \dfrac{B}{(x-1)} + \dfrac{C}{(x-3)}$.

Hence evaluate $\displaystyle\int_{-3}^0 g(x)\,\mathrm{d}x$.

8 Express $f(x) = \dfrac{x(x+4)}{(1-4x)(x^2+1)}$ in the form $\dfrac{A}{(1-4x)} + \dfrac{Bx+C}{(x^2+1)}$. Hence find $\displaystyle\int_{-1}^0 f(x)\,\mathrm{d}x$.

12.3 Technique

1 Use integration by parts to find:

a $\displaystyle\int xe^x\,\mathrm{d}x$ **b** $\displaystyle\int x^9 \ln x\,\mathrm{d}x$ **c** $\displaystyle\int x\cos 3x\,\mathrm{d}x$

2 Find:

a $\displaystyle\int (4x+3)\sin x\,\mathrm{d}x$ **b** $\displaystyle\int x\sec^2 2x\,\mathrm{d}x$

c $\displaystyle\int x(\cos x - \sin x)\,\mathrm{d}x$ **d** $\displaystyle\int x^2 \ln 2x\,\mathrm{d}x$

3 Find:

a $\displaystyle\int (t+1)e^{-t}\,\mathrm{d}t$ **b** $\displaystyle\int \frac{1}{t}\ln t\,\mathrm{d}t$

c $\displaystyle\int t\sin 2t\,\mathrm{d}t$

4 Evaluate:

a $\displaystyle\int_{\frac{\pi}{2}}^{\pi} x\sin 4x\,\mathrm{d}x$ **b** $\displaystyle\int_0^{\pi} (x-\pi)\cos x\,\mathrm{d}x$

c $\displaystyle\int_{\frac{1}{2}}^{1} \ln 2x\,\mathrm{d}x$

5 Use integration by parts twice to find:

a $\displaystyle\int x^2\cos x\,\mathrm{d}x$ **b** $\displaystyle\int \left(1+x^2\right)e^x\,\mathrm{d}x$

c $\displaystyle\int e^x\cos 2x\,\mathrm{d}x$

12.3 Contextual

1 Sunil is finding $\int x^2\cos 2x\,\mathrm{d}x$ by integration by parts. He stops after a few lines of working and is unsure whether he has tackled the question correctly. His last line of working is shown below.

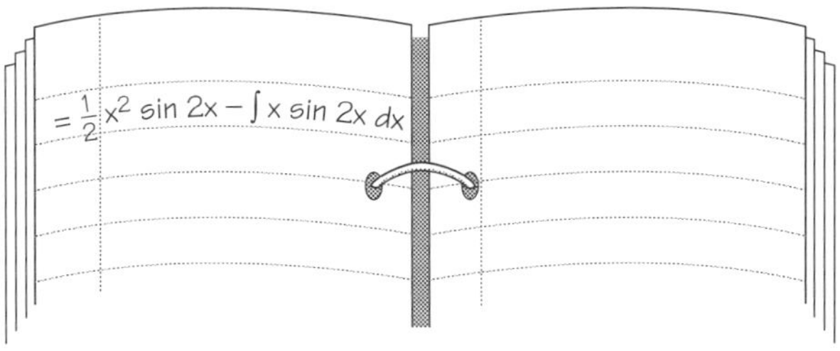

a How has Sunil arrived at this line of working?

b What advice would you offer Sunil?

c If he followed your advice, what would his final answer be?

2 The graph shows part of the curve $y = (x-1)\sin x$. Find the area of the shaded region, correct to two decimal places.

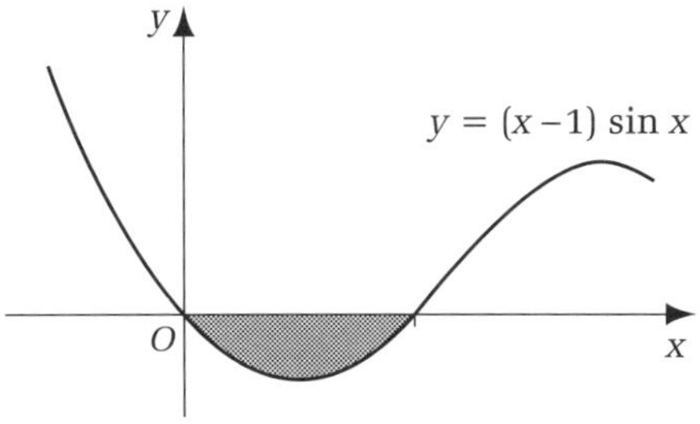

3 The graph shows part of the curve $y = g(x)$, where $g(x) = 5xe^x$. Find the area of the shaded region.

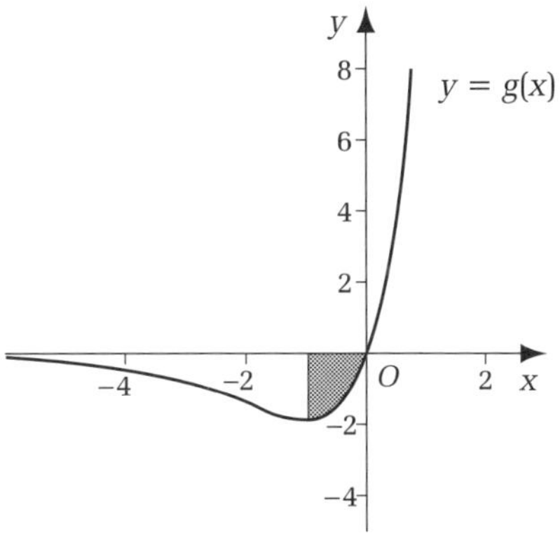

4 **a** Find $\int x\cos x\ \mathrm{d}x$.

b Use your answer to **a** to calculate the area of the shaded region correct to two decimal places.

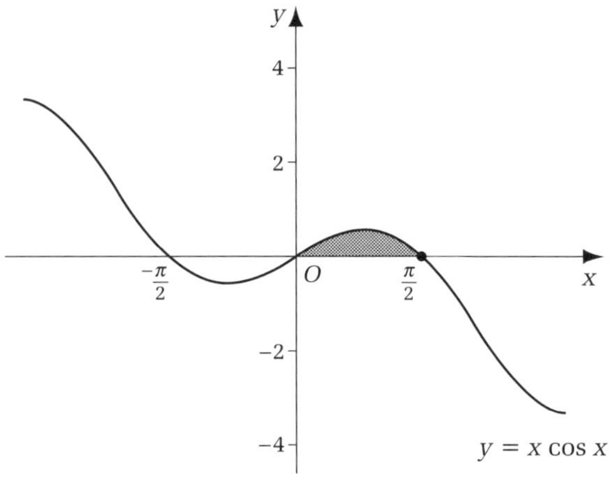

12.4 Technique

1 Use substitution to find the following:

a $\int (1-x)^3\ \mathrm{d}x$

b $\int \sqrt{4x+3}\ \mathrm{d}x$

c $\int (2x+1)^7\ \mathrm{d}x$

d $\int \frac{2}{(1-3x)^2}\ \mathrm{d}x$

e $\int \frac{1}{\sqrt{1+x}}\ \mathrm{d}x$

f $\int (6x-5)^{\frac{1}{3}}\ \mathrm{d}x$

Check your answers using the standard result for

$\int (ax+b)^n\ \mathrm{d}x$.

2 By making suitable substitutions, evaluate:

a $\int_0^{\frac{1}{3}} (1-3x)^4\ \mathrm{d}x$

b $\int_0^1 \sqrt{1-x}\ \mathrm{d}x$

3 Use the substitution $u = ax + b$ to show that

$$\int (ax+b)^n\ \mathrm{d}x = \frac{1}{a} \times \frac{(ax+b)^{n+1}}{(n+1)}.$$

Hence use the formula to:

a evaluate $\int_{-1}^{0} (3+4x)^3\ \mathrm{d}x$

b find $\int \frac{1}{\sqrt{4+x}}\ \mathrm{d}x$

c find $\int \frac{1}{(4-x)^2}\ \mathrm{d}x$

4 Use the suggested substitutions to find or evaluate:

a $\int_{-3}^{2} x\sqrt{7+x}\ \mathrm{d}x;\quad p = 7 + x$

b $\int \frac{x}{\sqrt{x^2+2}}\ \mathrm{d}x;\quad t = x^2 + 2$

c $\int_0^1 x\sqrt{1-x}\ \mathrm{d}x;\quad t^2 = 1 - x$

d $\int \frac{x}{\sqrt{3-x^2}}\ \mathrm{d}x;\quad u^2 = 3 - x^2$

e $\int x\mathrm{e}^{-\frac{1}{2}x^2}\ \mathrm{d}x;\quad t = x^2$

f $\int \sin\theta\cos^4\theta\ \mathrm{d}\theta;\quad u = \cos\theta$

g $\int \frac{2+\mathrm{e}^x}{2-\mathrm{e}^x}\ \mathrm{d}x;\quad z = e^x$

h $\int \frac{x}{4+x^4}\ \mathrm{d}x;\quad w = x^2$

5 Use a suitable substitution to integrate the following:

a $\int_1^2 x\sqrt{5+x^2}\ \mathrm{d}x$

b $\int \frac{x^2}{\sqrt{2+x^3}}\ \mathrm{d}x$

c $\int \cos\theta\sin^3\theta\ \mathrm{d}\theta$

d $\int xe^{2x^2}\ \mathrm{d}x$

e $\int \frac{\cos x}{\sin x}\ \mathrm{d}x$

f $\int \frac{e^{2x}}{1+e^{2x}}\ \mathrm{d}x$

g $\int_0^6 x\sqrt{6-x}\ \mathrm{d}x$

h $\int \frac{1}{\sqrt{2+x}}\ \mathrm{d}x$

i $\int_0^{\pi} \frac{\sin\theta}{1+\cos^2\theta}\ \mathrm{d}\theta$

12.4 Contextual

1 The graph shows part of the curve $y = \frac{2}{2-x}$.

Calculate the volume generated when the shaded region is rotated through one complete turn about the x-axis. Write your answer in terms of π.

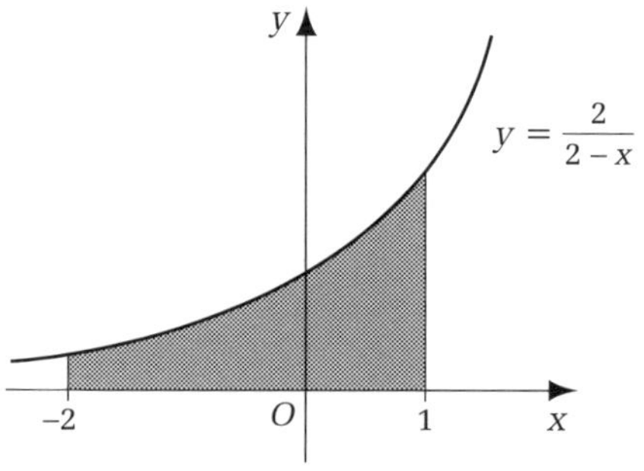

2 **a** Express $f(x) = \frac{1}{(x-1)(x-2)^2}$ in the form

$$\frac{A}{(x-1)} + \frac{B}{(x-2)} + \frac{C}{(x-2)^2}.$$

b Hence find the values of a and b if

$$\int_3^4 f(x)\ \mathrm{d}x = \frac{1}{2} + \ln\left(\frac{a}{b}\right).$$

3 Use the substitution $z = e^x$ to find $\int \frac{1}{2 + e^x}\,dx$.

4 The curve shows part of the graph of $y = \sin x \cos^2 x$.

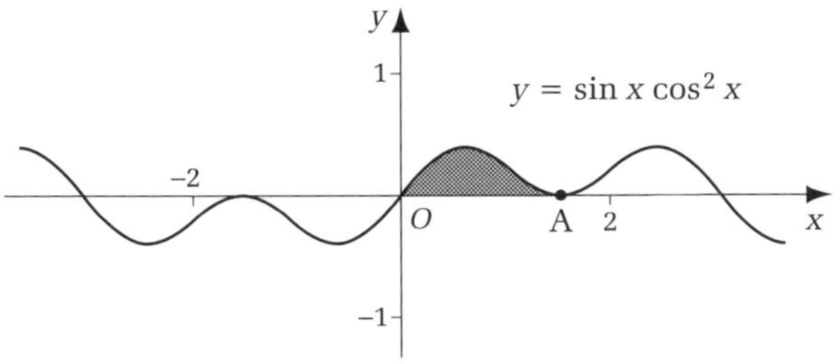

a Write down the exact coordinates of A.

b Use the substitution $u = \cos x$ to find $\int \sin x \cos^2 x\,dx$.

c Find the area of the shaded region.

5 **a** Use the substitution $u = \ln x$ to find $\int \frac{(\ln x)^2}{x}\,dx$.

b The graph shows the curve $y = \frac{(\ln x)^2}{x}$. Use your answer to **a** to find the area of the shaded region.

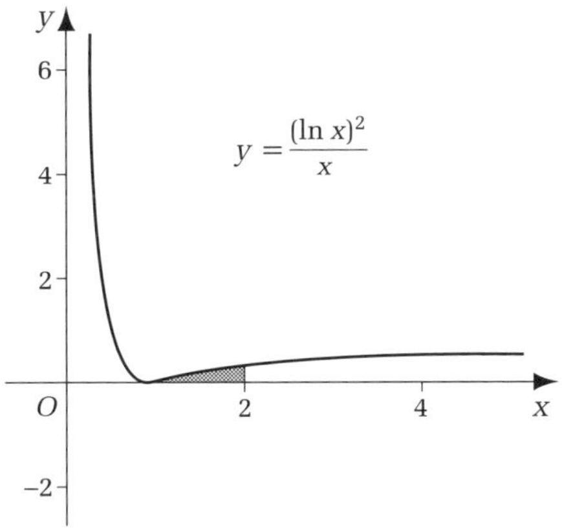

6 Use the substitution $x = \sin\theta$ to find $\int \sqrt{1 - x^2}\,dx$.

7 Use the substitution $u = \sin^{-1}x$ to find $\int \frac{\sin^{-1}x}{\sqrt{1 - x^2}}\,dx$.

8 Express $g(x) = \frac{2}{(x - 3)(x - 1)^2}$ in partial fractions.

Hence evaluate $\int_{-1}^{0} g(x)\,dx$.

12.5 Technique

1 An ellipse is defined parametrically as
$x = 5\cos\theta$, $y = 8\sin\theta$, $0 \le \theta \le 2\pi$.
Calculate the area of the ellipse above the x-axis.

2 The diagram shows an ellipse centred on the origin.

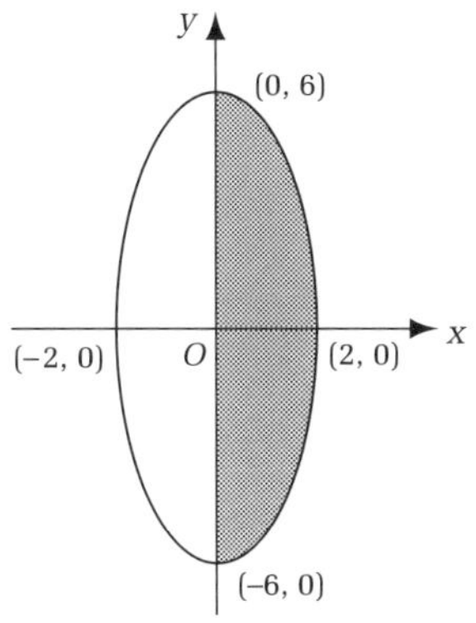

Find the area of the shaded region.

3 **a** Find the area of a region R bounded by the curve with parametric equations $x = t^3 - 1$, $y = 1 + 2t$, the line $x = 7$ and the x- and y-axes.

b Calculate the volume generated when R is rotated completely about the x-axis.

4 The diagram shows a sketch of the curve $x = 5\sqrt{t} - 2$, $y = \frac{2}{t}$, where t is a parameter and $t > 0$.

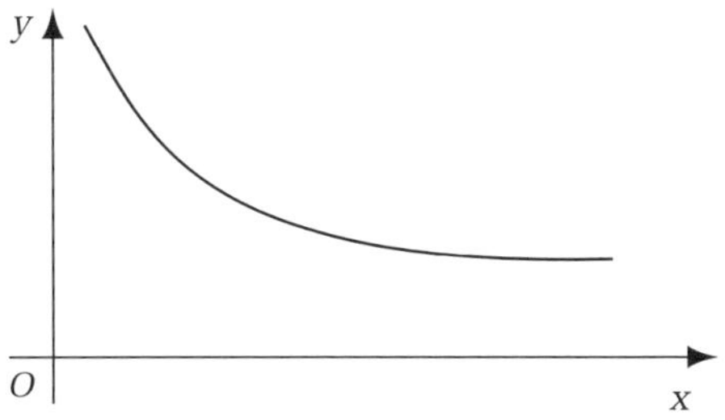

a Find the area bounded by the curve, the x-axis and the lines $x = 3$ and $x = 8$.

b Calculate the volume of the solid of revolution formed when this area is rotated through 2π about the x-axis.

12.6 Technique

1 Find the general solution to the following differential equations:

a $\frac{dy}{dx} - x = 1$

b $x\frac{dy}{dx} = 1$

c $\frac{dy}{dx} = x^2y^2$

d $x\frac{dy}{dx} - y = x^2y$

e $\frac{dy}{dx} - y^2e^x = 0$

f $x\frac{dy}{dx} + y = 1$

g $\frac{dy}{dx} = \frac{y}{x + 1}$

h $x^2\frac{dy}{dx} - y = 0$

2 **a** Find $\int \frac{1}{x(x + 1)}\,dx$.

b Use your result to **a** to solve the differential equation $r\frac{dT}{dr} = \frac{1}{r + 1}$, where $T = 2$ when $r = 1$.

3 The variables x and y satisfy the differential equation $\frac{dy}{dx} = x\sec y$.

a Find the general solution.

b Find the particular solution where $y = 0$ when $x = 0$.

4 **a** The variables h and t satisfy the differential equation $\frac{dh}{dt} = \frac{1}{h}$ for $h > 0$. Find the general solution.

b Find the particular solution if $h = 3$ when $t = 1$.

12.6 Contextual

1 The temperature θ°C of a cup of tea that is cooling can be modelled by a differential equation. The tea cools at a rate that is proportional to the temperature difference between the tea and room temperature.

a If the room temperature is 20°C write down a differential equation that models this situation.

b By solving the differential equation show that $\theta = Ae^{-kt} + 20$, where A and k are positive constants.

c If initially (at time $t = 0$) the temperature of the tea is 72°C, calculate the value of A.

d If the tea took five minutes to cool from 72°C to 55°C, calculate the value of k correct to two decimal places. Use this value to calculate how long it takes to cool from 72°C to 40°C.

2 In an experiment on cell division, the rate of increase of the number of cells, N, in a specimen after t days was believed to be proportional to the number of cells at that time.

a Write down a differential equation for N that models this situation.

The solution of the differential equation is of the form $N = N_0e^{\lambda t}$, where $\lambda > 0$.

b Sketch the graph of N against t and hence explain why the differential equation cannot hold over a long period.

c Given, further, that the number of cells was initially 10 000 and was 11 000 after one day, find the value of λ.

(SMP)

3 A cylindrical tank has a radius of 1 m and a height of 2.25 m as shown in the diagram. Water runs out through a hole in the base at a rate that is proportional to the square root of the depth of the water in the tank.

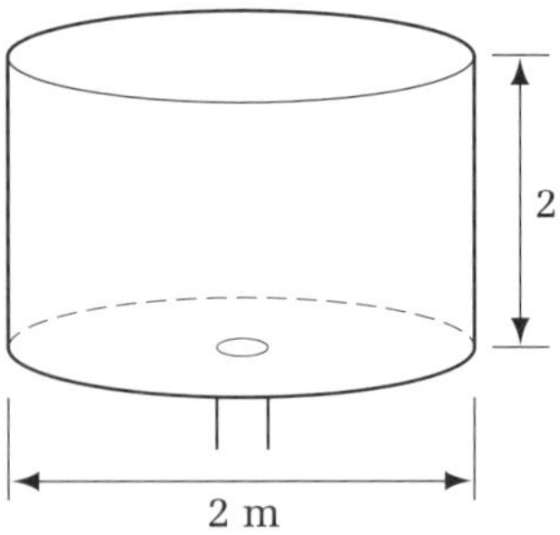

a If V represents the volume of water in the tank and h represents the depth of water at time t minutes, write down a differential equation to model this situation.

b Use your answer to **a** to show that $\frac{dh}{dt} = -\frac{k}{\pi}\sqrt{h}$, where k is a positive constant.

c If the tank is full initially ($t = 0$), show that $2\sqrt{h} = -\frac{k}{\pi}t + 3$.

d After four minutes, the depth of the water is 1.96 m. Show that $k = 0.05\pi$.

e How long does it take for the tank to empty?

4 The value of a car depreciates at a rate that is proportional to its value at that time.

a Write down a differential equation that models the value, V, of the car after t years.

b When the car is new its value is £9000. By solving the differential equation in **a**, show that the value of the car after t years can be written as $V = 9000e^{-kt}$.

c After one year the value of the car is £8000. Find the value of k correct to two decimal places.

d Find the value of the car after three years.

5 The gradient of a curve at any point (x, y) is inversely proportional to the square of the y-coordinate.

a Write down a differential equation to model this situation.

b Solve the differential equation to find the equation of the curve if it passes through (0, 1) and (2, −1).

Consolidation C

1 Find:

a $\int 2\cos(3x + 4)\,dx$

b $\int e^{2x}\,dx$

c $\int 10e^{\frac{1}{2}x}\,dx$

d $\int \frac{x^3}{x^4 + 2}\,dx$

e $\int \frac{1}{\sqrt{100 - x^2}}\,dx$

f $\int \frac{1}{x^2 + 100}\,dx$

g $\int xe^{x^2}\,dx$

h $\int \frac{1 + x}{1 + x^2}\,dx$

2 Evaluate:

a $\int_0^{\frac{\pi}{4}} \sec^2 x\,dx$

b $\int_0^{\frac{\pi}{2}} (2\sin 2x - \cos 4x)\,dx$

c $\int_0^{\frac{3\pi}{2}} \sin x \sin 2x\,dx$

d $\int_0^{\frac{\pi}{2}} \sin x \cos^2 x\,dx$

3 **a** Evaluate $\int_1^2 \frac{1 + x^2}{x^3}\,dx$.

b Find the area of the shaded region if the curve has equation $y = \frac{x}{1 + x^2}$.

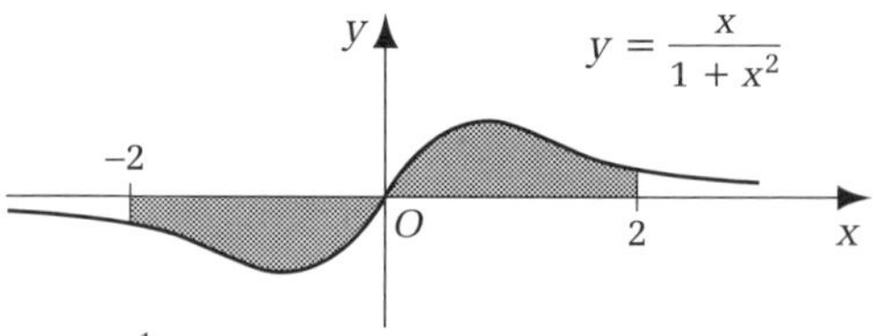

4 Evaluate $\int_0^1 xe^{-2x}\,dx$ leaving your answer in terms of e.

(ULEAC)

5 Evaluate:

a $\int_0^{\frac{\pi}{2}} (1 - x)\sin x\,dx$

b $\int_1^e \ln x\,dx$

6 Given that $f(x) = \dfrac{7}{(1+3x)(1-4x)}$, express $f(x)$ in partial fractions and find $\displaystyle\int_0^{0.2} f(x)\,dx$. (*ULEAC*)

7 A tank has water leaking from it such that the rate of change of the depth of water is proportional to the depth of the water remaining.

a Write down a differential equation to model this.

b Initially the depth of water is 6 m. After five minutes the depth of the water in the tank is 3 m. By solving the differential equation, find how much longer it will be before there is a depth of only 1 m remaining in the tank, giving your answer to the nearest minute. (*NICCEA*)

8 At any time, t hours, the rate of increase in the number of bacteria in a controlled environment is proportional to the number, n, of the bacteria present.

a Write down a differential equation to model this information.

b At $t = 0$ hours the number of bacteria present is N. After six hours the number of bacteria present is $\dfrac{3N}{2}$. By solving the differential equation find how much longer it will take for the number of bacteria present to become $2N$. (*NICCEA*)

9 **a** The rate of change of the radius (r cm) of a spherical pebble with time (t years), is modelled by the differential equation $\dfrac{dr}{dt} = k$, where k is a constant.

If the radius is initially 5 cm and after 20 years it is 4 cm, solve the differential equation, and hence find an expression for r in terms of t.

b Assuming the model $\dfrac{dr}{dt} = k$, use the chain rule to show that the rate of loss of the volume of the pebble is proportional to its surface area at time t. (Surface area of a sphere is $4\pi r^2$ and volume is $\frac{4}{3}\pi r^3$.) (*SMP*)

10 The curve C is given by the equations

$x = 2t, y = t^2$

where t is a parameter.

a Find an equation of the normal to C at the point P on C where $t = 3$.

The normal meets the y-axis at the point B. The finite region R is bounded by the part of the curve C between the origin O and P, and the lines OB and BP.

b Show the region R, together with its boundaries, in a sketch.

The region R is rotated through 2π about the y-axis to form a solid S.

c Using integration, and explaining each step in your method, find the volume of S giving your answer in terms of π. (*ULEAC*)

Review

1 Rearrange each of the following equations into the form $f(x) = 0$:

a $e^x = 5 - 2x$ **b** $\dfrac{6}{x+1} = x^2$ **c** $\sqrt[3]{x-4} = x$

2 Show how the equation $x^3 - 4x^2 + 3 = 0$ can be rearranged into each of the following forms:

a $x = \sqrt[3]{4x^2 - 3}$ **d** $x = 4 - \dfrac{3}{x^2}$

b $x = \sqrt{\dfrac{x^3+3}{4}}$ **e** $x = \frac{1}{4}\left(x^2 + \dfrac{3}{x}\right)$

c $x = \dfrac{3}{4x - x^2}$ **f** $x = \sqrt{\dfrac{3}{4-x}}$

3 Rearrange each of the following equations into the form $x = g(x)$:

a $e^x - x^2 + 2 = 0$ **b** $x + 7 - \sin x = 0$

c $x^2 - 9x + 5 = 0$

4 Find $f'(x)$ for each of the following functions:

a $f(x) = x^2 + \ln(2x+1)$ **b** $f(x) = 3x\tan x$

c $f(x) = \ln(\cos x) + 5x$ **d** $f(x) = xe^x - 6x^2$

5 Find the exact value of each of the following definite integrals:

a $\displaystyle\int_0^3 \frac{2}{x+3}\,dx$ **b** $\displaystyle\int_{-1}^1 xe^{-x}\,dx$ **c** $\displaystyle\int_0^{\frac{\pi}{2}} x^2\cos x\,dx$

13.1 Technique

1 **a** Show that the equation $10 + 8x - x^3 = 0$ has a root in the interval $3 < x < 4$.

b Use linear interpolation to find a first approximation for this root and state the interval that contains the root. Give a reason for your answer.

2 **a** Find the consecutive integer values of x between which the equation $x^5 - 3x - 7 = 0$ has a root.

b Use the interval bisection method to find the value of this root, correct to one decimal place.

c Taking a suitable value for the first approximation x_0, use the Newton–Raphson method to find the value of this root correct to four decimal places.

3 The sequence of values generated by the iterative formula $x_{n+1} = \ln(2x_n^2 + 5)$ with $x_0 = 3$ converges to a value α. Use this formula to find the value of α correct to three decimal places. State an equation satisfied by α.

4 **a** If $f(x) = x\sin x + 2x - 7$, find an expression for $f'(x)$.

b Show that the equation $x\sin x = 7 - 2x$ has a root in the interval $5 < x < 6$.

c Use linear interpolation to find a first approximation for the value of this root. Give your answer correct to two decimal places.

d Taking your answer in **c** as x_0, use the Newton–Raphson method to find this root correct to four decimal places.

5 **a** Show that the equation $x^5 - 3x - 8 = 0$ has a root in the interval $1 < x < 2$.

b Decide which two of the following iterative formulas could be used to find the root of this equation:

$$x_{n+1} = \sqrt[5]{3x_n - 8} \qquad x_{n+1} = \frac{8}{x_n^4 + 3}$$

$$x_{n+1} = \sqrt[5]{3x_n + 8} \qquad x_{n+1} = \frac{8}{x_n^4 - 3}$$

c Taking $x_0 = 2$, as an initial approximation, determine which of these two iterative formulae converge towards a value for this root. Find its value correct to three decimal places.

13.1 Contextual

1 **a** Show that the equation $\ln x - e^x + 5x = 0$ has a root in the interval $0 < x < 1$ and another in the interval $2 < x < 3$.

b Taking $x_0 = 2$ as an initial approximation, use the iterative formula $x_{n+1} = \ln(\ln x_n + 5x_n)$ to find the second of these two roots, correct to two decimal places.

c Describe what happens if this iterative formula is used to find the root in the interval $0 < x < 1$, taking:

i $x_0 = 0$ **ii** $x_0 = 1$

as a first approximation.

d Taking $x_0 = 1$ as a first approximation, use Newton's method to calculate the value of the other root, correct to three decimal places.

2 The diagram shows the graph of $y = e^x - x^5 + 2x + 3$ as it crosses the axes.

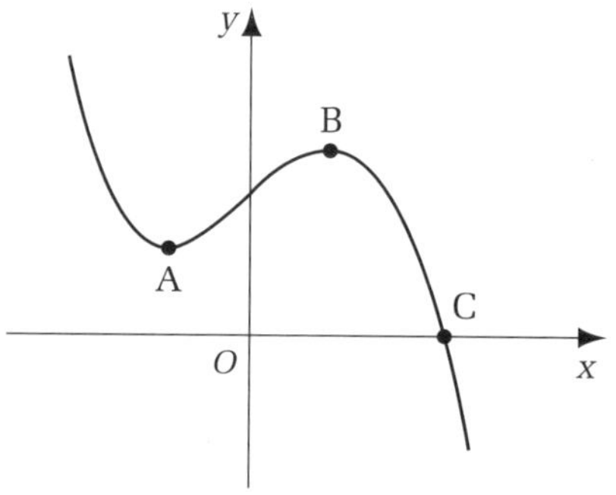

a Show that point C is located between $x = 1$ and $x = 2$.

b Using the Newton–Raphson method with $x = 2$ as a first approximation, calculate the x-coordinate of C, correct to three decimal places.

c Explain why the x-coordinates of A and B satisfy the equation $e^x - 5x^4 + 2 = 0$. By taking $x_0 = -1$ as a first approximation, use the Newton–Raphson method to solve this equation. Find the coordinates of A correct to three decimal places.

d Show that $e^x - 5x^4 + 2 = 0$ can also be solved numerically using the iterative formula

$x_{n+1} = \sqrt[4]{\dfrac{e^{x_n} + 2}{5}}$. Taking $x_0 = 1$ as a first

approximation, find the coordinates of B correct to three decimal places.

3

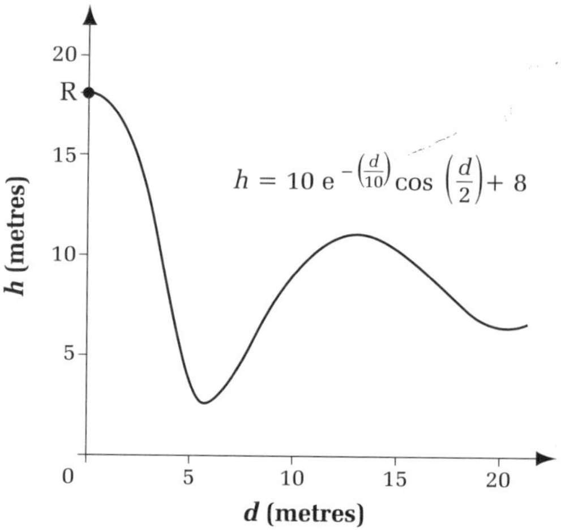

The diagram shows the track of a fairground ride. The height h metres of the track above the ground at a horizontal distance of d metres from the point R of release, can be modelled by $h = 10e^{-\left(\frac{d}{10}\right)}\cos\left(\frac{d}{2}\right) + 8$.

a Show that the track first falls to a height of 5 metres when $4 < d < 5$.

b Use Newton's method to find this value of d correct to the nearest centimetre.

13.2 Technique

1 Using the mid-ordinate rule with

a five strips and **b** ten strips,

calculate the value of $\int_1^2 2^x \, dx$ correct to three decimal places. Remember to show five decimal places of working.

2 Using

a the mid-ordinate rule **b** the trapezium rule

with ten strips, find the area under the curve $y = \sqrt{5 - x}$ between $x = 0$ and $x = 5$. Give your answers correct to two decimal places.

3 Find, giving your answers correct to four decimal places, the approximate value of $\int_{-1}^{1} \frac{4}{2 + \sin x} \, dx$ using:

a the trapezium rule with eleven ordinates

b Simpson's rule with eleven ordinates.

4 Given that $\frac{\sin x}{x} = 1$ when $x = 0$, use the trapezium rule with

a five strips **b** eight strips

to calculate the area under the graph of $y = \frac{\sin x}{x}$ between $x = 0$ and $x = 2$, correct to three decimal places.

5 Use Simpson's rule with

a four strips **b** eight strips

to evaluate $\int_{-1}^{1} e^{-x} \cos x \, dx$, giving your answers correct to four decimal places.

6 Find, correct to five decimal places, the value of

$$\int_2^5 x \ln x \, dx,$$ using:

a integration by parts

b Simpson's rule with seven ordinates.

c Calculate the percentage error of the result using Simpson's rule.

13.2 Contextual

1 An ornamental bottle with a perpendicular height of 30 cm can be modelled by rotating area A through 360° about the x-axis, as shown below, where A is the area bounded by the x-axis and the curve $y = 5e^{-\frac{(x-5)^2}{30}} + 3$ between $x = 0$ and $x = 30$.

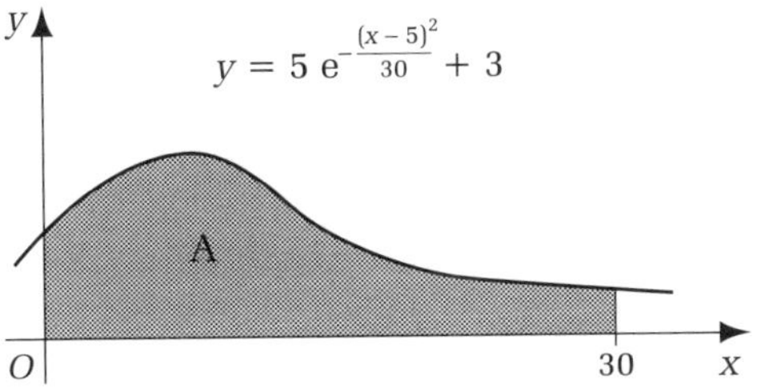

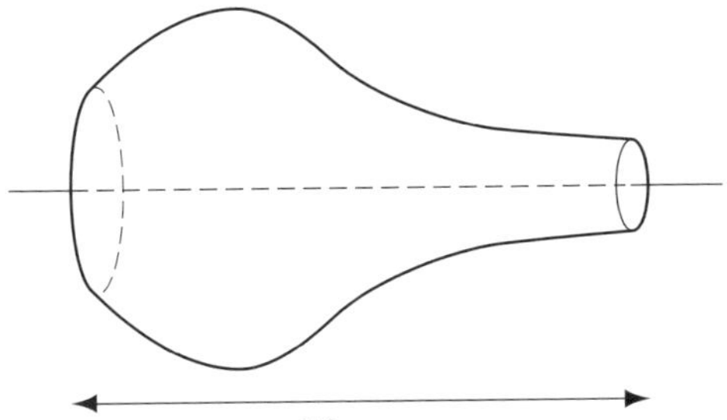

Use Simpson's rule with ten strips to calculate the volume of the bottle. Write your answer correct to one decimal place.

2 The number of vehicles per minute, N, passing through a particular point on a busy section of motorway are counted at five-minute intervals. Over a period of an hour, the results are tabulated as follows:

t (min)	0	5	10	15	20	25	30	35	40	45	50	55	60
N (vehicles per minute)	159	178	192	215	203	251	276	267	260	249	216	199	183

a Plot time t horizontally against the rate of vehicle flow N vertically.

b The total number of vehicles passing this point during this 60-minute period can be estimated by calculating the area under this graph. Use the trapezium rule with twelve strips to find an estimate for the total number of vehicles.

3 A train travelling at $40\,\text{m s}^{-1}$ begins to decelerate as it approaches a station. Its speed $v\,\text{m s}^{-1}$ at 15-second intervals are given in the table below.

t (s)	0	15	30	45	60	75	90	105	120	135	150
v (m s^{-1})	40	36	31	29	23	17	10	12	7	4	0

a Plot time t horizontally against speed v vertically.

b The total distance travelled by the train as it slows down in approaching the station can be estimated by calculating the area under this graph. Estimate this distance using:

i the trapezium rule with ten strips

ii Simpson's rule with ten strips.

Consolidation C

1 Show that the gradient of the curve $y = x^3 + 3x - 15$ is always positive. Deduce that the equation $x^3 + 3x - 15 = 0$ has just one real root.

For any given value of x_1, the nearest sequence defined by

$$x_{n+1} = (15 - 3x_n)^{\frac{1}{3}}, \quad n = 1, 2, 3, \ldots$$

converges to a value α. Show that α is a root of the equation $x^3 + 3x - 15 = 0$.

Use this sequence, starting with $x_1 = 2$, to obtain α to three decimal places, tabulating $x_2, x_3, \ldots$ to four decimal places as you proceed.

(NEAB)

2 **a** Use Simpson's rule, with five ordinates, to find, correct to two decimal places, an approximation for the area between the curve $y = xe^x$, the x-axis and the lines $x = 0$ and $x = 2$.

b Using integration by parts, find the exact value of the area.

(NICCEA)

3 **a** Show that the equation $x^3 - 5x + 1 = 0$ has a root α that lies between 0 and 1.

b Rearrangements of the above equation give the following iterative formulas:

$$x_{n+1} = \tfrac{1}{5}\left(x_n^3 + 1\right) \qquad \text{and} \qquad x_{n+1} = \sqrt[3]{5x_n - 1}$$

Using $x_0 = 0.5$, show that only one of these formulae will enable you to find α and determine the value of α correct to four decimal places. Record the values of $x_1, x_2, x_3, \ldots$ as accurately as your calculator will allow.

(WJEC)

4

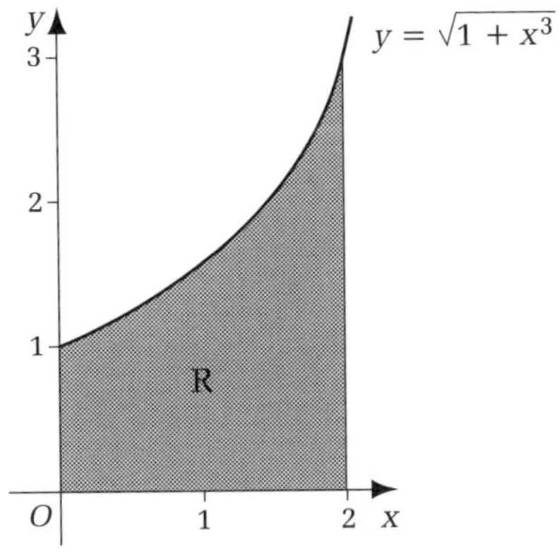

The diagram shows the region R bounded by the curve $y = \sqrt{1 + x^3}$, the axes and the line $x = 2$. Use the trapezium rule with four intervals to obtain an approximation for the area of R, showing your working and giving your answer to a suitable degree of accuracy. Explain, with the aid of a sketch, whether the approximation is an over-estimate or an under-estimate.

(UCLES)

5 The equation $e^x - x^2 - 2 = 0$ has exactly one root. Show that the root lies in the interval $1.3 < x < 1.4$. Use one application of Newton's method to find an approximate value of the root, giving two decimal places in your answer.

(NEAB)

6 The sequence given by the iteration formula

$x_{n+1} = \dfrac{3}{x_n^2} + \frac{2}{3}x_n$ with $x_1 = 2$, converges to α. Use this formula to find α correct to two decimal places, showing the result of each iteration. State an equation satisfied by α and hence find α in surd form.

(UCLES)

7 Show that the equation $x = 3 \sin x$ has a solution lying between 2.1 and 2.4. Use the Newton–Raphson method, showing the result of each iteration, to obtain this solution correct to three decimal places.

(UCLES)

8 **a** Show that the equation $e^x \sin x = 1$ has a root between $x = 0$ and $x = 1$.

b By repeated use of the Newton–Raphson approximation method, obtain the root correct to two decimal places.

(NICCEA)

9 **a** Sketch on one pair of axes the graphs of $y = \cot x$ and $y = \frac{1}{3}x$ for $0 < x \leq \dfrac{\pi}{2}$.

b Give a reason why the equation $\cot x = \frac{1}{3}x$ must have exactly one root in the interval $1 < x < \dfrac{\pi}{2}$.

c An iterative process is defined by $u_1 = 1$, $u_{n+1} = \tan^{-1}\left(\dfrac{3}{u_n}\right)$.

i Carry out the iteration as far as u_6, recording each value of u_n to three decimal places.

ii Given that the iteration converges to a limit α, show that α will be a root of the equation $\cot x = \frac{1}{3}x$.

iii State the value of α correct to two decimal places.

(NEAB)

10 It is given that the equation $f(x) = 0$, where $f(x) = \dfrac{1}{4+x} - \ln(1 + 4x), x \geq 0$ has exactly one root and that this root is close to zero. Apply Newton's method once, starting with $x = 0$, to find a second approximation to the root.

(NEAB)

Review

1 The coordinates of G are (1, 2, 2). Write down the coordinates of the points A, B, C, D, E and F.

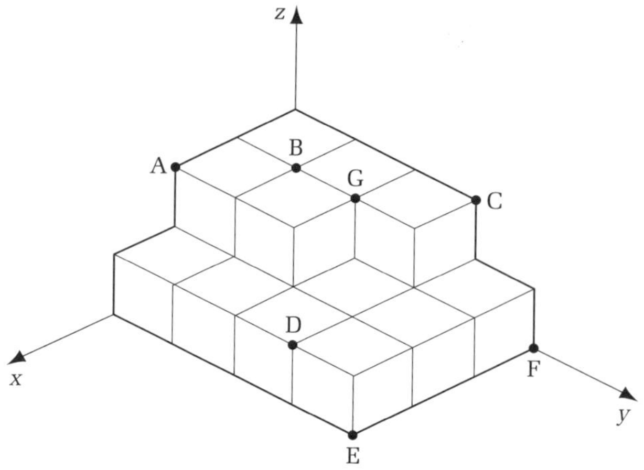

2 Use the cosine rule to find the angle θ in the following triangles:

a

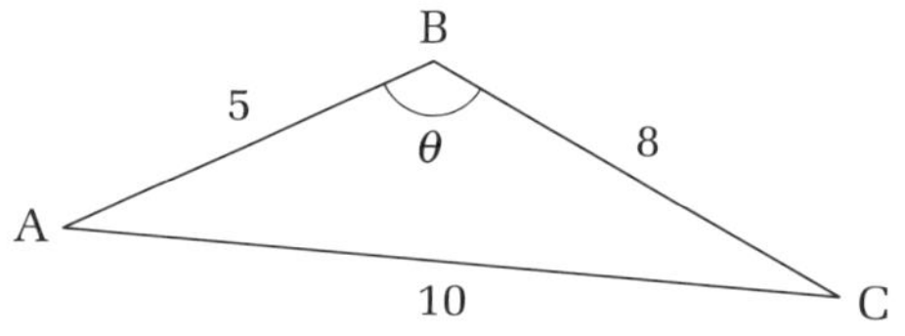

b

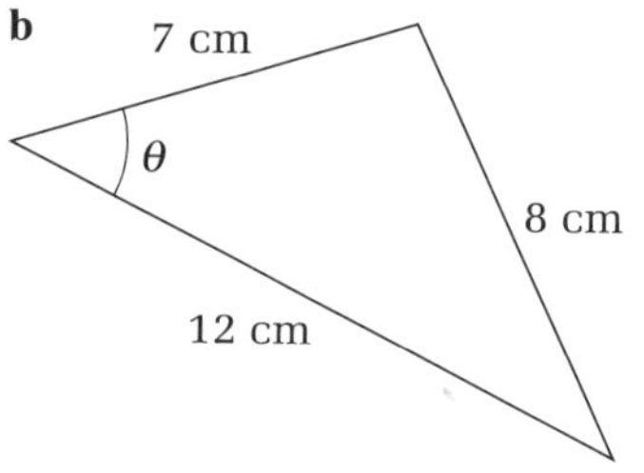

3 Use the formula area $= \frac{1}{2} ab\sin C$ to find the area of the following triangles:

a

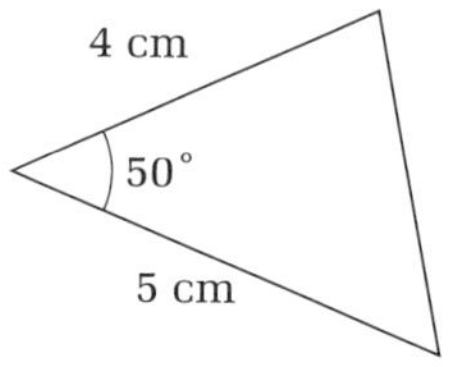

b

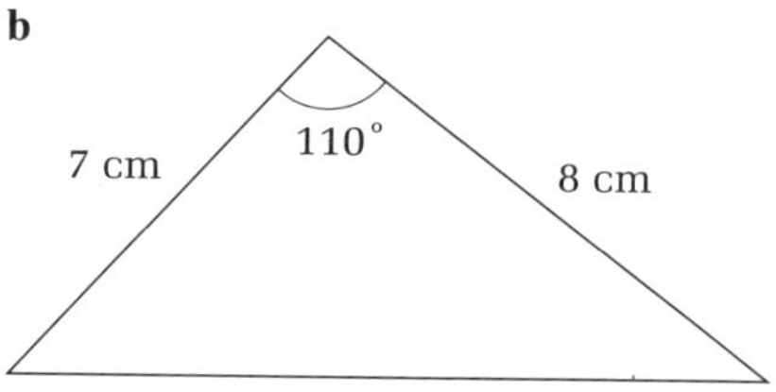

14.1 Technique

1 The grid shows vectors **a** and **b**. On squared paper, draw vectors **a** and **b**. Now draw line segments to represent the vectors:

a $\mathbf{a} + \mathbf{b}$ **b** $\mathbf{a} - \frac{1}{2}\mathbf{b}$ **c** $\frac{3}{2}\mathbf{a} + 2\mathbf{b}$

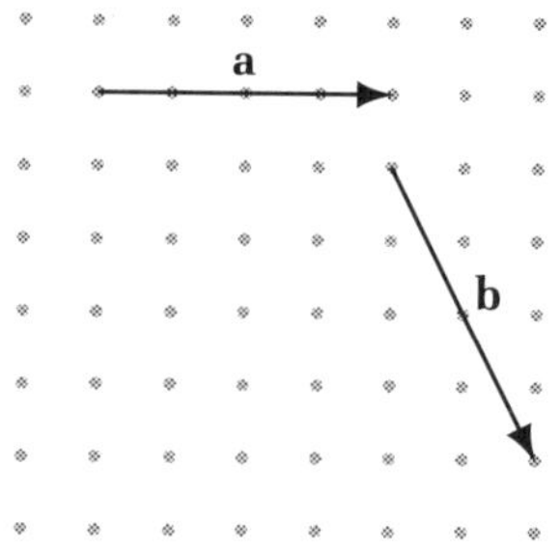

2 The grid shows vectors **u**, **v** and **w**. On squared paper, draw the vector $-\frac{1}{2}\mathbf{u} + 2\mathbf{v} + \frac{1}{3}\mathbf{w}$.

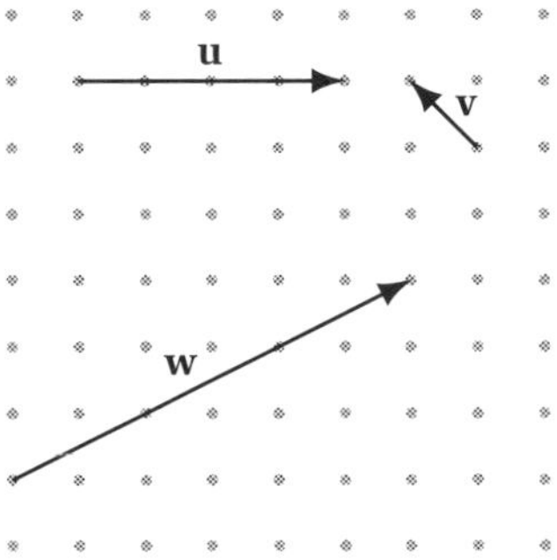

3 The grid shows vectors **a**, **b**, **c** and **d**. By drawing vectors, find:

a **c** in terms of **a** and **b** **b** **d** in terms of **a** and **b**.

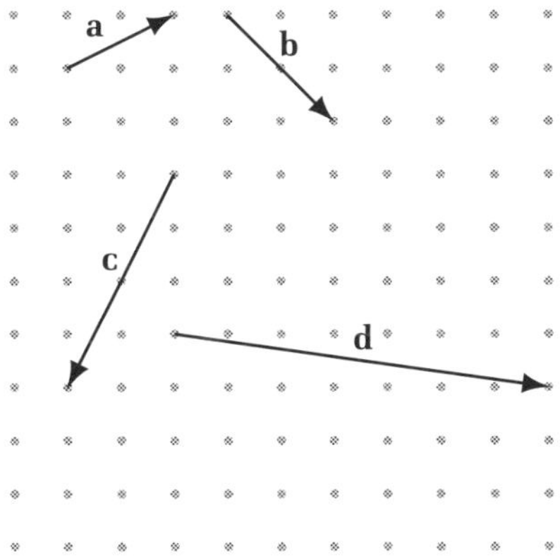

4

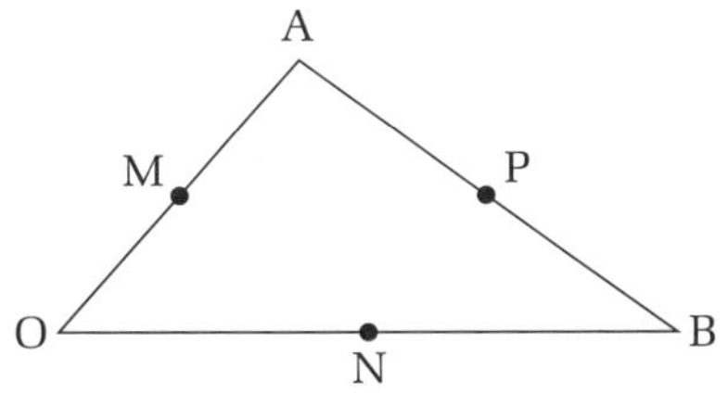

In the triangle OAB, $\overrightarrow{OA} = \mathbf{a}$ and $\overrightarrow{OB} = \mathbf{b}$ and M, N and P are the midpoints of OA, OB and AB, respectively. Write in terms of **a** and **b** the vectors:

a $\overrightarrow{OM}$ **b** $\overrightarrow{ON}$ **c** $\overrightarrow{AB}$
d $\overrightarrow{OP}$ **e** $\overrightarrow{MN}$ **f** $\overrightarrow{MP}$

Comment on your answers to **e** and **f**.

5

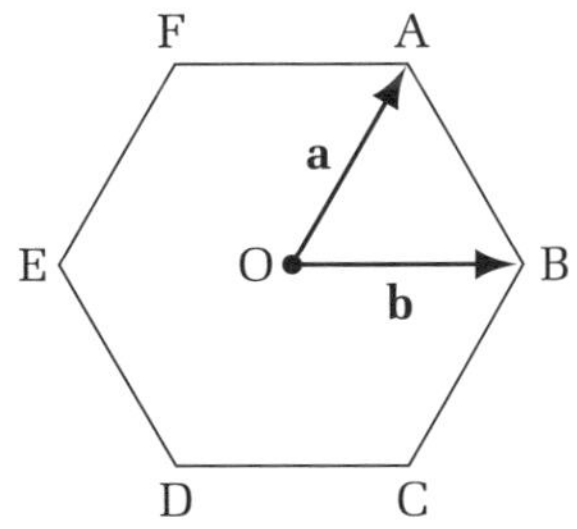

ABCDEF is a regular hexagon with centre O; $\overrightarrow{OA} = \mathbf{a}$ and $\overrightarrow{OB} = \mathbf{b}$. Express the following vectors in terms of **a** and **b**:

a $\overrightarrow{AF}$
b $\overrightarrow{OD}$
c $\overrightarrow{CB}$
d $\overrightarrow{AB}$
e vector $\overrightarrow{MN}$, if N and M are the midpoints of EF and CB, respectively.

6

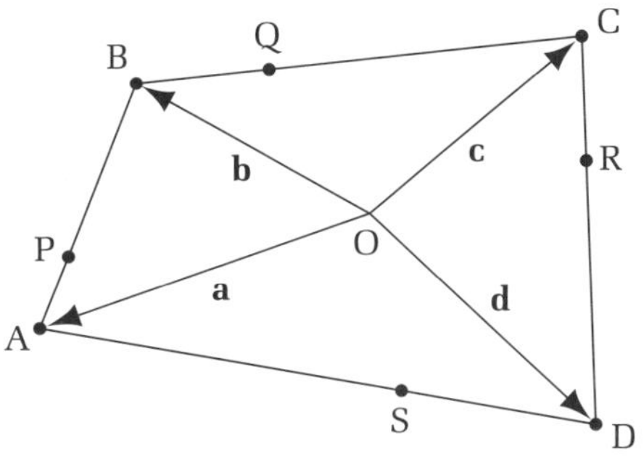

The diagram shows the quadrilateral ABCD. Points P, Q, R and S divide the lines AB, BC, CD and DA respectively in the ratio 1 : 2.

a Write expressions for the vectors $\overrightarrow{OP}$, $\overrightarrow{OQ}$, $\overrightarrow{OR}$ and $\overrightarrow{OS}$ in terms of **a**, **b**, **c** and **d**.

b Write down an expression for $\overrightarrow{OP} + \overrightarrow{OQ} + \overrightarrow{OR} + \overrightarrow{OS}$.

14.2 Technique

1 The diagram shows the x–y plane.

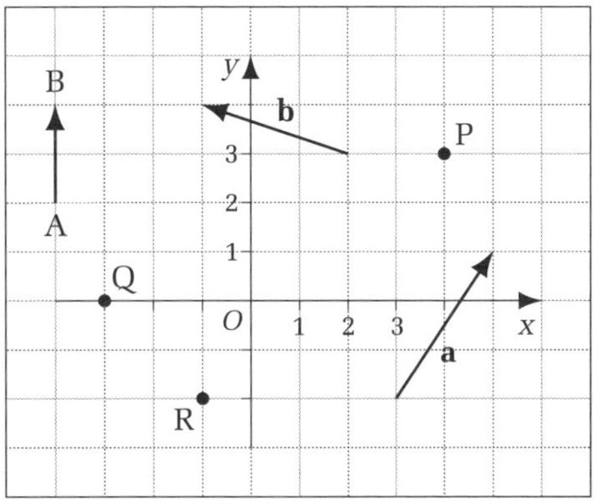

a Describe vectors **a**, **b** and $\overrightarrow{AB}$ using the base vectors **i** and **j**.

b Write vectors **a**, **b** and $\overrightarrow{AB}$ as column vectors.

c Find $|\mathbf{a}|$.

d Find the unit vector in the direction of vector **a**.

e Find the position vectors of points P, Q and R.

f Find, in column vector form, the vectors $3\mathbf{a}$, $\frac{1}{2}\mathbf{b}$, $\mathbf{a} + \mathbf{b}$ and $\mathbf{a} - \mathbf{b}$.

2 The diagram shows vectors in three-dimensional space.

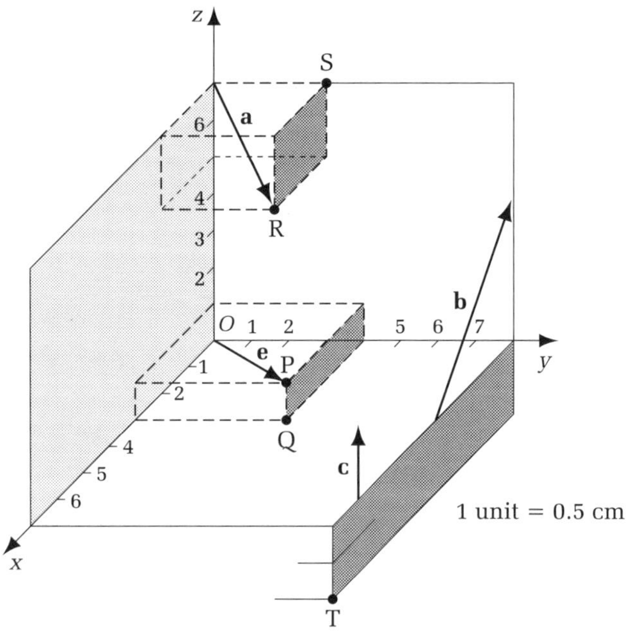

a Use both types of notation to describe vectors **a**, **b**, **c** and $\overrightarrow{OP}$.

b Write down, as column vectors, the position vectors of points P, Q, R, S and T.

c Write down the coordinates of points P, Q and T.

d Find *OP* and $|\mathbf{a}|$.

e Find a unit vector in the direction of **a**.

3 The points A and B have position vectors $\begin{pmatrix} -5 \\ 3 \\ 6 \end{pmatrix}$ and $\begin{pmatrix} 1 \\ 3 \\ -2 \end{pmatrix}$ respectively.

a Find the vector $\overrightarrow{AB}$.

b If M is the midpoint of AB, find the position vector of M.

c Find *AB*.

d How far is M from A?

4 Points R and S have coordinates (8, 6, –3) and (–4, 0, 1) respectively.

a Find vector $\overrightarrow{RS}$.

b Find vector $\overrightarrow{SR}$.

c Find $\left|\overrightarrow{RS}\right|$.

d Write down a unit vector in the direction of $\overrightarrow{RS}$.

5 Points A and B have position vectors $2\mathbf{j} + 3\mathbf{k}$ and $15\mathbf{i} - 3\mathbf{j} + 13\mathbf{k}$, respectively.

a Find $\overrightarrow{AB}$.

b Point C divides the line AB in the ration 2 : 3. Find the coordinates of C.

c How far is C from the origin?

6 The points A and B have position vectors $\begin{pmatrix} 7 \\ -3 \\ -4 \end{pmatrix}$ and $\begin{pmatrix} -5 \\ 7 \\ 0 \end{pmatrix}$ respectively.

a Find vector $\overrightarrow{AB}$.

b If M is the midpoint of AB, find the position vector of point M.

c Find the exact lengths of OA, OB and AB. What type of triangle is OAB?

d Find *OM* and the area of triangle OAB.

14.3 Technique

1 The diagram shows vectors $\mathbf{a} = \begin{pmatrix} -4 \\ 0 \\ 4 \end{pmatrix}$ and $\mathbf{b} = \begin{pmatrix} -2 \\ 0 \\ 0 \end{pmatrix}$.

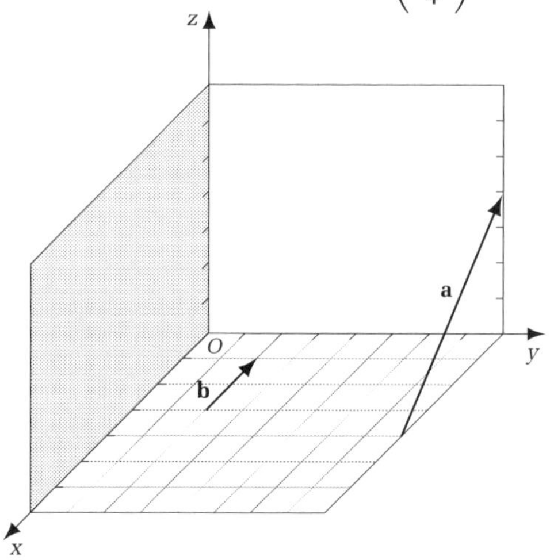

a Write down the angle between vectors **a** and **b**.

b Find $|\mathbf{a}|$ and $|\mathbf{b}|$.

c Use $\mathbf{a}.\mathbf{b} = ab\cos\theta$ to evaluate **a**.**b**.

d Use the components to find **a**.**b**.

2 If $\mathbf{u} = \begin{pmatrix} -1 \\ 8 \\ 7 \end{pmatrix}$ $\mathbf{v} = \begin{pmatrix} 2 \\ 3 \\ -1 \end{pmatrix}$, and $\mathbf{w} = \begin{pmatrix} 5 \\ 0 \\ 8 \end{pmatrix}$, find:

a $2\mathbf{u}$

b $\mathbf{v}.\mathbf{u}$

c $\mathbf{v}.(2\mathbf{u})$

d $\mathbf{u} - \mathbf{w}$

e $\mathbf{v}.(\mathbf{u} - \mathbf{w})$

f $\mathbf{v}.\mathbf{w}$

g $\mathbf{v}.\mathbf{u} - \mathbf{v}.\mathbf{w}$

Comment on your answers to **b** and **c**, and to **e** and **g**.

3 If $\mathbf{a} = 2\mathbf{i} - \mathbf{j} + 2\mathbf{k}$ and $\mathbf{b} = \mathbf{i} - \mathbf{j} + 5\mathbf{k}$, find:

a $|\mathbf{a}|$

b $|\mathbf{b}|$

c $\mathbf{a}.\mathbf{a}$

d $\mathbf{b}.\mathbf{b}$

e the connection between $|\mathbf{a}|$ and $\mathbf{a}.\mathbf{a}$

f $(\mathbf{a} + \mathbf{b}).(\mathbf{b} - \mathbf{a})$

4 Use the dot product to decide whether the angle between the following pairs of vectors is acute, obtuse or a right angle:

a $\begin{pmatrix} 2 \\ -7 \\ 3 \end{pmatrix}$ and $\begin{pmatrix} -3 \\ -1 \\ 1 \end{pmatrix}$

b $2\mathbf{j} - \mathbf{k}$ and $\mathbf{i} - \mathbf{j} - \mathbf{k}$

c $\begin{pmatrix} -3 \\ 2 \\ 6 \end{pmatrix}$ and $\begin{pmatrix} -2 \\ -9 \\ 2 \end{pmatrix}$

d $7\mathbf{i} - 3\mathbf{j} + \mathbf{k}$ and $\mathbf{i} - \mathbf{j} - 5\mathbf{k}$

5 If $\mathbf{u} = \begin{pmatrix} 24 \\ \lambda \\ 12 \end{pmatrix}$ and $\mathbf{v} = \begin{pmatrix} 6 \\ -2 \\ 3 \end{pmatrix}$ find the value of λ when:

a $\mathbf{u}$ and $\mathbf{v}$ are parallel

b $\mathbf{u}$ and $\mathbf{v}$ are perpendicular

6 Find the angle between the following pairs of vectors:

a $\mathbf{a} = \begin{pmatrix} 7 \\ -3 \\ 2 \end{pmatrix}$ and $\mathbf{b} = \begin{pmatrix} 2 \\ 4 \\ -1 \end{pmatrix}$

b $\mathbf{v} = \begin{pmatrix} 1 \\ 0 \\ -2 \end{pmatrix}$ and $\mathbf{u} = \begin{pmatrix} -3 \\ 1 \\ 1 \end{pmatrix}$

c $\mathbf{i} + \mathbf{k}$ and $\mathbf{i} - \mathbf{j} + \mathbf{k}$

7 Find the angle between the vector $3\mathbf{i} - \mathbf{j} - \mathbf{k}$ and the x-axis.

8 Find the values of λ if $\begin{pmatrix} \lambda \\ \lambda \\ 1 \end{pmatrix}$ and $\begin{pmatrix} \lambda \\ -5 \\ 4 \end{pmatrix}$ are perpendicular vectors.

9

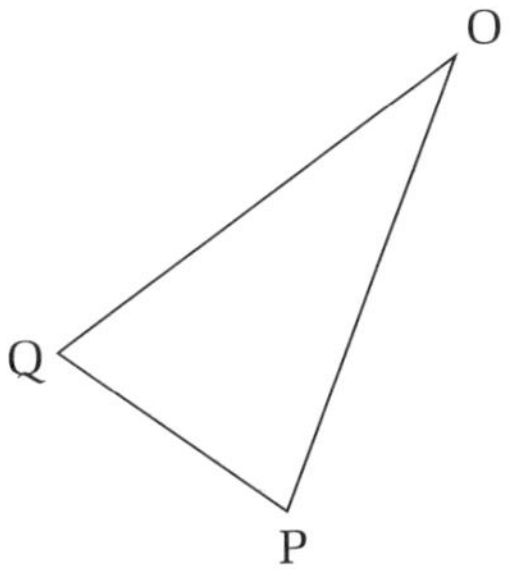

The diagram shows the triangle OPQ where O is the origin and P and Q have position vectors $\begin{pmatrix} 11 \\ -2 \\ -13 \end{pmatrix}$ and $\begin{pmatrix} 7 \\ -37 \\ -11 \end{pmatrix}$, respectively. Find:

a $\overrightarrow{PQ}$

b $\overrightarrow{OP}.\overrightarrow{PQ}$

c the area of the triangle.

10 The points A (0, 4, −1), B (3, 2, 0) and C (1, 1, 4) form a triangle ABC. Find the exact value of:

a lengths AB and AC

b the cosine of the angle between vectors $\overrightarrow{AB}$ and $\overrightarrow{AC}$

c the area of triangle ABC.

11 The angle between vectors $\begin{pmatrix} 1 \\ \lambda \\ 2 \end{pmatrix}$ and $\begin{pmatrix} 1 \\ 0 \\ 1 \end{pmatrix}$ is $\frac{2\pi}{3}$.

Find the two possible values of λ.

12 Find a vector that is perpendicular to both $2\mathbf{i} - 3\mathbf{j} + \mathbf{k}$ and $\mathbf{i} - \mathbf{j} - \mathbf{k}$.

14.3 Contextual

1 The diagram shows a triangular prism where ABE and OCD are isosceles triangles.

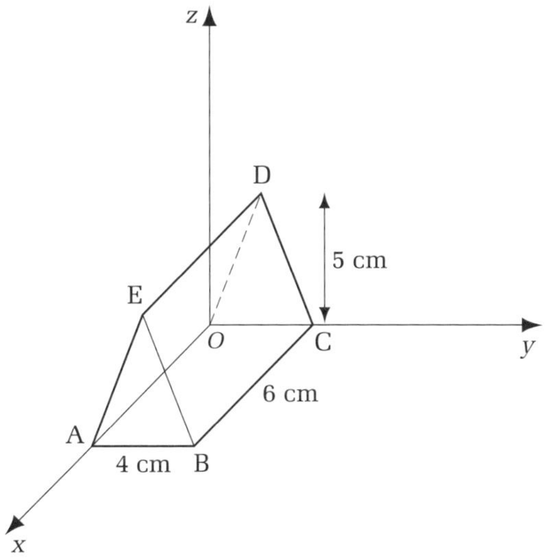

Use vectors to find the angle between:

a EC and BD

b CD and CO

2

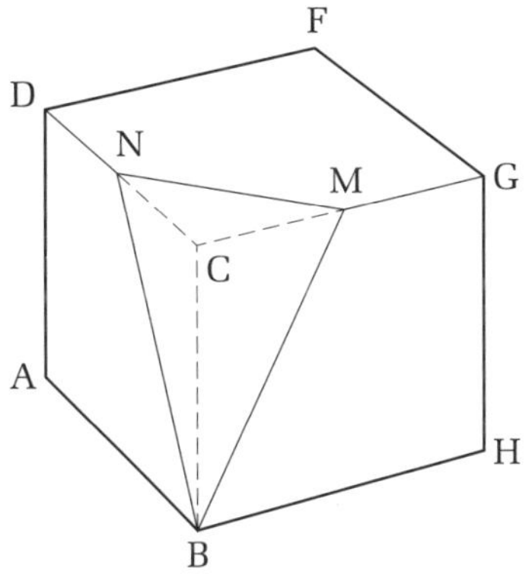

The diagram shows a cube of side length 4 units. Let point A be the origin. Use vectors to find the angle NBM, where N and M are the midpoints of DC and CG respectively.

14.4 Technique

1 A straight line l_1 passes through point (0, 1, 6) and is parallel to the vector $\begin{pmatrix} -2 \\ 5 \\ 1 \end{pmatrix}$.

a Find the equation of line l_1.

b Which point, P (4, –9, 5) or Q (6, –14, 3), lies on the line?

2 Straight line M has vector equation $\begin{pmatrix} x \\ y \\ z \end{pmatrix} = \begin{pmatrix} 1+2t \\ 2t \\ 3 \end{pmatrix}$.

a Write the vector equation in the form $\mathbf{r} = \mathbf{a} + t\mathbf{b}$.

b Write down the vector equation of the straight line parallel to M that passes through the point (–7, 1, 6).

c Another straight line passes through the origin and is also parallel to M. Write down the vector equation of this line.

3 The vector equation of a straight line L is

$\mathbf{r} = \begin{pmatrix} 2 \\ -7 \\ 3 \end{pmatrix} + t\begin{pmatrix} 3 \\ 1 \\ -2 \end{pmatrix}$. Find the values of x and y if P $(x, y, 5)$ lies on the line L.

4 Points A, B and C have position vectors $(\mathbf{i} + 3\mathbf{j} - 2\mathbf{k})$, $(2\mathbf{i} + 2\mathbf{j})$ and $(5\mathbf{i} + 3\mathbf{j} - \mathbf{k})$ respectively.

a Write down the vector equation of the straight line passing through points A and B.

b Write down the vector equation of the straight line passing through points B and C.

c Find the angle between these two lines.

5 The vector equation of a straight line N is $\mathbf{r} = \mathbf{a} + t\mathbf{b}$,

where $\mathbf{a} = \begin{pmatrix} 1 \\ -2 \\ 3 \end{pmatrix}$ and $\mathbf{b} = \begin{pmatrix} 3 \\ 0 \\ -4 \end{pmatrix}$.

a Find $|\mathbf{b}|$.

b Show that Q (7, –2, –5) lies on the line N.

c Find the coordinates of the two points that lie on the line N, and are 10 units from Q.

6 **a** Find the vector equation of a straight line passing through the points (2, 0, 7) and (4, 3, 3).

b Show that the three points (2, 0, 7), (4, 3, 3) and (–8, –15, 27) are collinear.

c Show that this line intersects with the line

$$\begin{pmatrix} x \\ y \\ 7 \end{pmatrix} = \begin{pmatrix} 4 \\ 5 \\ -1 \end{pmatrix} + \lambda\begin{pmatrix} 1 \\ 2 \\ -3 \end{pmatrix}$$

Write down the coordinates of their common point.

7 The straight line N is given by $\mathbf{r} = \begin{pmatrix} 26 \\ -4 \\ 17 \end{pmatrix} + \lambda\begin{pmatrix} 8 \\ -1 \\ 2 \end{pmatrix}$.

The straight line M passes through points (4, 2, 7) and (0, –4, 15).

a Find the vector equation of the straight line M.

b Show that lines M and N intersect and find the coordinates of the point of intersection.

c Find the angle between these two straight lines.

8 Straight lines L_1 and L_2 are given by

$$\mathbf{r} = \begin{pmatrix} 2-\lambda \\ -4+2\lambda \\ \lambda \end{pmatrix} \text{ and } \mathbf{r} = \begin{pmatrix} \mu \\ 3 \\ 1+2\mu \end{pmatrix}.$$

a Show that L_1 and L_2 are skew.

b Find the angle between these two lines.

9 The straight line L passes through points A and B with position vectors $(6\mathbf{i} + 3\mathbf{j} + 3\mathbf{k})$ and $(10\mathbf{i} + 5\mathbf{j} + 8\mathbf{k})$ respectively.

a Find the vector equation of the line L.

b Q is a point on L such that $\overrightarrow{OQ}$ is perpendicular to the line. Find the coordinates of Q.

c Find the shortest distance between the line L and the origin.

d Find the perpendicular distance from point D (–9, –7, –17) to line L.

10 Straight line L has vector equation

$$\begin{pmatrix} x \\ y \\ z \end{pmatrix} = \begin{pmatrix} 2 \\ 1 \\ 12 \end{pmatrix} + \lambda\begin{pmatrix} -1 \\ 2 \\ 6 \end{pmatrix}.$$

Point Q (0, 13, 35) is not on line L.

a Find the coordinates of point M on line L, such that $\overrightarrow{QM}$ is perpendicular to line L.

b Find the perpendicular distance from point Q to the line L.

c Find the coordinates of Q′, the reflection of Q in line L.

14.4 Contextual

1

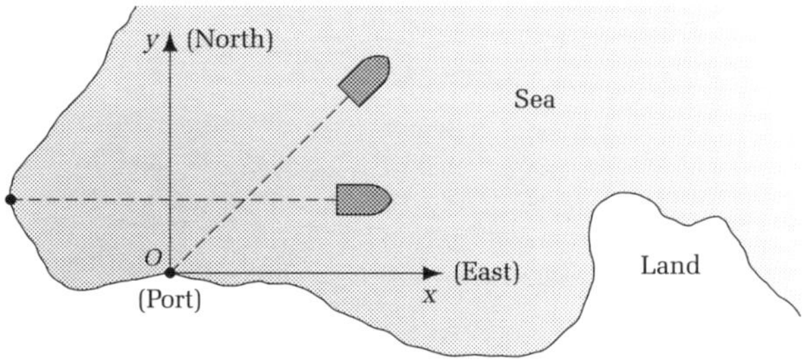

A ship leaves port O and its position t seconds later is given by the vector equation $\mathbf{r} = \begin{pmatrix} 2t \\ t \end{pmatrix}$. At the same time, a power boat sets off from point (–300, 200) and travels due east at a constant speed of $4\,\text{m}\,\text{s}^{-1}$.

a Write down the position vector of the power boat after t seconds.

b Write down an expression for the distance between the two vessels.

c If there is no change of course for the first five minutes, find the minimum distance between the two vessels.

2 Straight lines L and M are given by the vector equations

$$\mathbf{r}(\lambda) = \begin{pmatrix} 2 \\ 3 \\ 7 \end{pmatrix} + \lambda \begin{pmatrix} -2 \\ 2 \\ 1 \end{pmatrix} \text{ and}$$

$$\mathbf{r}(\mu) = \begin{pmatrix} -7 \\ 20 \\ 9 \end{pmatrix} + \mu \begin{pmatrix} -3 \\ 5 \\ 2 \end{pmatrix} \text{ respectively.}$$

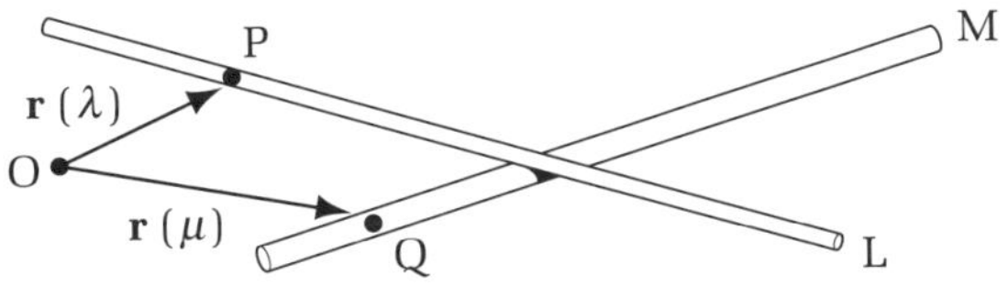

a P is a point on L, and Q is a point on M. Find an expression, in terms of λ and μ, for $\overrightarrow{PQ}$.

b Find $\overrightarrow{PQ}$ if this vector is perpendicular to both lines.

c Find the shortest distance between these two lines.

14.5 Technique

1

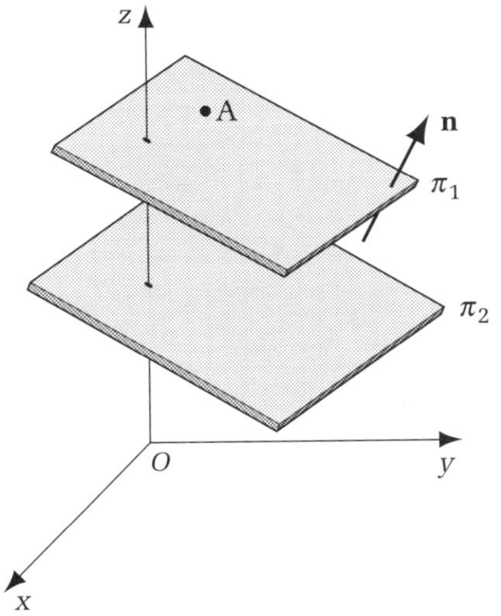

The diagram shows two planes π_1 and π_2. The vector $\mathbf{n} = \begin{pmatrix} 2 \\ -1 \\ 3 \end{pmatrix}$ is perpendicular to both planes, and the plane π_1 passes through point A $(-2, 1, 9)$.

a Find the vector equation and Cartesian equation of the plane π_1.

b Plane π_2 cuts the z-axis at $z = 5$. Find the vector equation and Cartesian equation of π_2.

c Find the value of x if Q $(x, 3, -1)$ lies on π_1.

d Find the value of x if Q $(x, 3, -1)$ lies on π_2.

2 The vector equation of a plane π_1 is $\mathbf{r}.\begin{pmatrix} 3 \\ 1 \\ -5 \end{pmatrix} = 4$.

a Does the point A $(2, -12, -2)$ lie on the plane?

b Find the vector equation and Cartesian equation of the plane that is parallel to π_1 and passes through the origin.

3 For each diagram, write down the vector perpendicular to the given plane, and the vector equation of the plane:

a

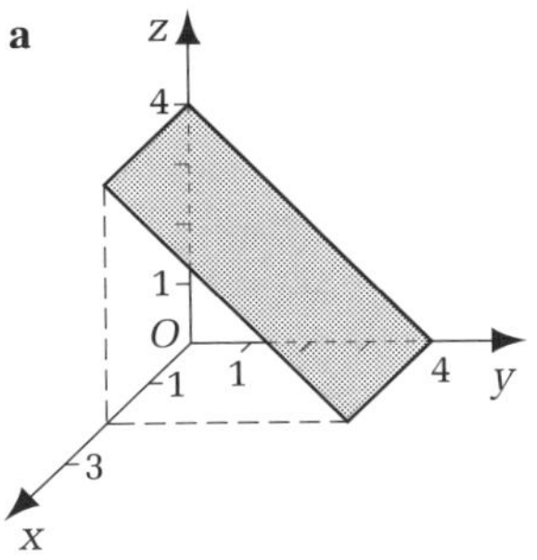

b

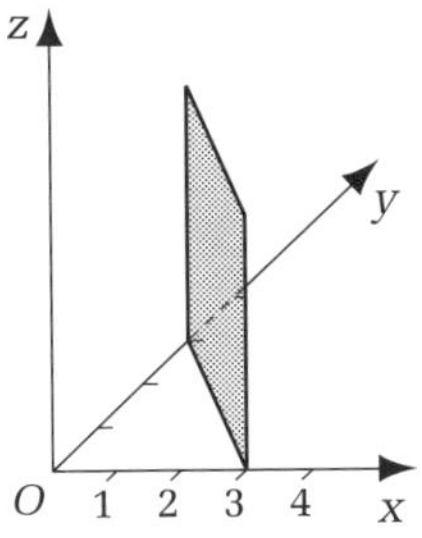

4 The Cartesian equation of the plane π is $x + 2y + 3z = 4$.

a Write the equation of the plane π in the vector form $\mathbf{r}.\mathbf{n} = d$.

b Find the value of y if Q $(4, y, 2)$ is a point on the plane π.

5 The diagram shows a plane passing through the points A, B and C with coordinates $(3, 0, 0)$, $(0, 3, 0)$ and $(0, 0, 2)$ respectively.

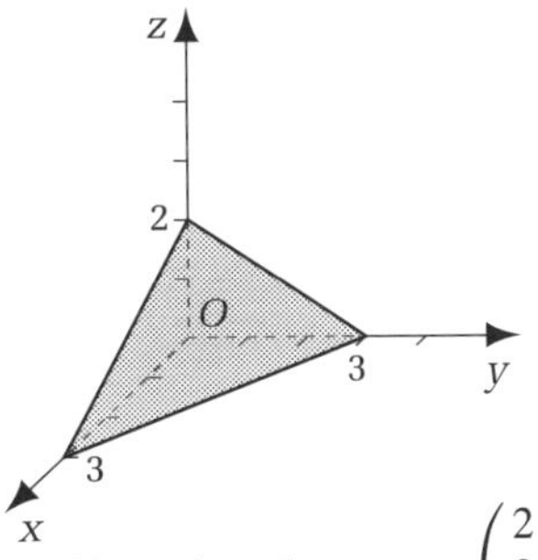

a Show that the vector $\begin{pmatrix} 2 \\ 2 \\ 3 \end{pmatrix}$ is perpendicular to the plane.

b Find the vector equation of the plane in the vector form $\mathbf{r}.\mathbf{n} = d$.

c Find the Cartesian equation of this plane.

6 Straight line L has vector equation

$$\begin{pmatrix} x \\ y \\ z \end{pmatrix} = \begin{pmatrix} 2 \\ -1 \\ 3 \end{pmatrix} + \lambda \begin{pmatrix} 7 \\ -2 \\ 5 \end{pmatrix}.$$

a Show that Q $(-12, 3, -7)$ lies on L.

b Find the vector equation of the plane that is perpendicular to L, and passes through Q.

7 Straight line M has vector equation

$$\mathbf{r} = \begin{pmatrix} 0 \\ 4 \\ -20 \end{pmatrix} + t \begin{pmatrix} 1 \\ -1 \\ 7 \end{pmatrix}.$$

The plane π has vector equation $\mathbf{r}.\begin{pmatrix} 2 \\ 3 \\ -4 \end{pmatrix} = 5$.

a Find the coordinates of the point where the plane π and line M intersect.

b Find the angle between the straight line M and plane π.

8 Find the angle between the two planes $2x - y + z = 1$ and $x - 3y + 2z = 0$.

9 The Cartesian equation of a plane π is $x - y + 2z = 6$.

a Find the perpendicular distance from the origin to the plane π.

b Find the perpendicular distance from the point $(1, -2, 1)$ to the plane π.

10 A plane π passes through the points P, Q and R with position vectors $\begin{pmatrix}3\\-1\\3\end{pmatrix}$, $\begin{pmatrix}4\\1\\7\end{pmatrix}$ and $\begin{pmatrix}0\\-2\\6\end{pmatrix}$ respectively.

a Find the vectors $\overrightarrow{PQ}$ and $\overrightarrow{PR}$.

b Find the vector equation of the plane π using two parameters.

c Show that the point S (5, −2, −4) lies on π.

11 The vector equations of two straight lines are given by

$$\mathbf{r} = \begin{pmatrix}1\\0\\3\end{pmatrix} + s\begin{pmatrix}2\\-1\\3\end{pmatrix} \text{ and } \mathbf{r} = \begin{pmatrix}-3\\2\\7\end{pmatrix} + t\begin{pmatrix}2\\-1\\1\end{pmatrix}.$$

a Show that these two lines intersect at the point (7, −3, 12).

b Find, using two parameters, the vector equation of the plane containing these two lines.

c Find the coordinates of the point where the plane cuts the x-axis.

14.5 Contextual

1 The points A, B and C have coordinates (1, −2, 6), (5, −1, −1) and (0, 1, −2) respectively.

a Find vectors $\overrightarrow{AB}$ and $\overrightarrow{AC}$.

b The vector perpendicular to the plane ABC can be written as $\mathbf{n} = \begin{pmatrix}s\\t\\1\end{pmatrix}$. Find the values of s and t if $\mathbf{n}$ is perpendicular to both $\overrightarrow{AB}$ and $\overrightarrow{AC}$.

c Find the vector equation of the plane ABC in the form $\mathbf{r.n} = d$.

d Point D (3, −5, 2) does not lie on the plane. Find point P on the plane such that $\overrightarrow{DP}$ is perpendicular to the plane.

e Find the coordinates of D′, the reflection of D in the plane.

2

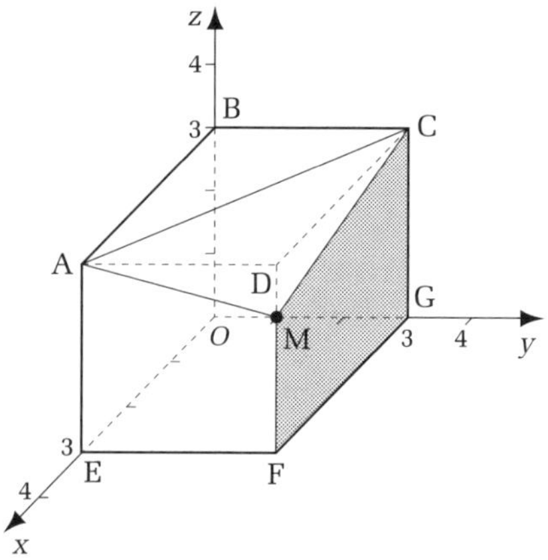

The diagram shows a cube OBADCGFE, where O is the origin, with sides of 3 cm. The point M is 1 cm below the point D.

a Find vectors $\overrightarrow{AC}$ and $\overrightarrow{AM}$.

b Find the vector equation of the plane ACM using two parameters λ and μ.

c By eliminating the parameters find the Cartesian equation of the plane ACM.

d Find the shortest distance from O to the plane ACM.

3 The diagram shows part of the plane $x + y + 3z = 24$ and the line l, which has vector equation

$$\begin{pmatrix}x\\y\\z\end{pmatrix} = \begin{pmatrix}1\\4\\-1\end{pmatrix} + t\begin{pmatrix}-2\\-3\\9\end{pmatrix}.$$

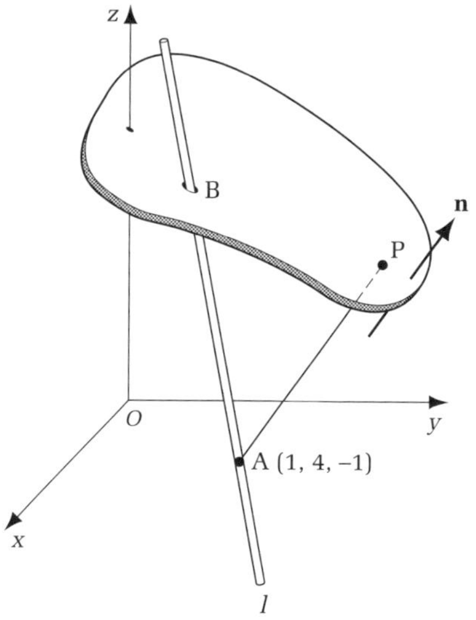

a Write down the vector $\mathbf{n}$, which is perpendicular to the plane.

b The line l meets the plane at the point B. Find the position vector of B.

c Line AP is the perpendicular from A to the plane. The position vector of P is $\overrightarrow{OP} = \overrightarrow{OA} + \overrightarrow{AP}$. Explain why $\overrightarrow{AP} = \lambda\begin{pmatrix}1\\1\\3\end{pmatrix}$, where λ is a constant. Hence write down the coordinates of P in terms of λ.

d Point P also lies on the plane and therefore must satisfy the equation $x + y + 3z = 24$. Hence find the value of λ and the vector $\overrightarrow{AP}$.

e Write down the coordinates of P and find the coordinates of A′, the image of A under a reflection in the plane $x + y + 3z = 24$. Hence find the vector equation of the line l', the reflection of the line l in the plane $x + y + 3z = 24$.

Consolidation C

1

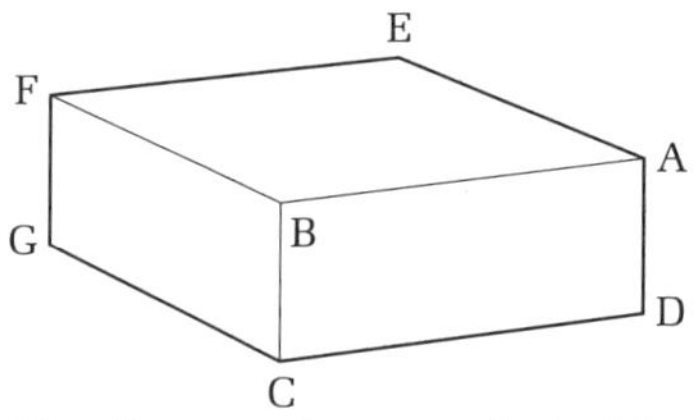

The diagram shows a cuboid. The position vectors of A, B and D are $\begin{pmatrix}5\\2\\3\end{pmatrix}$, $\begin{pmatrix}9\\-8\\6\end{pmatrix}$ and $\begin{pmatrix}13\\7\\9\end{pmatrix}$ respectively.

a Find the position vector of C.

b Show that $\mathbf{v} = \begin{pmatrix}-3\\0\\4\end{pmatrix}$ is perpendicular to $\overrightarrow{AD}$ and $\overrightarrow{AB}$.

c Find the unit vector in the direction of $\mathbf{v}$.

d AE is 10 units long in the direction of $\mathbf{v}$. Find the position vectors of E and F.

2 State the geometrical relationship between two non-zero vectors whose scalar product is zero. The position vectors of A and B are $\mathbf{i} + \mathbf{k}$ and $2\mathbf{j} - \mathbf{k}$ respectively. Find the position vector of C, where C is the foot of the perpendicular from the origin, O, to AB. Hence find the area of the triangle OAB.

(*AEB*)

3 With respect to an origin, O, the position vectors of points L and M are $2\mathbf{i} - 3\mathbf{j} + 3\mathbf{k}$ and $5\mathbf{i} + \mathbf{j} + c\mathbf{k}$ respectively, where c is a constant. Point N is such that OLMN is a rectangle.

a Find the value of c.

b Write down the position vector of N.

c Find, in the form $\mathbf{r} = \mathbf{p} + t\mathbf{q}$, an equation of the line MN.

(*ULEAC*)

4

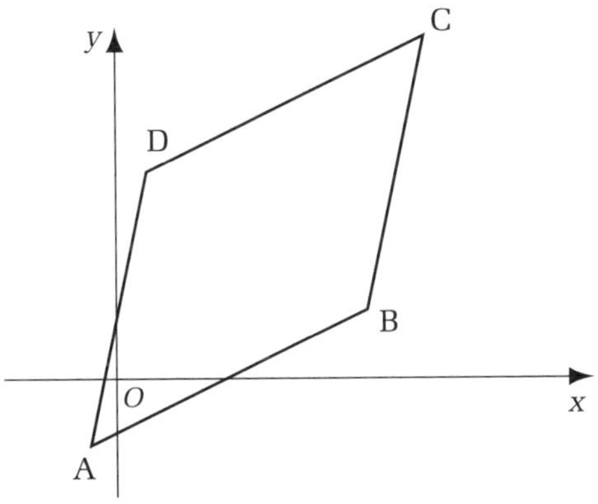

In the diagram, ABCD is a rhombus of area 100 square units. The diagonals AC and BD meet at the point (5, 5) and $|\overrightarrow{AC}| = 2|\overrightarrow{BD}|$.

a Calculate the length of AC.

b The equation of AC is $\mathbf{r} = \begin{pmatrix} 5 \\ 5 \end{pmatrix} + t\begin{pmatrix} 3 \\ 4 \end{pmatrix}$. Show that the vector $\begin{pmatrix} -4 \\ 3 \end{pmatrix}$ is perpendicular to AC.

c Deduce the vector equation of BD.

d Find the coordinates of A, B, C and D. (Hint: The diagonals of a rhombus bisect each other at right angles.)

(*OCSEB*)

5 Points A, B and C have position vectors **a**, **b** and **c** respectively, where $\mathbf{a} = -7\mathbf{i} + 6\mathbf{j} + 4\mathbf{k}$, $\mathbf{b} = 3\mathbf{i} - \mathbf{j} - 4\mathbf{k}$ and $\mathbf{c} = 8\mathbf{i} + \mathbf{j}$. The line l has vector equation $\mathbf{r} = 4\mathbf{i} + \mathbf{k} + s(2\mathbf{i} + \mathbf{j} + 3\mathbf{k})$, where s is a scalar parameter.

a Write down a vector equation of the line BC using a scalar parameter t.

b Show that the line l intersects the line BC and write down the position vector of the point of intersection.

c The midpoint of BC is M. Calculate the acute angle between AM and BC, giving your answer to the nearest degree.

d Determine the position vector of the point on the line l that is closest to the point A.

(*AEB*)

6 The position vectors of three points A, B and C on a plane ski-slope are $\mathbf{a} = 4\mathbf{i} + 2\mathbf{j} - \mathbf{k}$, $\mathbf{b} = -2\mathbf{i} + 26\mathbf{j} + 11\mathbf{k}$ and $\mathbf{c} = 16\mathbf{i} + 17\mathbf{j} + 2\mathbf{k}$, where the units are metres.

a Show that the vector $2\mathbf{i} - 3\mathbf{j} + 7\mathbf{k}$ is perpendicular to $\overrightarrow{AB}$ and also perpendicular to $\overrightarrow{AC}$. Hence find the equation of the plane of the ski-slope.

The track for an overhead railway lies along the straight line DEF, where D and E have position vectors $\mathbf{d} = 130\mathbf{i} - 40\mathbf{j} + 20\mathbf{k}$ and $\mathbf{e} = 90\mathbf{i} - 20\mathbf{j} + 15\mathbf{k}$, and F is a point on the ski-slope.

b Find the equation of the straight line DE.

c Find the position vector of point F.

d Find the length of the track DF.

(*MEI*)

7 A pyramid has a square base OABC and vertex V. The position vectors of A, B, C and V relative to the fixed origin O are $\mathbf{a} = 2\mathbf{i}$, $\mathbf{b} = 2\mathbf{i} + 2\mathbf{j}$, $\mathbf{c} = 2\mathbf{j}$ and $\mathbf{v} = \mathbf{i} + \mathbf{j} + 3\mathbf{k}$ respectively.

a Calculate the length of OV.

b The line l passes through A and the midpoint of VC.

i Find a vector equation of l in terms of parameter t.

ii Show that the line through O and the midpoint of BV intersects l and find the position vector of the point of intersection.

c By proving that the vector $\mathbf{x} = 3\mathbf{j} - \mathbf{k}$ is perpendicular to both OV and AV, show that **x** is perpendicular to the plane OAV. Hence, by considering the scalar product $\mathbf{x}.\overrightarrow{VB}$, determine the angle between the line VB and the plane OAV, giving your answer to the nearest tenth of a degree.

(*AEB*)

8

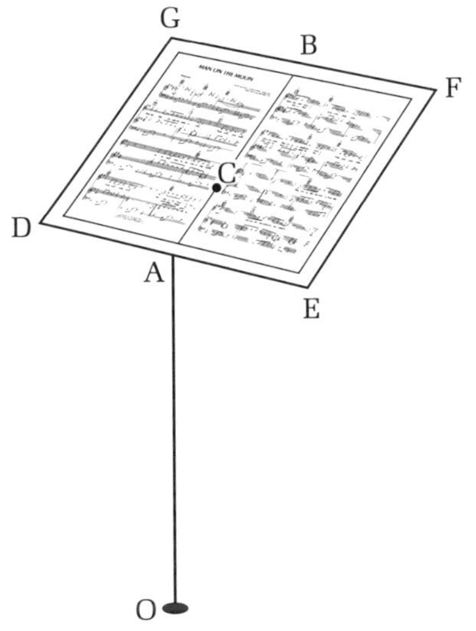

The drawing shows an ordinary music stand, which consists of a rectangle DEFG with a vertical support OA. Relative to axes through the origin O, which is on the floor, the coordinates of various points are given (with dimensions in metres) as A (0, 0, 1), D (−0.25, 0, 1) and F (0.25, 0.15, 1.3); DE and GF are horizontal, A is the midpoint of DE, B is the midpoint of GF and C lies on AB such that $AC = \frac{1}{3}AB$.

a Write down the vector $\overrightarrow{AD}$ and show that $\overrightarrow{EF}$ is $\begin{pmatrix} 0 \\ 0.15 \\ 0.3 \end{pmatrix}$.

b Calculate the coordinates of C.

c Find the equations of lines DE and EF in vector form.

d When the stand is set up, the conductor arranges it so that the line from his right eye to C is normal to the plane DEFG. His right eye is 1.6 metres above the floor, that is at $z = 1.6$ (but not at the point (0, 0, 1.6)). The equation of the plane is $-2y + z = 1$. Write down the equation of the normal through C, and hence find the distance of C from the conductor's right eye.

(MEI)

9 The equation of line L_1 through point P (2, 6, −5) is $\mathbf{r} = \begin{pmatrix} 2 \\ 6 \\ -5 \end{pmatrix} + t\begin{pmatrix} 0 \\ 2 \\ -1 \end{pmatrix}$. Find the numerical values of x_1 and y_1 for which the point Q $(x_1, y_1, -1)$ lies on L_1. Write down, in terms of a parameter s, a vector equation of the line L_2 passing through the points A (5, −3, 1) and B (9, −8, −1). Show that L_1 and L_2 are skew lines.

Given that N is the point on L_2 at which $s = \lambda$, calculate the scalar product $\overrightarrow{PN}.\overrightarrow{AB}$ in terms of λ. Find the coordinates of the foot of the perpendicular from P to L_2 and hence, or otherwise, determine whether P is nearer to AB than it is to Q.

(JMB)

10 Point P lies on the line $\mathbf{r}(\lambda) = -3\mathbf{i} + 2\mathbf{j} + 7\mathbf{k} + \lambda(\mathbf{i} - \mathbf{k})$ and point Q lies on the line $\mathbf{r}(\mu) = 2\mathbf{i} + 5\mathbf{j} + 2\mathbf{k} + \mu(\mathbf{j} - 2\mathbf{k})$. Find an expression for the vector $\overrightarrow{PQ}$.

If $\overrightarrow{PQ}$ is perpendicular to both lines, find the coordinates of P and Q, and hence calculate the shortest distance between these lines.

Review

1 Write each of the following mathematical statements using summation notation and find an expression for the total.

a $1^2 + 2^2 + 3^2 + \ldots + n^2$

b $a^n + a^{n-1}b + a^{n-2}b^2 + \ldots + ab^{n-1} + b^n$

c $\log\left(\frac{1}{a}\right) + \log\left(\frac{1}{a^2}\right) + \ldots + \log\left(\frac{1}{a^n}\right)$

2 What is a prime factor?

3 Factorise the following numbers into their prime factors:

a 56 **b** 128 **c** 233
d 457 **e** 597 **f** 1691

4 Use integration notation to write an expression for each of these shaded areas:

a

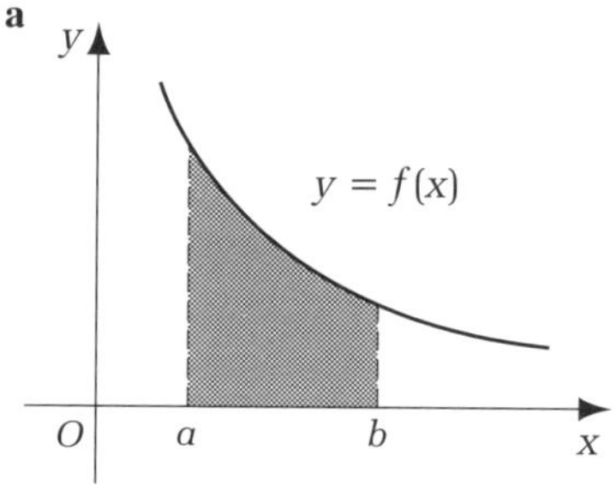

b

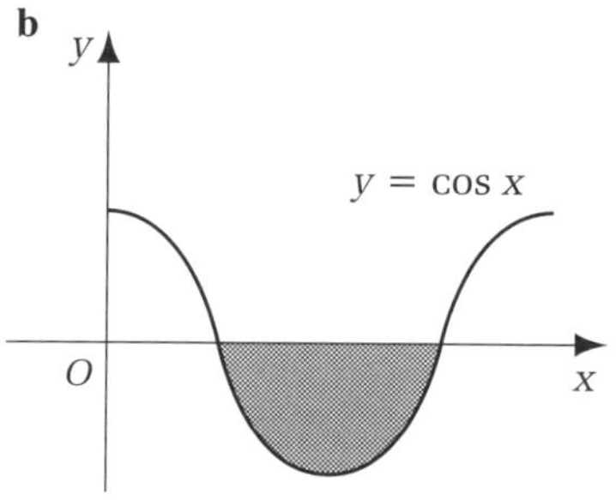

15.1 Technique

1 The following mathematical arguments are invalid. In each case identify the statement that produces the paradox and explain the inconsistency.

a To show that $1 = -3$:

$$\text{Let } x = 1 \qquad [1]$$
$$8x = 8 \qquad [2]$$
$$6x + 2x = 9 - 1 \qquad [3]$$
$$2x + 1 = -6x + 9 \qquad [4]$$
$$x^2 + 2x + 1 = x^2 - 6x + 9 \qquad [5]$$
$$(x + 1)^2 = (x - 3)^2 \qquad [6]$$
$$x + 1 = x - 3 \qquad [7]$$
$$1 = -3 \qquad [8]$$

b Deducing that $\frac{1}{9} > \frac{1}{3}$:

$$2 > 1 \qquad [1]$$
$$2\log(\tfrac{1}{3}) > \log(\tfrac{1}{3}) \qquad [2]$$
$$\log(\tfrac{1}{3})^2 > \log(\tfrac{1}{3}) \qquad [3]$$
$$(\tfrac{1}{3})^2 > \tfrac{1}{3} \qquad [4]$$
$$\tfrac{1}{9} > \tfrac{1}{3} \qquad [5]$$

c Showing that $\sin x = 0$ for $0 \leq x \leq 2\pi$.

$$\sin 0 = \sin 2\pi = 0 \qquad [1]$$

The area between the curve $y = \sin x$ and the x-axis between $x = 0$ and $x = 2\pi$ is given by

$$\int_0^{2\pi} \sin x \, dx \qquad [2]$$

$$\int_0^{2\pi} \sin x \, dx = [-\cos x]_0^{2\pi} \quad [3]$$

$$\int_0^{2\pi} \sin x \, dx = -(\cos 2\pi - \cos 0) \quad [4]$$

$$\int_0^{2\pi} \sin x \, dx = 0 \quad [5]$$

If there is no area between $y = \sin x$ and the x-axis, the curve must coincide with the axis. [6]

So $\sin x = 0$ for $0 \leq x \leq 2\pi$ [7]

15.1 Contextual

1 Prove that the exterior angle of a triangle is equal to the sum of the two opposite interior angles: $A + B = C$. State clearly any axioms used.

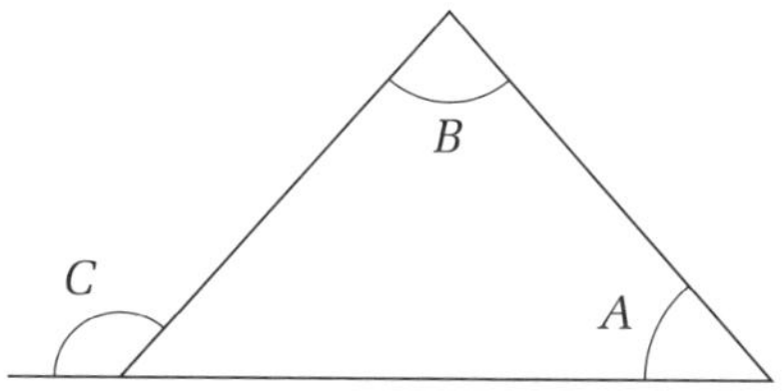

2 Without using a calculator find:

a the units figure of 15!

b the tens and units figures of 101!

15.2 Technique

1 Prove that a number is divisible by 9 if and only if the sum of its digits is divisible by 9.

2 Find a counter example for the following propositions.

a $n^2 + n + 17$ produces prime numbers for all $n \in \mathbb{N}$.

b $n^2 + n + c$, where $c > 1, c \in \mathbb{N}$ produces prime numbers for all $n \in \mathbb{N}$.

c If $n^3 - n$ is divisible by 4, then n must be divisible by 4.

3 Is there a three-digit number abc such that the difference between abc and cba is a prime number?

4 Prove that if a^2 is odd then a must be an odd number.

15.2 Contextual

1 **Fermat Numbers**

Pierre de Fermat (1601–1665) made the following conjecture in 1640.

The formula $2^{2^n} + 1$ produces prime numbers.

Prove, or disprove by counter example, Fermat's statement.

2 Prove, or disprove by counter example, the following conjecture: For $n \geq 1$, $n! + 1$ is either prime or a perfect square.

15.3 Technique

1 Prove that $\sqrt{5}$ is irrational.

2 Prove that $\sqrt{11}$ is irrational.

3 Prove that $\sqrt{p}$ is irrational, where p is prime.

4 Prove that for all $a, b \geq 0$ the following statement is true: $\frac{1}{2}(a + b) \geq \sqrt{ab}$.

15.3 Contextual

1 Prove that factorisation into prime factors is unique; that is that no number can be factorised into prime factors in two different ways.

15.4 Technique

1 Prove that $3^n > 2^n$ for all natural numbers n.

2 Prove that $\sum_{r=1}^{n} r^3 = \frac{1}{4}n^2(n + 1)^2$.

3 Prove the following by induction, where n is a positive integer in each case:

a $7^n + 5$ is divisible by 6

b $2^n + 2^{n+1}$ is divisible by 6

c $3^{4n} - 1$ is divisible by 80.

15.4 Contextual

1 Using the Binomial theorem, with a positive integer n, show that $(1 + x)^n$ is 1 more than a multiple of x, if x is an integer. Hence:

a prove that $3^{4n} - 1$ is divisible by 80

b prove that $4^{3n} - 1$ is divisible by 63.

Consolidation C

1 The following mathematical argument is invalid. Which statement is invalid?

$$(-1)^2 = (1)^2 \quad [1]$$

$$\log(-1)^2 = \log(1)^2 \quad [2]$$

$$2\log(-1) = 2\log(1) \quad [3]$$

$$2\log(-1) = 0 \quad [4]$$

$$\log(-1) = 0 \quad [5]$$

2 Prove that the highest common factor of consecutive integers n and $(n + 1)$ must always be 1.

3 If $\sqrt{n}$ and $\sqrt{m}$ are irrational then $\sqrt{n} - \sqrt{m}$ is also irrational. Prove, or disprove, this conjecture.

4 Find a counter example for the conjecture: 'numbers of the form 29, 229, 2229, . . . are always prime'.

5 Prove by induction that, for all positive integer values of n, $11^n - 6$ is divisible by 5.

(NICCEA)

QUESTION BANK

MIRROR EXERCISES

ANSWERS

Review

1 a y^3 b 5^3 c y^5 d a^3b^2 e $6x^2$ f $10a^3$

2 a 7 b 5 c 12 d 6 e y f b

3 a $7x$ b $5x^2$ c $2y$ d $7xy + 3x$ e $2x^2 + 2x$ f $16x - 2y$

4 a $3x + 12$ b $5x - 10$ c $-3x + 3$ d $-4x + 12$ e $-2x - 4y + 6$ f $15x - 5y - 15$ g $x^2 + x - 2$ h $x^2 + 5x + 6$ i $x^2 - 5x + 4$ j $x^2 + 5x - 6$ k $x^2 - x - 6$ l $x^2 - 6x + 9$

5 a $x = 7$ b $x = 2$ c $x = 2$ d $x = 4$ e $x = 4$ f $x = 3$

6 a $5(x + y)$ b $8(x - y)$ c $3(x - 4y)$ d $5(2x + 3y)$ e $7x(3x + y)$ f $5xy(3y^2 + 7x^2)$

7 a 912 b 342 c $611\frac{2}{13}$ d $732\frac{3}{14}$ e $422\frac{5}{16}$ f $545\frac{2}{17}$

1.1 Technique

1 a $2\sqrt{3}$ b $5\sqrt{2}$ c $3\sqrt{11}$ d $11\sqrt{3}$ e $2\sqrt{7}$ f $4\sqrt{2}$ g $4\sqrt{5}$ h $7\sqrt{2}$

2 a $4\sqrt{3}$ b $9\sqrt{2}$ c $2\sqrt{5}$ d $7\sqrt{5}$ e $5\sqrt{7}$ f $5\sqrt{5}$

3 a 3 b 2 c 95 d 5 e 27 f 44 g $15 - 4\sqrt{11}$ h $53 - 12\sqrt{11}$

4 a $\dfrac{\sqrt{7}}{7}$ b $\dfrac{2}{11}\sqrt{11}$ c $2\left(\sqrt{5} - \sqrt{2}\right)$ d $\sqrt{7} - 1$ e $2 + \sqrt{3}$ f $\dfrac{a - 2\sqrt{a} + 1}{a - 1}$ g $\dfrac{b + 2\sqrt{b} + 1}{b - 1}$ h $6 - \sqrt{35}$

1.2 Technique

1 a 5 b 3^6 c 4^2 d 8^7 e $5xy$ f $3y^2$

2 a x^{12} b x^{10} c $4x^{10}$ d $5x^{20}$ e $4x^4$ f $9x^6$

3 a $5x^{-3}$ b $\frac{3}{5}x^{-2}$ c 7^{-3} d 9^{-5} e $\frac{1}{3}x^2y^{-5}$ f $\frac{1}{4}x^{-1}$

4 a x^{-2} b x^3 c $\dfrac{4x^3}{y}$ d $\frac{1}{3}x^3$ e $3x^2$ f x^2

5 a $\frac{1}{4}$ b 125 c $\frac{1}{9}$ d 1 e $\frac{3}{2}$ f $\frac{3}{2}$ g $\frac{1}{125}$ h $\frac{3}{8}$

6 a 6 b 3 c -2 d -1 e 0 f $\frac{3}{2}$ g -1 h -5

1.3 Technique

1 a i 3 ii 2
b i 5 ii 3
c i 2 ii 12
d i 8 ii 0
e i 4 ii -2
f i not a polynomial, since it contains x^{-1} and x^{-3} terms ii 3

2 a $7x^2 + 4x$ b $5x^2 + 10x - 4$ c $3x^3 - 3x^2 - 2x + 14$ d $6x^4 - 3x^3 + 3x^2 + 1$

3 a $3x^2 - 2x + 7$ b $3x^2 + x - 7$ c $-9x^3 + 11x^2 - 8x + 7$ d $9x^4 - 2x^3 - 5x^2 - 2$

4 a $4x^2 + 4x + 1$ b $9x^2 - 12x + 4$ c $3x^3 + x^2 - 15x - 5$ d $x^3 - x^2 + x + 3$ e $-x^3 - x^2 - x + 14$ f $2x^3 + 6x^2 - 7x + 4$

5 a $3x^3 - x^2 + 2x - 4$
b $5x^3 + 7x^2 - 10x + 9$
c $7x^3 + x^2 - 4x + 3$
d $12x^3 - 12x^2 + 3x + 3$
e $6x^3 - 4x^2 + 4x - 4$
f $5x^4 + 11x^3 - 3x^2 + 2x + 8$

6 a $9x^2 + 6x + 1$
b $2x^3 - 5x^2 - 4x + 12$
c $x^4 - 8x^3 + 24x^2 - 32x + 16$
d $1 + 12x + 54x^2 + 108x^3 + 81x^4$
e $x^4 + 8x^3 + 24x^2 + 32x + 16$
f $16x^4 + 32x^3 + 24x^2 + 8x + 1$

7 a $1 - \dfrac{1}{x + 4}$ b $2 - \dfrac{3}{x + 3}$ c $4 + \dfrac{11}{x - 1}$ d $3 + \dfrac{4}{2x + 1}$ e $3 - \dfrac{10}{x + 3}$ f $3 + \dfrac{8}{x - 2}$

1.4 Technique

1 a $(x - 2)(x + 3)$ b $(x - 5)(x + 2)$ c $(x - 1)(x + 2)$ d $(x - 6)(x - 1)$ e $(x - 5)(x + 3)$ f $(x - 6)(x - 2)$ g $(x - 2)(x + 1)$ h $(x - 1)(x - 13)$

2 a $(x-5)(x+5)$
b $(y-10)(y+10)$
c $(9x-2)(9x+2)$
d $(3y-1)(3y+1)$
e $(\tan\theta-1)(\tan\theta+1)$
f $(2\sin\theta-1)(2\sin\theta+1)$
g $(3x-4y)(3x+4y)$
h $(11x-12y)(11x+12y)$

3 a $2(x-6)(x+6)$
b $3(y-9)(y+9)$
c $10(x-5)(x+5)$
d $7(y-2)(y+2)$
e $5t(t-1)(t+1)$
f $3(\cos\theta-1)(\cos\theta+1)$

4 a $(x-3)(2x-5)$
b $(2x+1)(x-3)$
c $(3x+2)(2x+5)$
d $(3x+2)(x+1)$
e $(4x+3)(x+5)$
f $(11x-1)(x+2)$
g $(3x-2)(x+4)$
h $(3x-4)(4x+3)$

5 a $3(x-7)(x+1)$
b $2(x-3)(x+3)$
c $2(x+2)(x-5)$
d $5(x-2)(x+4)$
e $4(x-1)(x-1)$
f $7(x+3)(x+2)$
g $3y(3x+2)(x+7)$
h $5y(2x-1)(x-2)$

1.5 Technique

1 a $x=-5, 2$
b $x=-1, -8$
c $x=3, -4$
d $x=2, -6$
e $x=\frac{1}{2}, -\frac{7}{2}$
f $x=1, -1$
g $x=0, 3$
h $x=-2, \frac{4}{3}$

2 a $x=2, -3$
b $x=-1, 13$
c $x=2, 7$
d $x=2, 5$
e $x=-\frac{4}{3}, \frac{2}{3}$
f $x=-\frac{2}{3}, 2$

3 a $x=1\pm\sqrt{6}$
b $x=-\frac{1}{2}(3\pm\sqrt{5})$
c $x=-\frac{1}{2}(7\pm\sqrt{29})$
d $x=3(1\pm\sqrt{2})$
e $x=-\frac{1}{2}(1\pm\sqrt{7})$
f $x=-\frac{1}{4}(3\pm\sqrt{41})$

4 a $x=\dfrac{-1\pm\sqrt{7}}{2}$
b $x=\dfrac{-9\pm\sqrt{17}}{4}$
c $x=\dfrac{-5\pm\sqrt{61}}{6}$
d $x=\dfrac{5\pm\sqrt{21}}{2}$
e $x=1\pm\sqrt{6}$
f $x=\dfrac{-1\pm\sqrt{6}}{5}$

5 a $x=-\frac{1}{2}, \frac{3}{2}$
b $x=\frac{1}{3}\left(9\pm\sqrt{51}\right)$
c $x=\dfrac{3\pm\sqrt{73}}{4}$
d $x=2\pm 7$
e $x=\frac{1}{3}$ (twice)
f $x=-5, 3$

6 a $(x-1)^2-6$
b $\left(x+\frac{3}{2}\right)^2-1\frac{1}{4}$
c $-\left(x-\frac{3}{2}\right)^2+5\frac{1}{4}$
d $\left(x+\frac{3}{2}\right)^2-10\frac{1}{4}$
e $-2\left(x-\frac{1}{2}\right)^2+8\frac{1}{2}$
f $3\left(x+\frac{1}{3}\right)^2-1\frac{1}{3}$

1.5 Contextual

1 $t=0.48$ s, 3 s; two answers because the body is at this height going up and coming down

2 4.2π cm^2

3 15

4 $2200m$

5 21

6 $\alpha=18.4°, 78.7°$

1.6 Technique

1 a (3, 2)
b (3, 2)
c (1, −2)
d (3, 2)
e (−8, −5)
f $\left(\frac{4}{3}, \frac{2}{3}\right)$

2 a (3, 5), (−2, 0)
b (1, 0), (0, 1)
c (−3, −10), (3, 20)
d (−2, 12), (4, 6)
e (5, −9), (−2, 5)
f (3, 4), (−4, −3)

3 a $a=\dfrac{10}{3}, T=40$
b (5, 2)
c (2, 3)
d $\left(\frac{3}{2}, 8\right), \left(-8, -\frac{3}{2}\right)$

1.6 Contextual

1 A (10, 60), B (−1, −6)

2 30 000 paid £25 (19 000 paid £40)

3 a CD costs £23 and a tape costs £11

4 $m=1, c=8$

5 Sîan is 17 and Helen is 14

Consolidation C

1 $t=0.8$ s, 2.9 s

2 A (−1, 7), B (5, −5)

3 a $x+y=900$
$24x+19y=17\,500$
b 80 paid at the door and 820 paid in advance

4 a area of outer circle $=\pi R^2$
area of inner circle $=\pi r^2$
cross-sectional area $=\pi R^2-\pi r^2$
$=\pi(R^2-r^2)$
$=\pi(R-r)(R+r)$
b volume $=8400\pi$ cm^3 ($8400\,000\pi$ mm^3)

5 $\theta=\tan^{-1}\frac{1}{3}=18.4°$; $\theta=\tan^{-1}6=80.5°$

6 a A (−2, 16), B (0, 4)
b $p=-3, q=16$

Review

1 **a** $y = x + 9$ **d** $y = 2x + 7$
b $y = -x + 6$ **e** $y = 3x + 7$
c $y = 3x - 6$ **f** $y = \frac{1}{3}x - 4$

2 **a** $x = 3\sqrt{5}$ **b** $x = \sqrt{181}$
c $x = \sqrt{15}$

3 **a** $x^2 + 8x + 16$ **d** $9x^2 + 6x + 1$
b $x^2 + 2x + 1$ **e** $25x^2 + 10x + 1$
c $x^2 + 10x + 25$ **f** $4x^2 + 12x + 9$

4 **a** $x^2 - 8x + 16$ **d** $9x^2 - 6x + 1$
b $x^2 - 10x + 25$ **e** $16x^2 - 8x + 1$
c $c^2 - 2cd + d^2$ **f** $25x^2 - 20x + 4$

2.1 Technique

1 **a** $\sqrt{2}$ **e** $\sqrt{8} = 2\sqrt{2}$
b $\sqrt{5}$ **f** $\sqrt{17}$
c $\sqrt{45} = 3\sqrt{5}$ **g** $\sqrt{56} = 2\sqrt{14}$
d $\sqrt{82}$ **h** $\sqrt{45} = 3\sqrt{5}$

2 **a** $\sqrt{29}$ **e** $\sqrt{90} = 3\sqrt{10}$
b $\sqrt{20} = 2\sqrt{5}$ **f** $\sqrt{29}$
c $\sqrt{109}$ **g** 7
d $\sqrt{58}$ **h** $\sqrt{30}$

3 **a** 4 **b** 4 **c** $-\frac{3}{2}$
d $2\frac{1}{3}$ **e** $\frac{1}{4}$ **f** $\frac{4}{0}$ (vertical line)

4 **a** **i** 3 **d** **i** 9
ii $-\frac{1}{3}$ **ii** $-\frac{1}{9}$
b **i** $-\frac{1}{2}$ **e** **i** $-\frac{1}{3}$
ii 2 **ii** 3
c **i** $-\frac{5}{7}$ **f** **i** $1\frac{1}{2}$
ii $1\frac{2}{5}$ **ii** $-\frac{2}{3}$

5 **a** perpendicular **d** parallel
b perpendicular **e** parallel
c parallel **f** perpendicular

2.1 Contextual

1 **a** $\sqrt{29}$ **b** 0.539 km

2 $\sqrt{106}$

3 38.6; 9.65 km

4 $AC = \sqrt{20}$; $BC = \sqrt{20}$

5 **a** S (10, 0)
b show all sides equal and PQ perpendicular to PS (or equivalent)

6 $AF = \sqrt{77}$

2.2 Technique

1 **a** $y = -2x - 6$ **e** $y = 3x - 3$
b $y = 5x + 2$ **f** $y = -2x + 7$
c $y = x + 8$ **g** $y = 10x + 6$
d $y = 4x - 5$ **h** $y = 3x + 4$

2 **a** gradient 4, intercept (0, 3)
b gradient −5, intercept (0, 1)
c gradient −4, intercept (0, 7)
d gradient $-\frac{1}{3}$, intercept (0, 0)

3 **a** gradient −1, intercept (0, −3)
b gradient 2, intercept (0, −5)
c gradient −3, intercept (0, −7)
d gradient −2, intercept $\left(0, -\frac{3}{2}\right)$
e gradient $-\frac{1}{5}$, intercept $\left(0, \frac{1}{5}\right)$
f gradient $\frac{2}{3}$, intercept $\left(0, -\frac{8}{3}\right)$

4 gradient $\frac{2}{3}$; A (−9, 0); B (0, 6)

5 $x + 3y - 14 = 0$

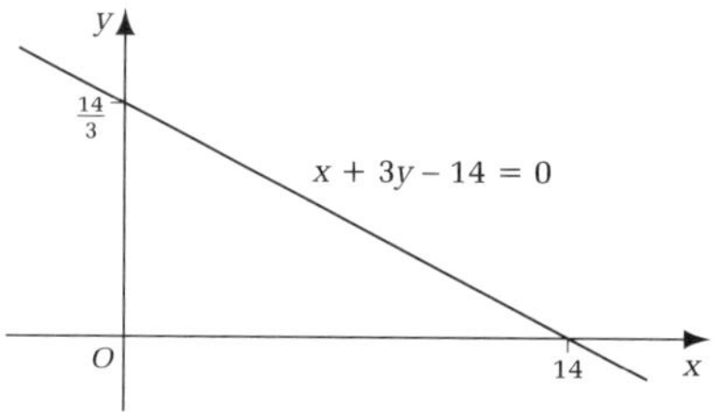

6 (1, 3)

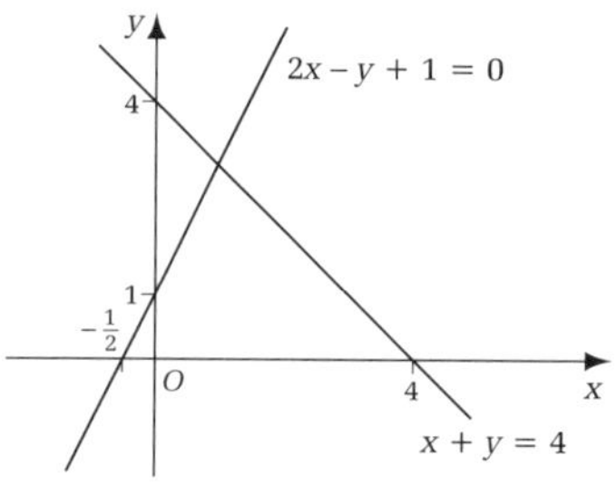

2.3 Technique

1 **a** $y = 2x + 1$ **d** $y = -2x + 6$
b $y = 4x - 11$ **e** $y = \frac{1}{3}x - 3$
c $y = x + 5$ **f** $y = -\frac{1}{4}x + \frac{15}{4}$

2 a $y = 2x - 3$
b $y = 5x + 1$
c $y = -2x - 1$
d $y = -3x - 1$
e $y = \frac{1}{2}x + 3$
f $y = -\frac{1}{3}x - 2$

3 a $y = -\frac{1}{4}x + \frac{19}{4}$
b $y = 2x + \frac{5}{2}$
c $y = x - 5$
d $y = -x - 7$

2.3 Contextual

1 $p = 3$; $y = 2x + 1$

2 a (2, 8)
b $\frac{2}{3}$
c $-\frac{3}{2}$
d $3x + 2y - 22 = 0$

3 M (3, 1); $y = -2x + 7$

4 a $x - y + 2 = 0$
b gradient AB $= -1$; gradient $x - y - 4 = 0$ is 1; hence perpendicular
c $3\sqrt{2}$

2.4 Technique

1 a $x > 2$
b $x \le -1$
c $x \ge 1$
d $x > 9$

2 a $x < -5, x > 5$
b $-10 \le x \le 10$
c $x < -11, x > 11$
d $-15 < x < 15$

3 a $x < -2, x > 5$
b $x \le -3, x \ge 4$
c $-3 < x < 2$
d $-5 < x < 7$
e $-3 \le x \le 8$
f $x < 2, x > 4$

4 a $x \le -4, x \ge 5$
b $x < -12, x > 9$
c $x < -4, x > 7$
d all values of x except $x = -1$
e all values of x
f $-3 \le x \le 7$

5 a $x < -\frac{7}{2}, x > 3$
b $\frac{2}{3} \le x \le \frac{3}{2}$
c $-\frac{1}{2} < x < 2$
d $x < -5, x > \frac{3}{2}$
e $x \le -\frac{1}{3}, x \ge \frac{1}{2}$
f $x < -\frac{5}{4}, x > 2$

2.4 Contextual

1 $x \ge 1 + \sqrt{6}, x \le 1 - \sqrt{6}$

2 $x^2 - 5x + 4 < 0$

3 $-1 < x < 3$

4 $x = -5\sqrt{3}$ or $2\sqrt{3}$

$x \le -5\sqrt{3}, x \ge 2\sqrt{3}$

2.5 Technique

1 a centre (3, 2), radius 10
b centre (−5, −2), radius 9
c centre (7, 0), radius 1
d centre (−3, 6), radius $\sqrt{10}$

2 a $x^2 + y^2 + 4x - 8y + 11 = 0$
b $x^2 + y^2 - 6x + 8 = 0$
c $x^2 + y^2 - 10x + 2y + 24 = 0$
d $x^2 + y^2 - 4x + 4y - 10 = 0$

3 a centre (7, 1), radius 2
b centre (0, 3), radius $\sqrt{3}$
c centre (4, −2), radius $\sqrt{2}$
d centre (−3, −2), radius $\sqrt{5}$

2.5 Contextual

1 $k = -2$ or $k = 8$

2 $k = 11$ or $k = -1$

3 (0, 5) or (0, −9)

4 tangent $y = \frac{1}{2}x + \frac{9}{2}$; normal $y = -2x + 7$

5 $y = 4x - 13$; perpendicular $y = -\frac{1}{4}x + \frac{3}{2}$

6 $x^2 + y^2 - 4x + 2y - 3 = 0$

7 $2\sqrt{15}$

Consolidation C

1 C (3, 1); gradient AB $= -1$; $y = x - 2$

2 $x - 2y - 8 = 0$; $(\frac{14}{5}, -\frac{13}{5})$; $3\sqrt{5}$

3 $\sqrt{15}$

4 $-5 \le x \le 2$

5 a $k = -14$
b $y = -\frac{1}{2}x$
c $\sqrt{405} = 9\sqrt{5}$

6 a centre (−3, 4), radius = 5
b $k = -3 + \sqrt{21}$ or $k = -3 - \sqrt{21}$

7 a $y = 5x - 14$
b $5y + x - 34 = 0$
c (4, 6)
d 13
e $(x - 4)^2 + (y - 6)^2 = 13^2$

Review

1 a 15
b 15
c 12
d 4.24

2 a $(p - 2q)(p + 2q)$
b $(r + 1)(r + 4)$
c $(y + 6)(y - 3)$
d $(m - 5)(m - 6)$
e $k = 7, -4$
f $p = \frac{3}{2}, 4$

3 **a** $\frac{3}{5}, \frac{4}{5}, \frac{3}{4}$ **b** $\frac{5}{13}, \frac{12}{13}, \frac{5}{12}$ **c** $\frac{60}{61}, \frac{11}{61}, \frac{60}{11}$

4 **a** 640 cm^2 **b** 27.5 m^2
c 55.4 m^2 **d** 660 cm^2

5 **a** A, 45°; a, 7.07 cm; c, 7.07 cm
b A, 53.1°; B, 36.9°; c, 17.5 m
c C, 40°; a, 10.7 cm; b, 14 cm
d A, 67.4°; B, 67.4°; C, 45.2°

6 **a** 048° **b** 128°
c 231° **d** 342°

3.1 Technique

1 **a** −cos 51° **d** −sin 52° **g** sin 55°
b tan 35° **e** tan 28° **h** −tan 60°
c cos 50° **f** −cos 55°

2 **a** $-\frac{1}{\sqrt{3}}$ **d** $\frac{\sqrt{3}}{2}$ **g** $\frac{\sqrt{3}}{2}$
b $-\frac{1}{\sqrt{2}}$ **e** $-\frac{1}{2}$ **h** $-\sqrt{3}$
c $-\frac{\sqrt{3}}{2}$ **f** $\frac{1}{2}$

3 **a** 38.2°, 141.8° **b** 117.4°, 242.6° **c** 60.7°, 240.7°
d 60.3°, 240.3° **e** 63.4°, 243.4° **f** 65.6°, 245.6°

4 **a** 1.167 **d** −1.095 **g** −3.420
b 3.236 **e** −1.143 **h** 6.392
c −0.4452 **f** −0.017 46

5 **a** −44.4° **b** −81.9° **c** 78.5°
d 56.3° **e** −76.0° **f** 23.6°

6 **a** −11.5°, 191.5°, 348.5°
b −25.8°, 25.8°, 334.2°, 385.8°
c 77.5°, 257.5°, 437.5°

7 **a** ±72.5°
b no solutions in this range

3.1 Contextual

1 **a**

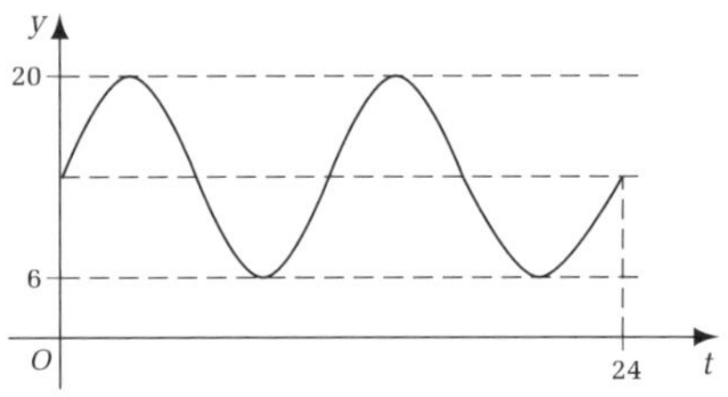

b high tide 0300, 1500
low tide 0900, 2100
c from 0131 to 0429, i.e. 2 h 58 min
and 1331 to 1629, i.e. 2 h 58 min

2 $a = 13, b = 15, c = 15$

3 **a** 7 m
b 13 m

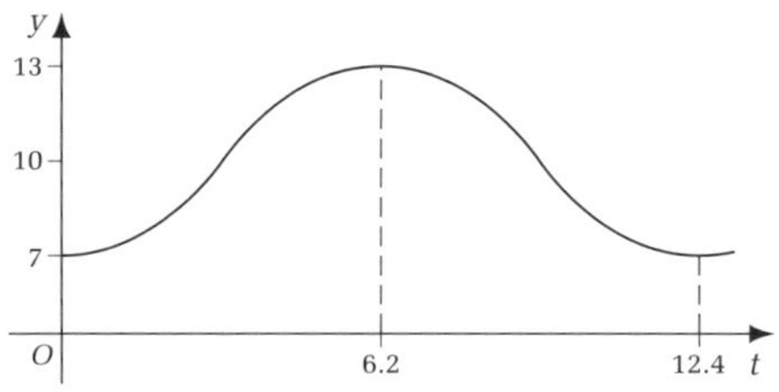

$k = 29.03$

3.2 Technique

1 **a** 68.2°, 248.2°
b 172.9°, 352.9°
c 11.3°, 191.3°
d 54.7°, 125.3°, 234.7°, 305.3°
e 30°, 150°, 210°, 330°
f 45°, 135°, 225°, 315°

2 **a** −108.4°, −101.3°; 71.6°, 78.7°
b −150°, −30°, 11.5°, 168.5°
c ±75.5°, ±104.5°
d ±45°, ±135°
e 41.8°, 138.2°
f −71.6, 0°, 108.4°±180°

3 **a** 9.7°, 80.3°, 189.7°, 260.3°
b 80.8°, 260.8°
c 40.7°, 179.3°, 220.7°, 359.3°
d 22.1°, 82.1°, 142.1°, 202.1°, 262.1°, 322.1°
e 120°, 240°
f 13.3°, 103.3°, 193.3°, 283.3°

4 **a** 19.5°, 90°, 160.5°
b ±60°
c −153.4°, −59.0°, 26.6°, 121.0°
d −160.5°, −19.5°, 30°, 150°
e ±78.7°, ±101.3°

3.2 Contextual

1 42.3° or 49.2°

2 The golfer should understand that it is possible for the ball to pass through a given point (X, Y) in more than one way. One club would pitch the ball high and another would keep it low. The stroke used (and so the speed of the ball) would be the same.

3.3 Technique

1 **a** $\frac{\sqrt{3}+1}{\sqrt{3}-1} = 2+\sqrt{3}$ **d** $\frac{1-\sqrt{3}}{1+\sqrt{3}} = \sqrt{3}-2$
b $\frac{\sqrt{2}}{4}(\sqrt{3}-1)$ **e** $\frac{\sqrt{2}}{4}(1-\sqrt{3})$
c $-\frac{\sqrt{2}}{4}(\sqrt{3}+1)$ **f** $\frac{\sqrt{3}-1}{\sqrt{3}+1} = 2-\sqrt{3}$

2 **a** $\cos 90° = 0$ **d** $\sin 30° = \frac{1}{2}$
b $\sin 45° = \frac{\sqrt{2}}{2}$ **e** $\cos 60° = \frac{1}{2}$
c $\tan 45° = 1$ **f** $\tan 135° = -1$

3 **a** 66.2°, 246.2° **b** 106.8°, 286.8°
c 90°, 270° **d** 97.4°, 277.4°

5 **a** $\frac{207}{305}$ **e** $-\frac{207}{224}$ **i** $\frac{3479}{3721}$
b $-\frac{273}{305}$ **f** $\frac{273}{136}$ **j** $-\frac{7}{25}$
c $-\frac{224}{305}$ **g** $\frac{1320}{3721}$ **k** $\frac{1320}{3479}$
d $-\frac{136}{305}$ **h** $-\frac{24}{25}$ **l** $\frac{24}{7}$

3.3 Contextual

1 **a** **ii** $\theta_{max} = 50.4°$
iii At θ_{max}, B is 1.75 m above the ground, which is not a practical height at which to move the object. C should just then pass under the top of the door frame.
b New inequality becomes $\sin(\theta + 7.7°) \leq \frac{2}{3}$,
$\theta_{max} = 34.1°$. At θ_{max}, B is 1.67 m above the ground.

3.4 Technique

1 **a** 13.3 cm **b** 22.3 m **c** 5.99 cm

2 **a** A = 48.6°, B = 24.2°, C = 107.2°
b A = 34.6°, B = 14.6°, C = 130.8°
c A = 40.6°, B = 123.2°, C = 16.2°

3 **a** 8.8 cm **b** 6.78 cm
c 25.1 cm **d** 15.9 cm

4 **a** A = 86°, B = 38.9°, C = 55.1°
b A = 105.7°, C = 28.3°, a = 22.7 cm
c B = 8.7°, C = 18.3°, a = 42 cm
d B = 70.2°, C = 97.8°, c = 90.5 cm
or B = 109.8°, C = 58.2°, c = 77.7 cm

3.4 Contextual

1 AC = 13.7 nautical miles; bearing of 133.4°

2 27 miles; 285°

3 11.3 km; 222°

4 26.7 m (3 s.f.)

3.5 Technique

1 **a** 30 cm² **b** 30 cm² **c** 30 cm²

2 **a** 96 m² **b** 8.02 m²
c 0.21 m² **d** 48 cm²

3.5 Contextual

1 10 cm

2 0.685 km²

3 **a** yes **b** 272 cm² **c** 1120 cm³
In practice, **c** would be much less, since it has ignored the thickness of the icing.

3.6 Technique

1 **a** $\frac{\pi}{18}$ **b** $\frac{7\pi}{9}$ **c** $\frac{17\pi}{18}$
d $\frac{11\pi}{9}$ **e** $\frac{31\pi}{18}$ **f** $\frac{35\pi}{18}$

2 **a** 0.0873 **b** 0.995 **c** 1.62
d 2.81 **e** 4.00 **f** 5.25

3 **a** 20° **b** 75° **c** 70°
d 84° **e** 330° **f** 380°

4 **a** arc length 3.6 cm, sector area 16.2 cm²
b arc length 7.2 cm, sector area 21.6 cm²
c arc length 29.4 cm, sector area 166.0 cm²
d arc length 25.4 cm, sector area 78.8 cm²

5 $r\theta = 20$; $\frac{1}{2}r^2\theta = 40$, $r = 4$ cm, $\theta = 5$

6 16.3 cm (3 s.f.)

7 **1** **a** 1.19, 4.33
b 3.02, 6.16
c 0.197, 3.34
d 0.955, 2.19, 4.10, 5.33
e 0.526, 2.62, 3.67, 5.76
f 0.785, 2.36, 3.93, 5.50

2 **a** −1.89, −1.77, 1.25, 1.37
b −2.62, −0.524, 0.201, 2.94
c ±1.32, ±1.82
d ±0.785, ±2.36
e 0.730, 2.41
f −1.25, 0, 1.89, ±3.14

4 **a** 0.340, 1.57, 2.80
b ±1.05
c −2.68, −1.03, 0.464, 2.11
d −2.80, −0.340, 1.05, 2.09
e ±1.37, ±1.77

3.6 Contextual

1 36 cm³

2 **a** $\frac{5}{3}$ rad **b** 750 cm³

3 **a** AOB = $\frac{2\pi}{3}$ rad = 2.09 (3 s.f.)
b 170 000 cm³

4 2520 cm²

Consolidation C

1 150°, 330°

2 **a** 10°, 50°, 130°, 170° **b** 38.7°

3 **a** January, 4.18 h (i.e. 4:11 p.m.)
May, 8.07 h (i.e. 8:04 p.m.)
b February and October
c $L = 6.125 + 2.25\sin\left(\frac{\pi t}{6}\right)$

4 **a** 35.5 cm^2 **b** 18.8 cm^2

5 **b** $\cos\alpha = \frac{3}{5}$, $AC^2 = 260$ **d** 13.9°

6 **a** A = (56.8°, 0), B = (123.2°, 0)
b P = (53.1°, 0.6), Q = (120°, −0.5)

7 **b** $\dfrac{2\pi}{3\sqrt{3}}$

8 **b** 0.85, 2.29, 4.71 **c** 1.70, 4.59

9 **a** 120 cm^2 **b** 2.16 **c** 161.07 cm^2

10 **a** ±60 **b** $-22\frac{1}{2}$, $37\frac{1}{2}$

Review

1 **a** $x = \dfrac{3y - 1}{4}$ **d** $x = (y + 3)^4$
b $x = \sqrt{\dfrac{5 - y}{2}}$ **e** $x = 1 + \dfrac{3}{2y}$
c $x = (y + 1)^2 - 5$ **f** $x = \dfrac{1}{6 - y}$

2 **a** $(x - 3)^2 + 7$ **d** $\left(x + \frac{5}{2}\right)^2 + \frac{3}{4}$
b $(x + 4)^2 - 5$ **e** $2(x - 5)^2 - 14$
c $\left(x - \frac{1}{2}\right)^2 + \frac{7}{4}$ **f** $3(x + 2)^2 - 1$

3 **a** $x = -5, x = 7$ **d** $x = 2 \pm \sqrt{5}$
b $x = \pm\dfrac{\sqrt{7}}{2}$ **e** $x = -4 \pm \sqrt{11}$
c $x = -4, x = -\frac{1}{2}$ **f** $x = \dfrac{-5 \pm \sqrt{2}}{3}$

4 **a** $1 - \dfrac{1}{x - 3}$ **d** $4 + \dfrac{13}{x - 5}$
b $1 + \dfrac{1}{x - 4}$ **e** $2 + \dfrac{3}{3x + 2}$
c $3 - \dfrac{4}{x + 2}$ **f** $5 - \dfrac{10}{x + 2}$

4.1 Technique

1 **a i**

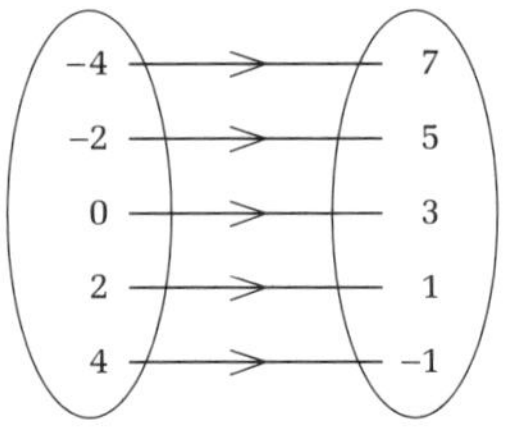

ii range is $\{-1, 1, 3, 5, 7\}$
iii one-to-one function

b i

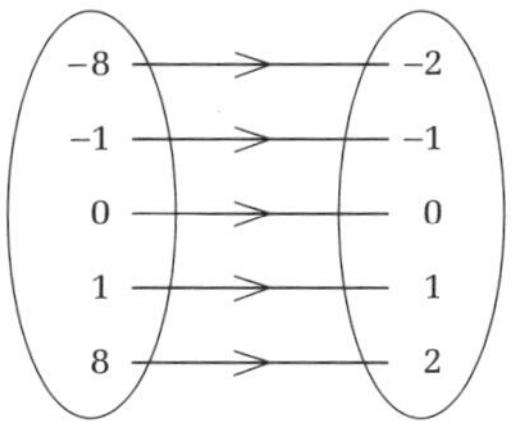

ii range is $\{-2, -1, 0, 1, 2\}$
iii one-to-one function

c i

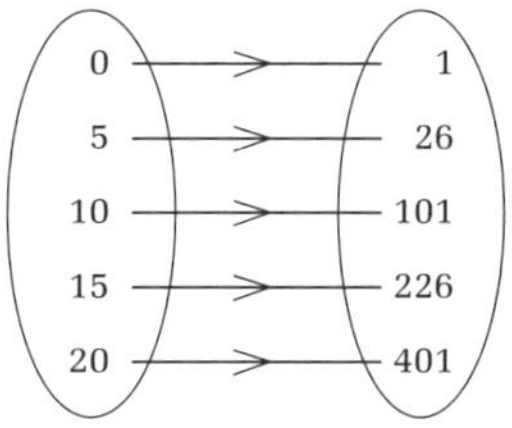

ii range is $\{1, 26, 101, 226, 401\}$
iii one-to-one function

d i

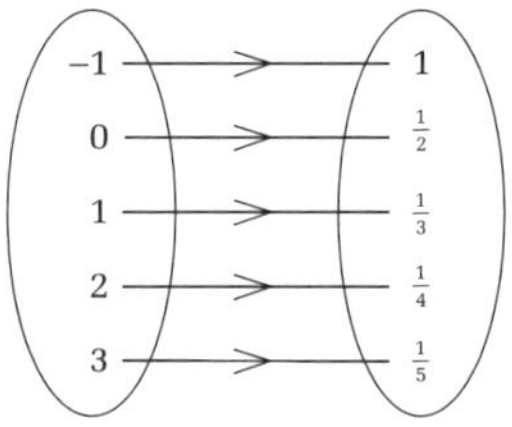

ii range is $\left\{\frac{1}{5}, \frac{1}{4}, \frac{1}{3}, \frac{1}{2}, 1\right\}$
iii one-to-one function

2 **a i**

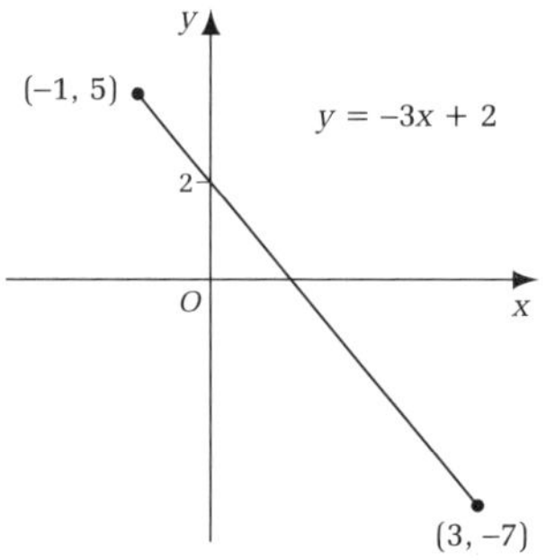

ii range is $\{y \in \mathbb{R}: -7 \leq y \leq 5\}$
iii one-to-one function

b i

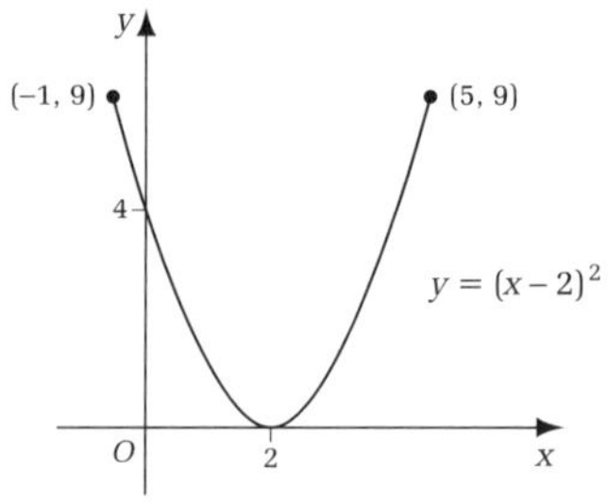

ii range is $\{y \in \mathbb{R}: 0 \leq y \leq 9\}$
iii many-to-one function

c i

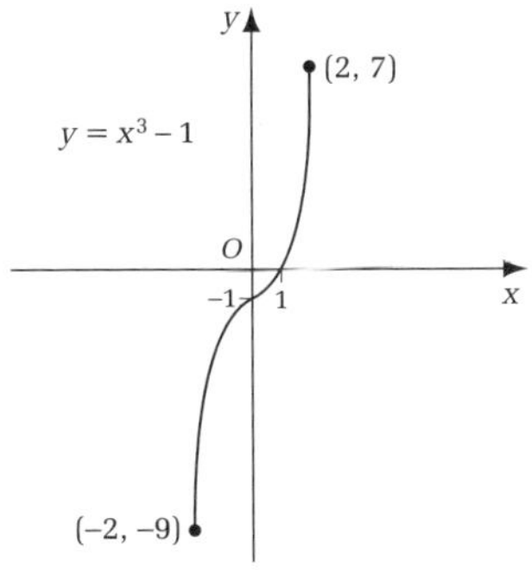

ii range is $\{y \in \mathbb{R}: -9 \le y \le 7\}$
iii one-to-one function

d i

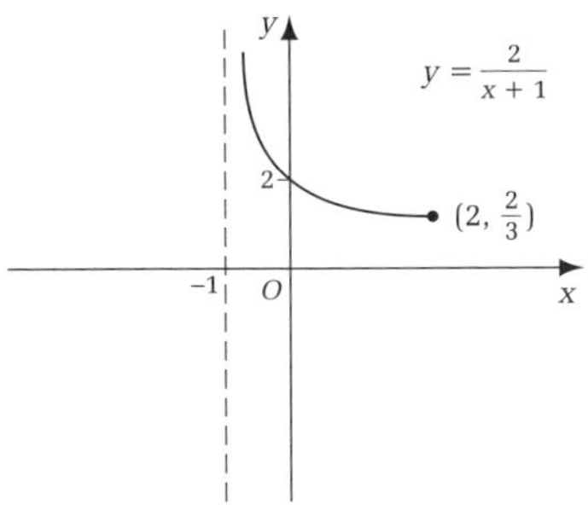

ii range is $\left\{y \in \mathbb{R}: y \ge \frac{2}{3}\right\}$
iii one-to-one function

3 **a** graph does not represent a function; relationship is one-to-many
b graph does represent a function; relationship is one-to-one
c graph does not represent a function; relationship is one-to-many
d graph does represent a function; relationship is many-to-one

4 f is a function; one-to-one mapping throughout the domain
g is not a function because $x^3 + 1 = 2$ and $\dfrac{2}{x+1} = 1$ when $x = 1$, so g is not uniquely defined for all values of x in its domain

5 **a** $a = 1$ or $a = 2$
b $b = -1$ or $b = 11$
c $c = -\frac{2}{3}$ or $c = 6$

6 $a = \frac{1}{2}, b = -5, c = 28$

7 $a = -\frac{3}{4}, b = \frac{3}{2}, c = 5, d = -2$ or 4

4.2 Technique

1 **i** inverse functions exist for **b** and **c**; both are one-to-one
no inverse functions exist for **a** and **d**; both are many-to-one

ii b

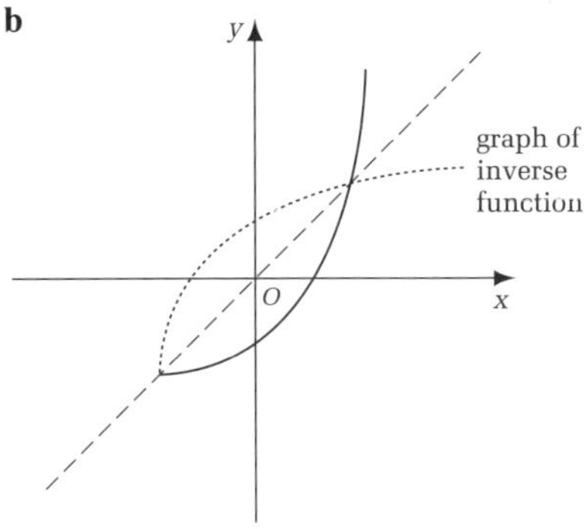

c

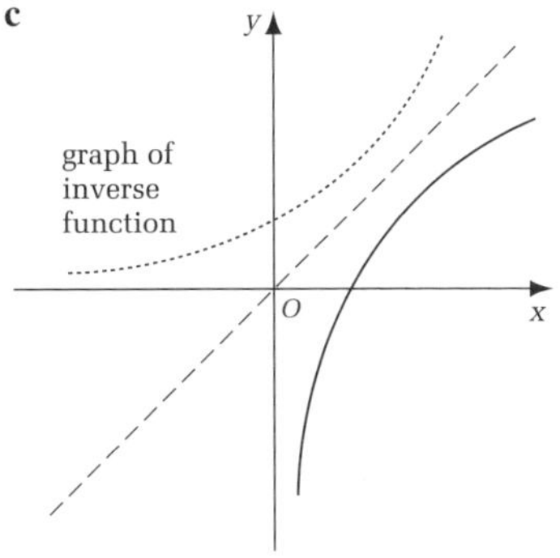

2 **a** $f^{-1}(x) = \dfrac{x-11}{5}$
b $f^{-1}(x) = \frac{1}{3}\left(\dfrac{10}{x} - 1\right)$
c $f^{-1}(x) = \sqrt[3]{\dfrac{x-3}{4}}$
d $f^{-1}(x) = x^3 - 2$
e $f^{-1}(x) = \dfrac{9}{x^2} + 4$
f $f^{-1}(x) = 1 + \sqrt{\dfrac{5}{x}}$

3 **a** $f(x)$ is undefined for $x = 3$; $g(x)$ is undefined for $x = \frac{1}{2}$; $h(x)$ is undefined for $x = -\frac{2}{3}$
b $f^{-1}(x) = 3 - \dfrac{5}{x}$; domain is $\{x \in \mathbb{R}: x \ne 0\}$
$g^{-1}(x) = \dfrac{x}{2x-3}$; domain is $\left\{x \in \mathbb{R}: x \ne \frac{3}{2}\right\}$
$h^{-1}(x) = \dfrac{4+2x}{1-3x}$; domain is $\left\{x \in \mathbb{R}: x \ne \frac{1}{3}\right\}$

4 **a** $f^{-1}(x) = \dfrac{-ax-15}{4x-1}$, $g^{-1}(x) = \dfrac{-10x}{x-b}$
b $a = -1, b = -10$

5 **a** $(x+3)^2 - 5$
b $q = -3$
c range is $f(x) \ge -5$
d $f^{-1}(x) = \sqrt{x+5} - 3$
e $x = -1$

6 **a** $f^{-1}(x) = 2\sin^{-1}\left(\dfrac{x}{4}\right)$; domain is $-4 \le x \le 4$
b $f^{-1}(2\sqrt{3}) = \dfrac{2\pi}{3}$

7 **a** $0 \le x \le 2\pi$
b i $f^{-1}(x) = 2\cos^{-1}x$
ii domain of f^{-1} is $-1 \le x \le 1$

iii

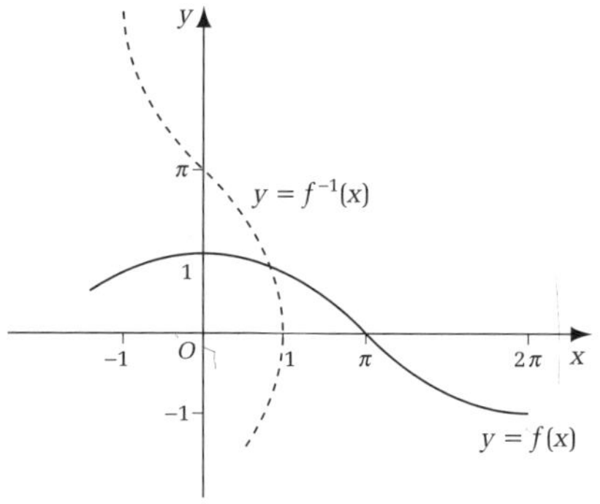

c $0 \leq x \leq \dfrac{\pi}{a}$

4.3 Technique

1 **a** $fg(x) = 2\sqrt{x} - 3$ **f** $hg(x) = x + 1$
b $gf(x) = \sqrt{2x - 3}$ **g** $f^2(x) = 4x - 9$
c $fh(x) = 2x^2 - 1$ **h** $g^2(x) = \sqrt[4]{x}$
d $hf(x) = 4x^2 - 12x + 10$ **i** $h^2(x) = x^4 + 2x^2 + 2$
e $gh(x) = \sqrt{x^2 + 1}$ **j** $fgh(x) = 2\sqrt{x^2 + 1} - 3$

2 **a** $hg(x)$ **b** $gh(x)$ **c** $f^2(x)$
d $hf(x)$ **e** $fh(x)$ **f** $hfg(x)$

3 **a** $fgh(x) = \dfrac{4}{x + 1} - 2$ **b** $x = -2$ or $x = \frac{1}{3}$

4 **b** since $fg(x) = gf(x)$, functions f and g are the inverse of one another

5 **a** range of f is $0 < f(x) \leq 1$
range of g is $g(x) \geq -3$

b $fg(x) = \dfrac{1}{2x - 2}$; domain of fg is $x \geq \frac{3}{2}$

c $gf(x) = \dfrac{2}{x + 1} - 3$; domain of gf is $x \geq 0$

6 **a** $fg(x) = 2 - \dfrac{1}{(2x + 4)^2}$; undefined for $x = -2$

b $gf(x) = \dfrac{1}{8 - 2x^2}$; undefined for $x = -2$ and $x = 2$

c $f^2(x) = -2 + 4x^2 - x^4$; defined for all values of x

d $g^2(x) = \dfrac{x + 2}{4x + 9}$; undefined for $x = -2$ and $x = -\dfrac{9}{4}$

7 **a** $f^{-1}(x) = \dfrac{x - 1}{3}, g^{-1}(x) = 2 - x, h^{-1}(x) = \dfrac{1}{\sqrt[3]{x}}$

b $fg(x) = 7 - 3x, (fg)^{-1} = g^{-1}f^{-1} = \dfrac{7 - x}{3}$

c $hg(x) = \dfrac{1}{(2 - x)^3}, (hg)^{-1} = g^{-1}h^{-1} = 2 - \dfrac{1}{\sqrt[3]{x}}$

d $fh(x) = \dfrac{3}{x^3} + 1, (fh)^{-1} = h^{-1}f^{-1} = \sqrt[3]{\dfrac{3}{x - 1}}$

4.4 Technique

1 **a i** translation of −3 units parallel to the y-axis
ii $y = -2x - x^2$
iii

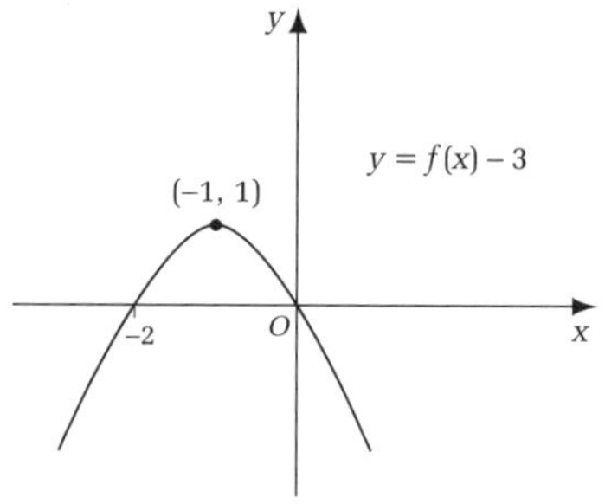

b i translation of +1 unit parallel to x-axis
ii $y = 4 - x^2$
iii

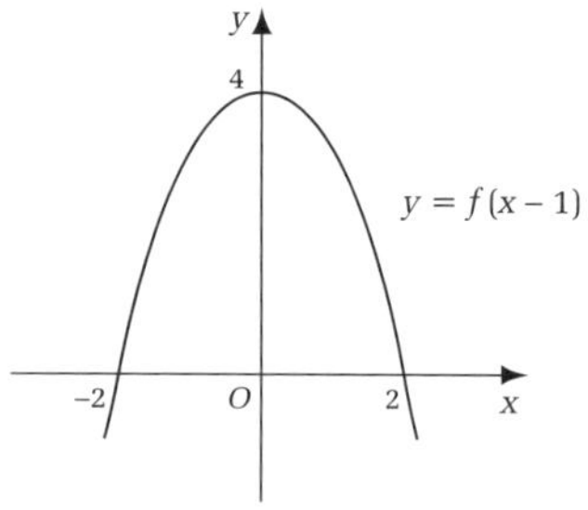

c i reflection in the x-axis and stretch of factor $\frac{1}{2}$ parallel to the y-axis
ii $y = \dfrac{x^2}{2} + x - \frac{3}{2}$
iii

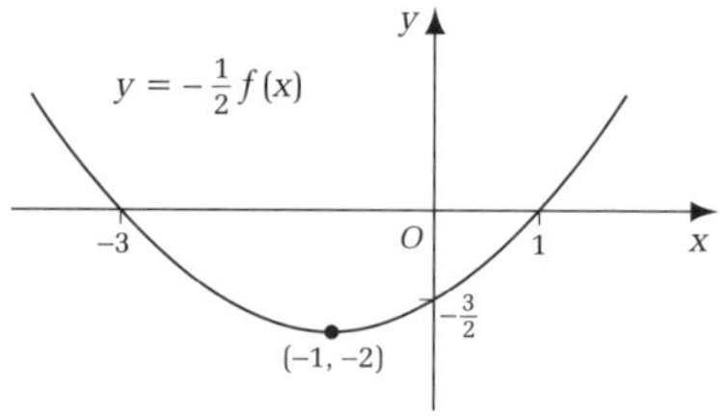

d i translation of +2 units parallel to the x-axis and −3 units parallel to the y-axis
ii $y = 2x - x^2$
iii

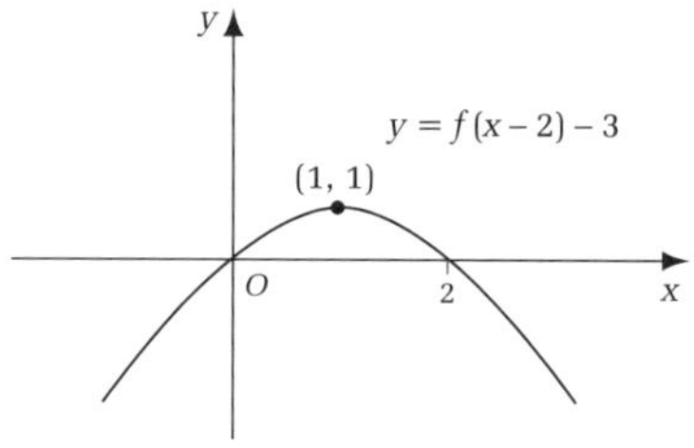

e i reflection in the y-axis
ii $y = 3 + 2x - x^2$
iii

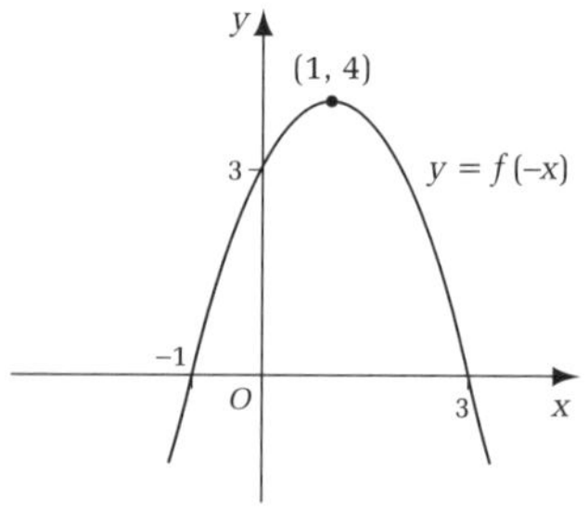

f i stretch of factor 2 parallel to the y-axis and translation of +3 units parallel to the x-axis
ii $y = 8x - 2x^2$

iii

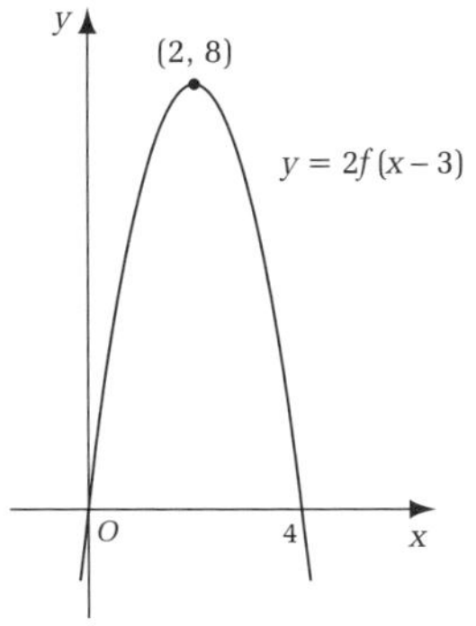

2 **a** $f(x) = (x - 1)^2 + 2$

b translations of +1 unit parallel to the x-axis and +2 units parallel to the y-axis

3 **a** $g(x) = 3(x + 1)^2 - 2$

b translation of −1 unit parallel to the x-axis and a stretch of factor 3 parallel to the y-axis, followed by a translation of −2 units parallel to the y-axis

4 **a** translation of −2 units parallel to the y-axis

b stretch of factor $\frac{1}{2}$ parallel to the y-axis

c translation of +7 units parallel to the x-axis

d reflection in the x-axis followed by a translation of +1 unit parallel to the y-axis; or a translation of −1 unit parallel to the y-axis followed by a reflection in the x-axis

e stretch of factor $\frac{1}{3}$ parallel to the x-axis and a translation of $+\frac{1}{3}$ unit parallel to the x-axis

f translations of −5 units parallel to the x-axis and −3 units parallel to the y-axis

5 **a i** translations of +2 units parallel to the x-axis and −1 unit parallel to the y-axis

ii $y = x - 3 + \dfrac{1}{x - 2}$

b i reflection in the y-axis followed by a translation of −1 unit parallel to the x-axis; or a translation of +1 unit parallel to the x-axis followed by a reflection in the y-axis

ii $y = -x - 1 - \dfrac{1}{x + 1}$

c i stretch of factor 2 parallel to the x-axis

ii $y = \dfrac{x}{2} + \dfrac{2}{x}$

6 **a i** translation by −180° parallel to the x-axis

ii

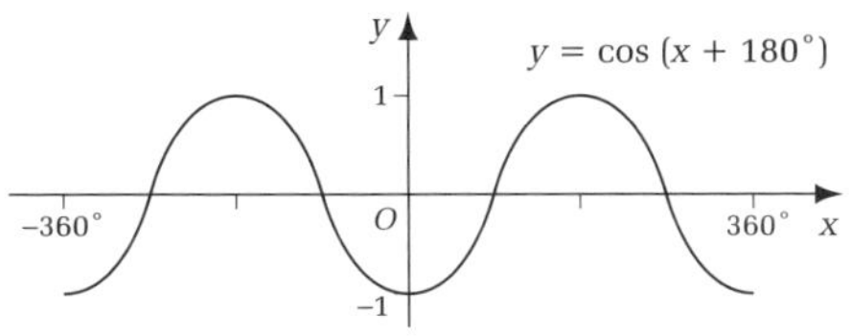

b i reflection in the y-axis

ii

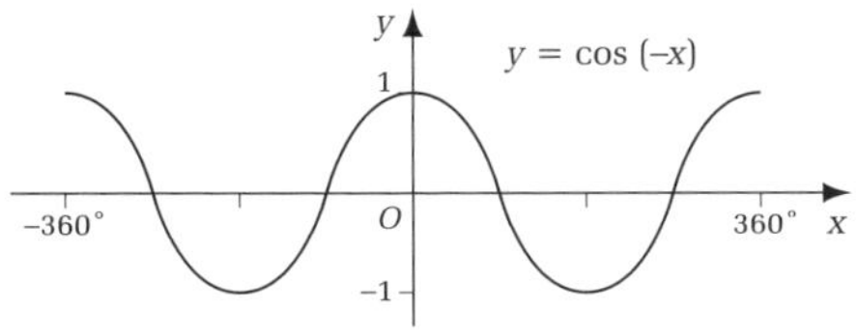

c i stretch of factor 3 parallel to the y-axis

ii

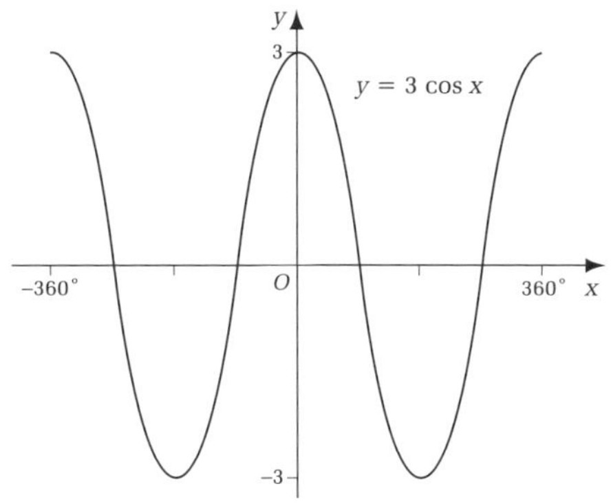

d i reflection in the x-axis followed by a translation of +2 units parallel to the y-axis; or a translation of −2 units parallel to the y-axis followed by reflection in the x-axis

ii

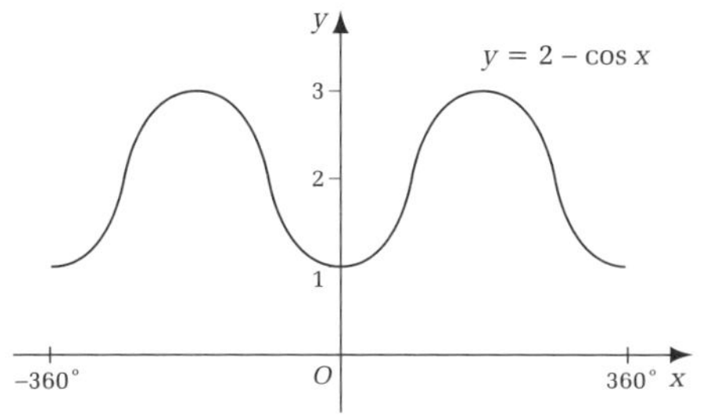

7 **a i** $g: x \to \dfrac{2x}{2x + 1}$

ii undefined for $x = -\frac{1}{2}$

b i $g: x \to \dfrac{x + 5}{x + 6}$

ii undefined for $x = -6$

c i $g: x \to \dfrac{-2x}{x + 1}$

ii undefined for $x = -1$

d i $g: x \to \dfrac{3x + 2}{x + 1}$

ii undefined for $x = -1$

4.5 Technique

1 **a** odd **b** even **c** neither

d neither **e** neither **f** odd

2 **a** neither **d** neither **g** odd

b even **e** neither **h** odd

c even **f** even

3 **a**

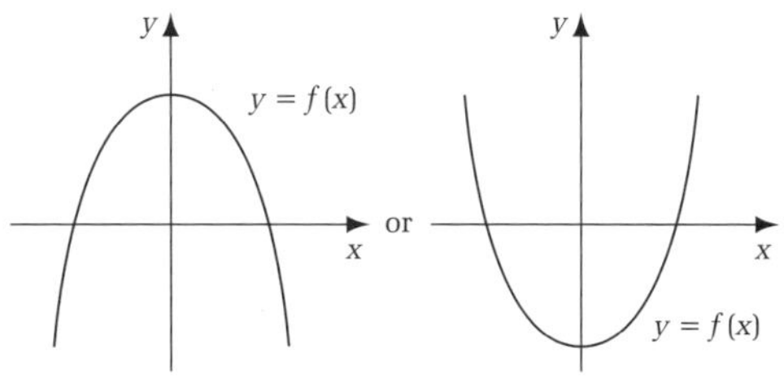

b Since $f(x)$ is a quadratic function, if the x^2 coefficient is negative, $f(x)$ is represented by the left graph. If the x^2 coefficient is positive, $f(x)$ is represented by the right graph.

4 **a**

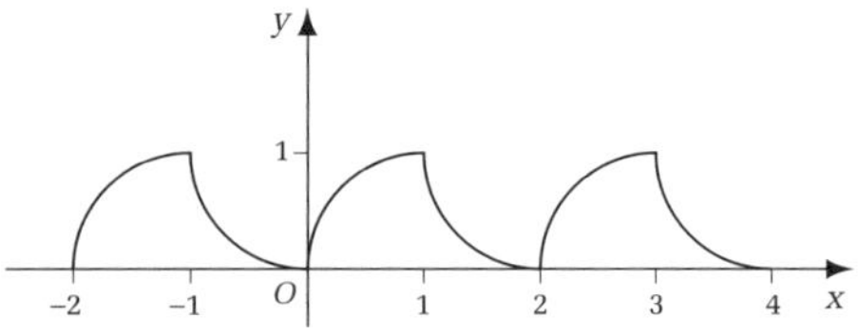

b $f(-1.875) = 0.5$; $f(5.5) = 0.25$

Consolidation C

1 **a**

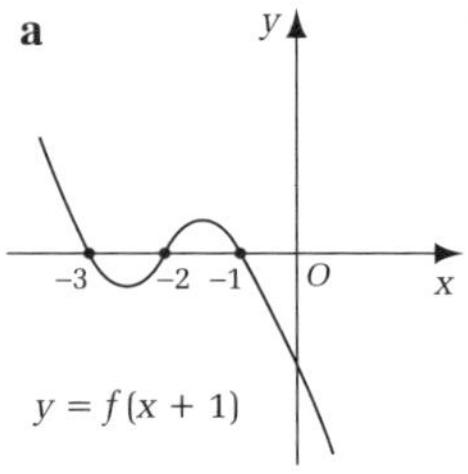

b

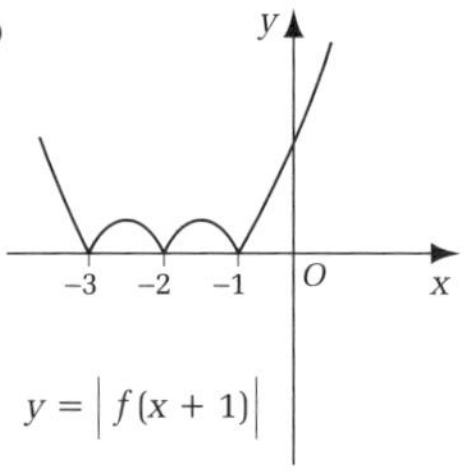

2 **a** $fg: x \to (2x + 1)^2 + 3 = 4x^2 + 4x + 4$

b range of fg is $fg(x) \geq 3$

c solution of $f(x) = 12g^{-1}(x)$ is $x = 3$

3 **a**

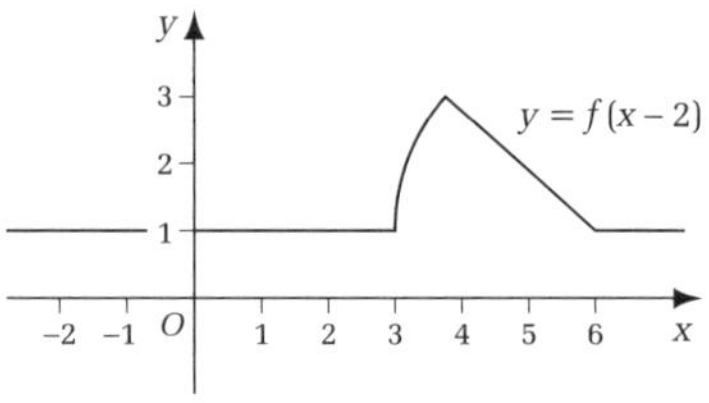

b

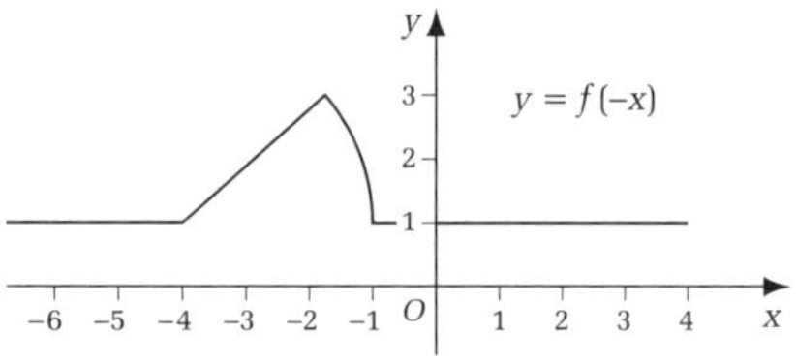

c

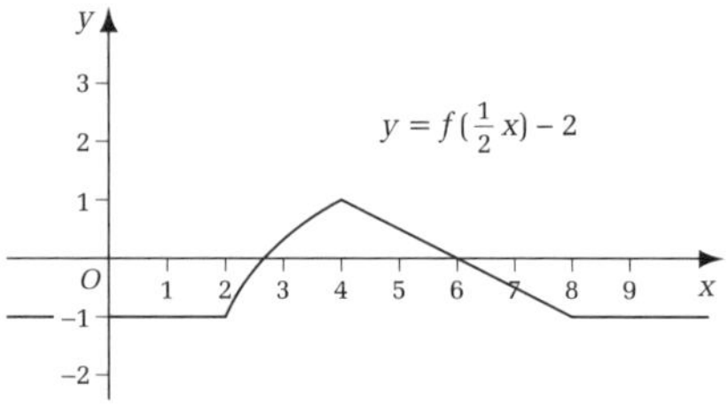

4 **a** range of f is $f(x) \geq -1$

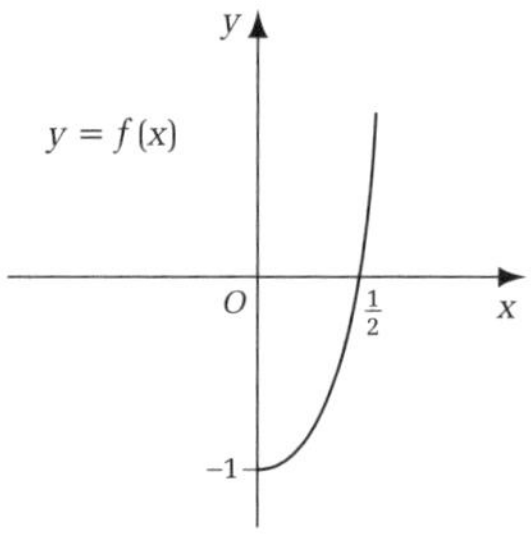

b inverse function f^{-1} exists because function f is one-to-one for the given domain

$$f^{-1}(x) = \sqrt{\frac{x+1}{4}}$$

c $fg(x) \geq f(x)$ for $0 \leq x \leq 3$

5 **a** $fg(x) = (2 - 3x)^3$ **b** $(fg)^{-1}(x) = \dfrac{2 - \sqrt[3]{x}}{3}$

6 **a** $f^{-1}(x) = x - 1$; function g does not have an inverse because, for the given domain, it is not a one-to-one function

b

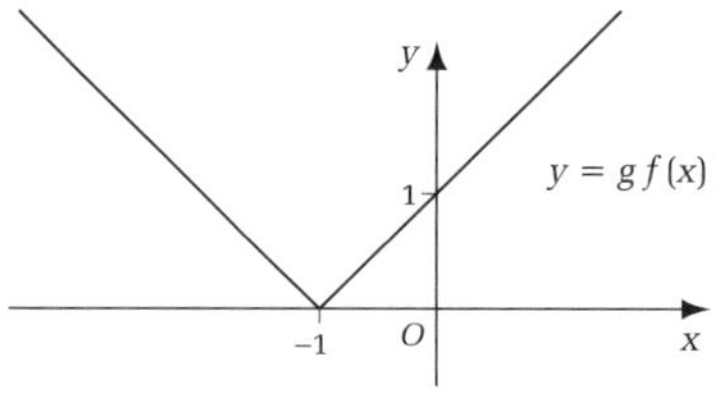

c solution set of the equation $gf(x) = fg(x)$ is $x \geq 0$

7 **a i** $f^{-1}(x) = \dfrac{2}{x-1}$

ii domain of f^{-1} is $x > 1$

b i $g(x) = \dfrac{3x+2}{x+2} = 3 - \dfrac{4}{x+2}$

ii range of g is $g(x) \in \mathbb{R} : \frac{5}{3} < g(x) < 3$

8 **a** range of g is $g(x) \geq 0$

b solutions to $gf(x) = 48$ are $x = 0$ and $x = 8$

c

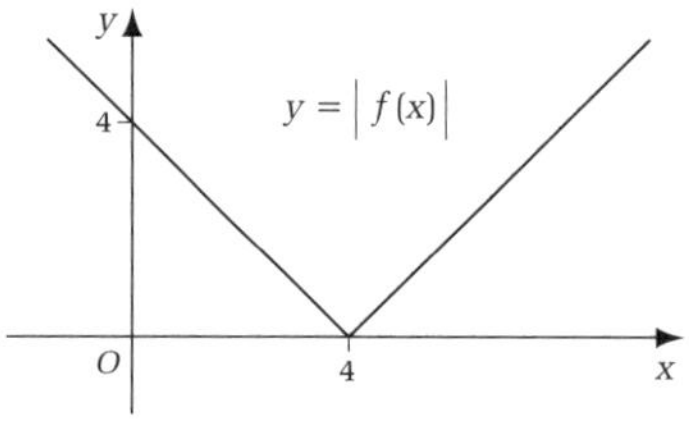

$|f(x)| = 2$ when $x = 2$ and $x = 6$

9 **a**

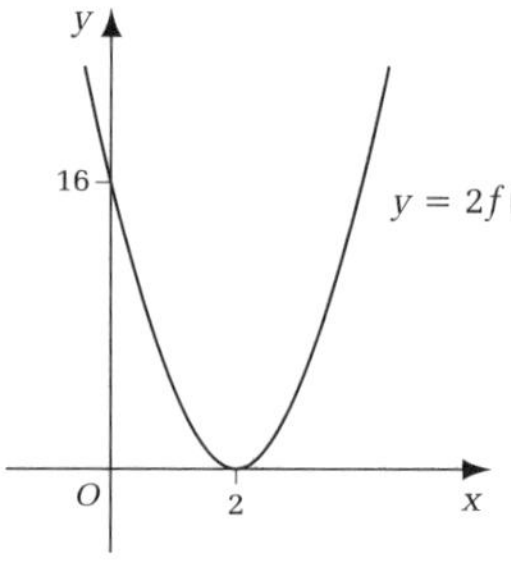

b

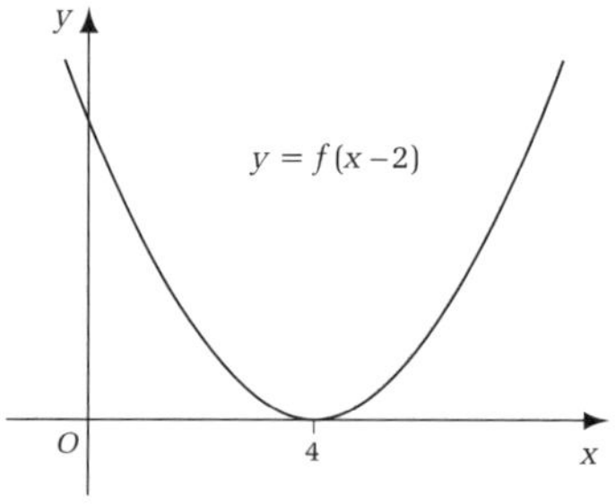

c

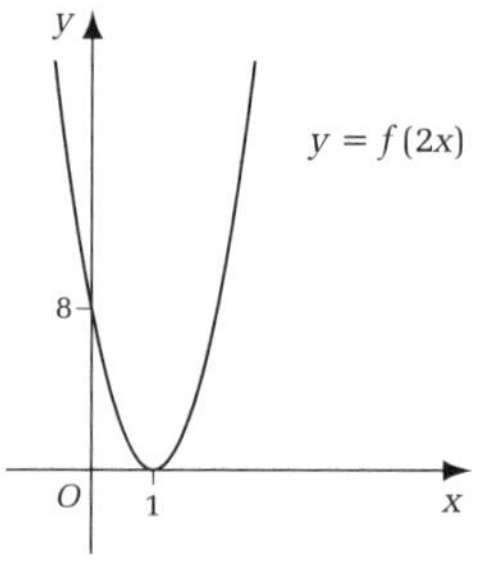

10 **a** range of f is $f(x) \geq -1$
b $f^{-1}(x) = \sqrt{x+1}$; domain is $x \in \mathbb{R}, x \geq -1$
c function g does not have an inverse because, for the given domain, it is not a one-to-one function
d domain of g for which g has an inverse is $x \in \mathbb{R}, x \geq 1$

Review

1 **a** 4 **b** -5
c $\frac{7}{3}$ **d** $-\frac{4}{3}$

2 **a** 15 **b** $\frac{13}{6}$ **c** $\sqrt{3}$

3 **a** $y = 4x - 5$ **b** $y = \frac{1}{3}x + 2$
c $y = -5x + 19$ **d** $y = -\frac{4}{3}x - \frac{5}{3}$

4 **a** $-\pi, 0, \pi$ **b** $-2.159, 0.983$ (3 d.p.)
c $-\frac{2\pi}{3}, -\frac{\pi}{2}, -\frac{\pi}{3}, \frac{\pi}{2}$ **d** $-\frac{2\pi}{3}, \frac{2\pi}{3}$

5 **a** $4x^{-2}$ **b** $x^{\frac{2}{3}}$ **c** $2(x-5)^{-2}$
d $9x^{-\frac{1}{3}}$ **e** $x^{\frac{3}{4}}$

5.1 Technique

1 **a** $\frac{dy}{dx} = 4x^3$ **b** $\frac{dy}{dx} = 4x + 7$
c $\frac{dy}{dx} = -\frac{2}{x^3}$

2 **a** $\frac{dy}{dx} = 9x^8$ **e** $\frac{dy}{dx} = 4 + \frac{2}{x^3}$
b $\frac{dy}{dx} = x^3 + x^2$ **f** $\frac{dy}{dx} = \frac{15}{x^4} - \frac{3}{x^2}$
c $\frac{dy}{dx} = 8x - 17$ **g** $\frac{dy}{dx} = -\frac{2}{x^{\frac{4}{3}}}$
d $\frac{dy}{dx} = \frac{3\sqrt{x}}{2}$ **h** $\frac{dy}{dx} = \frac{2 - 5x^2}{2\sqrt{x}}$

3 **a** $f'(x) = \frac{2}{5}x + \frac{3}{4}x^2$ **d** $h'(p) = -\frac{3}{p^2} - 4$
b $f'(t) = 1 - 2t$ **e** $\theta'(t) = 9\sqrt{t}$
c $f'(s) = 4s^3 + 6s^2 - 4$

4 **a** 1 **b** -20
c 0 **d** -6

5 **a** $(-2, 26)$ and $(4, -46)$ **b** $(-3, -18)$ and $(3, 24)$
c $(-1, 0)$ and $(1, 2)$ **d** $(-4, -25)$

6 **a** $\frac{dy}{dx} = -2$ at $(2, 0)$ **b** $\frac{dy}{dx} = -7$ at $(-\frac{1}{3}, 0)$
$\frac{dy}{dx} = 2$ at $(4, 0)$ $\frac{dy}{dx} = 7$ at $(2, 0)$
$\frac{dy}{dx} = -6$ at $(0, 8)$ $\frac{dy}{dx} = -5$ at $(0, -2)$

7 **a** $\frac{dy}{dx} = 5 - 18x + 9x^2 - 8x^3$
$\frac{d^2y}{dx^2} = -18 + 18x - 24x^2$
$\frac{d^3y}{dx^3} = 18 - 48x$
b $f'(x) = 2 + \frac{4}{x^2}$
$f''(x) = -\frac{8}{x^3}$
$f'''(x) = \frac{24}{x^4}$

5.2 Technique

1 **a i** $(3, 0)$, $(5, 0)$ and $(0, 15)$
ii $(4, -1)$; minimum point
iii

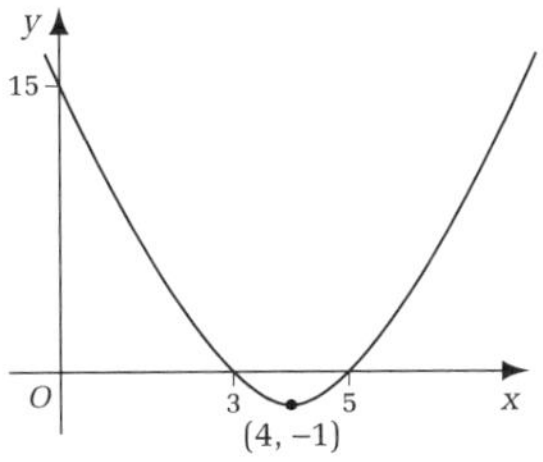

b i $(-1, 0)$, $(7, 0)$ and $(0, 7)$
ii $(3, 16)$; maximum point
iii

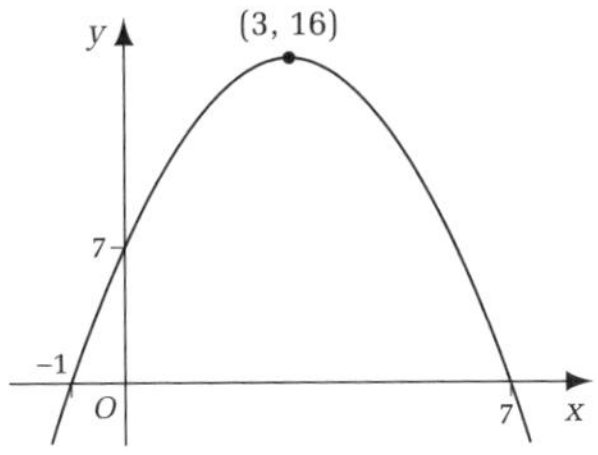

c i $(0, 5)$; graph does not cross x-axis
ii $(2, 1)$; minimum point
iii

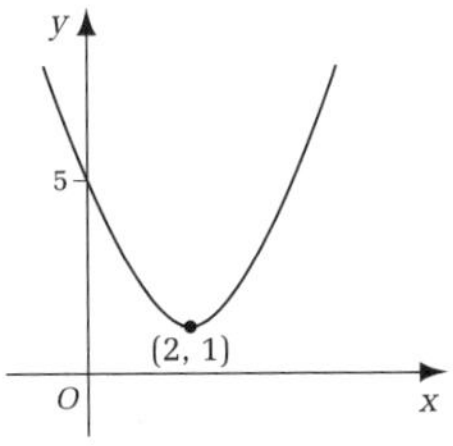

d i $(2, 0)$, $(5, 0)$ and $(0, 20)$
ii $\left(3\frac{1}{2}, -4\frac{1}{2}\right)$; minimum point
iii

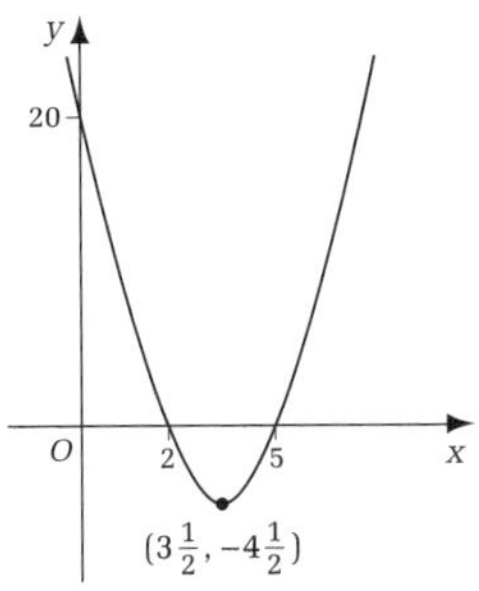

2 **a** point of inflexion at (3, 5)
b no stationary points
c point of inflexion at $\left(-\frac{1}{2}, -2\right)$
d maximum point at (−2, 48); minimum point at (4, −60)

3 **a** (−4, 0) and (0, 0)
b $4x^3 + 12x^2$; minimum point at (−3, −27), point of inflexion at (0, 0)
c

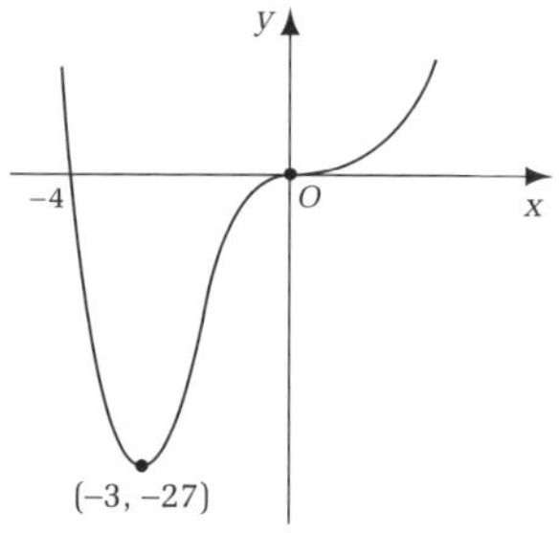

4 **a** $\frac{1}{2} - \frac{8}{x^3}$; (−4, −4) is a maximum point, (4, 4) is a minimum point
b $1 - \frac{64}{x^3}$; (4, 6) is a minimum point
c $2x - \frac{32}{x^3}$; (−2, 8) is a minimum point, (2, 8) is a minimum point
d $-\frac{2}{x^2} + \frac{6}{x^3}$; $\left(3, \frac{1}{3}\right)$ is a maximum point

5 $A = -4$ when $r = -\frac{1}{2}$

6 $W = 20\frac{1}{4}$ when $t = -\frac{1}{2}$

7 $x > 6\frac{1}{2}$

8 $-\frac{1}{3} < u < 4$

5.2 Contextual

1 **a** $V = 4y^3 - 610y^2 + 15\,600y$
b $\frac{dV}{dy} = 12y^2 - 1220y + 15\,600$; $y = 15$ cm, $y = 86\frac{2}{3}$ cm
c 110 250 cm^3; 210 cm × 35 cm × 15 cm

2 **a** $V = 36xy$ m^3; $y = \frac{375}{x}$
b $A = 45x + 4500 + \frac{18\,000}{x}$
c $x = 20$ m, $y = 18.75$ m
d 6300 m^2

3 **a** 380 pence; 428 pence
b 488 pence; 2 p.m.
c 64 pence fall

4 **c** $S = \frac{6000}{\sqrt{3}x} + \frac{3\sqrt{3}x^2}{2}$
d $x = 8.7$ cm, $h = 7.6$ cm, $S = 594.8$ cm^2

5 **a** $a = 350$, $b = -50$
c £4.50; £312.50; 125

5.3 Technique

1 **a** tangent, $y = 3x + 1$; normal, $x + 3y + 7 = 0$
b tangent, $y = 11x - 25$; normal, $x + 11y - 91 = 0$
c tangent, $-4x - 20$; normal, $x - 4y - 46 = 0$
d tangent, $y = \frac{5}{4}x - 2$; normal, $4x + 5y - 31 = 0$
e tangent, $x - 12y - 16 = 0$; normal, $y = -12x - 98$
f tangent, $7x - 4y - 1 = 0$; normal, $32x + 56y - 51 = 0$

2 $y = 8x + 40$ at (−5, 0); $y = -8x + 24$ at (3, 0)

3 (−6, 125); $y = -2x + 113$

4 $y = -\frac{1}{2}x + 1$; $\left(-\frac{4}{3}, \frac{5}{3}\right)$

5.3 Contextual

1 **a** 5 s **b** 2.5 s; 31.25 m

2 **a** $v = 3t^2 - 4t + 3$ m s^{-1}, $a = 6t - 4$ m s^{-2} **c** 18 m s^{-1}

3 **a** 16.2π cm^3 **b** 3.6π cm^2

4 **a** 6% decrease **b** 9% decrease

5.4 Technique

1 **a** $5(x + 7)^4$
b $18(3x - 4)^5$
c $-8(5 - 2x)^3$
d $9x^2(x^3 - 8)^2$
e $2(2x + 5)(x^2 + 5x - 2)$
f $\frac{-6}{(3x - 11)^2}$
g $\frac{2}{\sqrt{4x - 5}}$
h $6x^4(x^5 - 1)^{\frac{1}{5}}$
i $\frac{8}{(9 - 4x)^3}$
j $\frac{-x}{(16 + x^2)^{\frac{3}{2}}}$

2 **a** $30(6t + 1)^4$
b $-3(c - 6)^{-4}$
c $\frac{2x^3}{\sqrt{x^4 + 1}}$
d $\frac{4}{(12a - 6)^{\frac{2}{3}}}$
e $\frac{-3}{(\theta + 1)^{\frac{3}{2}}}$
f $\frac{-40t^3}{(t^4 + 1)^2}$
g $3\left(w - \frac{1}{w}\right)^2\left(1 + \frac{1}{w^2}\right)$
h $\frac{2(6 + \sqrt{\theta})^3}{\sqrt{\theta}}$

3 **a** 12 **b** 15 **c** 432
d $-\frac{2}{3}$ **e** $-\frac{4}{3}$

4 **a** (5, −32), (9, 32) **b** (−1, 1)
c (6, 4)

5 $\frac{dh}{dy} = 2y - 5$; $\frac{dy}{dx} = -2x$; $\frac{dh}{dx} = 4x^3 - 14x$

6 $6(2t + 13)^2$

5.4 Contextual

1 5π cm s^{-1}; 10π cm^2 s^{-1}

2 **a** 0.3 cm s^{-1}
b 15π cm^2 s^{-1}

3 (2, 3) is a maximum point; (3, 7) is a minimum point

4 **a** $V = 45x^3$; $\dfrac{dV}{dx} = 135x^2$
b 0.05 cm s^{-1}
c $S = 126x^2$; $\dfrac{dS}{dt} = 100.8$ cm^2 s^{-1}

5 **a** The dimensions of the cone are such that, numerically, height = diameter. Therefore, when the height of the unfilled flask is $20 - h$, the radius r of the top of the sand is $\dfrac{20-h}{2}$.
b $V = \dfrac{2000}{3}\pi - \dfrac{\pi}{12}(20-h)^3$ cm^3
d 0.45 cm s^{-1}

5.5 Technique

1 **a** $4\cos 4x$
b $-9\sin 9x + 7\sin 7x$
c $\dfrac{5}{2}\cos\dfrac{5x}{2}$
d $9\sin 3x$
e $\cos(x + \pi)$
f $-4\cos^3 x \sin x$
g $15\sin^2 5x \cos 5x$
h $8\cos 8x - 4\cos 4x$
i $2\sin\left(\dfrac{2x}{3}\right)$
j $ap\sin(2px) - bq\sin(2qx)$

2 **a** 2 **b** -1
c $3\pi^2 - 2$ **d** 2

3 **a** $\left(\dfrac{\pi}{2}, 0\right)$
b $(0, 0)$, $\left(\dfrac{\pi}{2}, 1\right)$, $(\pi, 0)$
c $(\pi, 2\pi)$
d $\left(\dfrac{\pi}{12}, \dfrac{\sqrt{3}}{2}\right)$, $\left(\dfrac{5\pi}{12}, -\dfrac{\sqrt{3}}{2}\right)$

4 **a** $f'(x) = -\frac{3}{2}\sin\left(\dfrac{3x}{4}\right)$; maximum point at (0, 2), minimum point at $\left(\dfrac{4\pi}{3}, -2\right)$
b $f'(x) = \frac{1}{2} - \cos x$; minimum point at $\left(\dfrac{\pi}{3}, \dfrac{\pi}{6} - \dfrac{\sqrt{3}}{2}\right)$, maximum point at $\left(\dfrac{5\pi}{3}, \dfrac{5\pi}{6} + \dfrac{\sqrt{3}}{2}\right)$
c $f'(x) = 1 + \sin 2x$; points of inflexion at $\left(\dfrac{3\pi}{4}, \dfrac{3\pi}{4} + \frac{1}{2}\right)$ and $\left(\dfrac{7\pi}{4}, \dfrac{7\pi}{4} + \frac{1}{2}\right)$

5 **a i** $\left(\dfrac{\pi}{6}, 0\right)$, $\left(\dfrac{7\pi}{6}, 0\right)$, (0, 1)
ii minimum point $\left(\dfrac{2\pi}{3}, -2\right)$, maximum point at $\left(\dfrac{5\pi}{3}, 2\right)$
iii

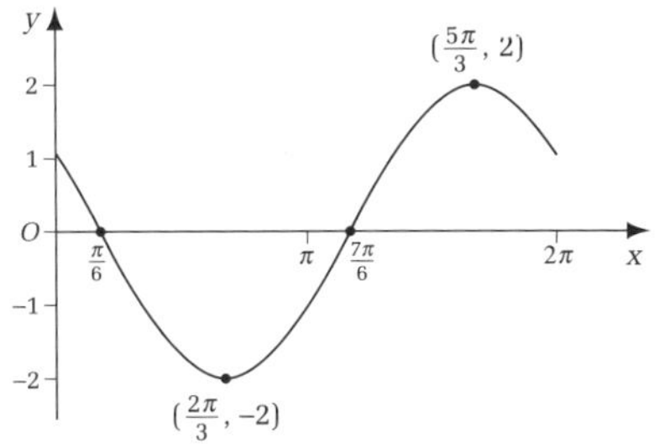

b i (1.816, 0), (4.957, 0), (0, 4)
ii maximum point at (0.245, 4.123), minimum point at (3.387, −4.123)
iii

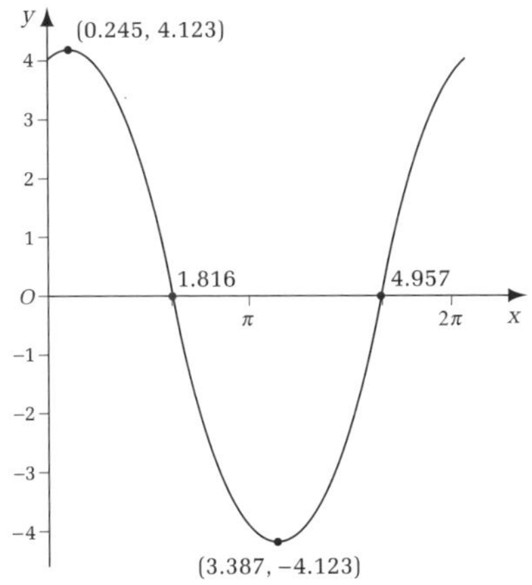

c i (1.190, 0), (4.332, 0), (0, −5)
ii maximum point at (2.761, 5.385), minimum point at (5.903, −5.385)
iii

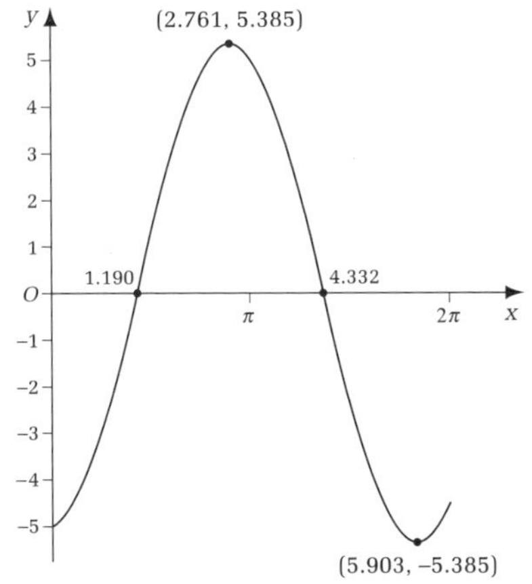

6 **a** tangent, $y = x - \dfrac{\pi}{2}$; normal, $y = -x + \dfrac{\pi}{2}$
b tangent, $y = 2x + 3 - 2\pi$; normal, $y = -\frac{1}{2}x + 3 + \dfrac{\pi}{2}$

5.5 Contextual

1 **a** 5 m
b $\dfrac{dD}{dt} = \dfrac{2\pi}{13}\cos\left(\dfrac{2\pi t}{13}\right) + \dfrac{6\pi}{13}\sin\left(\dfrac{2\pi t}{13}\right)$
c 0.48 m h^{-1} (rising)
d -1.32 m h^{-1} (falling)
e 5:50 a.m.

2 a $v = 8\cos 2t + 3\sin t$ m s^{-1}
$a = 3\cos t - 16\sin 2t$ m s^{-2}
b -3 m, 8 m s^{-1}, 3 m s^{-2}
c When $t = 0.94$ s, $x = 2.04$ m;
$t = 2.20$ s, $x = -2.04$ m;
$t = 3.81$ s, $x = 6.25$ m;
$t = 5.62$ s, $x = -6.25$ m

3 a $(x+1)(x-2)(x-3)$ d $(x+1)(x+2)(x-6)$
b $(x+1)(x+2)(x-3)$ e $(x+1)(x+2)(3x-2)$
c $(x+1)(x-1)(x+3)$ f $(x+2)(3x-2)(x+3)$

4 a $-7 < x < 2, x > 3$ d $-4 \leq x \leq 0, x \geq 3$
b $x < -1, 5 < x < 6$ e $-2 < x < 0, x > 5$
c $x \leq -3, 2 \leq x \leq 6$ f $x < -3, -\frac{1}{2} < x < 8$

Consolidation C

1 a $6x^2 + 6x - 72$ b 18
c 48; $y = 48x - 174$ d $(-4, 338)$, $(3, -5)$

2 a $4\pi r^2$ b 280 cm^3 s^{-1}

3 $5 - \dfrac{8}{x^2}$; $3p$

4 a 10 s b 200 m

5 b $\dfrac{dS}{dx} = 2x - \dfrac{250}{x^2}$ c $x = 5$ m, $S = 75$ m^2

6 a $a = -1$; $b = 150$ b £$(S - 100)$
c £125; £625 000

7 b $40\sqrt{\pi}$ cm^2 or 70.9 cm^2 (1 d.p.)

8 $w = 5.66$ cm, $V = 758.08$ cm^3 (2 d.p.)

9 a i 25 cm b $\frac{1}{2}$ and 2
ii 114.6° (1 d.p.)

10 a 2 m s^{-1} and -4 m s^{-2} b $\dfrac{\pi}{6}, \dfrac{\pi}{2}, \dfrac{5\pi}{6}$ and $\dfrac{3\pi}{2}$ s
c 3 m; $\dfrac{3\pi}{2}$ s

Review

1 a $96\frac{5}{13}$ b $445\frac{6}{7}$
c $1157\frac{1}{4}$ d $6478\frac{3}{11}$

2 a $\frac{8}{15}$ b $\frac{13}{21}$
c $\frac{11}{20}$ d $\frac{9}{35}$

3 a $1 + \dfrac{2}{x+1}$ b $1 - \dfrac{5}{x+4}$
c $3 + \dfrac{8}{x-2}$ d $-5 + \dfrac{5}{1-x}$

4 a 3 d 1
b 5 e 4
c 2

6.1 Technique

1 a $x^2 + 3x + 2$ b $x^2 + 5x + 3$
c $x^2 - 3x + 8$ d $x^2 - 2x + 3$

2 a 3 b 11
c 1 d -2

6.1 Contextual

1 a $(x+1)(x+2)(x-3)(x-5)$
b $-2 < x < -1$ and $3 < x < 5$

2 $a = 3$, $b = -23$; $(2x+1)$

3 $\left(-\frac{5}{2}, 0\right)$, $(-1, 0)$, $(3, 0)$, $(5, 0)$

4 $a = 2$, $b = -11$

5 $x = -\frac{1}{2}, x = \frac{1}{2}, x = 2$

6 $-\frac{1}{4} \leq x \leq \frac{1}{2}, x \geq 1$

6.2 Technique

1 a $\dfrac{5x+1}{(x-1)(x+1)}$ d $\dfrac{6(2x+5)}{(x+2)(x+3)}$
b $\dfrac{2x-11}{(x-1)(x+2)}$ e $\dfrac{2(5x-2)}{(x-1)(x+1)}$
c $\dfrac{3x+2}{(x+1)(x+2)}$ f $\dfrac{2(x+10)}{(x-2)(x+2)}$

2 a $\dfrac{2(4x-3)}{(2x+1)(x-2)}$ d $\dfrac{19x+7}{(x-2)(7x+1)}$
b $\dfrac{3x+11}{(x-1)(4x+3)}$ e $\dfrac{x}{(1-x)(3x-2)}$
c $\dfrac{2x-11}{(2x-3)(2x+1)}$ f $\dfrac{1}{(2x-1)(3x-2)}$

3 a $\dfrac{2x+9}{x^2-9}$ d $\dfrac{7}{(3-x)(x-2)(x+4)}$
b $\dfrac{2x^2+11x-7}{(x+2)(x-1)(3x-1)}$ e $\dfrac{2(x-1)}{3(x+2)^2}$
c $\dfrac{7x^2+10x-6}{(x+2)(2x-1)(x+1)}$ f $\dfrac{2-x}{(x+1)^2}$

4 a $\frac{1}{9}$ d $\dfrac{x}{10(x-10)}$
b $\dfrac{x+1}{12x}$ e $\dfrac{2}{x+3}$
c $\dfrac{2(x+6)}{5(x+2)}$ f $\dfrac{x-5}{2x}$

5 a $\frac{5}{24}$ d $\dfrac{4x}{3(x-5)}$
b $\dfrac{2x(x+3)}{15}$ e $\dfrac{1}{2(x-2)}$
c $\dfrac{1}{2(x-1)}$ f $\dfrac{x+4}{7x}$

6.3 Technique

1 a $\dfrac{3}{(x+3)} - \dfrac{1}{(x+1)}$

b $\dfrac{3}{x+3} - \dfrac{2}{x+2}$

c $\dfrac{2}{x-3} + \dfrac{5}{x-7}$

d $\dfrac{16}{x+2} + \dfrac{40}{x-5}$

e $\dfrac{1}{5(x-3)} - \dfrac{1}{5(x+2)}$

f $\dfrac{5}{6(x-2)} - \dfrac{5}{6(x+4)}$

2 a $\dfrac{1}{x-1} + \dfrac{3}{x^2+2x+3}$

b $\dfrac{2}{x-3} - \dfrac{3}{x^2+3}$

c $\dfrac{3}{x-2} - \dfrac{1-x}{x^2+1}$

d $\dfrac{3}{x-1} + \dfrac{x-3}{x^2+x+4}$

e $\dfrac{7}{x-2} - \dfrac{3x-2}{x^2+x+1}$

f $\dfrac{3}{x-5} + \dfrac{1-3x}{x^2+1}$

3 a $\dfrac{1}{x+3} - \dfrac{8}{(x+3)^2}$

b $\dfrac{1}{x-3} + \dfrac{4}{(x-3)^2}$

c $\dfrac{1}{x+5} - \dfrac{3}{(x+5)^2}$

d $\dfrac{3}{x-1} - \dfrac{2}{(x-1)^2}$

e $\dfrac{3}{16(x-3)} - \dfrac{3}{16(x+1)} + \dfrac{5}{4(x-3)^2}$

f $\dfrac{1}{16(x+1)} + \dfrac{5}{4(x-3)^2} - \dfrac{1}{16(x-3)}$

4 a $3x-3+\dfrac{8}{x+2}+\dfrac{4}{x-1}$

b $x+1+\dfrac{3}{x-2}+\dfrac{5}{x-5}$

c $3x+2+\dfrac{7}{x-1}+\dfrac{1}{x^2+2}$

d $4+\dfrac{3}{x-2}+\dfrac{2x+1}{x^2+2}$

e $1+\dfrac{2}{x+3}-\dfrac{1}{(x+3)^2}$

f $1-\dfrac{1}{4(x-1)}-\dfrac{19}{4(x+3)}+\dfrac{3}{(x+3)^2}$

5 a $\dfrac{7x+2}{x^2+x+5}+\dfrac{4}{x-5}$

b $x-1-\dfrac{2}{x+3}-\dfrac{1}{x+1}$

c $1+\dfrac{1}{x-5}-\dfrac{2}{x^2+4}$

d $\dfrac{7}{x-3}-\dfrac{2}{x+2}+\dfrac{2}{(x+2)^2}$

e $\dfrac{2}{x-7}-\dfrac{3}{x+2}+\dfrac{3}{(x+2)^2}$

f $\dfrac{2}{x-5}+\dfrac{3x+4}{x^2+x+4}$

6.4 Technique

1 a

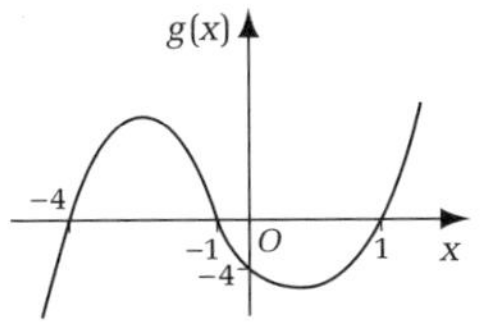

b

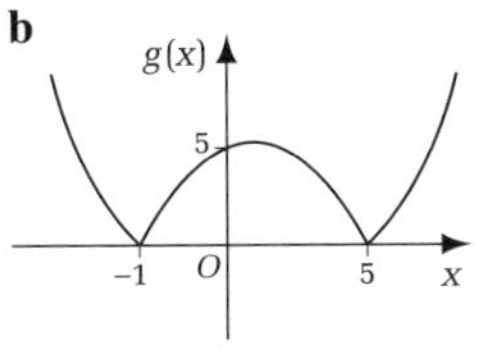

c

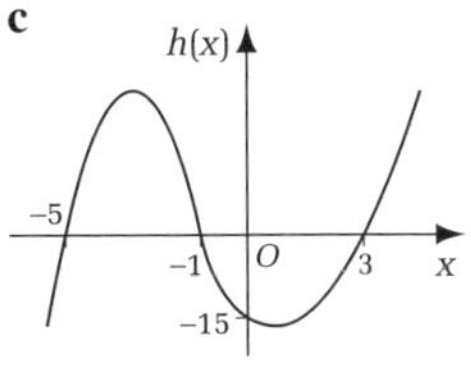

d

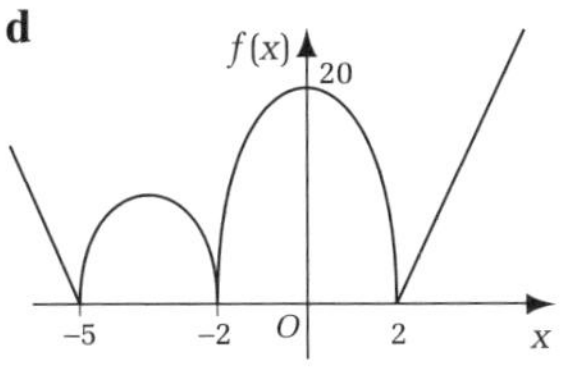

2 a

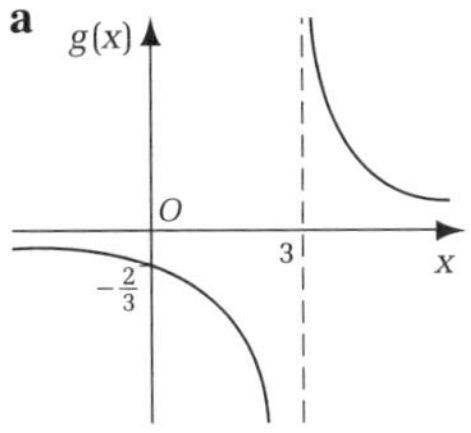

b

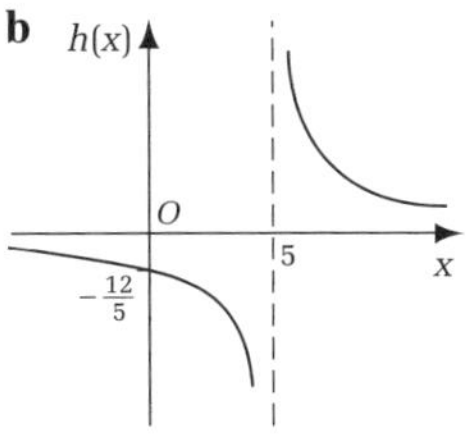

c

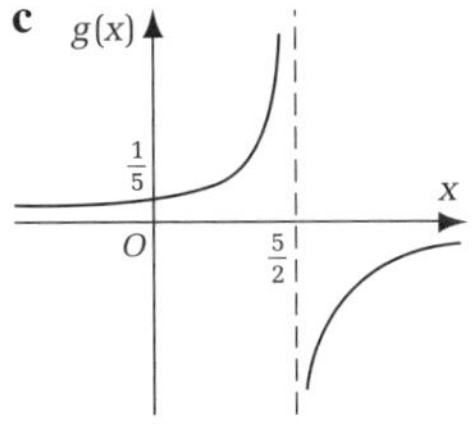

d

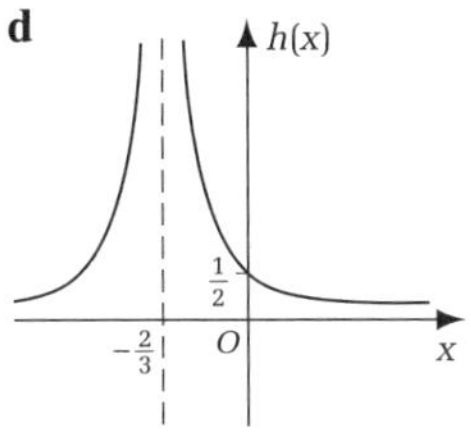

3 a

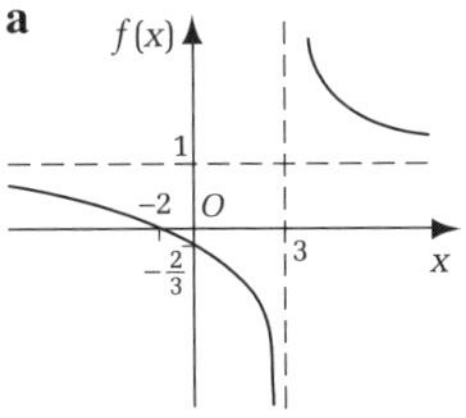

b

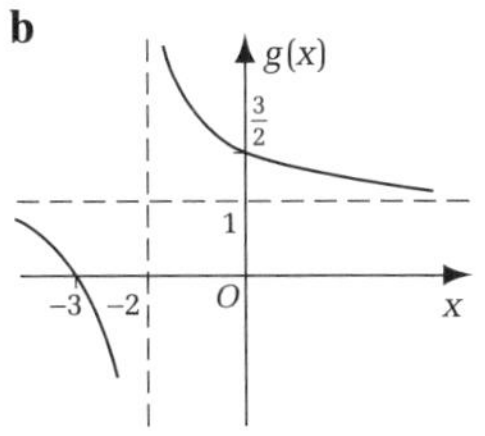

c

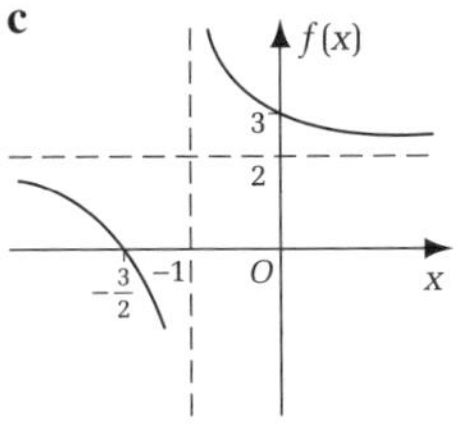

d

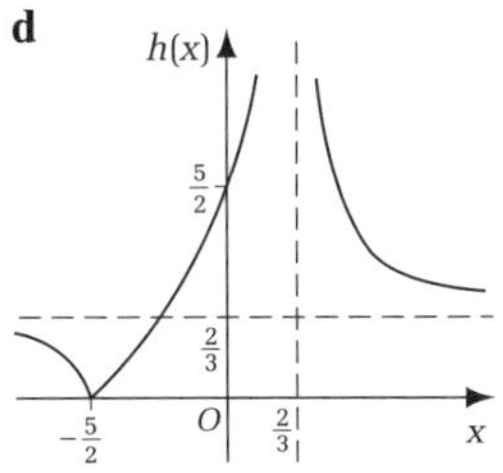

4 a $x=-5, x=-3$

b $x=6, x=18$

c $x=\frac{5}{3}, x=4$

d $2 \le x \le 4$

e $2 \le x \le \frac{5}{2}$

f $-1 \le x \le 9$

6.4 Contextual

1 a $f(x)=(x-1)(2x+3)(3x-2)$

b

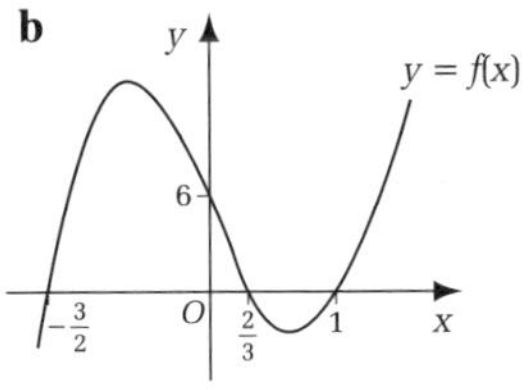

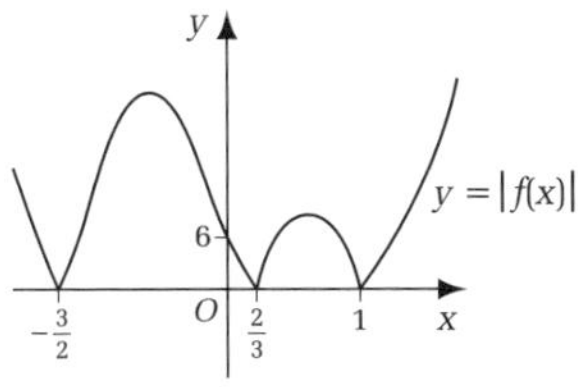

2 a $x > \frac{1}{3}$ b $x > -2$ c $x > \frac{1}{3}, x < -2$

3

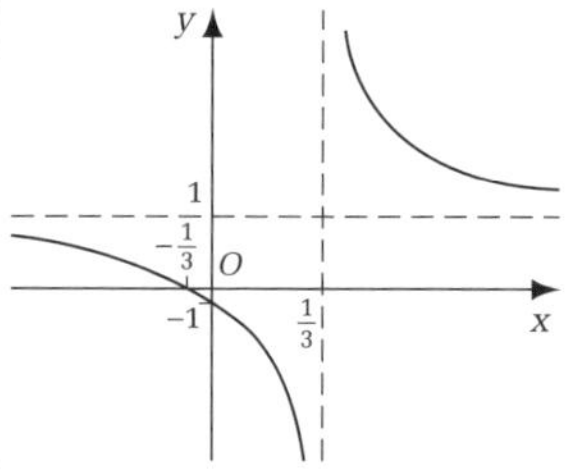

$-\frac{1}{3} < x < \frac{1}{3}$

4 **a**

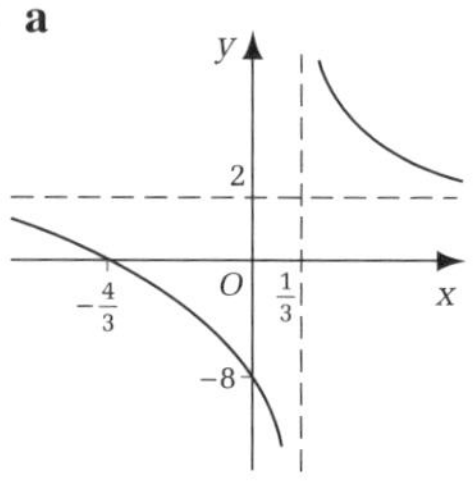

b

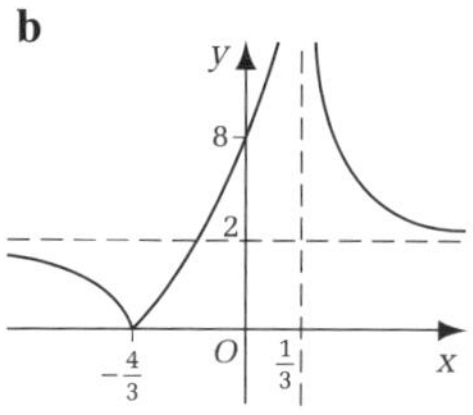

Consolidation C

1 $a = -11, b = -6; (x + 3)$

2 $a = 2, b = 5$

3 **a** $5x^3 - 14x^2 + 7x + 2$ **b** $\dfrac{2}{x-1} - \dfrac{1}{x-2} - \dfrac{3}{5x+1}$
c $-\frac{1}{5} < x < 1, x > 2$

4 $\frac{3}{4} < x < 1, x > 3$

5 **a** $\dfrac{1}{2(x-1)} - \dfrac{1}{2(x+2)}$
b $\dfrac{dy}{dx} = \dfrac{1}{2(x+2)^2} - \dfrac{1}{2(x-1)^2}$
c both are correct; Andy and Lynn have different forms of the same result

6

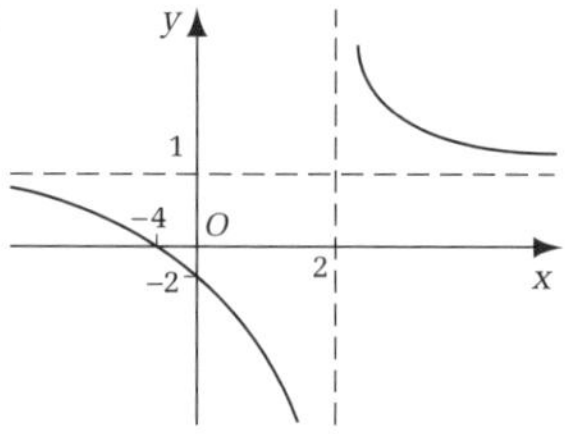

$y > 0$ when $x < -4$ and when $x > 2$

7

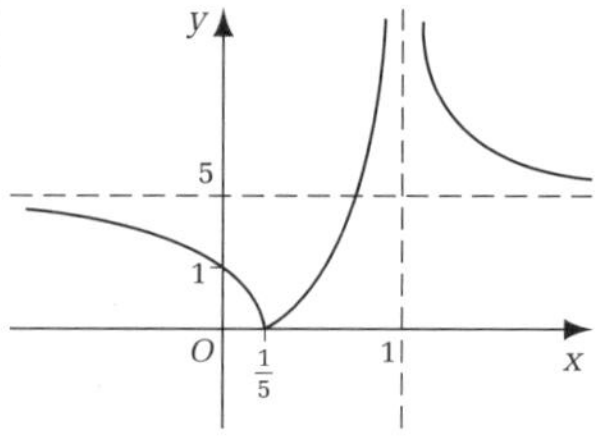

8 $\dfrac{3}{x-2} - \dfrac{2}{x-1} + \dfrac{1}{2x+3}$

Review

1 **a** 7.2×10^7 **b** 7.35×10^{-5}
c 5×10^6 **d** 6×10^{-3}

2 **a** a^8 **d** a^{x-y}
b a^9 **e** 1
c x^{15} **f** $a^{2x+2y} = a^{2(x+y)}$

3 **a** 3 **d** 1
b $\frac{6}{5} = 1\frac{1}{5}$ **e** $\frac{1}{27}$
c 4 **f** $\frac{5}{4} = 1\frac{1}{4}$

4 **a** 3.46 **b** 0.0693
c 0.0316 **d** 0.000 416

7.1 Technique

1 **a** 20.1 **b** 0.135
c 6.13 **d** 1.40

2 **a** translation by -3 units parallel to the x-axis
b translation by 3 units parallel to the y-axis
c one way stretch, scale factor 2, in the y direction
d one way stretch, scale factor $\frac{1}{2}$, in the x direction

3 **a** 5.2 **b** 1.9(3)
c 0.5(8) **d** 0.1(1)

4 **a** 1.61 **b** 1.96
c 0.405 **d** 1.14

7.1 Contextual

1 initial mass is 9.2 g; exceeds 100 g after $3\frac{1}{2}$ days

2 $t = 0.22$ days (about 5 h 16 min)

3 **a** 80°C **b** $k = 0.2$

4 **a** 0.487% **b** 2.96%
c 41.9% doubles after 12 years

5 **a**

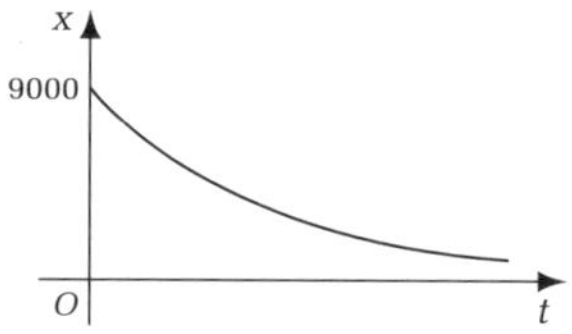

b i after 4.3 years
ii after $13\frac{1}{2}$ years (approx)

7.2 Technique

1 **a** 2 **b** -3
c 6 **d** -8

2 **a** 1.45 **b** 0.477
c -0.699 **d** -0.155

3 **a** 3 **b** 2
c -2 **d** -2

4 **a** 9 **b** 6 **c** 256 **d** 2

5 **a** 1.39 **b** 4.79 **c** 2.08 **d** 3.00

6 **a** 1 **b** $\frac{1}{3}$ **c** $\frac{4}{5}$ **d** $-\frac{5}{4}$

7

Exponent form	Logarithmic form
$2^3 = 8$	$\log_2 8 = 3$
$3^{-1} = \frac{1}{3}$	$\log_3(\frac{1}{3}) = -1$
$3^4 = 81$	$\log_3(81) = 4$
$4^3 = 64$	$\log_4(64) = 3$
$8^2 = 64$	$\log_8(64) = 2$
$5^{-1} = \frac{1}{5}$	$\log_5(\frac{1}{5}) = -1$
$25^{\frac{1}{2}} = 5$	$\log_{25}(5) = -\frac{1}{2}$

7.2 Contextual

1 **a** 1.58 : 1 **b** 20 : 1

2 **a** 8.45 dB **b** $10^6 = 1\,000\,000$ times **c** 3.8 dB **d** 3.01 dB

3 **a** 8.72 **b** 1.995×10^{-6} mol l^{-1} **c** 1000 times

7.3 Technique

1 **a** ln 28 **b** log 6 **c** $\log\left(\frac{14}{9}\right)$ **d** $\ln\left(\frac{1}{360}\right) = -\ln 360$ **e** ln 16 **f** $\log\left(\frac{3}{4}\right)$

2 **a** 1.2 **b** 1.4 **c** 1.5 **d** 0.4 **e** −0.3 **f** 2

3 **a** $\log(xyz^2)$ **b** $\ln\left(\frac{x^3}{y}\right)$ **c** $\ln\left(x^3y^{\frac{1}{2}}\right) = \ln\left(x^3\sqrt{y}\right)$ **d** $\log\left(\frac{xy}{z}\right)$ **e** $\log\left(\frac{1}{ab}\right) = -\log(ab)$ **f** $\log(10x^2)$

4 **a** $\ln x + \ln y + 2\ln z$ **b** $\ln x + 2\ln y - 2\ln z$ **c** $\ln x + \frac{1}{2}\ln y$ **d** $\ln x + \frac{1}{3}\ln y$ **e** $-\ln x - \ln y$ **f** $\ln x + \ln y - \frac{1}{2}\ln z$

5 **a** $x = 256$ **b** $x = \frac{1}{2}$ **c** $x = e^4 = 54.6$ **d** $x = 20$

7.3 Contextual

1 $xy = 1\,000\,000$
$\log_{10}(xy) = \log_{10}(1\,000\,000)$
$\log_{10} x + \log_{10} y = \log_{10} 10^6$
$\log_{10} x + \log_{10} y = 6$

2 transformation is reflection in the x-axis

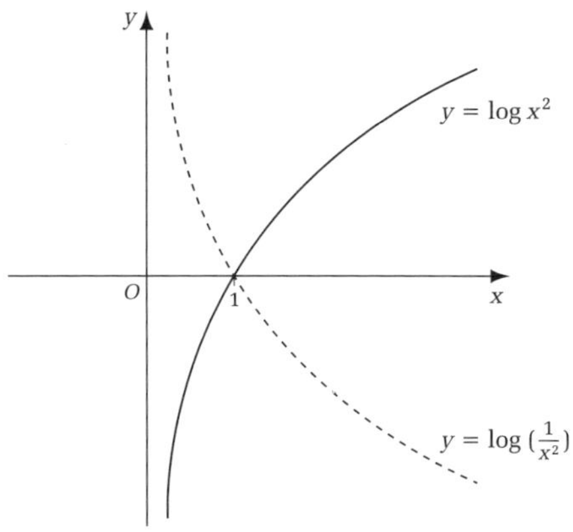

inverse function is $y = 10^{-x/2}$

3 **b** 0

4 $g(x) = 2\ln x$

7.4 Technique

1 **a** 2.21 **b** 2.82 **c** 3.06 **d** 0.946

2 **a** 2.46 **b** −0.637 **c** 4.32 **d** 0.864

3 **a** 1.19 **b** −0.828 **c** −4.5 **d** 4.26

4 **a** $x = \ln 1 = 0$
b $x = \ln\left(\frac{3}{2}\right) = 0.405$
c $x = \ln 2 = 0.693$
d $x = \frac{1}{2}\ln\left(\frac{1}{2}\right) = -0.347$ or $x = \frac{1}{2}\ln\left(\frac{7}{3}\right) = 0.424$

5 **a** $p = 9$ **b** $p = 7$ **c** $p = 18$ **d** $p = 4$

6 **a** $y = 2 \times 3^x$ **b** $y = 2x^3$ **c** $y = 13x^5$ **d** $y = 2 \times \left(\frac{4}{5}\right)^x$

7 **a** 0, log 9 **b** log 11 **c** $\log\left(\frac{1}{2}\right)$, $\log\left(\frac{19}{2}\right)$ **d** $\log\left(\frac{4}{3}\right)$

7.4 Contextual

1 $a = 5.98$, $b = 1.60$; population exceeds 250 thousand after 8 years

2 only one solution since the exponential function e^{2x} cannot be negative

3 $a = 3.8$, $b = 2.5$

4 **a** $A = 70.1$, $\lambda = -0.5$ (01) **b** 1 min 42 s

Consolidation C

1 **a** $\log\left(\frac{v}{u}\right)$ **b** $2\log(uvw) = \log(uvw)^2$

2 **a** $x > 4.29$ (3 s.f.) **b** $x < -1.31$ (3 s.f.)

3 a 32 000 b 0.144

4 b $t = \dfrac{\log V - \log 8500}{\log 0.9}$ c 4 years (approx.)

5 $x = -1, x = 2; y = 0.631$

6 52.1 s

7 a 3 b 1.79 (3 s.f.) c $\frac{1}{4}$; 8

8 c $\alpha = 1.8, \beta = 1.3$
d $P = 291$; prediction not reliable because model is valid for $5 \le t \le 25$ only.

9 $\ln\left(\dfrac{x^4}{y}\right), y = x^4$

10 a i $2\log\left(\dfrac{x}{y}\right)$ b $x = 0.210$ (3 s.f.)
ii 0
iii $\frac{2}{3}$

Review

1 a $3a + 2d$ b $4a + nd$
c $a + 3ar$ d $1 - x^2 - x^3$

2 a $n^2 + 5n + 6$ b $r^2 + r - 6$
c $r^3 + 3r^2 + 2r$ d $1 - 2x + x^2$

3 a 11 b 13 c 10
d 10 e 24 f 10

4 a $\dfrac{6}{r} - \dfrac{6}{r+1}$ b $\dfrac{7}{r+1} - \dfrac{7}{r+2}$
c $\dfrac{8}{r+4} - \dfrac{8}{r+5}$ d $\dfrac{11}{n+1} - \dfrac{11}{n+2}$

8.1 Technique

1 a 5, 12, 19, 26 d 0, 4, 10, 18
b 2, 1, 0, −1 e 4, 8, 16, 32
c 2, 5, 10, 17 f 2, 10, 30, 68

2 a $4n - 3$ d $-5n + 4$ g $\dfrac{2n+1}{2n+3}$
b $5n - 1$ e n^2 h $n^3 - 1$
c $-4n + 27$ f $n(n + 2)$

3 a 3, 5, 7, 9 b 1, 5, 13, 29 c 3, 8, 23, 68
d 1, 3, 5, 7 e 1, 2, 7, 23 f 8, 7, $6\frac{1}{2}$, $6\frac{1}{4}$

4 a $\sum_{r=1}^{4}(r^2 + 1)$ b $\sum_{r=1}^{4}(r^2 + 3)$ c $\sum_{r=1}^{7}(r + 4)$
d $\sum_{r=1}^{6}(3r + 1)$ e $\sum_{r=1}^{5}r(r + 2)$ f $\sum_{r=1}^{5}r^3$

5 a 14 b 29 c 219
d 11 e 68 f 355

6 a $\sum_{r=3}^{6}(2r - 3)$ b $\sum_{r=3}^{8}(2r - 3)$ c $\sum_{r=4}^{7}2r$
d $\sum_{r=3}^{8}(5r - 4)$ e $\sum_{r=2}^{5}(7r - 11)$ f $\sum_{r=5}^{9}(4r - 12)$

8.2 Technique

1 a 11, 13; $2n + 1$ b 23, 27; $4n + 3$
c 21, 19; $-2n + 31$ d −9, −11; $-2n + 1$

2 a 20 b 59 c 42
d 122 e 51 f 29

3 a 192 b 216
c 154 d 210

4 a 1392 b 210 c 45
d 168 e 75 f −77

5 a 124 b 128
c 245 d 520

8.2 Contextual

1 $6n + 4$

2 109

3 2910

4 2016

5 a 5 b 4 c 11

6 648

7 11, 18; 35 050

8 £9000

9 5, 13, 21

10 20

8.3 Technique

1 a i 1000, 10 000 c i 20.25, 30.375
ii 10^{n-1} ii $6 \times 1.5^{n-1}$
b i 243, 729 d i 0.75, −0.375
ii 3^n ii $12 \times \left(-\frac{1}{2}\right)^{n-1}$

2 a 192 b 10 935 c 1215
d 1.5625 e 3.125 f 96

3 a 255 b 9.30
c 215.3 d 53.6

4 a $S_{11} = 10\,235$ b $S_9 = 9841$ c $S_7 = 1397.375$
d $S_6 = -555.75$ e $S_9 = 976\,562$ f $S_{11} = 1535.25$

8.3 Contextual

1 1.37

2 $r = 1.5$

3 22 143

4 $4, \frac{8}{9}, \frac{16}{81}$
a 4 **b** $\frac{2}{9}$ **c** 5.14

5 **a** 4766.4 **b** 149.9

8.4 Technique

1 **a** 600
b $91\frac{3}{7}$
c 1000
d 700
e does not exist; $r > 1$
f 2500

2 **a** oscillates
b oscillates
c converges $r < 1$
d diverges, $r > 1$
e converges
f converges

3 **a** 405 **b** $202\frac{1}{2}$ **c** $341\frac{1}{3}$
d 1250 **e** $357\frac{1}{7}$ **f** 300

4 **a** $a = 40$ **b** $r = \frac{2}{11}$ **c** $r = \frac{9}{10}$
d $a = 4$ **e** $r = \frac{4}{5}$ **f** $a = 42$

8.4 Contextual

1 49

2 51

3 $\frac{2}{3}$

4 364

5 $\frac{15}{16}$

6 $r = \frac{3}{5}, a = 50$

7 **a** $r > 1$ (diverges) **b** $r = \frac{6}{7}$

8.5 Technique

1 **a** $1 + 5x + 10x^2 + 10x^3 + 5x^4 + x^5$
b $27 - 27x + 9x^2 - x^3$
c $256 + 256x + 96x^2 + 16x^3 + x^4$
d $64 - 192x + 240x^2 - 160x^3 + 60x^4 - 12x^5 + x^6$

2 **a** 10 **b** 36
c 35 **d** 15

3 **a** $1 + 4x + 6x^2 + 4x^3 + x^4$
b $625 - 500x + 150x^2 - 20x^3 + x^4$
c $1 - 12x + 60x^2 - 160x^3 + 240x^4 - 192x^5 + 64x^6$
d $x^4 + 12x^2 + 54 + \dfrac{108}{x^2} + \dfrac{81}{x^4}$

4 **a** $1 + 11x + 55x^2 + 165x^3 + \ldots$
b $81 + 108x + 54x^2 + 12x^3 + \ldots$
c $625 - 1000x + 600x^2 - 160x^3 + \ldots$
d $243 + 810x + 1080x^2 + 720x^3 + \ldots$

5 **a** $|x| < \frac{5}{3}$ **b** $|x| < \frac{1}{3}$
c $|x| < \frac{5}{2}$ **d** $|x| < \frac{7}{5}$

6 **a** $1 - 10x + 60x^2 - 280x^3 + \ldots$
b $\sqrt{2}\left(1 - \frac{3}{4}x - \frac{9}{32}x^2 - \frac{27}{128}x^3 + \ldots\right)$
c $1 + 3x - 3x^2 + 7x^3 + \ldots$
d $1 - 8x + 30x^2 - 80x^3 + \ldots$

8.5 Contextual

1 $243 + 405x + 270x^2 + 90x^3 + 15x^4 + x^5$; valid for all x

2 $295\,245x^8$

3 $-745\,472$

4 1120

5 0.9321

6 $1 - \frac{2}{3}x - \frac{4}{9}x^2 - \frac{40}{81}x^3 + \ldots$; valid for $|x| < \frac{1}{2}$

Consolidation C

1 2100

2 **a i** 5
ii $S_{12} = 366$
b i $r = 2.5$
ii $a = 4$
iii 25 400

3 **a** $a = 25, d = -3$ **b** $S_{60} = -3810$

4 **b** $S_{39} = \dfrac{105(1.05^{39} - 1)}{0.05} = 11\,980$; $a = 105$ $r = 1.05$
at start of fortieth year = 12080

5 **a** $u_1 = 1, u_2 = 3, u_3 = 1, u_4 = 3$; oscillates
b $u_1 = 6, u_2 = 2, u_3 = 2$, etc.; converges
c $u_1 = 2$

6 **a** $u_1 = 30, u_2 = 18, u_3 = 10.8$
b $u_8 = 0.840$
c $S_{10} = 74.5$
d $S_\infty = 75$

7 **a i** $1 + 2x + 4x^2 + 8x^3 + \ldots$
ii $|x| < \frac{1}{2}$
b $1 + 5x + 10x^2 + 20x^3$

8 **a** $\dfrac{2}{2x-1} - \dfrac{3}{x+2}$
b $1 + 2x + 4x^2$; $|x| < \frac{1}{2}$
c $\frac{1}{2} - \frac{1}{4}x + \frac{1}{8}x^2$; $|x| < 2$

Review

1 **a** $6x^2 - 12x$
b $x^2 - 2x + 1$
c $-\dfrac{20}{x^3}$
d $5 \cos x$
e $-5 \sin 5x$
f $6 \cos 3x - 10 \sin 2x$

2 a $\frac{3}{5}x^{-2}$
b $4x^{-\frac{1}{2}}$
c $10x^{-3} - 3x^{-1}$
d $7x^{-\frac{1}{2}}$
e $x^{\frac{3}{2}} - x$
f $\frac{3}{2}x^3$

3 a 6 b $-\frac{10}{3}$ c $\frac{1}{5}$

4 a

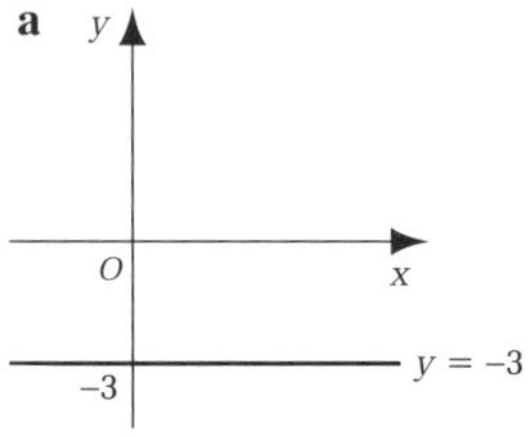

b

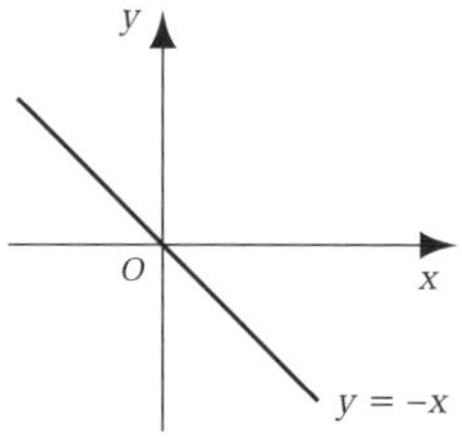

c

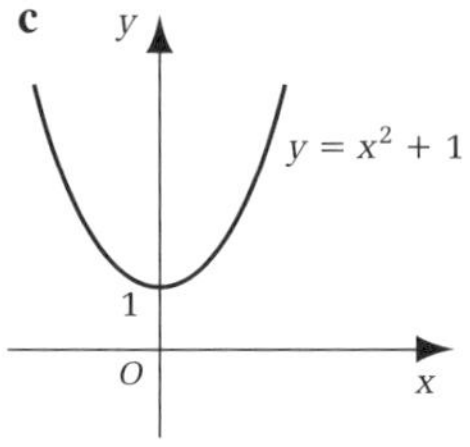

d

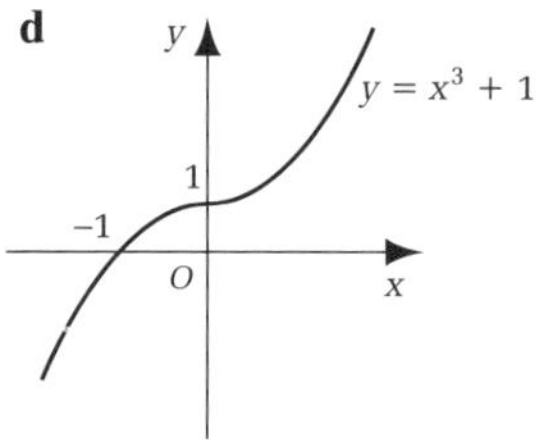

e

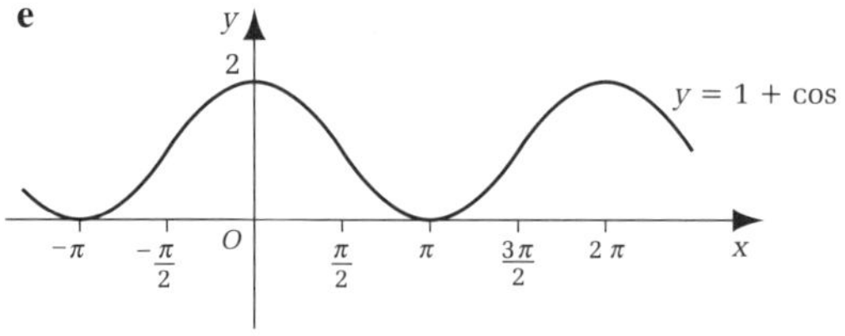

f

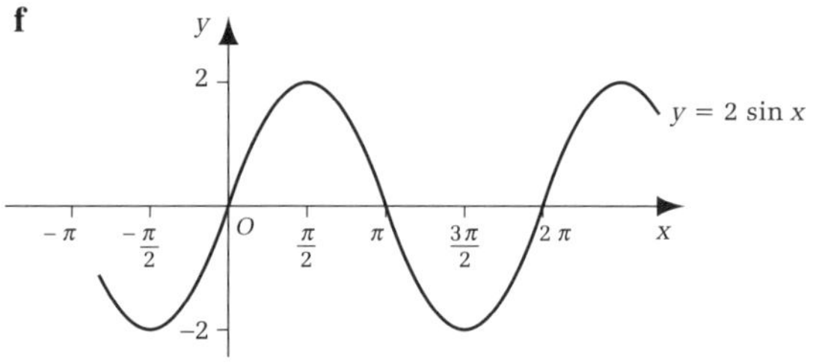

5 a $\sum_{r=1}^{5}(2r-1)$
b $\sum_{r=1}^{4}\left(\frac{1}{r}\right)$
c $\sum_{r=1}^{7}2^{r-1}$
d $\sum_{r=1}^{3}(2r+1)$

9.1 Technique

1 a $\int\left(t^2 - 2t + 1\right)\,dt$
b $\int e^{-2x}\,dx$
c $\int \cos 2\theta\,d\theta$

2 a $5x + c$
b $10x - \frac{3}{2}x^2 + c$
c $7x - \frac{1}{2}x^2 + \frac{1}{3}x^3 + c$
d $\frac{5}{2}x^2 - \frac{2}{3}x^3 + c$
e $\frac{1}{3}x^3 - \frac{5}{2}x^2 - 14x + c$
f $\frac{4}{3}x^3 - 6x^2 + 9x + c$

3 a $t + c$
b $32x + c$
c $\frac{1}{3}t^3 - \frac{1}{2}t^2 + c$
d $\frac{3}{4}x^{\frac{4}{3}} + c$
e $-\frac{1}{2x^2} + c$
f $\frac{x^{2.2}}{2.2} + c$
g $4t^4 + c$
h $-\frac{2}{t^5} + c$
i $20t^{2.5} + c$
j $24x^{\frac{1}{2}} + c$
k $\frac{3}{2}x^{\frac{2}{3}} + c$

4 a $x - x^2 - x^3 + c$
b $x^2 - \frac{2}{3}x^3 + c$
c $4t - 2t^2 + \frac{1}{3}t^3 + c$
d $\frac{1}{4}t^4 - \frac{1}{2}t^2 + c$
e $\frac{1}{3}r^3 + r^2 - 3r + c$
f $\frac{1}{2}r^2 - \frac{4}{3}r^{\frac{3}{2}} + c$

5 a $-\frac{1}{x^2} + \frac{1}{x} + c$
b $10x^{\frac{1}{2}} - \frac{2}{3}x^{\frac{3}{2}} + c$
c $-\frac{12}{t} - \frac{1}{2}t^2 + c$
d $t - \frac{4}{3}t^{\frac{3}{2}} + \frac{1}{2}t^2 + c$
e $\frac{2}{3}x^{\frac{3}{2}} - 2x^{\frac{1}{2}} + c$
f $\frac{2}{5}x^{\frac{5}{2}} + \frac{2}{3}x^{\frac{3}{2}} + c$

6 $g(x) = -\frac{8}{x} - 1$

7 $s = \frac{1}{3}t^3 + 2t^2$

9.1 Contextual

1 a $\frac{d}{dx}(x^4) = 4x^3;\ \int x^3\,dx = \frac{1}{4}x^4 + c$
b all $10x^9 + 1$

2 a $\int\left(x^4 - 6x^3 + 4\right)\,dx = \frac{1}{5}x^5 - \frac{3}{2}x^4 + 4x + c$
b $\int\left(6 - 2t^2\right)\,dt = 2\int\left(3 - t^2\right)\,dt$
$= 2\left(3t - \frac{1}{3}t^3\right) + c$
$= 6t - \frac{2}{3}t^3 + c$
c $\int dr = r + c$
d $\int 2\,dy = 2y + c$

3 $y = x^3 - x + 3$

4 $g(x) = x + \frac{1}{x} + 1$

5 $y = -4x^{\frac{3}{2}} + 37$

6 a $s = \frac{1}{2}t^2 + t$ b 1.5 metres

7 a $s = 0.9t^2 - 0.01t^3$ m b 1080 m

8 a $v = 7t - 3t^2 + 13\,\text{m s}^{-1}$
b $s = \frac{7}{2}t^2 - t^3 + 13t$ m

9.2 Technique

1 a $7x - 2x^2$
b $\frac{1}{3}x^3 - \frac{1}{2}x^2 - x$
c θ
d $15t^2 - 14t + 1$

2 a $\int_1^4\left(-x^2 + 5x - 4\right)\,dx$ b $\int_0^2 x(2-x)\,dx$

3 $\left[\frac{1}{3}x^3 - 2x^2 + 4x\right]_0^2 = \left(\frac{8}{3} - 8 + 8\right) - (0)$
$= \frac{8}{3}$

4 **a**

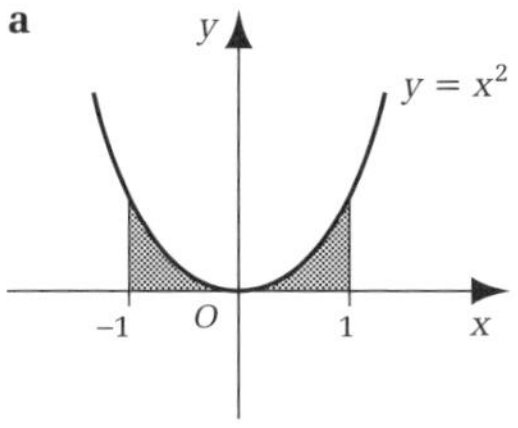

b

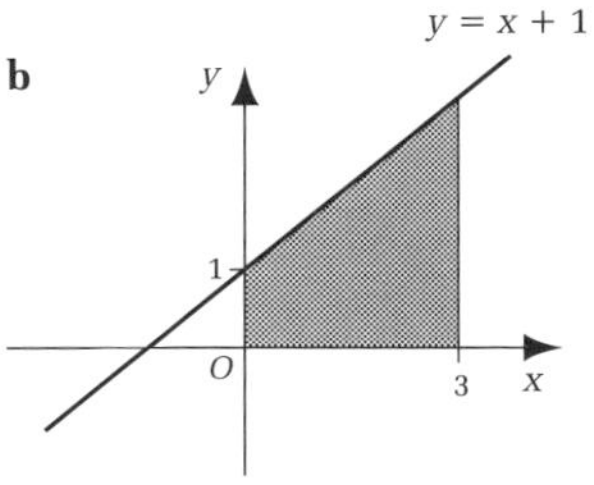

c

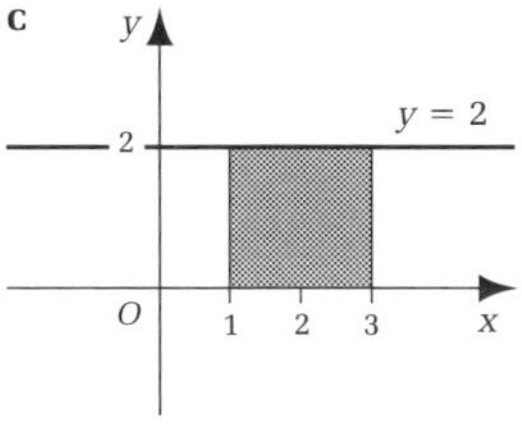

5 **a** 36 **b** $\frac{9}{2}$

6 **a** $\frac{125}{6}$ **b** $\frac{3}{4}$

7 **a** (1, 1) **b** 1

8 **a** $\frac{16}{3}$ **b** $\frac{128}{3}$

9 **a** $\frac{1}{42}$ **b** 12

9.2 Contextual

1 $15\frac{1}{3}$ square units

2 35.324 square units

3 69 m

4 $\dfrac{\sqrt{10}}{3}$

5 2

9.3 Technique

1 **a** $x = y + 2$
b $4\frac{1}{2}$ (minus sign indicates area is to left of vertical axis)

2 $x = \pm 2\sqrt{y}$; $\frac{64}{3}$

3 $x = 2y^{\frac{1}{3}}$; 8.02

4 $x = y^2 - 2y + 2$; $\frac{4}{3}$

9.4 Technique

1 $\frac{9}{2}$

2 9

3 $\frac{9}{8}$

9.4 Contextual

1 **a** (2, 4) **b** $\frac{8}{3}$

2 **a** P(−1, 4), Q(3, 0) **b** $10\frac{2}{3}$

3 **b** (4, 0) **c** 385.07

4 32

5 **a** $\frac{192}{7}$ **b** $\frac{1}{2}$

9.5 Technique

1 **a**

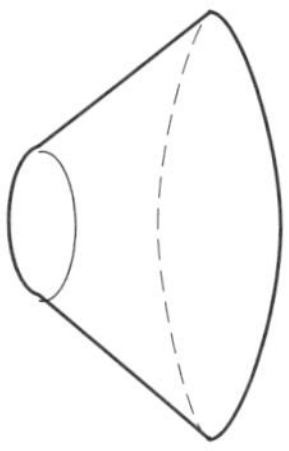

volume $= 21\pi$ cubic units

b

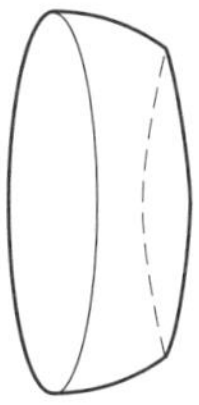

volume $= \frac{203}{15}\pi$ cubic units

c

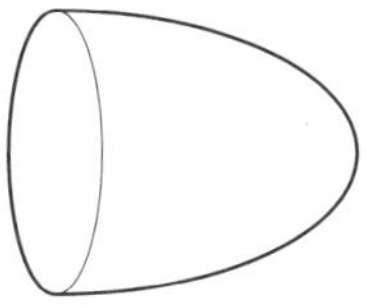

volume $= 2\pi$ cubic units

d

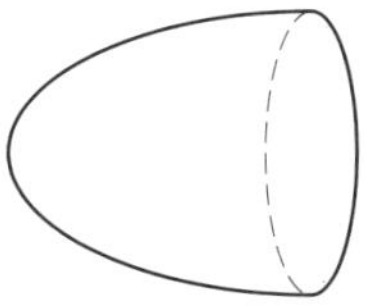

volume $= 27.14$ cubic units

2 **a**

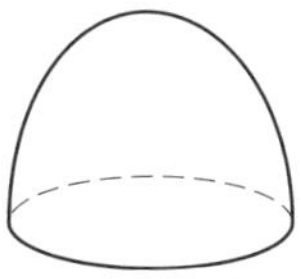

volume $= \dfrac{\pi}{2}$ cubic units

b

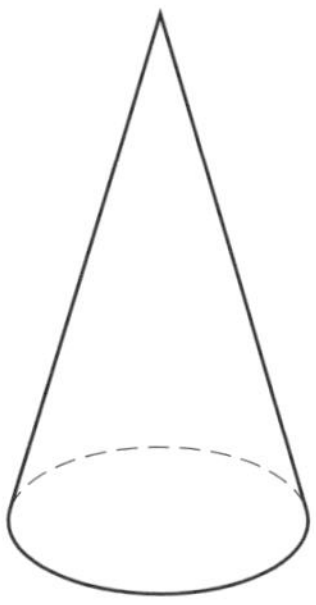

volume $= \frac{16}{3}\pi$ cubic units

3 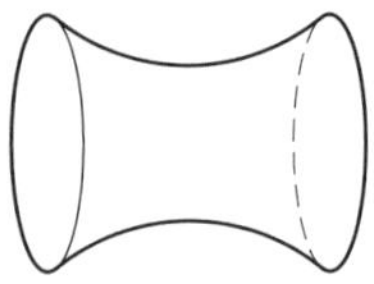

volume $= \frac{83}{30}\pi$ cubic units

9.5 Contextual

1 $\frac{500}{3}\pi$ cubic units, $\frac{1000}{3}\pi$ cubic units

2 **a** OP is a straight line; equation is $y = mx + c$, with $m = \dfrac{r}{h}$ and $c = 0$ $\therefore y = \dfrac{r}{h}x$
b $\frac{1}{3}\pi r^2 h$

3 $\frac{100}{3}\pi$ cubic units

9.6 Technique

1 **a** $2 \sin x + c$ **b** $2 \cos x + c$
c $2x + \sin x + c$ **d** $x + \sin x - \cos x + c$

2 **a** $\frac{1}{2}\sin 2x + c$ **d** $\frac{1}{2}x^2 + \cos 2x + c$
b $2 \sin 2x + c$ **e** $3x - 2\cos 5x + c$
c $x - 2\sin 2x + c$ **f** $\frac{1}{2}\sin 2x + \frac{1}{2}\cos 2x + c$

3 **a** 0 **b** $\frac{3}{2}(\sqrt{3} - 1)$

4 **a** 1.84 **b** 0.87

5 **a** $3\pi - 6$ **b** 2 **c** $\frac{\pi}{4} + \frac{3}{2}$

6 10.10

7 **a** $\dfrac{\sqrt{3}}{2}$ **b** $\pi - 1$
c 3 **d** $\dfrac{\pi}{4} + 1$

8 **a** $\dfrac{x}{2} - \dfrac{\sin 4x}{8} + c$ **b** $\dfrac{x}{2} + \dfrac{\sin x}{2} + c$

9.6 Contextual

1 **b** 2π

2 -1;

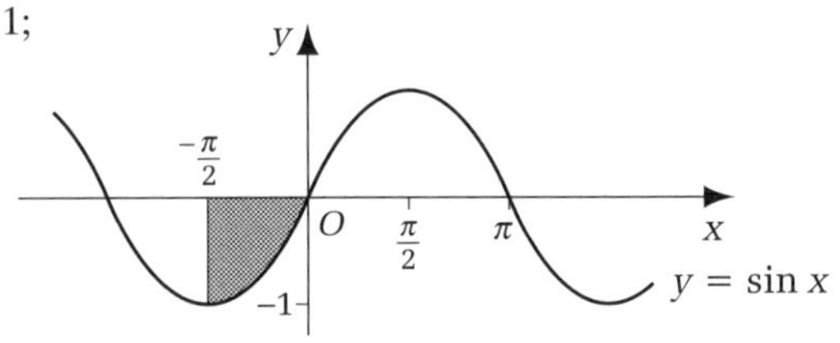

The minus sign indicates that the area is under the x-axis. The actual area is 1 square unit.

3 **a** $y = \sin 2x; \frac{1}{2}$ **b** $y = 2\cos x; 2$

4 **b** $\frac{1}{4}$

5 $\dfrac{\pi}{2}$

6 $\dfrac{\pi}{4}(9\pi + 8)$ cubic units

7 $\dfrac{\pi^2}{2}$ cubic units

Consolidation C

1 **a** $\frac{1}{3}x^3 + \frac{3}{2}x^2 - 10x + c$ **b** $x^2 - x^3 + c$

2 **a** $f'(x) = 3x^2 + 3\cos x$ **b** $g(x) = 3x^2 + \dfrac{4}{x} + 2$
c $\frac{1}{4}x^4 - \frac{1}{2}x^2 + 2x + c; 20$

3 **a** $v = 3t^2 - 24t + 21; t = 1$ and $t = 7$
b $s = t^3 - 12t^2 + 21t - 1$; 108 m
c $-15\,\text{m s}^{-1}$; 1 m

4 $\dfrac{16}{7}\sqrt{2}$

5 **b** 0

6 **a** (0, 0) and (4, 8) **b** $\frac{40}{3}$

7 **b** gradient $= 3$; $y = 17$; $y = 3x - 1$
c When $x = 1$ the equation of the tangent and the equation of the curve both give 2 for the y-coordinate.
d $52\frac{1}{12} = \frac{625}{12}$ square units

8 **a** $x = \dfrac{1}{y^2}; \frac{3}{2}$ **b** $\frac{21}{8}\pi$

9 **a** $\sin\left(\dfrac{5\pi}{4}\right) = \cos\left(\dfrac{5\pi}{4}\right) = -\dfrac{1}{\sqrt{2}}$
x-coordinate of P is $\dfrac{\pi}{4}$
b The area between two curves and the lines $x = a$ and $x = b$ is given by $\int_a^b [f(x) - g(x)]\,dx$ (if $f(x) \geq g(x)$ when $a \leq x \leq b$. For the shaded region, $\sin x > \cos x$. So area is $\int_a^b (\sin x - \cos x)\,dx$. Here $a = \dfrac{\pi}{4}$ and $b = \dfrac{5\pi}{4}$.
c $2\sqrt{2}$

10 **a** $10x + \frac{1}{2}\cos 2x + 2\sin 3x + c$
b 2.28

Review

1 **a** $\sin A \cos B - \cos A \sin B$
b $\cos A \cos B - \sin A \sin B$
c $\dfrac{\tan A - \tan B}{1 + \tan A \tan B}$
d $\frac{1}{2}(\sqrt{3}\cos x + \sin x)$
e $\dfrac{\sqrt{2}}{2}(\sin x + \cos x)$
f $\dfrac{\tan x + 1}{1 - \tan x}$

2 **a** **i** $\dfrac{\pi}{3}$ **ii** $\dfrac{5\pi}{6}$ **iii** $\dfrac{5\pi}{3}$
iv $\dfrac{5\pi}{4}$ **v** $\dfrac{41\pi}{18}$ **vi** $-\dfrac{7\pi}{18}$
b **i** $60°$ **ii** $112\frac{1}{2}°$ **iii** $10°$
iv $330°$ **v** $630°$ **vi** $-70°$

3 a $\frac{\sqrt{2}}{2}$ b $\sqrt{3}$ c $-\frac{\sqrt{3}}{2}$
d -1 e $\sqrt{3}$ f $\frac{\sqrt{3}}{2}$

10.1 Technique

1 a 15°, 30°, 75°, 90°, 150° b 20°, 90°, 100°, 140°
c 30°, 45°, 135°, 150° d 15°, 45°, 75°, 135°

2 a $\frac{\pi}{6}, \frac{\pi}{2}, \frac{5\pi}{6}$
b $0, \frac{\pi}{3}, \frac{7\pi}{12}, \frac{2\pi}{3}, \frac{11\pi}{12}, \pi$
c $0, \frac{\pi}{5}, \frac{\pi}{4}, \frac{3\pi}{5}, \frac{2\pi}{3}, \frac{3\pi}{4}, \pi$
d $\frac{\pi}{8}, \frac{\pi}{2}, \frac{5\pi}{8}$

4 a $\sin 5\theta - \sin 3\theta$ d $\cos 6\theta + \cos 4\theta$
b $\sin 7\theta + \sin 3\theta$ e $\cos 7\theta - \cos 3\theta$
c $\sin 7\theta - \sin 3\theta$ f $\cos 5\theta - \cos 9\theta$

5 a $\frac{\sqrt{6}}{2}$ b $\frac{\sqrt{6}}{2}$
c $\frac{(1+\sqrt{2})}{2}$ d $\frac{(\sqrt{2}-\sqrt{3})}{2}$

10.2 Technique

1 a $10\sin(x + 53.1°)$ b $13\sin(x + 22.6°)$
c $61\sin(x + 10.4°)$ d $\sqrt{34}\sin(x + 59°)$

2 a $2\sin(x - \frac{\pi}{3})$ b $\sqrt{2}\sin(x - \frac{\pi}{4})$
c $5\sin(x - 0.644)$ d $\sqrt{74}\sin(x - 0.951)$

3 a $13\cos(x + 22.6°)$ b $25\cos(x + 16.3°)$
c $\sqrt{13}\cos(x + 56.3°)$ d $\sqrt{2}\cos(x + 45°)$

4 a $5\cos(x - 0.927)$ b $61\cos(x - 1.39)$
c $2\cos(x - \frac{\pi}{6})$ d $\sqrt{65}\cos(x - 1.05)$

5 a 5, 306.9°; − 5, 126.9°
b $\sqrt{29}$, 68.2°; $-\sqrt{29}$, 248.2°
c $\sqrt{34}$, 149°; $-\sqrt{34}$, 329°
d 13, 22.6°; − 13, 202.6°

6 a 96.9°, 336.9° d 90°, 330°
b 126.9° e 158.2°, 338.2°
c 315° f 143.1°, 270°

7 a 0.363, 2.135 b −2.498, 1.571
c −1.139, 0.744 d −0.747, 3.007

8 a $1, \frac{1}{11}$ b $1, \frac{1}{11}$
c $1, \frac{1}{5}$ d $\frac{3}{7}, \frac{1}{11}$

10.2 Contextual

1 a 31 cm, 1 cm b 38.6 s (3 s.f.)

2 a 35 cm, 165 cm b lowest 07:08, Saturday
highest 01:08, Wednesday

10.3 Technique

1 a $360°n \pm 78.5°$
b $360°n \pm 139°$
c $180°n + (-1)^n 64.2°$
d $180°n + (-1)^n(-23.6°)$
e $180°n + 5.7°$
f $180°n - 71°$

2 a $2\pi n \pm \frac{\pi}{4}$ d πn
b $2\pi n \pm \frac{3\pi}{4}$ e $\pi n - \frac{\pi}{4}$
c $\pi n + (-1)^n\left(\frac{\pi}{4}\right)$ f $\pi n + \frac{\pi}{3}$

3 a $\pi n \pm \frac{\pi}{6}$ b $n\pi$, $2n\pi \pm \frac{\pi}{3}$
c $\frac{n\pi}{3}$ d $\frac{n\pi}{2} + \frac{\pi}{8}$

4 $\theta = 360°n + 121.9°$, $\theta = 360°n + 1.9°$

5 $2\pi n + 0.217$, $2\pi n - 1.39$

6 $\frac{n\pi}{3}$

Consolidation C

1 $0, \frac{\pi}{4}, \frac{\pi}{2}, \frac{3\pi}{4}, \pi$

2 b maximum value is 4 when $\theta = 60°$

3 a $65\cos(2\theta + 14.3)°$ b 26.1°, 139.7°

4 $R = \sqrt{5}$, $\alpha = 26.6°$ (3 s.f.);
maximum turning point at $(63.4, \sqrt{5})$

5 a $R = 13$, $\alpha = 1.176$ b $-13 \le f(x) \le 13$

6 a $R = 17$ b 61.9°
c 17 d 298.1°

7 $\sqrt{10}\cos(\theta + 0.322)$
$\left(2\pi - \cos^{-1}\sqrt{\frac{2}{5}}\right) - \tan^{-1}\left(\frac{1}{3}\right) = 5.075$ (3 d.p.)

8 $\frac{2\pi n}{5}$

9 $-2\pi n, \frac{\pi}{3}(2n + 1) + 0.618$

10 a $R = 25$ b -25
c 1.85 d 3.83 and 6.15

Review

1
- **a** $-3 - 2x$
- **b** $30x^4 - 6x^2 + 1$
- **c** $6\sqrt{x}$
- **d** $-2x^{-3} - 12x^{-5}$
- **e** $\dfrac{4}{\sqrt[3]{x}}$
- **f** $4x^3 + 6x^2 - 2$
- **g** $-5 \sin 5x$
- **h** $2 \cos\left(\dfrac{x}{2}\right)$
- **i** $-\sin\left(x + \dfrac{3\pi}{2}\right)$
- **j** $-3 \cos(\pi - 3x)$
- **k** $-6 \sin 6x - \sin 2x$
- **l** $3x^2 + 3\cos^2 x \sin x$

2
- **a** $5(x - 9)^4$
- **b** $20(5x + 6)^3$
- **c** $\dfrac{-1}{2\sqrt{3 - x}}$
- **d** $\dfrac{3x^2}{2\sqrt{x^3 - 27}}$
- **e** $\dfrac{-20x}{(x^2 - 9)^2}$
- **f** $\dfrac{3}{(2x - 8)^{\frac{3}{2}}}$

3 point of inflexion at $(-3, -17)$

4 tangent, $y = -20x - 15$; normal $x - 20y + 502 = 0$

5
- **a** $2 \ln 3 - 1$
- **b** $e^2 - 5$
- **c** $\frac{1}{2}\ln 5$
- **d** $\dfrac{3}{e^2}$

6
- **a** $x^2 + y^2 = 100$
- **b** $(x - 11)^2 + y^2 = 144$
- **c** $(x - 7)^2 + (y + 8)^2 = 81$
- **d** $(x + 5)^2 + (y - 5)^2 = 5$

7
- **a** $1 + 3x + 6x^2 + 10x^3$
- **b** $1 + 2x - x^2 + \frac{4}{3}x^3$
- **c** $1 - 5x + 25x^2 - 125x^3$
- **d** $1 - x - \frac{1}{2}x^2 - \frac{1}{2}x^3$

11.1 Technique

1
- **a** $6(2x - 1)(x - 3)^4$
- **b** $(10x - 3)(2x + 5)^3$
- **c** $\dfrac{3(x + 1)}{\sqrt{2x + 1}}$
- **d** $x \sin 2x + \sin^2 x$
- **e** $\dfrac{5}{(x + 1)^2}$
- **f** $\dfrac{x^2(8x + 3)}{(4x + 1)^2}$
- **g** $5 \cos x \cos 5x - \sin x \sin 5x$
- **h** $\dfrac{5 \cos x \cos 5x + \sin x \sin 5x}{\cos^2 x}$

2
- **a** -80
- **b** 28
- **c** -15
- **d** $\frac{16}{25}$
- **e** -2π
- **f** $\frac{5}{2}$

11.1 Contextual

1
- **a** tangent, $y = 10x - 13$; normal, $x + 10y + 29 = 0$
- **b** tangent, $y = -7x - 19$; normal, $x - 7y + 17 = 0$
- **c** tangent, $y = 2x$; normal, $x + 2y - 5\pi = 0$
- **d** tangent, $y = -24x - 80$; normal, $x - 24y + 388 = 0$

11.2 Technique

1
- **a** $3e^{3x+5}$
- **b** $4x^3 e^{x^4}$
- **c** $-e^{2-x}$
- **d** $-\dfrac{1}{x^2} e^{\frac{1}{x}}$
- **e** $2\cos 2x\, e^{\sin 2x}$
- **f** $x^2(x + 3)e^x$
- **g** $-e^{-x} \cos x (\cos x + 2 \sin x)$
- **h** $\dfrac{(x - 2)e^x}{(x - 1)^2}$
- **i** $e^x (\cos 2x - 2 \sin 2x)$
- **j** $\dfrac{\sin x(2 \cos x - \sin x)}{e^x}$

2
- **a** $\dfrac{1}{x}$
- **b** $\dfrac{1}{2x}$
- **c** $\dfrac{1}{3x}$
- **d** $\dfrac{2}{2x - 1}$
- **e** $\dfrac{2x}{x^2 - 1}$
- **f** $2 \cot x$
- **g** $x^3 (4 \ln x + 1)$
- **h** $\dfrac{1 - 3 \ln x}{x^4}$
- **i** $\dfrac{2x}{x^2 - 1}$
- **j** $\dfrac{1}{\ln 10}$

3
- **a i** $\dfrac{dy}{dx} = 3(x - 1)e^x, \quad \dfrac{d^2y}{dx^2} = (3x + 2)e^x$
  - **ii** $\left(\frac{1}{3}, -3e^{\frac{1}{3}}\right)$ is a minimum point
- **b i** $\dfrac{dy}{dx} = \dfrac{4(x - 3)}{(2x - 1)^2}, \quad \dfrac{d^2y}{dx^2} = \dfrac{4(11 - 2x)}{(2x - 1)^3}$
  - **ii** $(3, 1 + \ln 5)$ is a minimum point
- **c i** $\dfrac{dy}{dx} = -e^{-x}(\cos 2x + 2 \sin 2x)$,
  $\dfrac{d^2y}{dx^2} = e^{-x}(4 \sin 2x - 3 \cos 2x)$
  - **ii** $(-0.232, 1.13)$ is a maximum point
    $(1.34, -0.234)$ is a minimum point
- **d i** $\dfrac{dy}{dx} = 2(x + 2)(2x - 3)e^{4x}$,
  $\dfrac{d^2y}{dx^2} = 2\left(8x^2 + 8x - 23\right)e^{4x}$
  - **ii** $\left(-2, \dfrac{1}{e^8}\right)$ is a maximum point
    $\left(\frac{3}{2}, -\frac{3}{4}e^6\right)$ is a minimum point

11.2 Contextual

1
- **a** tangent, $y = \frac{2}{3}x - \frac{8}{3} + \ln 3$; normal, $y = -\frac{3}{2}x + 6 + \ln 3$
- **b** tangent, $y = -4ex - 7e$; normal, $y = \dfrac{1}{4e}x + \dfrac{1}{2e} + e$
- **c** tangent, $y = 4e^2x - 3e^3$; normal, $y = -\dfrac{1}{4e^2}x + e^3 + \dfrac{1}{4e}$
- **d** tangent, $y = -4ex - 2e$; normal, $y = \dfrac{1}{4e}x + \dfrac{1}{4e} + 2e$

2 **a** 6101; 6306
b 48 days
c 22 visitors per day
d the model takes no account of patterns of shopping during the week and at weekends

3 **a** $\frac{dS}{dt} = 0.45t^2 - 1.2t - 8$; £142.68
b $P = 0.15t^3 - 0.6t^2 - 8t + 180 - 170 \ln (0.1t + 3)$
£48.32 loss; £74.96 loss; £62.62 loss; £12.84 profit
c $\frac{dP}{dt} = 0.45t^2 - 1.2t - 8 - \frac{17}{(0.1t+3)}$
−£12.70 per month; −£3.72 per month; +£13.29 per month; +£38.35 per month

4 **a** 5 years; 51 000
b 48 000; increasing by 3000 per year

11.3 Technique

1 **a** $-2 \operatorname{cosec} 2x \cot 2x$
b $-6 \operatorname{cosec}^2 6x$
c $10 \tan 5x \sec^2 5x$
d $-3x^2 \operatorname{cosec}(x^3) \cot(x^3)$
e $6 \sec^2 3x \tan 3x$
f $\frac{7}{\sqrt{1-49x^2}}$
g $\frac{2}{4+x^2}$
h $\frac{1}{\sqrt{x(1-x)}}$
i $\frac{3\sqrt{x}}{2(1+x^3)}$
j $\frac{1}{5x^2-4x+1}$

2 **a** $\cot 2x - 2x \operatorname{cosec}^2 2x$
b $2(1+x) \tan x + (1+x)^2 \sec^2 x$
c $\frac{\sec x(x \tan x - 1)}{2x^2}$
d $3x^2 \sin^{-1} x + \frac{x^3}{\sqrt{1-x^2}}$
e $3 \tan^{-1}\left(\frac{x}{3}\right) + \frac{9x}{9+x^2}$

11.3 Contextual

1 **a** $2\sqrt{2}$; tangent, $y = 2\sqrt{2}x + \sqrt{2} - \frac{\pi\sqrt{2}}{4}$;
normal, $y = -\frac{x}{2\sqrt{2}} + \sqrt{2} + \frac{\pi}{16\sqrt{2}}$
b $\frac{1}{10}$; tangent, $y = \frac{x}{10} - \frac{1}{2} + \frac{\pi}{4}$;
normal, $y = -10x + 50 + \frac{\pi}{4}$
c 2π; tangent, $y = 2\pi x - 4\pi^2$; normal, $y = -\frac{x}{2\pi} + 1$

11.4 Technique

1 **a** **i** $f'(x) = 3x^2 + 12x - 36$; $f''(x) = 6x + 12$
ii $(-6, 206)$ is a maximum point; $(2, -50)$ is a minimum point
iii

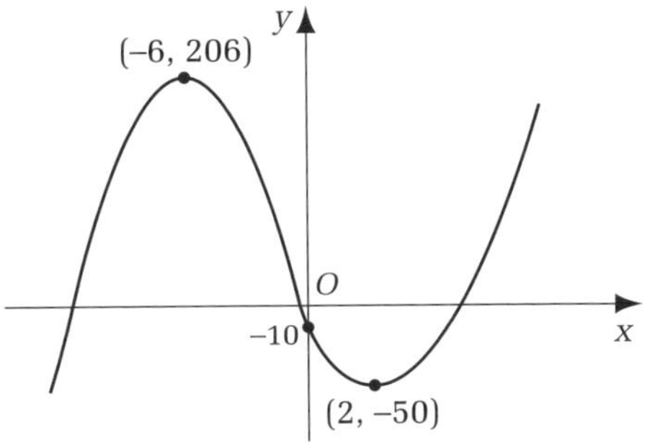

b **i** $f'(x) = 4x^2 - 4x + 1$; $f''(x) = 8x - 4$
ii $(\frac{1}{2}, 6\frac{1}{6})$ is a point of inflexion
iii

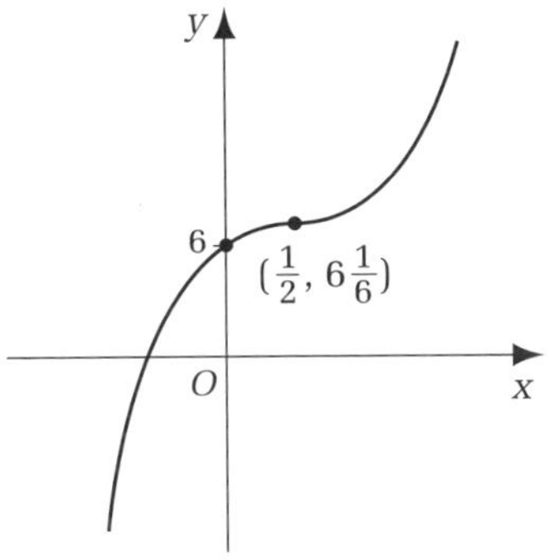

c **i** $f'(x) = 120 - 6x - 6x^2$; $f''(x) = -6 - 12x$
ii $(-5, -353)$ is a minimum point; $(4, 376)$ is a maximum point
iii

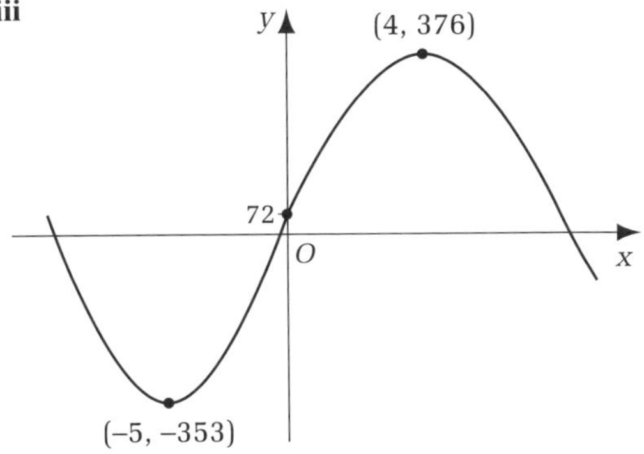

2 **a** $(-2, 0)$, $(4, 0)$, $(0, -32)$
b $x^4 + 2x^3 - 12x^2 - 40x - 32$
c $(-2, 0)$ is a point of inflexion; $(\frac{5}{2}, \frac{-2187}{16})$ is a minimum point
d $(1, -81)$; $\frac{dy}{dx} < 0$, $\frac{d^3y}{dx^3} > 0$
e

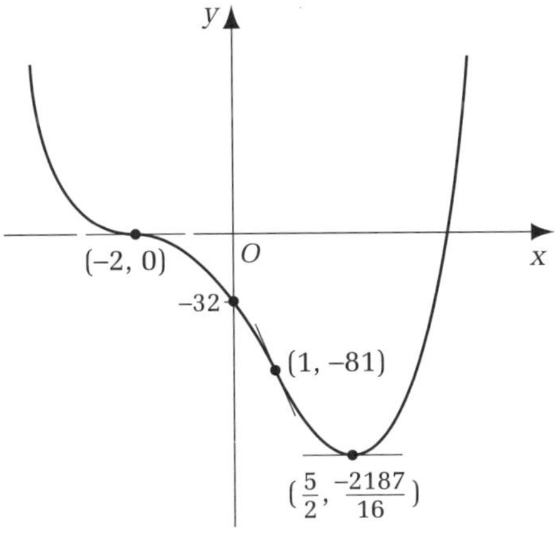

3 point of inflexion at (0, 0); minimum point at (3, −27); $a = 2$, $b = -16$

4 **a** points of inflexion at $\left(\frac{\pi}{4}, \frac{9}{2} - \frac{\pi}{4}\right)$ and $\left(\frac{5\pi}{4}, \frac{9}{2} - \frac{5\pi}{4}\right)$

b $\left(\frac{3\pi}{4}, \frac{9}{2} - \frac{3\pi}{4}\right)$ and $\left(\frac{7\pi}{4}, \frac{9}{2} - \frac{7\pi}{4}\right)$

c

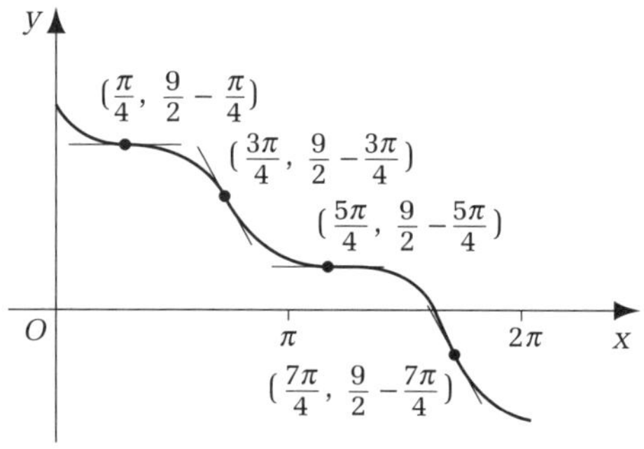

5 **a** **i** (0, 0), (2, 0)

ii none

iii $\frac{dy}{dx} = 3(2x - 1)(x - 2)^2$

iv $\left(\frac{1}{2}, -\frac{27}{16}\right)$ is a minimum point; (2, 0) is a point of inflexion

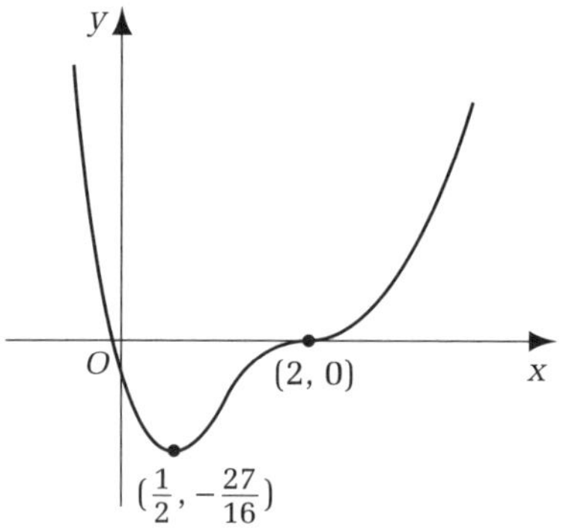

b **i** (−1, 0), (3, 0), (0, −3)

ii none

iii $\frac{dy}{dx} = 4(x - 2)(x + 1)^2$

iv (−1, 0) is a point of inflexion; (2, −27) is a minimum point

v

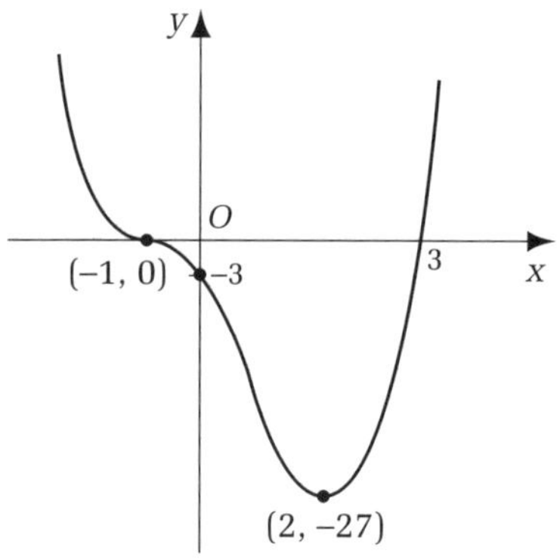

c **i** (−2, 0), (0, −4)

ii $x = 1$

iii $\frac{dy}{dx} = \frac{(x + 2)(x - 4)}{(x - 1)^2}$

iv (−2, 0) is a maximum point; (4, 12) is a minimum point

v

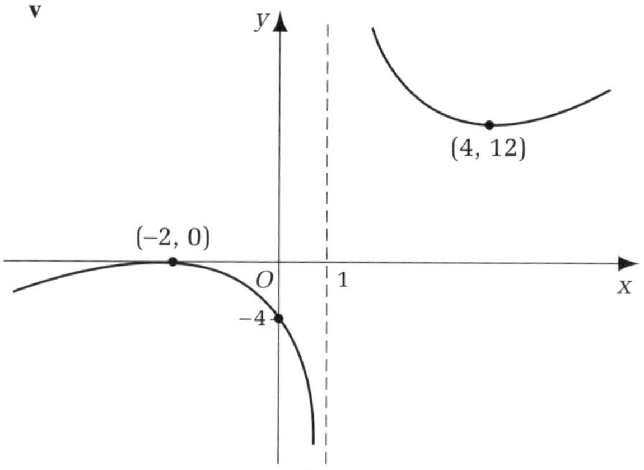

6 **a** **i** $\frac{dy}{dx} = -(x + 1)e^{-x}$; $\frac{d^2y}{dx^2} = xe^{-x}$

ii (−1, e) is a maximum point

b **i** $\frac{dy}{dx} = \frac{(x - 4)}{(x - 1)^2}$; $\frac{d^2y}{dx^2} = \frac{(7 - x)}{(x - 1)^3}$

ii (4, 1 + ln 3) is a minimum point

c **i** $\frac{dy}{dx} = e^x(\cos x - \sin x)$; $\frac{d^2y}{dx^2} = -2e^x \sin x$

ii $\left(-\frac{3\pi}{4}, -\frac{1}{\sqrt{2}e^{\frac{3\pi}{4}}}\right)$ is a minimum point; $\left(\frac{\pi}{4}, \frac{e^{\frac{\pi}{4}}}{\sqrt{2}}\right)$ is a maximum point

d **i** $\frac{dy}{dx} = \frac{1}{2}(x - 3)(x + 1)e^{\frac{x}{2}}$; $\frac{d^2y}{dx^2} = \frac{1}{4}(x^2 + 2x - 7)e^{\frac{x}{2}}$

ii $\left(-1, \frac{16}{\sqrt{e}}\right)$ is a maximum point; (3, 0) is a minimum point

11.4 Contextual

1 **a** 96 m; 120 m

b $-\frac{1}{24}(x - 10)(x - 58)$

$\frac{1}{576}(x^2 - 116x + 2212)\, e^{-(x+1)/24}$

c 10 km

d 58 km; 197 m

e 24.1 km; 70 m

11.5 Technique

1 **a** **i** $(x - 5)^2 + (y - 3)^2 = 100$

ii $\frac{dy}{dx} = \frac{x - 5}{3 - y}$

iii $\frac{3}{4}$

b **i** $(x + 9)^2 + (x + 4)^2 = 289$

ii $\frac{dy}{dx} = -\frac{x + 9}{y + 4}$

iii $-\frac{15}{8}$

c i $(x+6)^2+(y-7)^2=29$

ii $\dfrac{dy}{dx}=\dfrac{x+6}{7-y}$

iii $\frac{5}{2}$

2 **a** $\dfrac{dy}{dx}=\dfrac{5x-y}{x+y}$ **d** $\dfrac{dy}{dx}=\dfrac{5y-8}{3y^2-5x}$

b $\dfrac{dy}{dx}=\dfrac{2(3x^2+2y)}{1-4x}$ **e** $\dfrac{dy}{dx}=\dfrac{8xy+1}{5-4x^2}$

c $\dfrac{dy}{dx}=\dfrac{-6x}{4y+5}$ **f** $\dfrac{dy}{dx}=\dfrac{y^2-4}{2y(3-x)}$

3 **a** -5 **b** $\frac{2}{3}$ **c** -2

4 **a i** $\dfrac{dy}{dx}=\dfrac{2(x+y)}{y-2x}$

ii $(-2, 2), (2, -2)$

iii $\dfrac{d^2y}{dx^2}=\dfrac{\left[2+4\left(\frac{dy}{dx}\right)-\left(\frac{dy}{dx}\right)^2\right]}{y-2x}$

iv $(-2, 2)$ is a minimum point; $(2, -2)$ is a maximum point

b i $\dfrac{dy}{dx}=\dfrac{-5x}{3(y+1)}$

ii $(0, -3), (0, 1)$

iii $\dfrac{d^2y}{dx^2}=\dfrac{-\left[5+3\left(\frac{dy}{dx}\right)^2\right]}{3(y+1)}$

iv $(0, -3)$ is a minimum point; $(0, 1)$ is a maximum point

c i $\dfrac{dy}{dx}=\dfrac{2(xy-1)}{24-x^2}$

ii $\left(4, \frac{1}{4}\right), \left(6, \frac{1}{6}\right)$

iii $\dfrac{d^2y}{dx^2}=\dfrac{2\left[y+2x\frac{dy}{dx}\right]}{24-x^2}$

iv $\left(4, \frac{1}{4}\right)$ is a minimum point; $\left(6, \frac{1}{6}\right)$ is a maximum point

5 **a** tangent, $y=3x+6$; normal, $y=-\frac{1}{3}x+\frac{8}{3}$
b tangent, $y=5x-4$; normal, $y=-\frac{1}{5}x-4$
c tangent, $y=-\frac{5}{3}x+5$; normal, $y=\frac{3}{5}x-\frac{9}{5}$

11.6 Technique

1 **a** $y=\dfrac{x^3}{64}$

b $y=4x^2+2x+1$

c $x^2+y^2=49$

d $\dfrac{x^2}{4}+\dfrac{y^2}{25}=1$

e $\left(\dfrac{x+1}{2}\right)^2+\left(\dfrac{y-4}{3}\right)^2=1$

f $\left(\dfrac{x-p}{a}\right)^2+\left(\dfrac{y-q}{b}\right)^2=1.$

2 **a** $\dfrac{dy}{dx}=-2(t+1)$ **b** $\dfrac{dy}{dx}=3\sin\theta\cos\theta$

c $\dfrac{dy}{dx}=t^3(t+4)^2$

3 **a** $(-11, 21)$ **b** $(\frac{1}{2}, 2)$ and $(-\frac{1}{2}, -6)$
c $(2\ln 5, 5+\ln 5)$

4 tangent, $y=2x-2\pi-2$; normal, $y=-\frac{1}{2}x+\frac{\pi}{2}+\frac{1}{2}$

5 $y=3t^2x-3t^5-t+10$; $y=3x+6$

6 **a** $\dfrac{d^2y}{dx^2}=-\frac{1}{2}$; maximum point at $(13, 16)$

b $\dfrac{d^2y}{dx^2}=\dfrac{2(2+3t^2)}{(2-3t^2)^3}$; minimum point at $(0, 0)$

11.7 Technique

1 **a** $5x-\frac{125}{6}x^3+\frac{625}{24}x^5$ **d** $1+\frac{1}{2}x^2+\frac{1}{24}x^4$

b $2x-6x^2+9x^3$ **e** $x+\frac{1}{6}x^3+\frac{1}{120}x^5$

c $\frac{1}{2}x-\frac{1}{8}x^2+\frac{1}{24}x^3$ **f** $x+x^2+\frac{1}{3}x^3$

2 0.42

3 $-3x-\frac{9}{2}x^2-9x^3$; $4x-8x^2+\frac{64}{3}x^3$; $-7x+\frac{7}{2}x^2-\frac{91}{3}x^3$; -0.695

4 $\tan x\approx 1+2\left(x-\dfrac{\pi}{4}\right)$ near $x=\dfrac{\pi}{4}$; $\tan 48^\circ\approx 1.105$

5 **a** $e+2e(x-1)+\dfrac{3e}{2}(x-1)^2$
b 3.058

Consolidation C

1 **a** $\dfrac{dy}{dx}=-(x+2)e^{-x}$ **b** $(-2, e^2)$

c $-e^2$, maximum point

2 **a** $\dfrac{\pi}{4}$ **b** $2x+y=2\sqrt{2}$

3 **a** $-\frac{2}{3}$ **b** $y=-\frac{2}{3}x+7$

4 $\dfrac{dy}{dx}=2x\tan x\,(\tan x+x\sec^2 x)$

5 **b** $y=-2x-2$

6 $\ln 3+\frac{1}{3}x-\frac{1}{18}x^2$

7 $5x-13y+3=0$

8 **a** $e^{3x}(3\cos x-\sin x)$ **b** $2e^{6x}\cos x\,(3\cos x-\sin x)$

9 $x+3y-31=0$; $(q+1)(q-2)^2$; $(4, -1)$

10 $(-2, -4)$ is a maximum point; $(0, 0)$ is a minimum point

Review

1 $\dfrac{d}{dx}(\sin 2x)=2\cos 2x$, so $\displaystyle\int\cos 2x\,dx=\frac{1}{2}\sin 2x+c$

2 a $\frac{1}{6}x^6 + x^5 - \frac{1}{2}x^2 + 10x + c$ d $-\frac{1}{5}\cos 5x + c$
b $\frac{1}{3}x^3 + x^2 + x + c$ e $-\frac{1}{x} + c$
c $\sin x + c$ f $\frac{1}{a}\sin ax + c$

3

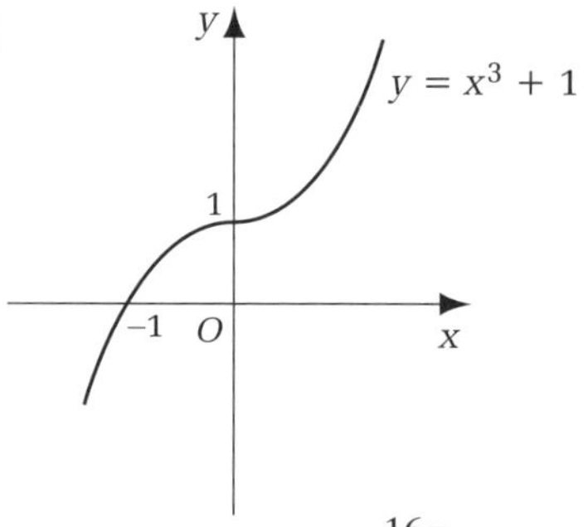

a 2 b $\frac{16\pi}{7}$

4 a $8\cos(8x + 1)$ f $-5e^{2-5x}$
b $-4\sin(4x - 1)$ g $\frac{1}{x}$
c $2e^{2x}$ h $\frac{2x + 1}{x^2 + x - 1}$
d $\frac{1}{2}e^{\frac{1}{2}x}$ i $\cot x$
e $2xe^{x^2}$ j $\frac{4x^3}{x^4 - 1}$

5 a i $1 + \frac{4}{x - 1}$
ii $1 - \frac{3}{x + 3}$
b $\frac{1}{4(x - 2)} - \frac{1}{4(x + 2)}$
c $\frac{1}{2(x - 3)} - \frac{x + 3}{2(x^2 + 1)}$
d $1 - \frac{2}{x + 1} + \frac{1}{x + 4}$

6 a $\frac{dy}{dx} = \frac{2x - 1}{2y}$ b $\frac{dy}{dx} = \frac{-\sin x}{2y}$

12.1 Technique

1 a $-\frac{1}{2}\cos 2x + c$ d $\frac{1}{3}\cos(1 - 3x) + c$
b $2\sin\frac{1}{2}x + c$ e $\frac{1}{2}\tan 2x + c$
c $\frac{1}{2}\sin(2x + 1) + c$ f $\tan x - x + c$

2 a 2 b $\frac{1}{2}$
c $1 - \frac{\pi}{4}$ d $\frac{\sqrt{3}}{4}$

3 a $-\frac{1}{8}\cos 4x - \frac{1}{4}\cos 2x + c$ b $\frac{1}{8}\sin 4x + \frac{1}{4}\sin 2x + c$
c $\frac{1}{12}$ d $-\frac{1}{2}$

4 a $\frac{1}{6}e^{6x} + c$ f $-3e^{-\frac{1}{3}x} + c$
b $-\frac{1}{2}e^{-2x} + c$ g $2e^{3x} + c$
c $\frac{-1}{4e^{4x}} + c$ h $18e^{\frac{2}{3}x} + c$
d $2e^{\frac{1}{2}x} + c$ i $25e^{0.2x} + c$
e $\frac{5}{2}e^{\frac{2}{5}x} + c$

5 a $x - \frac{1}{3}e^{-3x} + c$ e $20e - 10$
b $\frac{1}{2}e^{2x} - \frac{1}{2}e^{-2x} + c$ f $e + \frac{1}{2e^2} - \frac{3}{2}$
c $-2e^{1-\frac{1}{2}x} + c$ g $1 - \frac{1}{e}$
d $4x + 4e^{-x} - \frac{1}{2}e^{-2x} + c$

6 a $6\ln|x| + c$ e $\frac{1}{3}\ln|3x - 5| + c$
b $\frac{1}{10}\ln|x| + c$ f $-\frac{1}{3}\ln|5 - 6x| + c$
c $\ln|x - 3| + c$ g $2\ln|5x + 7| + c$
d $-\ln|3 - x| + c$ h $-\frac{1}{2}\ln|1 - 2x| + c$

7 a $\ln|3x + 5| + c$ d $\ln|x^2 - 9x| + c$
b $\frac{1}{2}\ln(1 + x^2) + c$ e $2\ln|3x^2 - 4x + 1| + c$
c $\frac{1}{4}\ln(1 + x^4) + c$ f $\ln(1 + e^x) + c$

8 a $-\ln|\cos x| + c$ b $-\frac{1}{2}\ln|\cos 2x| + c$
c $\ln|1 + \tan x| + c$

9 a $\sin^{-1}\left(\frac{x}{5}\right) + c$ d $\frac{1}{\sqrt{5}}\tan^{-1}\left(\frac{x}{\sqrt{5}}\right) + c$
b $\sin^{-1}\left(\frac{x}{10}\right) + c$ e $\sin^{-1}\left(\frac{x}{\sqrt{3}}\right) + c$
c $\frac{1}{2}\tan^{-1}\left(\frac{x}{2}\right) + c$ f $\sqrt{2}\tan^{-1}\left(\frac{x}{2\sqrt{2}}\right) + c$

12.1 Contextual

1

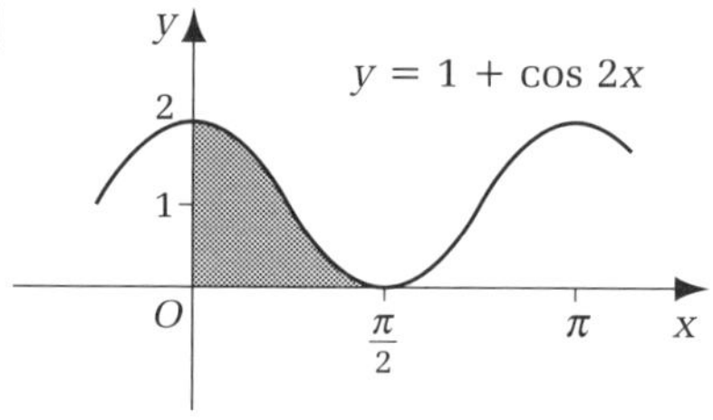

area $= \frac{\pi}{2}$

2 b 0.92

3 $2\pi\sqrt{3}$

4 a $\sec^2\left(\frac{x}{2}\right) - 1$ b $\frac{1}{2}\pi(4 - \pi)$

5 $\frac{2\sqrt{2}}{3}$

6 a $3 - e^{-2}$ b $\frac{\pi}{2}(9 - e^{-4} - 4e^{-2})$

7 a $P = x\ln a$ b $\frac{a^x}{\ln a} + c$

8 $\frac{1}{2}\ln 2$

9 $\frac{\pi}{3}$

10 $\frac{\pi}{2}$

11 a 4π b 2π

12.2 Technique

1 a $2x - \ln|x+1| + c$ b $x + 2\ln|x+5| + c$
c $\frac{1}{3}x - \frac{1}{9}\ln|3x+1| + c$ d $-x - \ln|1-x| + c$

2 a $6 + \ln(\frac{3}{7})$ b $2 - \frac{\pi}{2}$

3 a 1.76 b 0.08

4 a $3\ln|x| - 5\ln|x+9| + c$
b $\ln|x-1| - 5\ln|x+2| + c$
c $\ln\left|\frac{x+2}{x+3}\right| + c$
d $\frac{1}{8}\ln\left|\frac{4+x}{4-x}\right| + c$

5 a $\ln(\frac{10}{9})$ d $\ln(\frac{3}{2})$
b $-\ln 2$ e $-4\ln 5$
c $\frac{1}{2}\ln 2$ f $\frac{1}{2}\ln(\frac{8}{3})$

6 $\frac{1}{2(x-2)} - \frac{1}{2x} - \frac{1}{x^2}$; $\frac{1}{2}\ln(\frac{3}{2}) - \frac{1}{12}$

7 $3 - \frac{1}{x-4} + \frac{2}{x+3}$; $3x + \ln\left|\frac{(x+3)^2}{x-4}\right| + c$

8 $1 + \frac{2}{x-3} - \frac{1}{x+2}$; $x + \ln\left|\frac{(x-3)^2}{x+2}\right| + c$

9 $\frac{1}{5(x+1)} + \frac{1-x}{5(x^2+4)}$; $\frac{1}{10}\ln(\frac{9}{2}) + \frac{\pi}{40}$

12.2 Contextual

1 a $1 + \frac{5}{x-5}$ b $2 + 5\ln(\frac{5}{3})$

2 a $1 - \frac{6}{x+3}$ b area is positive; $6\ln 2 - 3$

3 $\frac{1}{3}\ln 10$

4 $-\frac{1}{x} + \frac{6}{x-8}$; $-7\ln 3$

5 $a = 2$

6 $\frac{2}{x^2} + \frac{3}{x} + \frac{9}{2-3x}$; $1 - 3\ln 2$

7 $2 - \frac{1}{x-1} + \frac{4}{x-3}$; $6 - 2\ln 2$

8 $\frac{1}{1-4x} - \frac{1}{x^2+1}$; $\frac{1}{4}(\ln 5 - \pi)$

12.3 Technique

1 a $(x-1)e^x + c$
b $\frac{1}{100}x^{10}(10\ln x - 1) + c$
c $\frac{1}{3}x\sin 3x + \frac{1}{9}\cos 3x + c$

2 a $-(4x+3)\cos x + 4\sin x + c$
b $\frac{1}{2}x\tan 2x + \frac{1}{4}\ln(\cos 2x) + c$
c $(x-1)\sin x + (x+1)\cos x + c$
d $\frac{1}{9}x^3(3\ln 2x - 1) + c$

3 a $-(t+2)e^{-t} + c$
b $\frac{1}{2}(\ln t)^2 + c$
c $-\frac{1}{2}t\cos 2t + \frac{1}{4}\sin 2t + c$

4 a $-\frac{\pi}{8}$ b -2
c $-\frac{1}{2} + \ln 2$

5 a $(x^2-2)\sin x + 2x\cos x + c$
b $(x^2 - 2x + 3)e^x + c$
c $\frac{1}{5}(2\sin 2x + \cos 2x)e^x + c$

12.3 Contextual

1 a use integration by parts and let $u = x^2$ and $\frac{dv}{dx} = \cos 2x$
b use integration by parts again to evaluate $\int x\sin 2x\,dx$
c $\frac{1}{4}(2x^2 - 1)\sin 2x + \frac{1}{2}x\cos 2x + c$

2 0.16

3 Area $= 5 - \frac{10}{e}$

4 a $x\sin x + \cos x + c$ b $\frac{\pi}{2} - 1$

12.4 Technique

1 a $-\frac{1}{4}(1-x)^4 + c$ d $\frac{2}{3(1-3x)} + c$
b $\frac{1}{6}(4x+3)^{\frac{3}{2}} + c$ e $2\sqrt{x+1} + c$
c $\frac{1}{16}(2x+1)^8 + c$ f $\frac{1}{8}(6x-5)^{\frac{4}{3}} + c$

2 a $\frac{1}{15}$ b $\frac{2}{3}$

3 a 5 b $2\sqrt{x+4} + c$
c $\frac{1}{4-x} + c$

4 a $-\frac{64}{15}$ e $-e^{-\frac{1}{2}x^2} + c$
b $\sqrt{x^2+2} + c$ f $-\frac{1}{5}\cos^5\theta + c$
c $\frac{4}{15}$ g $x - 2\ln|2-e^x| + c$
d $-\sqrt{3-x^2} + c$ h $\frac{1}{4}\tan^{-1}\left(\frac{x^2}{2}\right) + c$

5 a $9 - 2\sqrt{6}$ f $\frac{1}{2}\ln\left(e^{2x}+1\right) + c$
b $\frac{2}{3}\sqrt{2+x^3} + c$ g $\frac{48\sqrt{6}}{5}$
c $\frac{1}{4}\sin^4\theta + c$ h $2\sqrt{x+2} + c$
d $\frac{1}{4}e^{2x^2} + c$ i $\frac{\pi}{2}$
e $\ln|\sin x| + c$

12.4 Contextual

1 3π

2 a $\frac{1}{x-1} - \frac{1}{x-2} + \frac{1}{(x-2)^2}$ b $a = 3, b = 4$

3 $\frac{x}{2} - \frac{1}{2}\ln|e^x + 2| + c$

4 a A $\left(\frac{\pi}{2}, 0\right)$
b $-\frac{1}{3}\cos^3 x + c$
c $\frac{1}{3}$

5 a $\frac{1}{3}(\ln x)^3 + c$
b $\frac{(\ln 2)^3}{3}$

6 $\frac{1}{2}\sin^{-1} x + \frac{1}{2}x\sqrt{1-x^2} + c$

7 $\frac{1}{2}\left(\sin^{-1} x\right)^2 + c$

8 $\frac{1}{2(x-3)} - \frac{1}{2(x-1)} - \frac{1}{(x-1)^2}$; $\frac{1}{2}\ln\left(\frac{3}{2}\right) - \frac{1}{2}$

12.5 Technique

1 20π square units

2 6π square units

3 a $29\frac{1}{2}$ square units
b $\frac{623}{5}\pi$ cubic units

4 a 5 square units
b $\frac{35}{6}\pi$ cubic units

12.6 Technique

1 a $y = \frac{1}{2}x^2 + x + c$
b $y = \ln|x| + c$
c $y = \frac{3}{c - x^3}$
d $y = Axe^{\frac{1}{2}x^2}$
e $y = \frac{1}{c - e^x}$
f $y = 1 - \frac{A}{x}$
g $y = A(x+1)$
h $y = Ae^{-\frac{1}{x}}$

2 a $\ln\left|\frac{x}{x+1}\right| + c$
b $T = \ln\left|\frac{2r}{1+r}\right| + 2$

3 a $\sin y = \frac{1}{2}x^2 + c$
b $\sin y = \frac{1}{2}x^2$ or $y = \sin^{-1}\left(\frac{1}{2}x^2\right)$

4 a $h = \sqrt{2t + c}$
b $h = \sqrt{2t + 7}$

12.6 Contextual

1 a $\frac{d\theta}{dt} = -k(\theta - 20)$
c $A = 52$
d $k = 0.08$; $t = 11.9$ min to cool to 40°C

2 a $\frac{dN}{dt} = \lambda N$
b

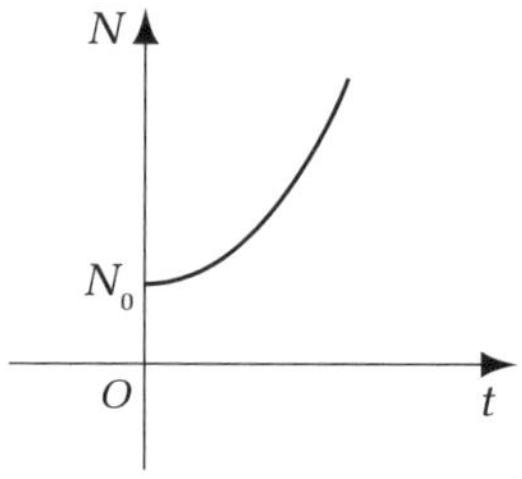

N would eventually increase to infinity
c $\lambda = \ln(1.1)$

3 a $\frac{dV}{dt} = -k\sqrt{h}$
e 60 min

4 a $\frac{dV}{dt} = -kV$
c $k = 0.12$
d £6279, to the nearest pound

5 a $\frac{dy}{dx} = \frac{k}{y^2}$
b $y = \sqrt[3]{1-x}$

Consolidation C

1 a $\frac{2}{3}\sin(3x+4) + c$
b $\frac{1}{2}e^{2x} + c$
c $20e^{\frac{1}{4}x} + c$
d $\frac{1}{4}\ln|x^4 + 2| + c$
e $\sin^{-1}\left(\frac{x}{10}\right) + c$
f $\frac{1}{10}\tan^{-1}\left(\frac{x}{10}\right) + c$
g $\frac{1}{2}e^{x^2} + c$
h $\tan^{-1} x + \frac{1}{2}\ln(x^2+1) + c$

2 a 1
b 2
c $-\frac{2}{3}$
d $\frac{1}{3}$

3 a $\frac{3}{8} + \ln 2$
b $\ln 5$

4 $\frac{1}{4}(1 - 3e^{-2})$

5 a 0
b 1

6 $\frac{3}{1+3x} + \frac{4}{1-4x}$; $3\ln 2$

7 a $\frac{dh}{dt} = -kh$
b $h = 6e^{-kt}$, where $k = \frac{1}{5}\ln 2$, or $k = 0.139$ (3 d.p.); a further 8 min

8 a $\frac{dn}{dt} = kn$; about 4.25 hours longer

9 a $r = -\frac{1}{20}t + 5$

10 **a** $y = -\frac{1}{3}x + 11$

b

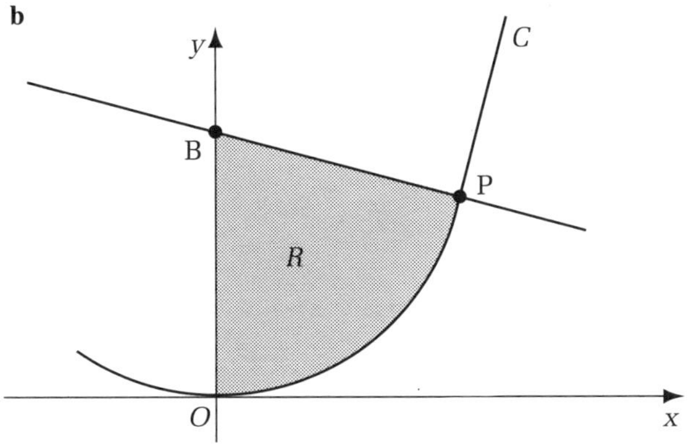

c 186π cubic units

Review

1 **a** $e^x + 2x - 5 = 0$

b $\dfrac{6}{x+1} - x^2 = 0$ or $x^3 + x^2 - 6 = 0$

c $\sqrt[3]{x-4} - x = 0$ or $x^3 - x + 4 = 0$

3 **a** $x = \sqrt{e^x + 2}$ or $x = \ln(x^2 - 2)$

b $x = \sin x - 7$ or $x = \sin^{-1}(x + 7)$

c $x = \frac{1}{9}(x^2 + 5)$ or $x = \pm\sqrt{9x - 5}$ or $x = \dfrac{5}{9-x}$ or $x = 9 - \dfrac{5}{x}$

4 **a** $f'(x) = 2x + \dfrac{2}{2x+1}$

b $f'(x) = 3\tan x + 3x\sec^2 x$

c $f'(x) = 5 - \tan x$

d $f'(x) = xe^x + e^x - 12x$

5 **a** $2\ln 2$ **b** $-\dfrac{2}{e}$

c $\dfrac{\pi^2}{4} - 2$

13.1 Technique

1 **b** 3.24 (2 d.p.)
Since the sign of $10 + 8x - x^3$ is positive when $x = 3.24$ and negative when $x = 4$, the root lies in the interval $3.24 < x < 4$.

2 **a** 1 and 2 **b** 1.6 **c** 1.6417

3 3.276; $e^x - 2x^2 - 5 = 0$

4 **a** $f'(x) = x\cos x + \sin x + 2$

c 5.35

d 5.4767

5 **b** $x_{n+1} = \sqrt[5]{3x_n + 8}$ and $x_{n+1} = \dfrac{8}{x_n^4 - 3}$

c $x_{n+1} = \sqrt[5]{3x_n + 8}$; 1.671

13.1 Contextual

1 **b** 2.66

c **i** The iterative formula fails immediately because, with $x_0 = 0$, $x_1 = \ln(\ln 0)$, and $\ln 0$ is undefined.

ii With $x_0 = 1$, the iterative formula converges towards the other root.

d 0.471

2 **b** 1.625 **c** A (−0.835, 2.170)

d B (0.983, 6.721)

3 **b** 4.08 m

13.2 Technique

1	**a** 2.883	**b** 2.885
2	**a** 7.47 square units	**b** 7.38 square units
3	**a** 4.3129	**b** 4.3084
4	**a** 1.600 square units	**b** 1.603 square units
5	**a** 1.9307	**b** 1.9333
6	**a** 13.481 68	**b** 13.481 75; 0.0005%

13.2 Contextual

1 2190.5 cm^3

2 **a**

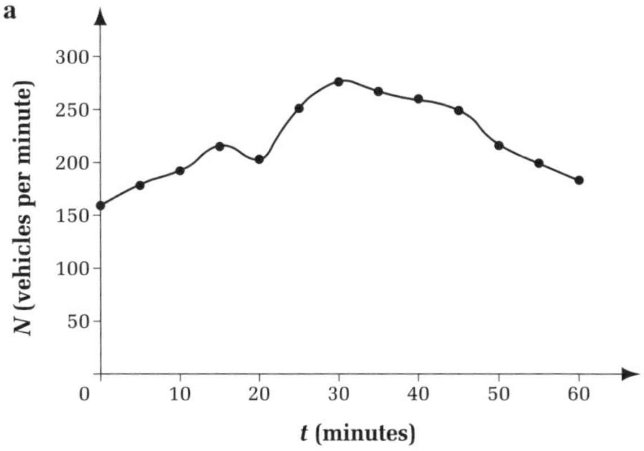

b 13 385 vehicles

3 **a**

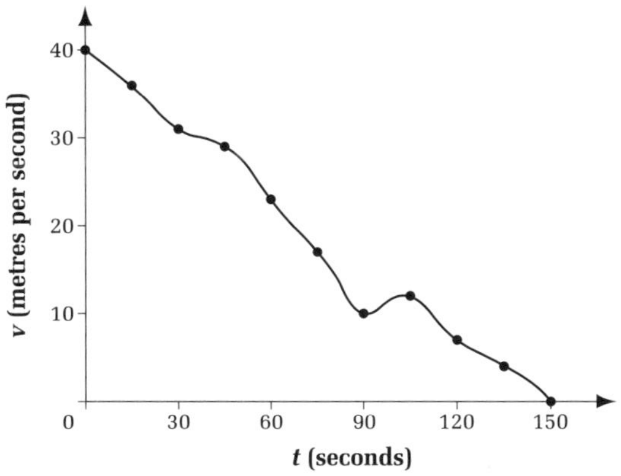

b **i** 2835 m

ii 2870 m

Consolidation C

1 $x_1 = 2, x_2 = 2.0801, x_3 = 2.0614, x_4 = 2.0658, x_5 = 2.0648,$ $x_6 = 2.0650;\ \alpha = 2.065$ (3 d.p.)

2 **a** 8.40 (2 d.p.) **b** $e^2 + 1$

3 **b** Using $x_{n+1} = \frac{1}{5}(x_n^3 + 1)$ and $x_0 = 0.5$
$x_1 = 0.225, x_2 = 0.202\,278, x_3 = 0.201\,655, x_4 = 0.201\,640,$ $x_5 = 0.201\,640;\ \alpha = 0.2016$ (4 d.p.)

4 3.283 (3 d.p.); overestimate because the tops of the trapezia are drawn above the curve

5 1.32

6 $x_1 = 2, x_2 = 2.083\,333, x_3 = 2.080\,089, x_4 = 2.080\,084;$ $\alpha = 2.08$ (2 d.p.); $\alpha^3 = 9;\ \alpha = \sqrt[3]{9}$

7 $x_0 = 2.1, x_1 = 2.294\,719, x_2 = 2.278\,958, x_3 = 2.278\,863,$ $x_4 = 2.278863$; solution is 2.279 (3 d.p.)

8 0.59 (2 d.p.)

9 **a**

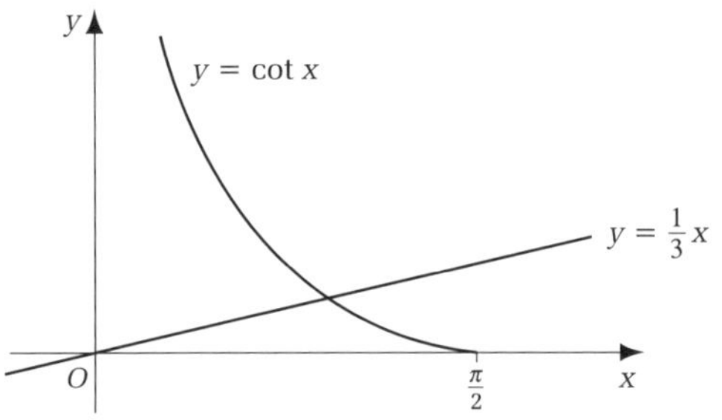

b The equation $\cot x = \frac{1}{3}x$ has exactly one root in the interval $1 < x < \frac{\pi}{2}$, because when $x = 1$, $\cot x > \frac{1}{3}x$, but when $x = \frac{\pi}{2}$, $\cot x < \frac{1}{3}x$. Therefore the curve $y = \cot x$ and the line $y = \frac{1}{3}x$ intersect between 1 and $\frac{\pi}{2}$.

c **i** $u_1 = 1, u_2 = 1.249, u_3 = 1.176, u_4 = 1.197,$ $u_5 = 1.191, u_6 = 1.193$

iii $\alpha = 1.19$

10 0.0615 (4 d.p.)

Review

1 A (2, 0, 2), B (1, 1, 2), C (0, 3, 2), D (3, 3, 1), E (3, 4, 0), F (0, 4, 0)

2 **a** 97.9° **b** 39.8°

3 **a** 7.66 cm^2 **b** 26.31 cm^2

14.1 Technique

1

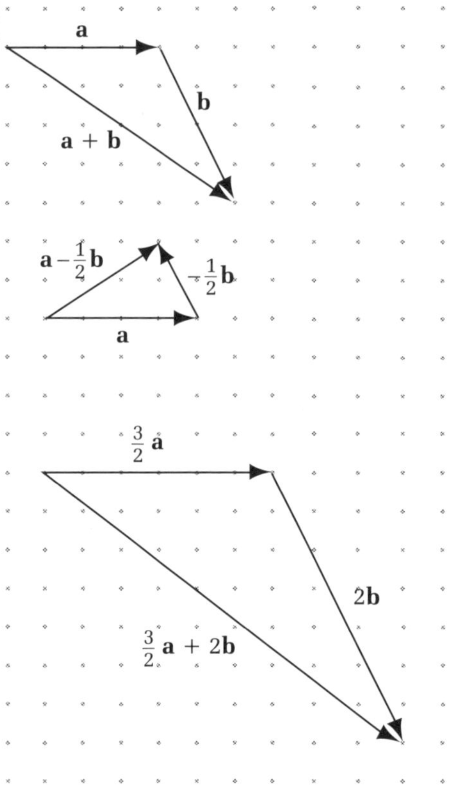

2

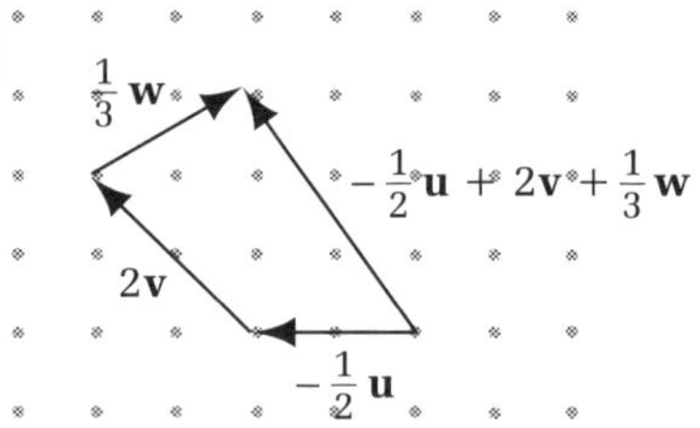

3

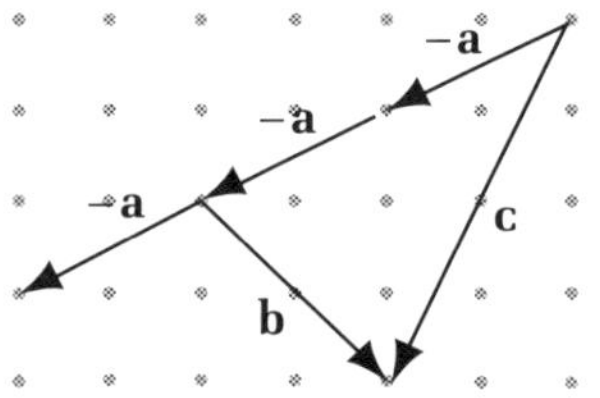

$\therefore\ \mathbf{c} = -2\mathbf{a} + \mathbf{b}$

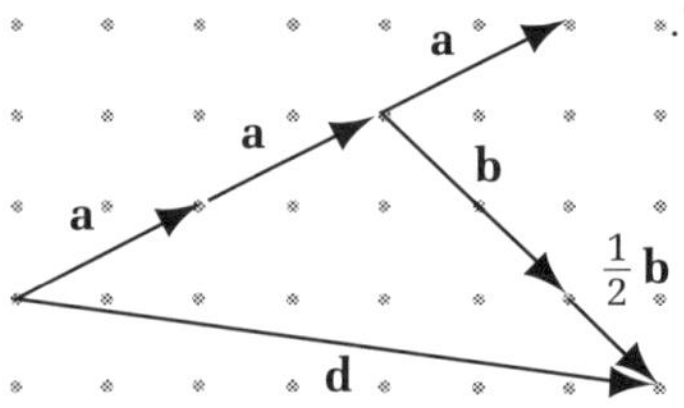

$\therefore\ \mathbf{d} = 2\mathbf{a} + \frac{3}{2}\mathbf{b}$

4 a $\overrightarrow{OM} = \frac{1}{2}\mathbf{a}$

b $\overrightarrow{ON} = \frac{1}{2}\mathbf{b}$

c $\overrightarrow{AB} = -\mathbf{a} + \mathbf{b} = \mathbf{b} - \mathbf{a}$

d $\overrightarrow{OP} = \overrightarrow{OA} + \overrightarrow{AP} = \mathbf{a} + \frac{1}{2}(\mathbf{b} - \mathbf{a}) = \frac{1}{2}(\mathbf{a} + \mathbf{b})$

e $\overrightarrow{MN} = \frac{1}{2}(\mathbf{b} - \mathbf{a})$

f $\overrightarrow{MP} = \frac{1}{2}\mathbf{b}$

The line MN is parallel to the line AB and the length of MN is half that of AB. The line MP is parallel to the line OB and is half the length of OB.

5 a $\overrightarrow{AF} = -\mathbf{b}$
b $\overrightarrow{OD} = -\mathbf{a}$
c $\overrightarrow{CB} = \mathbf{a}$
d $\overrightarrow{AB} = \mathbf{b} - \mathbf{a}$
e $\overrightarrow{MN} = \mathbf{a} - 2\mathbf{b}$

6 a $\overrightarrow{OP} = \frac{2}{3}\mathbf{a} + \frac{1}{3}\mathbf{b} = \frac{1}{3}(2\mathbf{a} + \mathbf{b})$
$\overrightarrow{OQ} = \frac{2}{3}\mathbf{b} + \frac{1}{3}\mathbf{c} = \frac{1}{3}(2\mathbf{b} + \mathbf{c})$
$\overrightarrow{OR} = \frac{2}{3}\mathbf{c} + \frac{1}{3}\mathbf{d} = \frac{1}{3}(2\mathbf{c} + \mathbf{d})$
$\overrightarrow{OS} = \frac{2}{3}\mathbf{d} + \frac{1}{3}\mathbf{a} = \frac{1}{3}(2\mathbf{d} + \mathbf{a})$
b $\mathbf{a} + \mathbf{b} + \mathbf{c} + \mathbf{d}$

14.2 Technique

1 a $\mathbf{a} = 2\mathbf{i} + 3\mathbf{j}$; $\mathbf{b} = -3\mathbf{i} + \mathbf{j}$; $\overrightarrow{AB} = 2\mathbf{j}$

b $\mathbf{a} = \begin{pmatrix} 2 \\ 3 \end{pmatrix}$; $\mathbf{b} = \begin{pmatrix} -3 \\ 1 \end{pmatrix}$; $\overrightarrow{AB} = \begin{pmatrix} 0 \\ 2 \end{pmatrix}$

c $|\mathbf{a}| = \sqrt{13}$

d $\hat{\mathbf{a}} = \frac{1}{\sqrt{13}}\begin{pmatrix} 2 \\ 3 \end{pmatrix} = \begin{pmatrix} \frac{2}{\sqrt{13}} \\ \frac{3}{\sqrt{13}} \end{pmatrix}$

e $\begin{pmatrix} 4 \\ 3 \end{pmatrix}$; $\begin{pmatrix} -3 \\ 0 \end{pmatrix}$; $\begin{pmatrix} -1 \\ -2 \end{pmatrix}$

f $3\mathbf{a} = \begin{pmatrix} 6 \\ 9 \end{pmatrix}$; $\frac{1}{2}\mathbf{b} = \begin{pmatrix} -\frac{3}{2} \\ \frac{1}{2} \end{pmatrix}$;

$\mathbf{a} + \mathbf{b} = \begin{pmatrix} -1 \\ 4 \end{pmatrix}$; $\mathbf{a} - \mathbf{b} = \begin{pmatrix} 5 \\ 2 \end{pmatrix}$

2 a $\mathbf{a} = 2\mathbf{i} + 3\mathbf{j} - 2\mathbf{k} = \begin{pmatrix} 2 \\ 3 \\ -2 \end{pmatrix}$; $\mathbf{b} = -3\mathbf{i} + 4\mathbf{k} = \begin{pmatrix} -3 \\ 0 \\ 4 \end{pmatrix}$;

$\mathbf{c} = 2\mathbf{k} = \begin{pmatrix} 0 \\ 0 \\ 2 \end{pmatrix}$; $\overrightarrow{OP} = 3\mathbf{i} + 4\mathbf{j} + \mathbf{k} = \begin{pmatrix} 3 \\ 4 \\ 1 \end{pmatrix}$

b $\overrightarrow{OP} = \begin{pmatrix} 3 \\ 4 \\ 1 \end{pmatrix}$; $\overrightarrow{OQ} = \begin{pmatrix} 3 \\ 4 \\ 0 \end{pmatrix}$; $\overrightarrow{OR} = \begin{pmatrix} 2 \\ 3 \\ 5 \end{pmatrix}$;

$\overrightarrow{OS} = \begin{pmatrix} 0 \\ 3 \\ 7 \end{pmatrix}$; $\overrightarrow{OT} = \begin{pmatrix} 7 \\ 8 \\ -2 \end{pmatrix}$

c P (3, 4, 1); Q (3, 4, 0); T (7, 8, –2)

d $OP = \sqrt{26}$, $|\mathbf{a}| = \sqrt{17}$

e $\frac{1}{\sqrt{17}}\begin{pmatrix} 2 \\ 3 \\ -2 \end{pmatrix}$ or $\begin{pmatrix} \frac{2}{\sqrt{17}} \\ \frac{3}{\sqrt{17}} \\ \frac{-2}{\sqrt{17}} \end{pmatrix}$

3 a $\begin{pmatrix} 6 \\ 0 \\ -8 \end{pmatrix}$ b $\begin{pmatrix} -2 \\ 3 \\ 2 \end{pmatrix}$

c 10 units d 5 units

4 a $\overrightarrow{RS} = \begin{pmatrix} -12 \\ -6 \\ 4 \end{pmatrix}$

b $\overrightarrow{SR} = \begin{pmatrix} 12 \\ 6 \\ -4 \end{pmatrix}$

c 14

d $\frac{1}{14}\begin{pmatrix} -12 \\ -6 \\ 4 \end{pmatrix}$ or $\frac{1}{7}\begin{pmatrix} -6 \\ -3 \\ 2 \end{pmatrix}$ or $\begin{pmatrix} -\frac{6}{7} \\ -\frac{3}{7} \\ \frac{2}{7} \end{pmatrix}$

5 a $15\mathbf{i} - 5\mathbf{j} + 10\mathbf{k}$
b (6, 0, 7)
c $\sqrt{85}$ units

6 a $\begin{pmatrix} -12 \\ 10 \\ 4 \end{pmatrix}$
b $\begin{pmatrix} 1 \\ 2 \\ -2 \end{pmatrix}$
c $\sqrt{74}$; $\sqrt{74}$; $2\sqrt{65}$; isosceles
d 3; $3\sqrt{65}$ square units

14.3 Technique

1 a $\frac{\pi}{4}$ or 45° b $|\mathbf{a}| = 4\sqrt{2}$; $|\mathbf{b}| = 2$
c 8 d 8

2 a $\begin{pmatrix} -2 \\ 16 \\ 14 \end{pmatrix}$ e 13
b 15 f 2
c 30 g 13
d $\begin{pmatrix} -6 \\ 8 \\ -1 \end{pmatrix}$

Comment: $\mathbf{v}.(2\mathbf{u}) = 2 \times (\mathbf{v}.\mathbf{u})$.
The dot product obeys the distributive law
$\mathbf{v}.(\mathbf{u} - \mathbf{w}) = \mathbf{v}.\mathbf{u} - \mathbf{v}.\mathbf{w}$

3 a 3 d 27
b $3\sqrt{3}$ e $\mathbf{a}.\mathbf{a} = |\mathbf{a}|^2$
c 9 f 18

4 a acute b obtuse
c right angle d acute

5 a –8 b 90

6 a $\frac{\pi}{2}$ or 90° b 2.31 rad or 132.4°
c 0.616 rad or 35.3°

7 0.440 rad or 25.2°

8 1 and 4

9 a $\begin{pmatrix} -4 \\ -35 \\ 2 \end{pmatrix}$ b 0

c $\frac{21}{2}\sqrt{830}$

10 a $\sqrt{14}$, $\sqrt{35}$ b $\dfrac{\sqrt{10}}{5}$

c $\dfrac{7\sqrt{6}}{2}$

11 $\pm\sqrt{13}$

12 $\begin{pmatrix} 4 \\ 3 \\ 1 \end{pmatrix}$ or $\begin{pmatrix} -4 \\ -3 \\ -1 \end{pmatrix}$

14.3 Contextual

1 a 1.46 rad or 83.8° b 1.19 rad or 68.2°

2 0.644 rad or 36.9°

14.4 Technique

1 a $\mathbf{r} = \begin{pmatrix} 0 \\ 1 \\ 6 \end{pmatrix} + \lambda \begin{pmatrix} -2 \\ 5 \\ 1 \end{pmatrix}$ b Q

2 a $\mathbf{r} = \begin{pmatrix} 1 \\ 0 \\ 3 \end{pmatrix} + t \begin{pmatrix} 2 \\ 2 \\ 0 \end{pmatrix}$

b $\mathbf{r} = \begin{pmatrix} -7 \\ 1 \\ 6 \end{pmatrix} + \lambda \begin{pmatrix} 2 \\ 2 \\ 0 \end{pmatrix}$

c $\mathbf{r} = \mu \begin{pmatrix} 2 \\ 2 \\ 0 \end{pmatrix}$ or $\mathbf{r} = \begin{pmatrix} 0 \\ 0 \\ 0 \end{pmatrix} + \mu \begin{pmatrix} 2 \\ 2 \\ 0 \end{pmatrix}$

3 $x = -1, y = -8$

4 a $\mathbf{r} = (\mathbf{i} + 3\mathbf{j} - 2\mathbf{k}) + \lambda(\mathbf{i} - \mathbf{j} + 2\mathbf{k})$

b $\mathbf{r} = (2\mathbf{i} + 2\mathbf{j}) + \mu(3\mathbf{i} + \mathbf{j} - \mathbf{k})$

c $\dfrac{\pi}{2}$ or 90°

5 a 5

c (1, −2, 3) and (13, −2, −13)

6 a $\mathbf{r} = \begin{pmatrix} 2 \\ 0 \\ 7 \end{pmatrix} + \lambda \begin{pmatrix} 2 \\ 3 \\ -4 \end{pmatrix}$

c (0, −3, 11)

7 a $\mathbf{r} = \begin{pmatrix} 4 \\ 2 \\ 7 \end{pmatrix} + \lambda \begin{pmatrix} -2 \\ -3 \\ 4 \end{pmatrix}$

b (2, −1, 11)

c 1.68 rad or 96.4°

8 b 1.39 rad or 79.5°

9 a $\mathbf{r} = 6\mathbf{i} + 3\mathbf{j} + 3\mathbf{k} + \lambda(4\mathbf{i} + 2\mathbf{j} + 5\mathbf{k})$

b (2, 1, −2)

c 3

d $\sqrt{5}$

10 a (−2, 9, 36) b $\sqrt{21}$

c (−4, 5, 37)

14.4 Contextual

1 a $\mathbf{r} = \begin{pmatrix} -300 + 4t \\ 200 \end{pmatrix}$

b $\sqrt{(300 - 2t)^2 + (t - 200)^2}$

c $\sqrt{2000}$ m (44.7 m)

2 a $\overrightarrow{PQ} = \begin{pmatrix} -9 + 2\lambda - 3\mu \\ 17 - 2\lambda + 5\mu \\ 2 - \lambda + 2\mu \end{pmatrix}$

b $\begin{pmatrix} -1 \\ 1 \\ -4 \end{pmatrix}$

c $3\sqrt{2}$

14.5 Technique

1 a $\mathbf{r}.\begin{pmatrix} 2 \\ -1 \\ 3 \end{pmatrix} = 22$; $2x - y + 3z = 22$

b $\mathbf{r}.\begin{pmatrix} 2 \\ -1 \\ 3 \end{pmatrix} = 15$; $2x - y + 3z = 15$

c 14

d $10\frac{1}{2}$

2 a yes

b $\mathbf{r}.\begin{pmatrix} 3 \\ 1 \\ -5 \end{pmatrix} = 0$; $3x + y - 5z = 0$

3 a $\begin{pmatrix} 0 \\ 1 \\ 1 \end{pmatrix}$; $\mathbf{r}.\begin{pmatrix} 0 \\ 1 \\ 1 \end{pmatrix} = 4$ b $\begin{pmatrix} 1 \\ 1 \\ 0 \end{pmatrix}$; $\mathbf{r}.\begin{pmatrix} 1 \\ 1 \\ 0 \end{pmatrix} = 3$

4 a $\mathbf{r}.\begin{pmatrix} 1 \\ 2 \\ 3 \end{pmatrix} = 4$ b −3

5 b $\mathbf{r}.\begin{pmatrix} 2 \\ 2 \\ 3 \end{pmatrix} = 6$

c $2x + 2y + 3z = 6$

6 b $\mathbf{r}.\begin{pmatrix} 7 \\ -2 \\ 5 \end{pmatrix} = -125$

7 a (3, 1, 1) b 0.854 rad or 48.9°

8 0.702 rad or 40.2°

9 a $\sqrt{6}$ b $\frac{\sqrt{6}}{6}$

10 a $\begin{pmatrix}1\\2\\4\end{pmatrix}, \begin{pmatrix}-3\\-1\\3\end{pmatrix}$

b $\mathbf{r} = \begin{pmatrix}3\\-1\\3\end{pmatrix} + \lambda\begin{pmatrix}1\\2\\4\end{pmatrix} + \mu\begin{pmatrix}-3\\-1\\3\end{pmatrix}$

11 a $s = 3$ and $t = 5$ gives the common point S.

b $\mathbf{r} = \begin{pmatrix}7\\-3\\12\end{pmatrix} + \lambda\begin{pmatrix}2\\-1\\3\end{pmatrix} + \mu\begin{pmatrix}2\\-1\\1\end{pmatrix}$

c (1, 0, 0)

14.5 Contextual

1 a $\begin{pmatrix}4\\1\\-7\end{pmatrix}, \begin{pmatrix}-1\\3\\-8\end{pmatrix}$

b $s = 1, t = 3$

c $\mathbf{r}.\begin{pmatrix}1\\3\\1\end{pmatrix} = 1$

d P (4, −2, 3)

e D′ (5, 1, 4)

2 a $\begin{pmatrix}-3\\3\\0\end{pmatrix}, \begin{pmatrix}0\\3\\-1\end{pmatrix}$

b $\mathbf{r} = \begin{pmatrix}3\\0\\3\end{pmatrix} + \lambda\begin{pmatrix}-3\\3\\0\end{pmatrix} + \mu\begin{pmatrix}0\\3\\-1\end{pmatrix}$

c $x + y + 3z = 12$

d $\frac{12\sqrt{11}}{11}$

3 a $\mathbf{n} = \begin{pmatrix}1\\1\\3\end{pmatrix}$

b $\begin{pmatrix}-1\\1\\8\end{pmatrix}$

c $P(1+\lambda, 4+\lambda, -1+3\lambda)$

d $\lambda = 2; \begin{pmatrix}2\\2\\6\end{pmatrix}$

e P (3, 6, 5); A′ (5, 8, 11); $\mathbf{r} = \begin{pmatrix}-1\\1\\8\end{pmatrix} + \mu\begin{pmatrix}6\\7\\3\end{pmatrix}$

Consolidation C

1 a $\begin{pmatrix}17\\-3\\12\end{pmatrix}$

c $\frac{1}{5}\begin{pmatrix}-3\\0\\4\end{pmatrix}$

d $\begin{pmatrix}-1\\2\\11\end{pmatrix}; \begin{pmatrix}3\\-8\\14\end{pmatrix}$

2 the vectors are perpendicular; $\overrightarrow{OC} = \frac{1}{3}\begin{pmatrix}2\\2\\1\end{pmatrix}$; $\frac{3}{2}$ square units

3 a $c = 5$

b $3\mathbf{i} + 4\mathbf{j} + 2\mathbf{k}$

c $\mathbf{r} = (5\mathbf{i} + \mathbf{j} + 5\mathbf{k}) + t(2\mathbf{i} - 2\mathbf{j} + 3\mathbf{k})$

4 a 20

c $\mathbf{r} = \begin{pmatrix}5\\5\end{pmatrix} + s\begin{pmatrix}-4\\3\end{pmatrix}$

d (−1, −3), (9, 2), (11, 13), (1, 8)

5 a $\mathbf{r} = 3\mathbf{i} - \mathbf{j} - 4\mathbf{k} + t(5\mathbf{i} + 2\mathbf{j} + 4\mathbf{k})$

b $-2\mathbf{i} - 3\mathbf{j} - 8\mathbf{k}$

c 75°

d $5\mathbf{i} - \frac{1}{2}\mathbf{j} - \frac{1}{2}\mathbf{k}$

6 a $\mathbf{r}.\begin{pmatrix}2\\-3\\7\end{pmatrix} = -5$

b $\mathbf{r} = \begin{pmatrix}130\\-40\\20\end{pmatrix} + \lambda\begin{pmatrix}-8\\4\\-1\end{pmatrix}$

c $\begin{pmatrix}10\\20\\5\end{pmatrix}$

d 135 metres

7 a $\sqrt{11}$

b i $\mathbf{r} = 2\mathbf{i} + t(-\mathbf{i} + \mathbf{j} + \mathbf{k})$

ii $\mathbf{i} + \mathbf{j} + \mathbf{k}$

c **x** is perpendicular to the plane because it is perpendicular to two lines in the plane; 34.9°

8 a $\overrightarrow{AD} = \begin{pmatrix}-0.25\\0\\0\end{pmatrix}$

b (0, 0.05, 1.1)

c $\mathbf{r} = \begin{pmatrix}0\\0\\1\end{pmatrix} + \lambda\begin{pmatrix}0.25\\0\\0\end{pmatrix}$; $\mathbf{r} = \begin{pmatrix}0.25\\0.15\\1.3\end{pmatrix} + \mu\begin{pmatrix}0\\0.15\\0.3\end{pmatrix}$

d $\mathbf{r} = \begin{pmatrix}0\\0.05\\1.1\end{pmatrix} + s\begin{pmatrix}0\\-2\\1\end{pmatrix}; \frac{\sqrt{5}}{2}$

9 $x_1 = 2, y_1 = -2$; $\mathbf{r} = \begin{pmatrix}5\\-3\\1\end{pmatrix} + s\begin{pmatrix}4\\-5\\-2\end{pmatrix}$

$45 + 45\lambda$; (1, 2, 3); P is nearer Q

10 $\overrightarrow{PQ} = \begin{pmatrix}5-\lambda\\3+\mu\\-5+\lambda-2\mu\end{pmatrix}$; (1, 2, 3); (2, 4, 4); $\sqrt{6}$

Review

1 **a** $\sum_{r=1}^{n} r^2 = \frac{1}{6}n(n+1)(2n+1)$

b $\sum_{r=0}^{n} a^{n-r}b^r$

c $-\log a \times \sum_{r=1}^{n} r = \frac{-n(n+1)}{2}\log a$

2 A prime factor is a factor that is also a prime number

3 **a** $2^3 \times 7$ **d** 457 is prime
b 2^7 **e** 3×199
c 233 is prime **f** 19×89

4 **a** $\int_a^b f(x)\,dx$ **b** $\int_{\frac{\pi}{2}}^{\frac{3\pi}{2}} \cos x\,dx$

15.1 Technique

1 **a** Statement [7]: this is produced by taking the square root of both sides of statement [6] – there should be both positive and negative values

b Statement [2]: $\log\left(\frac{1}{3}\right) < 0$ so the inequality should have been reversed

c Statement [6]: the area bounded by $y = \sin x$ and the x-axis consists of two regions, numerically equal but opposite in sign – the curve doesn't coincide with the axis

15.1 Contextual

1 Use the axioms:
1 angles on a straight line add to 180°
2 alternate angles on parallel lines are equal (Z angles).

2 **a** 0, since it is created by multiplying by 10

b both are zero since this expression is created by multiplication by 100

15.2 Technique

1 Break this statement down and show the case for two digits, then three digits. An argument can then be constructed using induction (Section 15.4). Alternatively accept a proof for three- and four-digit numbers only.

2 **a** $n = 16$, $n^2 + n + 17 = 289$ and $289 = 17^2$

b Let $n = (c - 1)$

Then $n^2 + n + c = n(n+1) + c$
$= (c-1)c + c$
$= c^2 - c + c$
$= c^2$

c^2 cannot be prime.

c $n = 3$, $n^3 - n = 27 - 3 = 24$; but $4 \nmid 3$

3 No. $(abc) - (cba) = (100a + 10b + c) - (100c + 10b + a)$
$= 100a + c - 100c - a$
$= 99(a - c)$,
which is always a multiple of 99 (not prime)

15.2 Contextual

1 Formula is true for $n = 0, 1, 2, 3, 4$.

When $n = 5$, $2^{2^n} + 1 = 2^{32} + 1$
$= 4\,294\,967\,297$
$= 641 \times 6\,700\,417$

This counter-example was first discovered by Euler in 1732.

2 $6! + 1 = 721 = 7 \times 103$. Conjecture is false.

15.3 Technique

1 Assume that $\sqrt{5} = \frac{a}{b}$ and show that this leads to a and b both being multiples of 5

2 Assume that $\sqrt{11} = \frac{a}{b}$ and show that this leads to a and b both being multiples of 11

3 Assume that $\sqrt{p} = \frac{a}{b}$ and show that this leads to p being a factor of both a and b

4 If $\frac{1}{2}(a+b) \le \sqrt{ab}$
then $\frac{1}{4}(a+b)^2 \le ab$
$(a+b)^2 \le 4ab$
$a^2 + 2ab + b^2 \le 4ab$
$a^2 - 2ab + b^2 \le 0$
$(a-b)^2 \le 0$, which is absurd

15.3 Contextual

1 Assume that a number can be factorised into primes in two distinct ways, then let N be the smallest number that can be factorised in two distinct ways. Then $N = a \times b \times c \times \ldots = x \times y \times z \times \ldots$ for primes $a, b, c, \ldots, x, y, z, \ldots$ with no factors common to both. Each factorisation must contain at least one prime, so let these be a and x. Since there is more than one factor in each expression, $N > a^2$ and $N > x^2$. This means that $N > ax$.

Now consider $N - ax$. $N - ax$ cannot have two factorisations, since $N - ax < N$ and N is assumed to be the smallest number. But a is a factor of $N - ax$ and x is a factor of $N - ax$, so ax is a factor of $N - ax$. This means ax is a factor of N, so a is a factor of $\frac{N}{x} = y \times z \times \ldots$. This is a contradiction!

15.4 Technique

1 $3^1 > 2^1$

Assume true for $n = k$: $3^k > 2^k$

$3^{k+1} = 3.3^k > 3.2^k > 2.2^k = 2^{k+1}$

2 $1^3 = \frac{1}{4}1^2(2)^2$

Assume true for $n = k$. Next term is $(k + 1)^3$

$\frac{1}{4}k^2(k + 1)^2 + (k + 1)^3 = \frac{1}{4}(k + 1)^2(k + 2)^2$

3 **a** $7^1 + 5 = 12; 12 \mid 6$

Assume true for $n = k$; $7^k + 5 = 6s$

$$\begin{aligned} 7^{k+1} + 5 &= 7(7^k) + 5 \\ &= 7(6s - 5) + 5 \\ &= 42s - 35 + 5 \\ &= 6(7s - 5) \end{aligned}$$

b $2^1 + 2^2 = 6; 6 \mid 6$

Assume true for $n = k$; $2^k + 2^{k+1} = 6s$

$$\begin{aligned} 2^{k+1} + 2^{k+2} &= 2(2^k + 2^{k+1}) \\ &= 2 \times 6s \\ &= 6(2s) \end{aligned}$$

c $3^{(4\times1)} - 1 = 80; 80 \mid 80$

Assume true for $n = k$; $3^{4k} - 1 = 80s$

$3^{4(k+1)} - 1 = 3^4(3^{4k}) - 1 = 3^4(80s + 1) - 1 =$

$3^4 \times 80s + 3^4 - 1 = 3^4 \times 80s + 80 = 80\,(3^4s + 1)$

15.4 Contextual

1 $$\begin{aligned} (1 + x)^n &= 1 + nx + \ldots + nx^{n-1} + x^n \\ &= 1 + x(n + \ldots + nx^{n-2} + x^{n-1}) \end{aligned}$$

Since n and x are integers this bracket contains only integers, so $(1 + x)^n = 1 + xt$, where t is an integer, and xt is therefore a multiple of x.

a Now put $x = 80$ into this result.

$$(1 + 80)^n = 1 + 80t$$
$$\Rightarrow \quad (81)^n = 1 + 80t$$
$$\Rightarrow \quad (3^4)^n = 1 + 80t$$
$$\Rightarrow 3^{4n} - 1 = 80t$$

So $3^{4n}-1$ is divisible by 80.

b Put $x = 63$ into the result.

$$(1 + 63)^n = 1 + 63t$$
$$\Rightarrow \quad 64^n = 1 + 63t$$
$$\Rightarrow \quad (4^3)^n = 1 + 63t$$
$$\Rightarrow 4^{3n} - 1 = 63t$$

So $4^{3n} - 1$ is divisible by 63.

Consolidation C

1 Logarithms of negative numbers are not defined in the real number system. Euler stated that $\log(-1) = \pi i$ where $i = \sqrt{-1}$. Error here is in line [3], since $\log(-1)$ is not defined in the real number system.

2 Use proof by contradiction. Suppose that $k > 1$ is the highest common factor of n and $(n + 1)$. So $n = kp$ and $(n + 1) = kq$ for some p, q. Subtracting gives $1 = kq - kp$, that is $1 = k(q - p)$. Since $k > 1$ this is impossible.

3 Subtracting is false; when $n = m$, $\sqrt{n} - \sqrt{m} = 0$, and 0 isn't irrational.

4 $2229 = 3 \times 743$

5 When $n = 1$, $11^n - 6 = 5$ and $5 \mid 5$.

Assume true for $n = k$, that is $11^k - 6 = 5t$.

$$\begin{aligned} 11^{(k+1)} - 6 &= 11(11^k) - 6 \\ &= 11(5t + 6) - 6 \\ &= 55t + 66 - 6 \\ &= 5(11t + 12) \end{aligned}$$

This is divisible by 5, so if true for $n = k$, it is also true for $n = (k + 1)$

CONSOLIDATION A

WORKED SOLUTIONS

(refer to *Complete Advanced Level Mathematics: Pure Mathematics*)

1 Algebra I (p. 44)

1 Substituting $u = 14$ and $g = -9.8$ into the equation $h = ut + \frac{1}{2}gt^2$ gives

$$6 = 14t - \tfrac{1}{2}9.8t^2$$

$$\Rightarrow \quad 6 = 14t - 4.9t^2$$

$$\Rightarrow \quad 4.9t^2 - 14t + 6 = 0$$

Using the quadratic formula with $a = 4.9, b = -14, c = 6$,

$$t = \frac{14 \pm \sqrt{196 - 117.6}}{9.8} = \frac{14 \pm \sqrt{78.4}}{9.8} = \frac{22.85}{9.8} \text{ or } \frac{5.15}{9.8} = 2.33 \text{ or } 0.53$$

The ball reaches a height of 6 m after 0.53 s on the way up, and after 2.33 s on the way down.

2 The coordinates of A and B simultaneously satisfy both $y = x^2 - 4x$ and $y + x = 10$. Substituting $y = x^2 - 4x$ into $y + x = 10$ gives

$$x^2 - 4x + x = 10$$

$$\Rightarrow \quad x^2 - 3x - 10 = 0$$

$$\Rightarrow \quad (x-5)(x+2) = 0$$

$$\Rightarrow \quad x - 5 = 0 \quad \text{or} \quad x + 2 = 0$$

$$\Rightarrow \quad x = 5 \quad \text{or} \quad x = -2$$

Substituting these values into $y + x = 10$,

when $x = 5,\ y + 5 = 10$

$$\Rightarrow \quad y = 5$$

when $x = -2,\ y + (-2) = 10$

$$\Rightarrow \quad y = 12$$

So the coordinates are A $(-2, 12)$ and B $(5, 5)$.

3 Let the number of people paying £31 be x and let the number of people paying £16 be y.

a Total attendance is 240 people
$\Rightarrow \quad x + y = 240$ [1]

Total box office receipts are £5595
$\Rightarrow \quad 31x + 16y = 5595$ [2]

b Using the elimination method, multiply equation [1] by 16 to give

$16x + 16y = 3840$ [3]

Subtracting equation [3] from equation [2],

$$15x = 1755$$

$$\Rightarrow \quad x = 117$$

Substituting this value into equation [1] gives

$$117 + y = 240$$

$$y = 240 - 117 = 123$$

So 117 people pay £31, and 123 people pay £16.

4 volume of concrete required = length × cross-sectional area

$$\text{cross-sectional area} = \pi R^2 - \pi r^2 = \pi(R^2 - r^2) = \pi(R - r)(R + r)$$

Substituting $R = 1$ and $r = 0.95$,

$$\text{cross-sectional area} = \pi(1 - 0.95)(1 + 0.95) = \pi(0.05)(1.95) = 0.0975\pi \text{ m}^2$$

$$\text{volume of concrete mix} = \text{length} \times \text{cross-sectional area} = 1200 \times 0.0975\pi = 117\pi \text{ m}^3$$

5 Rearranging the angle formula,

$$2 = 4\tan\theta - (1 + \tan^2\theta) = 4\tan\theta - 1 - \tan^2\theta$$

$$\Rightarrow \quad \tan^2\theta - 4\tan\theta + 3 = 0$$

$$\Rightarrow \quad (\tan\theta - 1)(\tan\theta - 3) = 0$$

So $\tan\theta - 1 = 0$ or $\tan\theta - 3 = 0$

$$\Rightarrow \quad \tan\theta = 1 \text{ or } \tan\theta = 3$$

$$\Rightarrow \quad \theta = \tan^{-1}(1) \text{ or } \theta = \tan^{-1}(3)$$

$$\Rightarrow \quad \theta = 45° \text{ or } \theta = 71.6°$$

So the angle the ball was kicked at could be either 45° or 71.6°.

6 The given equations are:

$\frac{1}{4}N_1 - N_2 = 0$ [1]

and $N_1 + \frac{1}{2}N_2 - 99 = 0$ [2]

Multiplying equation [1] by 4,

$N_1 - 4N_2 = 0$ [3]

Subtracting equation [3] from equation [2],

$$\tfrac{1}{2}N_2 - 99 + 4N_2 = 0$$

$$\Rightarrow \quad 4\tfrac{1}{2}N_2 - 99 = 0$$

$$\Rightarrow \quad 9N_2 = 198$$

$$\Rightarrow \quad N_2 = 22$$

Substituting this value into equation [1] to find N_1,

$$\tfrac{1}{4}N_1 - 22 = 0$$

$$\Rightarrow \quad N_1 = 88$$

So $N_1 = 88$ and $N_2 = 22$.

7

$x - 2y = 1$ [1]

$3xy - y^2 = 8$ [2]

Rearranging equation [1], $x = 1 + 2y$. Substituting this into equation [2] gives

$$3(1 + 2y)y - y^2 = 8$$

$$\Rightarrow \quad 3(y + 2y^2) - y^2 = 8$$

$$\Rightarrow \quad 3y + 6y^2 - y^2 - 8 = 0$$

$$\Rightarrow \quad 5y^2 + 3y - 8 = 0$$

Solve this quadratic equation by factorisation using PAFF.

P: $5 \times -8 = -40$

A: coefficient of y is 3

F: $-5, 8$

F: $$5y^2 + 3y - 8 = 5y^2 - 5y + 8y - 8$$
$$= 5y(y-1) + 8(y-1)$$
$$= (y-1)(5y+8)$$

So $$5y^2 + 3y - 8 = 0$$

can be written as $(y-1)(5y+8) = 0$

$\Rightarrow$ $$y - 1 = 0 \text{ or } 5y + 8 = 0$$

$\Rightarrow$ $$y = 1 \text{ or } y = -\tfrac{8}{5}$$

Substituting each of these values into equation [1],

$$x = 1 + 2y$$

When $y = 1$, $\quad x = 1 + 2 = 3$

When $y = -\frac{8}{5}$, $\quad x = 1 + (-\frac{16}{5}) = -\frac{11}{5}$

So the solutions of these equations are (3, 1) and $(-\frac{11}{5}, -\frac{8}{5})$

2 Coordinate Geometry (p. 83)

1 **a** Triangle ABC is isosceles if it has two sides of the same length. Using Pythagoras' theorem to calculate the three side lengths:

$$AB = \sqrt{(4-(-1))^2 + (7-2)^2} = \sqrt{50}$$

$$AC = \sqrt{(6-4)^2 + (1-7)^2} = \sqrt{40}$$

$$BC = \sqrt{(6-(-1))^2 + (1-2)^2} = \sqrt{50}$$

Since $AB = BC$, the triangle is isosceles.

The mid-point of AC, M, has coordinates

$$\left(\frac{4+6}{2}, \frac{7+1}{2}\right) = (5, 4)$$

Since M is the mid-point of AC, it follows that BM is the height of the triangle with base AC.

$$BM = \sqrt{(5-(-1))^2 + (4-2)^2} = \sqrt{40}$$

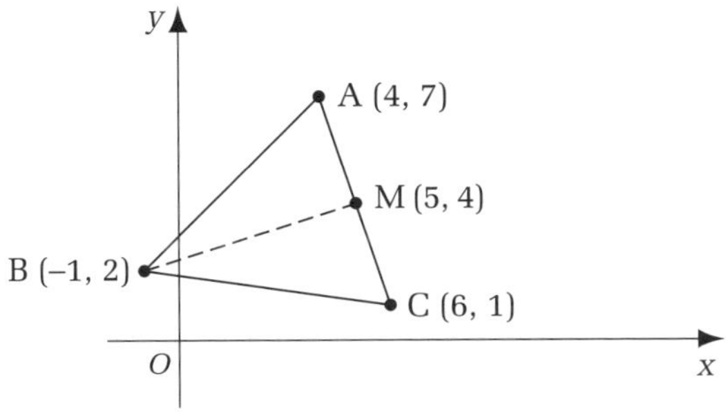

So $\quad$ area $= \frac{1}{2} \times$ base $\times$ perpendicular height

$$= \tfrac{1}{2} \times AC \times BM$$

$$= \tfrac{1}{2} \times \sqrt{40} \times \sqrt{40}$$

$$= 20$$

b The gradient of the line BM is given by

$$\frac{\text{change in } y\text{-coordinate}}{\text{change in } x\text{-coordinate}} = \frac{4-2}{5-(-1)} = \frac{2}{6} = \frac{1}{3}$$

Using $y - y_1 = m(x - x_1)$,

$$y - 4 = \tfrac{1}{3}(x-5)$$

$\Rightarrow$ $$3y - 12 = x - 5$$

$\Rightarrow$ $x - 3y + 7 = 0$

So the equation of BM is $x - 3y + 7 = 0$.

c Gradient of BC $= \dfrac{2-1}{-1-6} = -\frac{1}{7}$, and therefore the gradient of any line perpendicular to BC is 7.

Using $y - y_1 = m(x - x_1)$, the equation of the line perpendicular to BC that passes through A (4, 7) is given by

$$y - 7 = 7(x-4)$$

$\Rightarrow$ $$y - 7 = 7x - 28$$

$\Rightarrow$ $$7x - y - 21 = 0$$

So the equation of the perpendicular through A (4, 7) is $7x - y - 21 = 0$.

d Find the coordinates of H by solving the simultaneous equations

$$7x - y - 21 = 0 \qquad [1]$$

$$x - 3y + 7 = 0 \qquad [2]$$

Multiplying equation [1] by 3 gives

$$21x - 3y - 63 = 0 \qquad [3]$$

Subtracting equation [2] from equation [3],

$$20x - 70 = 0$$

$\Rightarrow$ $$20x = 70$$

$\Rightarrow$ $$x = 3.5$$

Substituting $x = 3.5$ into equation [2],

$$3.5 - 3y + 7 = 0$$

$\Rightarrow$ $$10.5 = 3y$$

$\Rightarrow$ $$y = 3.5$$

Check these answers by substituting them into equation [1].

The coordinates of H are (3.5, 3.5)

Now, the gradient of CH $= \dfrac{3.5-1}{3.5-6}$

$$= \frac{2.5}{-2.5}$$

$$= -1$$

and the gradient of AB $= \dfrac{7-2}{4-(-1)}$

$$= \tfrac{5}{5}$$

$$= 1$$

Since the product of the two gradients is -1, the lines CH and AB must be perpendicular.

2 length of AB $= \sqrt{(4-2)^2+(-3-3)^2}$

$$= \sqrt{2^2+(-6)^2}$$
$$= \sqrt{40}$$
$$= 2\sqrt{10}$$

The coordinates of the mid-point are

$$\left(\frac{2+4}{2}, \frac{3+(-3)}{2}\right) = (3, 0)$$

3

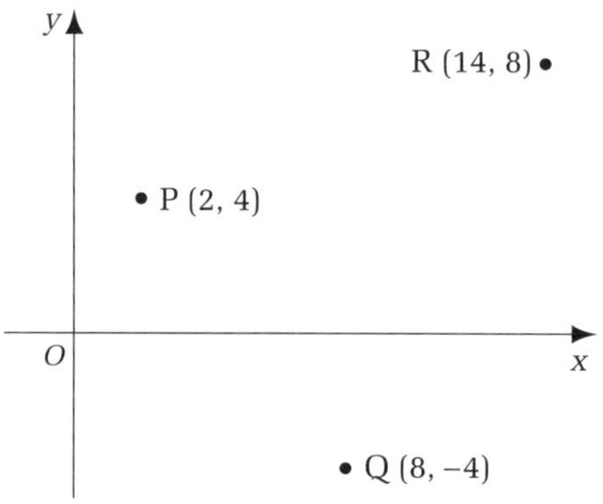

a gradient of PQ $= \frac{4-(-4)}{2-8}$

$$= \frac{8}{-6}$$
$$= -\tfrac{4}{3}$$

So the gradient of any line perpendicular to PQ is $\frac{3}{4}$.
The coordinates of the mid-point of PQ are

$$\left(\frac{2+8}{2}, \frac{4+(-4)}{2}\right) = (5, 0)$$

Using $y - y_1 = m(x - x_1)$, the equation of the perpendicular bisector of PQ is given by

$$y - 0 = \tfrac{3}{4}(x-5)$$
$$\Rightarrow \quad 4y = 3(x-5)$$
$$\Rightarrow \quad 4y = 3x - 15$$
$$\Rightarrow \quad 3x - 4y - 15 = 0$$

gradient of PR $= \frac{8-4}{14-2}$

$$= \tfrac{4}{12}$$
$$= \tfrac{1}{3}$$

So the gradient of any line perpendicular to PR is -3.
The coordinates of the mid-point of PR are

$$\left(\frac{14+2}{2}, \frac{8+4}{2}\right) = (8, 6)$$

Using $y - y_1 = m(x - x_1)$, the equation of the perpendicular bisector of PR is given by

$$y - 6 = -3(x-8)$$
$$\Rightarrow \quad y - 6 = -3x + 24$$
$$\Rightarrow \quad 3x + y - 30 = 0$$

b Find the coordinates of C by solving the simultaneous equations:

$$3x - 4y - 15 = 0 \qquad [1]$$
$$3x + y - 30 = 0 \qquad [2]$$

Use the elimination method to find y.
Subtracting equation [1] from equation [2],

$$5y - 15 = 0$$
$$\Rightarrow \quad y = 3$$

Substituting this into equation [2] to find x,

$$3x + 3 - 30 = 0$$
$$\Rightarrow \quad 3x = 27$$
$$\Rightarrow \quad x = 9$$

The coordinates of C are (9, 3). It follows that

$$CP^2 = (9-2)^2 + (3-4)^2$$
$$= 7^2 + (-1)^2$$
$$= 50$$

c

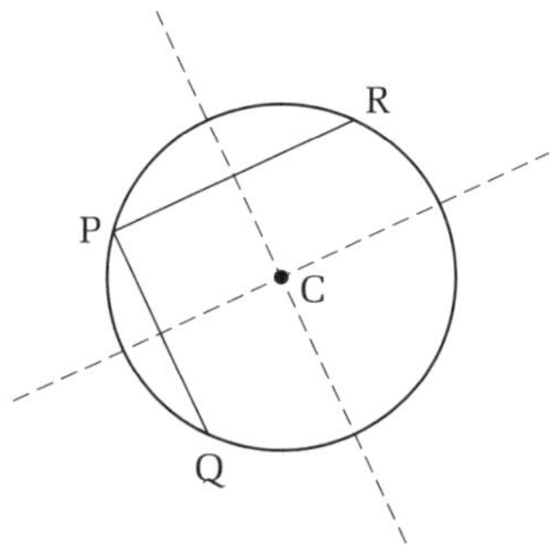

We know that the coordinates of C are (9, 3) and that $CP^2 = 50$. So we can deduce that the equation of the circle is

$$(x-9)^2 + (y-3)^2 = 50$$
$$\Rightarrow \quad (x^2 - 18x + 81) + (y^2 - 6y + 9) = 50$$
$$\Rightarrow \quad x^2 + y^2 - 18x - 6y + 40 = 0$$

4 $2(2m+1)^2 > 9(m^2+1)$

Multiplying out the brackets gives

$$2(4m^2 + 4m + 1) > 9m^2 + 9$$
$$\Rightarrow \quad 8m^2 + 8m + 2 > 9m^2 + 9$$
$$\Rightarrow \quad 0 > m^2 - 8m + 7$$
$$\Rightarrow \quad m^2 - 8m + 7 < 0$$

The graph of $y = m^2 - 8m + 7$ is shown below.

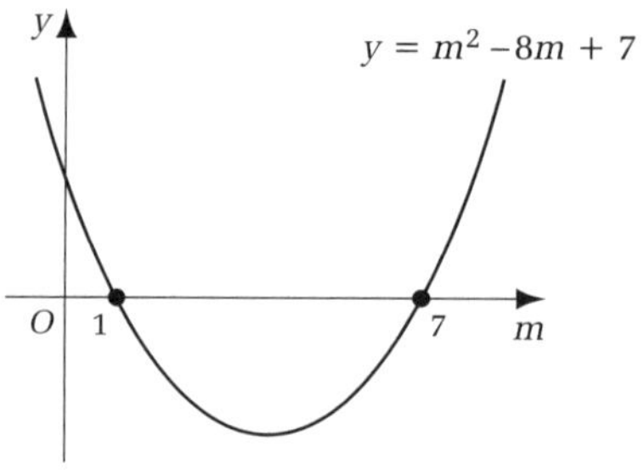

The solution set of the inequality corresponds to these values of m for which the curve lies below the m-axis, that is, between the critical values of $m = 1$ and $m = 7$. The solution set is therefore $1 < m < 7$.

5 To find the centre and radius of the circle, first rearrange its equation into form $(x-a)^2+(y-b)^2=r^2$.

$$x^2+y^2+2x=9$$
$$\Rightarrow \quad x^2+2x+1+y^2=9+1$$
$$\Rightarrow \quad (x+1)^2+y^2=10$$

The circle is therefore centred at $(-1, 0)$ and has a radius of $\sqrt{10}$.

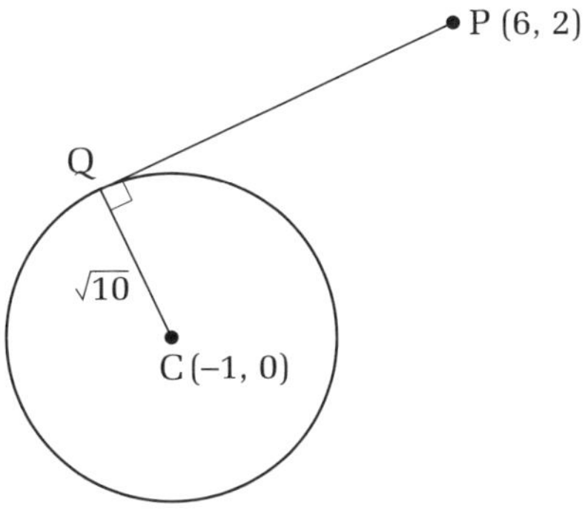

Let P be (6, 2) and Q be the point of contact of the tangent with the circle. We know that the coordinates of the centre of the circle are C $(-1, 0)$. So

$$PC=\sqrt{(6-(-1))^2+(2-0)^2}$$
$$=\sqrt{7^2+2^2}$$
$$=\sqrt{53}$$

Using Pythagoras' theorem in triangle CQP,

$$(PC)^2=(QC)^2+(QP)^2$$
$$\Rightarrow \quad (\sqrt{53})^2=(\sqrt{10})^2+(QP)^2$$
$$\Rightarrow \quad 53=10+(QP)^2$$
$$\Rightarrow \quad QP=\sqrt{43}$$

Therefore the length of the tangent is $\sqrt{43}$.

6 $(x-3)(x-6)>x+9$

Multiplying out the brackets.

$$x^2-9x+18>x+9$$
$$\Rightarrow \quad x^2-10x+9>0$$
$$\Rightarrow \quad (x-1)(x-9)>0$$

The critical values of x are 1 and 9.

	$x<1$	$1<x<9$	$x>9$
$(x-1)$	–ve	+ve	+ve
$(x-9)$	–ve	–ve	+ve
$(x-1)(x-9)$	+ve	–ve	+ve

So the inequality is satisfied if $x<1$ or $x>9$.

3 Trigonometry I (p. 132)

1 $\sin 2\theta=\cos^2\theta$

Using the double angle formula,

$$2\sin\theta\cos\theta=\cos^2\theta$$
$$2\sin\theta\cos\theta-\cos^2\theta=0$$

Factorising,

$$\cos\theta(2\sin\theta-\cos\theta)=0$$

So $\cos\theta=0$ or $2\sin\theta-\cos\theta=0$

Now $2\sin\theta-\cos\theta=0 \Rightarrow 2\sin\theta=\cos\theta$

$$\frac{\sin\theta}{\cos\theta}=\frac{1}{2}$$
$$\tan\theta=\frac{1}{2}$$

So $\cos\theta=0 \Rightarrow \theta=90°, 270°$

and $\tan\theta=\frac{1}{2} \Rightarrow \theta=27°, 207°$ (to the nearest degree)

2 Using $\text{area}=\frac{1}{2}ab\sin C$

$$12=\tfrac{1}{2}x(x+2)\sin 30°$$
$$\Rightarrow \quad 12=\tfrac{1}{2}x(x+2)\times\tfrac{1}{2}$$
$$\Rightarrow \quad 48=x(x+2)$$
$$\Rightarrow \quad 48=x^2+2x$$
$$\Rightarrow \quad x^2+2x-48=0$$
$$\Rightarrow \quad (x+8)(x-6)=0$$

Lengths cannot be negative, so discard the solution $x=-8$.
The only valid solution is $x=6$.

3

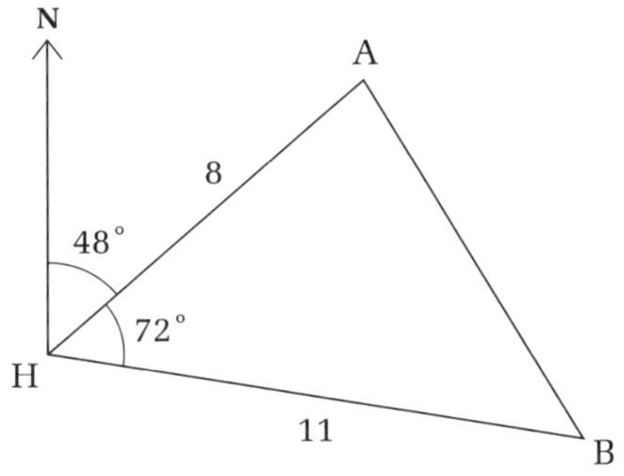

a Since bearing of B is 120°, angle H of the triangle is

$$120°-48°=72°$$

Using the cosine rule with $a=11$, $b=8$, $H=72°$

$$h^2=a^2+b^2-2ab\cos H$$
$$=11^2+8^2-2\times8\times11\times\cos 72°$$
$$=130.613\,009$$
$$h=\sqrt{130.613\,009}$$

So $AB=11.4$ km (3 s.f.)

b

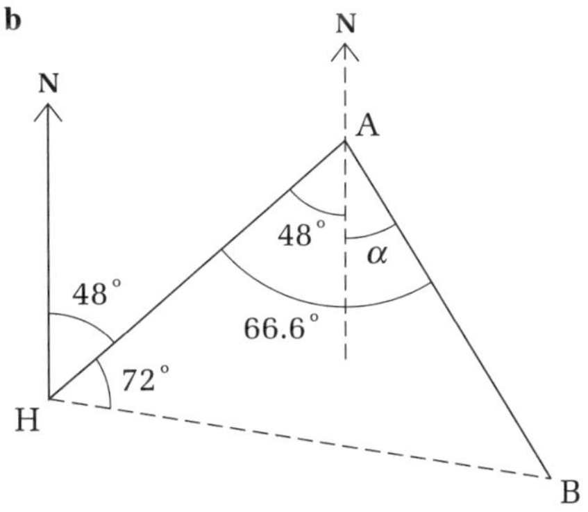

Required bearing is $180° - \alpha$. We can find α by finding angle A. Using the sine rule,

$$\frac{\sin A}{a} = \frac{\sin H}{h}$$

$$\frac{\sin A}{11} = \frac{\sin 72°}{11.4}$$

$$\Rightarrow \quad \sin A = \frac{11 \times \sin 72°}{11.4}$$

$$= 0.9177 \text{ (4 d.p.)}$$

So $\quad A = 66.6°$ (1 d.p.)

Now,

$$\alpha = 66.6° - 48° = 18.6°$$

So required bearing is

$$180° - \alpha = 180° - 18.6°$$
$$= 161.4°$$
$$= 161° \text{ (to the nearest degree)}$$

4 $4\cot^2\theta + 12\operatorname{cosec}\theta + 1 = 0$

Using the Pythagorean identities,

$$4(\operatorname{cosec}^2\theta - 1) + 12\operatorname{cosec}\theta + 1 = 0$$

$$4\operatorname{cosec}^2\theta + 12\operatorname{cosec}\theta - 3 = 0$$

$$\operatorname{cosec}^2\theta + 3\operatorname{cosec}\theta = \tfrac{3}{4}$$

This equation will not factorise so complete the square.

$$\operatorname{cosec}^2\theta + 3\operatorname{cosec}\theta + \tfrac{9}{4} = \tfrac{3}{4} + \tfrac{9}{4}$$

$$(\operatorname{cosec}\theta + \tfrac{3}{2})^2 = \tfrac{12}{4} = 3$$

$$\operatorname{cosec}\theta + \tfrac{3}{2} = \pm\sqrt{3}$$

$$\operatorname{cosec}\theta = -\tfrac{3}{2} \pm \sqrt{3}$$

So $\quad \operatorname{cosec}\theta = 0.232$ (3 d.p.) or $\operatorname{cosec}\theta = -3.232$ (3 d.p.)

$\Rightarrow \quad \sin\theta = 4.31$ or $\sin\theta = -0.3094$

$\sin\theta = 4.31$ has no solutions.

$\sin\theta = -0.3094 \Rightarrow \theta = 198°, 342°$ (nearest degree)

5 $\tan(2x + 30°) = -\sqrt{3}$

Tangent is negative in the second and fourth quadrants, and $\sqrt{3}$ is the tangent of a 'special angle'.

$$\tan(2x + 30°) = -\sqrt{3} \Rightarrow 2x + 30° = 120°, 300°, 480°, 660°$$

$$\Rightarrow \quad 2x = 90°, 270°, 450°, 630°$$

$$\Rightarrow \quad x = 45°, 135°, 225°, 315°$$

6 **a** If $\cos 2x < 0$, then $2x$ is in the second quadrant. So,

$$2x > 90°$$
$$\Rightarrow \quad x > 45°$$

b If $\sin 4y < 0$ and $\cos 4y > 0$, then $4y$ is in the fourth quadrant. So,

$$4y > 270°$$
$$\Rightarrow \quad y > 67\tfrac{1}{2}°$$

c We have $x > 45°$ and $y > 67\frac{1}{2}°$

i Given $x > 45°$ and $y > 67\frac{1}{2}°$, $(x + y) > 112\frac{1}{2}°$. And we know that x and y are acute, so $(x + y) < 180°$. Thus $(x + y)$ is in the second quadrant, and therefore $\tan(x + y)$ is negative.

ii Given $y > 67\frac{1}{2}°$, $3y > 202\frac{1}{2}°$. Since y is acute, $3y < 270°$, so $3y$ is in the third quadrant, and therefore $\tan 3y$ is positive.

7 **a** By substituting $\theta = \frac{\pi}{12}$, the formula

$$\tan 3\theta = \frac{3\tan\theta - \tan^3\theta}{1 - 3\tan^2\theta}$$

becomes

$$\tan\left(\tfrac{\pi}{4}\right) = \frac{3\tan\left(\frac{\pi}{12}\right) - \tan^3\left(\frac{\pi}{12}\right)}{1 - 3\tan^2\left(\frac{\pi}{12}\right)}$$

Using $\tan\left(\frac{\pi}{4}\right) = 1$ and putting $x = \tan\left(\frac{\pi}{12}\right)$,

$$1 = \frac{3x - x^3}{1 - 3x^2}$$

$$\Rightarrow \quad 1 - 3x^2 = 3x - x^3$$

$$\Rightarrow \quad x^3 - 3x^2 - 3x + 1 = 0$$

Therefore $x = \tan\left(\frac{\pi}{12}\right)$ is a root of the given cubic equation.

b $\sin 2y = 2 - 2\cos 2y$

Using the double angle formulas

$$2\sin y\cos y = 2 - 2(1 - 2\sin^2 y)$$
$$= 2 - 2 + 4\sin^2 y$$

$$\Rightarrow \quad 2\sin y\cos y - 4\sin^2 y = 0$$

Factorising,

$$2\sin y(\cos y - 2\sin y) = 0$$

So $\quad \sin y = 0$ or $\cos y - 2\sin y = 0$

Now $\quad \sin y = 0 \quad \Rightarrow \quad y = 0, \pi$

and $\quad \cos y - 2\sin y = 0 \quad \Rightarrow \quad \cos y = 2\sin y$

$$\Rightarrow \quad \frac{\sin y}{\cos y} = \tfrac{1}{2}$$

$$\Rightarrow \quad \tan y = \tfrac{1}{2}$$

$$\Rightarrow \quad y = 0.464$$

So the solutions are $y = 0, 0.464, 3.14$ radians (3 s.f.)

8 This identity is straightforward to show once we spot the difference of two squares in factorised form.

$$\text{LHS} = (\operatorname{cosec} x - 1)(\operatorname{cosec} x + 1)(\sec x - 1)(\sec x + 1)$$
$$= (\operatorname{cosec}^2 x - 1)(\sec^2 x - 1)$$
$$= (\cot^2 x)(\tan^2 x)$$
$$= \left(\frac{1}{\tan^2 x}\right) \times (\tan^2 x)$$
$$= 1 = \text{RHS}$$

9 Begin by drawing a diagram.

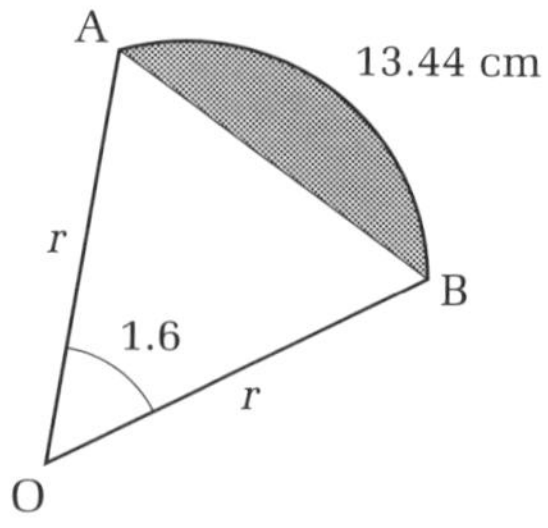

a Use the formula arc length $= r\theta$

$$13.44 = r \times 1.6$$

$$r = \frac{13.44}{1.6} = 8.4 \text{ cm}$$

b The required area is shaded in the diagram.

required area = area of sector OAB – area of triangle OAB

$$= \tfrac{1}{2}r^2\theta - \tfrac{1}{2}r^2 \sin\theta$$

$$= (\tfrac{1}{2} \times 8.4^2 \times 1.6) - (\tfrac{1}{2} \times 8.4^2 \times \sin 1.6)$$

$$= 56.448 - 35.265$$

$$= 21.183$$

$$= 21.2 \text{ cm}^2 \text{ (3 s.f.)}$$

10 **a** $\tan x = 5$
Tangent is positive in the first and third quadrants. Use a calculator to find the first solution and then add 180° (the period of the tangent function) to obtain the second solution.

$$x = \tan^{-1}(5)$$

$$= 78.7°, 258.7° \text{(1 d.p.)}$$

b $3\cos^2 y + 8\sin y = 0$
Using the Pythagorean identity, $\sin^2\theta + \cos^2\theta \equiv 1$ we get

$$3(1 - \sin^2 y) + 8\sin y = 0$$

$$\Rightarrow \quad 3 - 3\sin^2 y + 8\sin y = 0$$

$$\Rightarrow \quad 3\sin^2 y - 8\sin y - 3 = 0$$

$$\Rightarrow \quad (3\sin y + 1)(\sin y - 3) = 0$$

$$\Rightarrow \quad 3\sin y + 1 = 0 \text{ or } \sin y - 3 = 0$$

$$\Rightarrow \quad \sin y = -\tfrac{1}{3} \text{ or } \sin y = 3$$

$\sin y = 3$ has no solutions

$$\sin y = -\tfrac{1}{3} \Rightarrow y = 3.48, 5.94 \text{ (2 d.p.)}$$

4 Functions (p. 184)

1 **a** **i** This graph *does not* represent a function because the mapping is *one-to-many*.

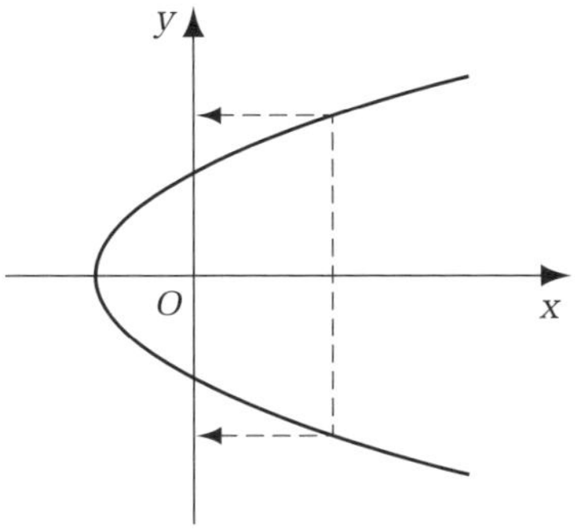

ii This graph *does* represent a function because the mapping is *one-to-one*.

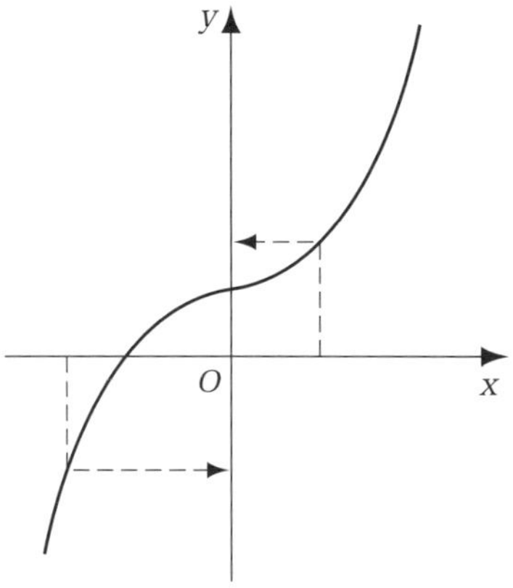

iii This graph *does* represent a function because the mapping is *many-to-one*.

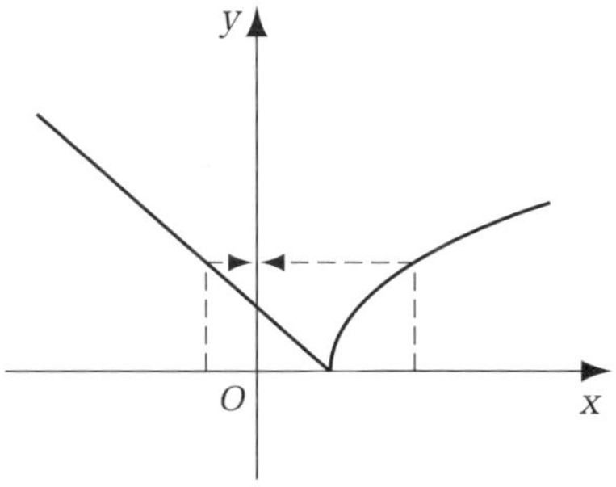

b The only function for which an inverse function exists is that represented by graph **ii**.

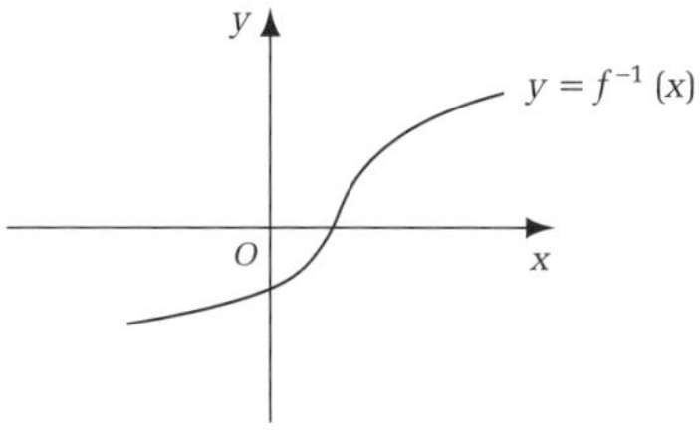

2 **a** $gf(x) = g(x^2) = 4 - 9(x^2) = 4 - 9x^2$

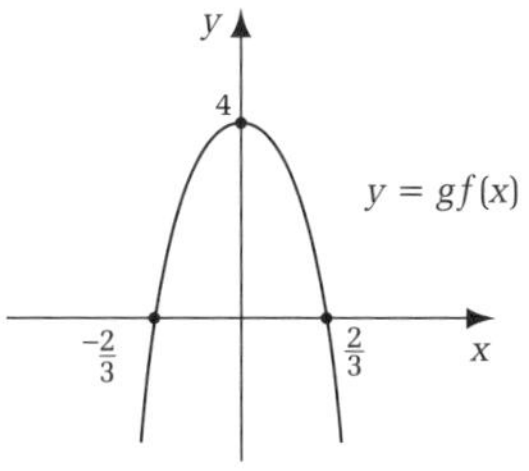

c The range of gf is $gf(x) \leq 4$.

d To find the inverse function $g^{-1}(x)$:

Express $g(x)$ in the form $\quad y = 4 - 9x$.

Interchange x and y: $\quad x = 4 - 9y$

Rearrange to make y the subject again:

$$y = \frac{4 - x}{9}.$$

Therefore $\quad g^{-1}(x) = \dfrac{4 - x}{9}$

If $g(x) = g^{-1}(x)$, then $\quad 4 - 9x = \dfrac{4 - x}{9}$

$$36 - 81x = 4 - x$$

$$80x = 32$$

$$x = \tfrac{2}{5}$$

3 **a** $f: x \to 9x^2 - 4$ for $x \geq 0$.
Since the minimum value of x^2 is zero in this domain, the range of f is $f(x) \geq -4$.

b The graph crosses the y-axis when $x = 0$

$\Rightarrow \quad y = -4$

The graph crosses the x-axis when $y = 0$

$$\Rightarrow \quad 9x^2 - 4 = 0$$
$$x^2 = \tfrac{4}{9}$$
$$x = \pm\tfrac{2}{3}$$

The solution $x = -\frac{2}{3}$ is outside the domain, so $x = \frac{2}{3}$.

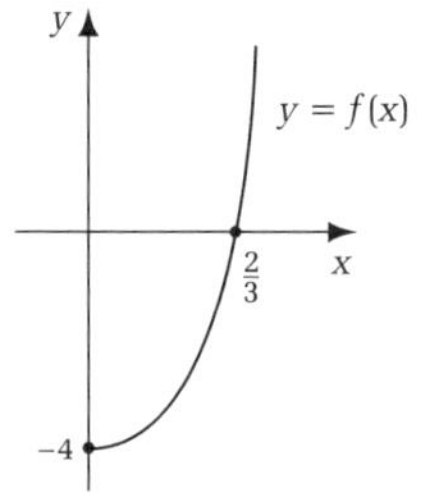

The inverse function f^{-1} exists because function f is one-to-one for the given domain.

To find the inverse function $f^{-1}(x)$:

Express $f(x)$ in the form $\quad y = 9x^2 - 4$

Interchange x and y: $\quad x = 9y^2 - 4$

Rearrange to make y the subject again: $\quad y = \sqrt{\dfrac{x+4}{9}}$

Therefore $f^{-1}(x) = \sqrt{\dfrac{x+4}{9}}$, and the domain of $f^{-1}(x)$ is $x \geq -4$.

c **i** Recall that $f \circ g(x)$ means 'function g followed by function f'.

$$f \circ g(x) = f(\sqrt{x+1})$$
$$= 9(\sqrt{x+1})^2 - 4$$
$$= 9(x+1) - 4$$
$$= 9x + 5$$

ii If $\quad f(x) = f \circ g(x)$

then $\quad 9x^2 - 4 = 9x + 5$

$\Rightarrow \quad 9x^2 - 9x - 9 = 0$

$\Rightarrow \quad x^2 - x - 1 = 0$

$\Rightarrow \quad x = \dfrac{1+\sqrt{5}}{2}$

Note that we can ignore the negative solution because it is outside the domain of f and g.

4

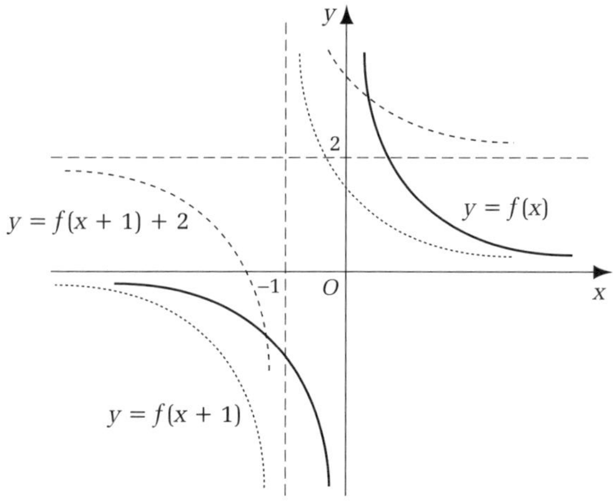

The graph of $y = f(x+1)$ is obtained by translating the graph of $y = f(x)$ by -1 unit parallel to the x-axis.

The graph of $y = f(x+1) + 2$ is obtained by translating the graph of $y = f(x)$ by -1 unit parallel to the x-axis and by $+2$ units parallel to the y-axis.

Writing $\quad y = \dfrac{2x+3}{x+1}$ in terms of $f(x) = \dfrac{1}{x}$,

$$y = \frac{2x+3}{x+1} = 2 + \frac{1}{x+1}$$
$$\equiv f(x+1) + 2, \text{ where } f(x) = \tfrac{1}{x}$$

Since the graph of $y = f(x)$ is symmetrical about the origin, the graph of $y = \dfrac{2x+3}{x+1}$ is symmetrical about the point $(-1, 2)$ as shown on the diagram.

5 To find the inverse function $f^{-1}(x)$:

Express $f(x)$ in the form $\quad y = \tan 2x$.

Interchange x and y: $\quad x = \tan 2y$

Rearrange to make y the subject again: $\quad 2y = \tan^{-1} x$

$$y = \tfrac{1}{2}\tan^{-1} x$$

Therefore $\quad f^{-1}(x) = \tfrac{1}{2}\tan^{-1} x$

The domain of f^{-1} is equivalent to the range of f. So the domain of f^{-1} is $0 \leq x \leq 1$.

6 **a** The range of g is $g(x) \geq 0$.

b
$$gf(x) = g(4-x)$$
$$= 3(4-x)^2$$
$$= 3(16 - 8x + x^2)$$
$$= 48 - 24x + 3x^2$$

If $gf(x) = 48$, then $\quad 48 - 24x + 3x^2 = 48$

$\Rightarrow \quad 3x^2 - 24x = 0$

$\Rightarrow \quad 3x(x-8) = 0$

$\Rightarrow \quad x = 0$ or $x = 8$

c

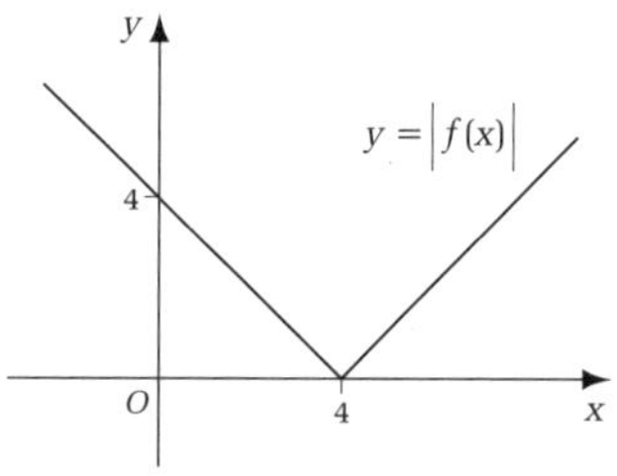

If $|f(x)| = 2$ then $|x - 4| = 2$

$\Rightarrow \quad x - 4 = 2$ or $x - 4 = -2$

$\Rightarrow \quad x = 6$ or $x = 2$

Hence, $|f(x)| = 2$ for $x = 2$ and $x = 6$

7 **a** The range of g is $g(x) \geq 1$.

b
$$gf(x) = g(3x-1)$$
$$= (3x-1)^2 + 1$$
$$= (9x^2 - 6x + 1) + 1$$
$$= 9x^2 - 6x + 2$$

$$fg(x) = f(x^2+1) = 3(x^2+1) - 1 = 3x^2 + 2$$

If $gf(x) = fg(x)$

then $9x^2 - 6x + 2 = 3x^2 + 2$

$\Rightarrow 6x^2 - 6x = 0$

$\Rightarrow 6x(x-1) = 0$

$\Rightarrow x = 0$ or $x = 1$

c If $|f(x)| = 8$ then $|3x - 1| = 8$

$\Rightarrow 3x - 1 = 8$ or $3x - 1 = -8$

$\Rightarrow x = 3$ or $x = -\frac{7}{3}$

d $h(x) = x^2 + 3x$

$= x^2 + 3x + \frac{9}{4} - \frac{9}{4}$

$= (x + \frac{3}{2})^2 - \frac{9}{4}$

So function h is a quadratic function, with a minimum value of $-\frac{9}{4}$ when $x = -\frac{3}{2}$. Since h is a one-to-one function, with domain $x \in \mathbb{R}, x \geq q$, only those values of x greater than $x = -\frac{3}{2}$ can be included. Hence, the least value of q is $-\frac{3}{2}$.

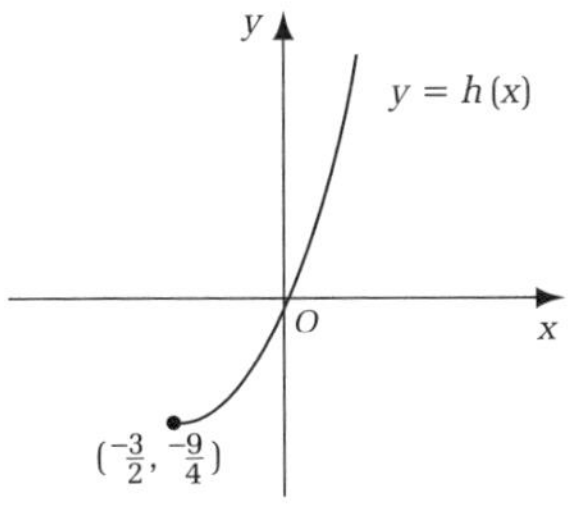

8 **a**

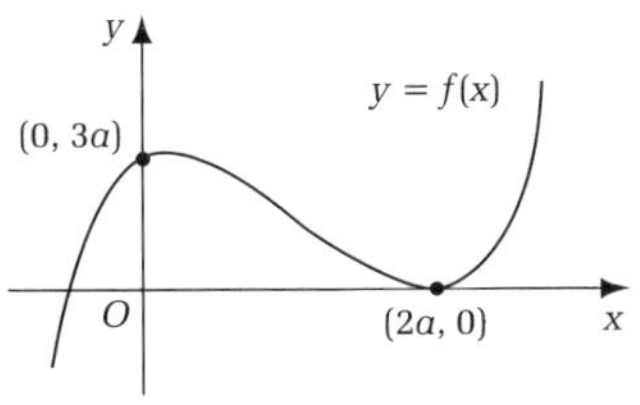

The graph of $y = f(x)$ is obtained by translating the graph of $y = f(x+a)$ by $+a$ units parallel to the x-axis. So the maximum point is (0, 3a) and the minimum point is (2a, 0).

b

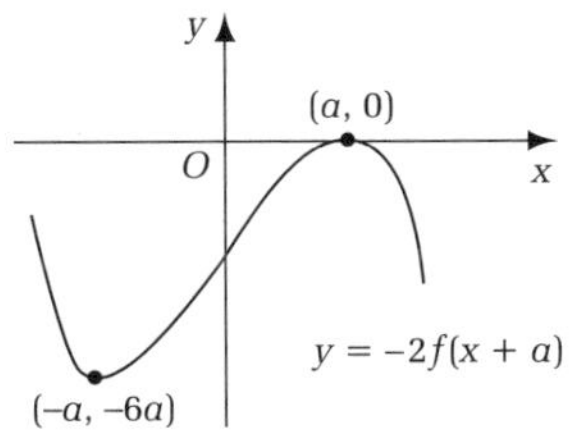

The graph of $y = -2f(x+a)$ is obtained by stretching the graph of $y = f(x+a)$ by a factor of 2 parallel to the y-axis, followed by a reflection in the x-axis. So the minimum point is $(-a, -6a)$ and the maximum point is $(a, 0)$.

9 **a** **i** To find the inverse function $f^{-1}(x)$:

Express $f(x)$ in the form $y = 1 + \dfrac{2}{x}$.

Interchange x and y: $x = 1 + \dfrac{2}{y}$

Rearrange to make y the subject again:

$$y = \frac{2}{x-1}.$$

Therefore $f^{-1}(x) = \dfrac{2}{x-1}$

ii The domain of f^{-1} is equivalent to the range of f, so the domain of f^{-1} is $x > 1$.

b **i** $g(x) = ff(x)$

$$= f\left(1 + \frac{2}{x}\right) = 1 + \frac{2}{(1+\frac{2}{x})} = \frac{(1+\frac{2}{x}) + 2}{(1+\frac{2}{x})} = \frac{\left(\frac{3x+2}{x}\right)}{\left(\frac{x+2}{x}\right)} = \frac{3x+2}{x+2}$$

Therefore $g(x) = \dfrac{3x+2}{x+2} = 3 - \dfrac{4}{x+2}$

ii We have $x > 0$, so the range of g is $1 < g(x) < 3$.

10 **a** $f(x) = 4 - 2x - x^2$

$= -(x^2 + 2x - 4)$

$= -(x^2 + 2x + 1) + 5$

$= 5 - (x+1)^2$

So $a = 5$ and $b = 1$.

b The graph of $y = f(x)$ can be obtained from part of the $y = -x^2$ graph by translations of $+5$ units parallel to the y-axis and of -1 units parallel to the x-axis.

c The coordinates on the points of the $y = -x^2$ graph which correspond to the end points of $y = f(x)$ are found by reversing the translations in **b** for the end points $(-1, 5)$ and $(2, -4)$. So the coordinates are $(0, 0)$ and $(3, -9)$.

d The graph of $y = f^{-1}(x)$ is obtained by reflecting the graph of $y = f(x)$ in the line $y = x$.

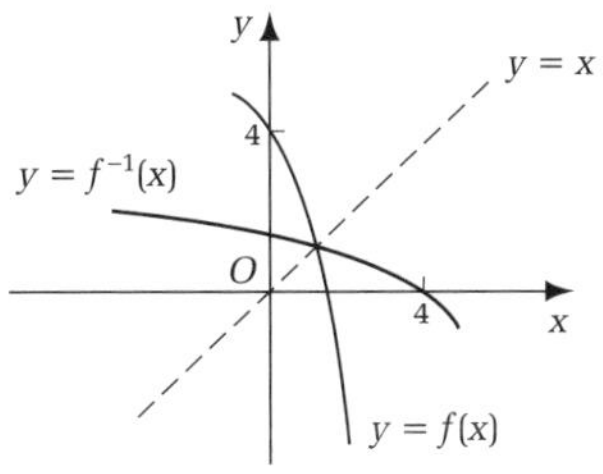

e From the above diagram, it is clear that the solution to $f(x) = f^{-1}(x)$ can alternatively be found by solving $f(x) = x$.

If $f(x) = x$ then $4 - 2x - x^2 = x$

$\Rightarrow x^2 + 3x - 4 = 0$

$\Rightarrow (x+4)(x-1) = 0$

$\Rightarrow x = -4$ (ignore) or $x = 1$

We can discard $x = -4$, because f is defined only on the domain $-1 \leq x \leq 2$.

Hence, $f(x) = f^{-1}(x)$ at $x = 1$.

5 Differentiation I (p. 249)

1 **a** When $x = 3$,

$$y = 2 \times (3)^3 - 3 \times (3)^2 - 36 \times (3) + 120$$
$$= 54 - 27 - 108 + 120$$
$$= 39$$

When $x = -2$,

$$y = 2 \times (-2)^3 - 3 \times (-2)^2 - 36 \times (-2) + 120$$
$$= -16 - 12 + 72 + 120$$
$$= 164$$

b $\frac{dy}{dx} = 6x^2 - 6x - 36$

c At stationary points on the curve, $\frac{dy}{dx} = 0$

$\Rightarrow \quad 6x^2 - 6x - 36 = 0$

$\Rightarrow \quad 6(x^2 - x - 6) = 0$

Factorising,

$$6(x + 2)(x - 3) = 0$$

$\Rightarrow \quad x = -2 \quad \text{or} \quad x = 3$

Using the y-coordinates calculated in **a**, the coordinates of the stationary points on the curve are $(-2, 164)$ and $(3, 39)$.

d To determine the nature of these stationary points, look at the values of $\frac{dy}{dx}$ on either side of each point.

When $x = -3$, $\frac{dy}{dx} = 6 \times (-3)^2 - 6 \times (-3) - 36$
$= 36$

When $x = -1$, $\frac{dy}{dx} = 6 \times (-1)^2 - 6 \times (-1) - 36$
$= -24$

Since the gradient changes from +ve to −ve as it passes through $x = -2$, the stationary point at $(-2, 164)$ is a maximum.

When $x = 2$, $\frac{dy}{dx} = 6 \times (2)^2 - 6 \times (2) - 36 = -24$

When $x = 4$, $\frac{dy}{dx} = 6 \times (4)^2 - 6 \times (4) - 36 = 36$

Since the gradient changes from −ve to +ve as it passes through $x = 3$, the stationary point at $(3, 39)$ is a minimum point.

e

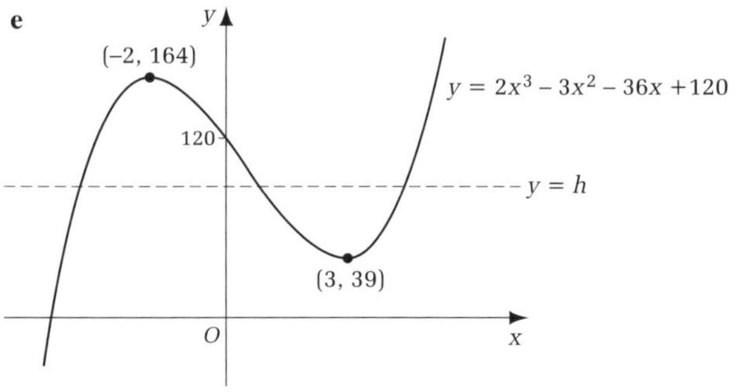

The cubic equation $2x^3 - 3x^2 - 36x + 120 = 0$ only has one root because, as the sketch shows, the graph of $y = 2x^3 - 3x^2 - 36x + 120$ only crosses the x-axis, or $y = 0$, at one point.

However, the graph of $y = 2x^3 - 3x^2 - 36x + 120$ does intersect with the line $y = h$ at three distinct values of x (as shown), for $39 < h < 164$ (that is, between the stationary points). These points of intersection are found by solving the equation $2x^3 - 3x^2 - 36x + 120 = h$. Therefore, the cubic equation $2x^3 - 3x^2 - 36x + 120 = h$ has three real roots, provided $39 < h < 164$.

2

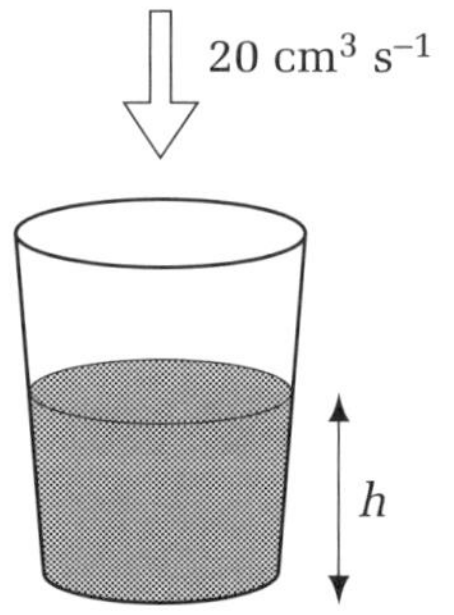

When the height of water in the cup is h cm, the volume of water is $V = \frac{1}{30}\pi h^3$ cm^3. The rate at which the volume changes with respect to height is given by

$$\frac{dV}{dh} = \frac{1}{10}\pi h^2 \text{ cm}^3 \text{ cm}^{-1}$$

Since the machine delivers water at a constant rate of 20 cm^3 s^{-1},

$$\frac{dV}{dt} = 20 \text{ cm}^3 \text{ s}^{-1}$$

Using the chain rule, the rate of change of height h is given by

$$\frac{dh}{dt} = \frac{dh}{dV} \times \frac{dV}{dt}$$
$$= \frac{10}{\pi h^2} \times 20$$
$$= \frac{200}{\pi h^2} \text{ cm s}^{-1}$$

When $h = 5$, $\frac{dh}{dt} = \frac{200}{\pi(5)^2} = 2.55$ cm s^{-1} (2 d.p.)

3 **a** Let $y = u^6$ where $u = 3 - x^2$.

Then $\frac{dy}{du} = 6u^5$ and $\frac{du}{dx} = -2x$.

By the chain rule,

$$\frac{dy}{dx} = \frac{dy}{du} \times \frac{du}{dx}$$
$$= 6u^5 \times -2x$$

Substituting $u = 3 - x^2$,

$$\frac{dy}{dx} = -12x\left(3 - x^2\right)^5$$

b When $x = 2$, $y = (3 - 2^2)^6 = 1$

and $\frac{dy}{dx} = -12 \times 2 \times (3 - 2^2)^5 = 24$

The gradient of the tangent to the curve at $(2, 1)$ is 24, so the gradient of the normal to the curve at this point is $-\frac{1}{24}$

Using $y - y_1 = m(x - x_1)$, the equation of the normal is given by

$$y - 1 = -\frac{1}{24}(x - 2)$$

$\Rightarrow \quad 24y - 24 = -x + 2$

$\Rightarrow \quad x + 24y - 26 = 0$

4 **a** The displacement of train T from point C t minutes past noon is $x = \frac{1}{20}(24t - 9t^2 - 2t^3)$ km.

The velocity of the train at this time is given by

$$v = \frac{dx}{dt}$$
$$= \tfrac{1}{20}(24 - 18t - 6t^2)$$

Taking out the common factor of 6:

$$v = \tfrac{6}{20}(4 - 3t - t^2)$$
$$= \tfrac{3}{10}(4 - 3t - t^2)\text{ km min}^{-1}$$

b At the times of departure of train from A, and its arrival at B, velocity v is zero.

$$\Rightarrow \quad \tfrac{3}{10}(4 - 3t - t^2) = 0$$
$$\Rightarrow \quad 4 - 3t - t^2 = 0$$

Factorising,

$$(4 + t)(1 - t) = 0$$
$$\Rightarrow \quad t = -4 \text{ or } t = 1$$

Since time t is measured in minutes *past* noon the train departs A at 11.56 a.m. and arrives at B at 12.01 p.m.

c At station A, when $t = -4$, the displacement of the train from point C is given by

$$x = \tfrac{1}{20}[24 \times (-4) - 9 \times (-4)^2 - 2 \times (-4)^3]$$
$$= -5.6 \text{ km}$$

At station B, when $t = 1$, the displacement of the train from point C is given by

$$x = \tfrac{1}{20}[24 \times (1) - 9 \times (1)^2 - 2 \times (1)^3]$$
$$= 0.65 \text{ km}$$

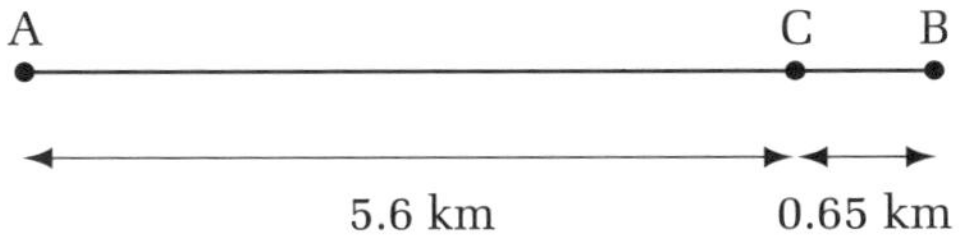

This means that the distance between the stations is

$$5.6 + 0.65 = 6.25 \text{ km}$$

d The average speed of the train for the journey from A to B is given by $\dfrac{\text{total distance travelled}}{\text{time taken}}$.

$$\frac{\text{total distance travelled}}{\text{time taken}} = \frac{6.25 \text{ km}}{5 \text{ min}}$$
$$= 1.25 \text{ km min}^{-1}$$
$$= 75 \text{ km h}^{-1}$$

e The acceleration of the train is given by

$$a = \frac{dv}{dt}$$
$$= \tfrac{3}{10}(-3 - 2t)$$
$$= -\tfrac{3}{10}(3 + 2t)$$

When the train reaches its greatest speed, acceleration is zero.

$$-\tfrac{3}{10}(3 + 2t) = 0$$
$$\Rightarrow \quad 3 + 2t = 0$$
$$\Rightarrow \quad t = -1.5 \text{ s}$$

Substituting $t = -1.5$ into the expression for v:

$$v = \tfrac{3}{10}[4 - 3 \times (-1.5) - (-1.5)^2]$$
$$= 1.875 \text{ km min}^{-1}$$
$$\equiv 112.5 \text{ km h}^{-1}$$

Therefore, the greatest speed of the train is 112.5 km h^{-1}

5 **a** $y = 4\cos x + 2\cos 2x$

$$\Rightarrow \quad \frac{dy}{dx} = -4\sin x - 4\sin 2x$$

When $x = \dfrac{\pi}{3}$,

$$\frac{dy}{dx} = -4\sin\frac{\pi}{3} - 4\sin\frac{2\pi}{3}$$
$$= -4 \times \frac{\sqrt{3}}{2} - 4 \times \frac{\sqrt{3}}{2}$$
$$= -4\sqrt{3}$$

So, the exact value of the gradient of the curve at $x = \dfrac{\pi}{3}$ is $-4\sqrt{3}$.

b When $x = \dfrac{\pi}{3}$,

$$y = 4\cos\frac{\pi}{3} + 2\cos\frac{2\pi}{3}$$
$$= (4 \times \tfrac{1}{2}) + (2 \times -\tfrac{1}{2})$$
$$= 1$$

So the gradient of the tangent to the curve at $\left(\dfrac{\pi}{3}, 1\right)$ is $-4\sqrt{3}$.

Using $y - y_1 = m(x - x_1)$, the equation of the tangent is given by

$$y - 1 = -4\sqrt{3}\left(x - \frac{\pi}{3}\right)$$
$$\Rightarrow \quad y = -4\sqrt{3}x + \frac{4\pi\sqrt{3}}{3} + 1$$

6 **a**

$$y = 3x - \frac{4}{x}$$
$$= 3x - 4x^{-1}$$
$$\frac{dy}{dx} = 3 + 4x^{-2}$$
$$= 3 + \frac{4}{x^2}$$

b When $x = 2$,

$$\frac{dy}{dx} = 3 + \frac{4}{2^2}$$
$$= 4$$

For small changes, δx, in the value of x,

$$\frac{\delta y}{\delta x} \approx \frac{dy}{dx}$$

Hence, the approximate increase in y as x increases from 2 to $2 + p$ is given by

$$\delta y \approx \frac{dy}{dx} \times \delta x$$

The small change in x, $\delta x = p$ and $\dfrac{dy}{dx} = 4$.

So $\quad \delta y \approx 4 \times p = 4p$

7 **a** surface area of lid $= \pi x^2 + 2\pi x$

surface area of tin $= \pi x^2 + 2\pi xh$

Since total surface area $= 80\pi \text{ cm}^2$, it follows that

$$2\pi x^2 + 2\pi x(h+1) = 80\pi$$
$$x^2 + xh + x = 40$$
$$h = \frac{40}{x} - 1 - x$$

volume of the tin, $V = \pi x^2 h$

Substituting for h,

$$V = \pi x^2\left(\frac{40}{x} - 1 - x\right)$$
$$= \pi(40x - x^2 - x^3)$$

b $\dfrac{dV}{dx} = \pi(40 - 2x - 3x^2)$

For values of x for which V is stationary, $\dfrac{dV}{dx} = 0$

So $\pi(40 - 2x - 3x^2) = 0$

$\Rightarrow \quad 40 - 2x - 3x^2 = 0$

Using the quadratic formula,

$$x = \frac{2 \pm \sqrt{4 + 480}}{-6}$$
$$= \frac{2 \pm 22}{-6}$$
$$= \frac{10}{3} \quad \text{or} \quad -4$$

The positive value of x for which V is stationary is $x = \frac{10}{3}$.

c To determine the nature of this stationary point, look at the value of $\dfrac{dV}{dx}$ on either side of $x = \frac{10}{3}$.

When $x = 3$,

$$\frac{dV}{dx} = \pi[40 - 2\times(3) - 3\times(3)^2]$$
$$= 7\pi$$

When $x = 4$,

$$\frac{dV}{dx} = \pi[40 - 2\times(4) - 3\times(4)^2]$$
$$= -16\pi$$

Since the gradient changes sign from +ve to −ve as x passes through $x = \frac{10}{3}$, this value of x gives a maximum value of V.

d When $x = \frac{10}{3}$,

$$V = \pi[(40 \times \tfrac{10}{3}) - (\tfrac{10}{3})^2 - (\tfrac{10}{3})^3]$$
$$= \frac{2300}{27}\pi \text{ cm}^3$$
$$= 267.62 \text{ cm}^3 \text{(2 d.p.)}$$

e When $x = \frac{10}{3}$, the surface area of the lid is

$$\pi \times (\tfrac{10}{3})^2 + 2\pi \times (\tfrac{10}{3}) = \tfrac{160}{9}\pi \text{ cm}^2$$

The percentage of the sheet metal used in the lid when V is a maximum is therefore

$$\frac{\frac{160}{9}\pi}{80\pi} \times 100\% = 22.2\%$$

8 **a** $\dfrac{dy}{dx} = 3x^2 - 12x$

b At the point (2, −8), the gradient of the tangent is given by

$$\frac{dy}{dx} = 3 \times (2)^2 - 12 \times (2)$$
$$= -12$$

Using $y - y_1 = m(x - x_1)$, the equation of the tangent is given by

$$y - (-8) = -12(x - 2)$$
$$\Rightarrow \quad y + 8 = -12x + 24$$
$$\Rightarrow \quad y = -12x + 16$$

c When $x = 3$, $\quad y = (3)^3 - 6(3)^2 + 8$

$$= -19$$

and $\quad \dfrac{dy}{dx} = 3 \times (3)^2 - 12 \times (3)$

$$= -9$$

The gradient of the normal to the curve at $x = 3$ is $\frac{1}{9}$.

Using $y - y_1 = m(x - x_1)$, the equation of this normal is given by

$$y - (-19) = \tfrac{1}{9}(x - 3)$$
$$\Rightarrow \quad y + 19 = \tfrac{1}{9}(x - 3)$$
$$\Rightarrow \quad 9y + 171 = x - 3$$
$$\Rightarrow \quad x - 9y - 174 = 0$$

9 **a** Using speed $= \dfrac{\text{distance travelled}}{\text{time taken}}$, it follows that

$$\text{time taken} = \frac{\text{distance travelled}}{\text{speed}}$$
$$= \frac{60}{v} \text{ hours}$$

b Total cost of the whole journey,

C = cost per hour × time taken

$$= £\left(v^2 + \frac{32\,000}{v}\right) \times \frac{60}{v}$$
$$= £\left(60v + \frac{1\,920\,000}{v^2}\right)$$

c $\dfrac{dC}{dv} = 60 - \dfrac{3\,840\,000}{v^3}$

At stationary values of C, $\dfrac{dC}{dv} = 0$

So $\quad 60 - \dfrac{3\,840\,000}{v^3} = 0$

$$\Rightarrow \quad v^3 = \frac{3\,840\,000}{60} = 64\,000$$
$$\Rightarrow \quad v = 40 \text{ km h}^{-1}$$

Look at the sign of $\dfrac{dC}{dv}$ on either side of $v = 40$ to determine the nature of this stationary point:

When $x = 35$, $\dfrac{dC}{dv} = 60 - \dfrac{3\,840\,000}{(35)^3}$

$$= -29.6$$

When $x = 45$, $\dfrac{dC}{dv} = 60 - \dfrac{3\,840\,000}{(45)^3}$

$$= 17.9$$

Since the sign of $\frac{dC}{dv}$ changes from –ve to +ve as v passes through $v = 40$ km h^{-1}, this is the speed at which the train should travel to minimise the cost of the journey.

d The total cost of the whole journey

$$C = \text{cost per hour} \times \text{time taken}$$
$$= £\left(v^2 + \frac{32\,000}{v}\right) \times \frac{\text{distance travelled}}{\text{speed}}$$
$$= £\left(v^2 + \frac{32\,000}{v}\right) \times \frac{\text{distance travelled}}{v}$$
$$= £\left(v + \frac{32\,000}{v^2}\right) \times \text{distance travelled}$$

Since, distance travelled is not a function of v, we can consider it to be a constant.

So $\frac{dC}{dv} = £\left(1 - \frac{64\,000}{v^3}\right) \times \text{distance travelled}$

The value of v for which $\frac{dC}{dv} = 0$ is therefore independent of the distance travelled.

10 **a** $c = 250n + 5000$

b £400 per machine will sell 5500 machines

$\Rightarrow \quad 5500 = a + 400b \qquad [1]$

£500 per machine will sell 3500 machines

$\Rightarrow \quad 3500 = a + 500b \qquad [2]$

Solve these simulataneous equations by eliminating a.

Equation [2] – equation [1] gives

$100b = -2000$

$\Rightarrow \quad b = -20$

Substituting this into equation [1] gives

$5500 = a - 8000$

$\Rightarrow \quad a = 13\,500$

Therefore $n = 13\,500 - 20s$

c total receipts = number sold × selling price

$$= n \times s$$
$$= (13\,500 - 20s)s$$
$$= 13\,500s - 20s^2$$

total costs $c = 250n + 5000$

$$= 250 \times (13\,500 - 20s) + 5000$$
$$= 3\,375\,000 - 5000s + 5000$$
$$= 3\,380\,000 - 5000s$$

total profit p = total receipts – total costs

$$= (13\,500s - 20s^2) - (3\,380\,000 - 5000s)$$
$$= 18\,500s - 20s^2 - 3\,380\,000$$

d $\frac{dp}{ds} = 18\,500 - 40s$

At stationary value of p, $\frac{dp}{ds} = 0$

$\Rightarrow \quad 18\,500 - 40s = 0$

$s = £462.50$

Look at the sign of $\frac{dp}{ds}$ on either side of $s = 462.50$ to decide whether this corresponds to a maximum or minimum profit:

When $s = 462$, $\frac{dp}{ds} = 18\,500 - 40 \times (462)$
$= 20$

When $s = 463$, $\frac{dp}{ds} = 18\,500 - 40 \times (463)$
$= -20$

Hence, $s = £462.50$ will *maximise* the importer's profit.

The number of machines she should purchase is

$$n = 13\,500 - 20s$$
$$= 13\,500 - 20 \times (462.50)$$
$$= 4250 \text{ machines}$$

The corresponding profit is

$$p = 18\,500s - 20s^2 - 3380000$$
$$= 18\,500 \times (462.50) - 20 \times (462.50)^2 - 3380000$$
$$= £898,125$$

6 Algebra II (p. 290)

1 Let $f(x) = x^3 - x - 6$

$f(1) = 1 - 1 - 6 \neq 0$

$f(2) = 2^3 - 2 - 6 = 0$

By the factor theorem, $(x - 2)$ is a factor.

Now use polynomial long division.

$$\begin{array}{r} x^2 + 2x + 3 \\ x - 2\,\big)\overline{x^3 \qquad\quad - x - 6} \\ \underline{x^3 - 2x^2} \qquad\qquad \\ 2x^2 - x \qquad \\ \underline{2x^2 - 4x} \qquad \\ 3x - 6 \\ \underline{3x - 6} \\ 0 \end{array}$$

So $x^3 - x - 6 = (x - 2)(x^2 + 2x + 3)$.

2 **a** $g(x) = (2x - 1)(x + 2)(x - 3)$

$$= (2x - 1)(x^2 - x - 6)$$
$$= 2x(x^2 - x - 6) - 1(x^2 - x - 6)$$
$$= 2x^3 - 2x^2 - 12x - x^2 + x + 6$$
$$= 2x^3 - 3x^2 - 11x + 6$$

This is in the required form, with $A = 2$, $B = -3$, $C = -11$, $D = 6$.

b Using the factor theorem, if $(x + 3)$ is a factor of $g(x) + ax$ then

$g(-3) - 3a = 0$

$\Rightarrow \quad (-7) \times (-1) \times (-6) - 3a = 0$

$\Rightarrow \quad -42 - 3a = 0$

$\Rightarrow \quad a = -14$

c $\dfrac{x-3}{g(x)} \equiv \dfrac{x-3}{(2x-1)(x+2)(x-3)}$

$= \dfrac{1}{(2x-1)(x+2)}$

Writing this as partial fractions,

$$\frac{1}{(2x-1)(x+2)} \equiv \frac{A}{(2x-1)} + \frac{B}{(x+2)}$$

Since the denominators are linear, we can use the cover-up rule.

$x = -2 \quad \Rightarrow \quad B = \dfrac{1}{-4-1} = -\frac{1}{5}$

$x = \frac{1}{2} \quad \Rightarrow \quad A = \dfrac{1}{\frac{1}{2}+2} = \frac{2}{5}$

So $\dfrac{x-3}{g(x)} = \dfrac{2}{5(2x-1)} - \dfrac{1}{5(x+2)}$

d

$g(x) > 3x(x+2)(x-3)$

$\Rightarrow \quad (2x-1)(x+2)(x-3) > 3x(x+2)(x-3)$

$0 > 3x(x+2)(x-3)$

$\Rightarrow \quad -(2x-1)(x+2)(x-3)$

$0 > [3x-(2x-1)] \times$

$\Rightarrow \quad (x+2)(x-3)$

$\Rightarrow \quad 0 > (x+1)(x+2)(x-3)$

$\Rightarrow \quad (x+1)(x+2)(x-3) < 0$

Consider the graph of $y = (x+1)(x+2)(x-3)$. It crosses the x-axis when $y = 0$ at $x = -2$, $x = -1$ and $x = 3$.

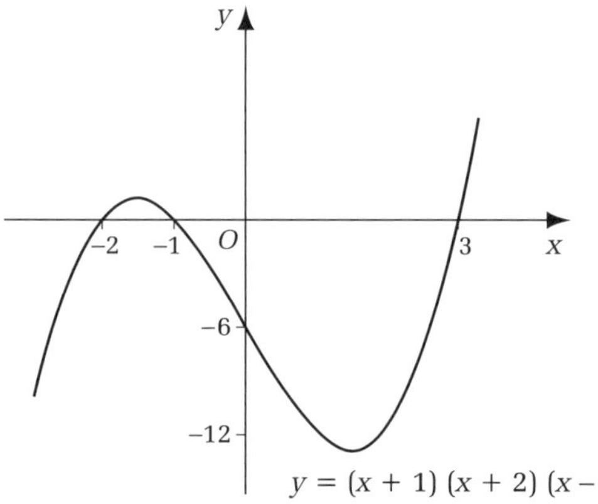

The inequality $(x+1)(x+2)(x-3) < 0$ is satisfied by values of x for which the curve lies below the x-axis.

So $g(x) > 3x(x+2)(x-3)$ for $x < -2$ and $-1 < x < 3$.

3 **a** Let $f(x) = x^3 + ax^2 - (a^2 - 3a + 17)x$

Since division of $f(x)$ by $(x-3)$ gives a remainder of 3, then using the remainder theorem,

$f(3) = 3$

$\Rightarrow \quad 27 + 9a - (a^2 - 3a + 17) \times 3 = 3$

$\Rightarrow \quad 27 + 9a - 3a^2 + 9a - 51 = 3$

$\Rightarrow \quad 3a^2 - 18a + 27 = 0$

$\Rightarrow \quad 3(a^2 - 6a + 9) = 0$

$\Rightarrow \quad 3(a-3)^2 = 0$

$\Rightarrow \quad a = 3$

b Let $f(x) = 3x^3 - x^2 - 75x + 25$

Using the factor theorem and substituting factors of 25 into $f(x)$.

$f(1) = 3 - 1 - 75 + 25 \neq 0$

$f(5) = 375 - 25 - 375 + 25 = 0$

This means that $(x-5)$ is a factor of $f(x)$.

Now $3x^3 - x^2 - 75x + 25 \equiv (x-5)(3x^2 + kx - 5)$

Find k by comparing coefficients of x^2.

$-1 = -15 + k$

$\Rightarrow \quad k = 14$

So $\quad 3x^3 - x^2 - 75x + 25 = (x-5)(3x^2 + 14x - 5)$

To factorise the quadratic, use PAFF.

P: $\quad -15$

A: $\quad 14$

F: $\quad 15, -1$

F: $\quad 3x^2 + 14x - 5 = 3x^2 + 15x - x - 5$

$= 3x(x+5) - 1(x+5)$

$= (x+5)(3x-1)$

So $\quad 3x^3 - x^2 - 75x + 25 = (x-5)(x+5)(3x-1)$

4 **a** $p(x) = x^3 + cx^2 + 7x + d$

Using the factor theorem, if $x + 2$ is a factor of $p(x)$ then

$p(-2) = 0$

$\Rightarrow \quad -8 + 4c - 14 + d = 0$

$\Rightarrow \quad 4c + d - 22 = 0 \qquad [1]$

Now using the remainder theorem, if division by $(x-1)$ leaves a remainder of 3 then

$\Rightarrow \quad p(1) = 3$

$\Rightarrow \quad 1 + c + 7 + d = 3$

$\Rightarrow \quad c + d + 5 = 0 \qquad [2]$

Subtracting equation [2] from equation [1],

$3c - 27 = 0$

$\Rightarrow \quad c = 9$

Substituting c into equation [2],

$9 + d + 5 = 0$

$d = -14$

So $c = 9$ and $d = -14$.

b $p(x) = x^3 + 9x^2 + 7x - 14$

We know that $(x+2)$ is a factor of $p(x)$ so

$x^3 + 9x^2 + 7x - 14 \equiv (x+2)(x^2 + kx - 7)$

Find k by comparing coefficients of x^2.

$9 = 2 + k$

$\Rightarrow \quad k = 7$

Try and factorise $(x^2 + 7x - 7)$

There are no rational factors, so use the quadratic formula.

This gives $\quad x = \dfrac{-7 \pm \sqrt{49+28}}{2} = \dfrac{-7 \pm \sqrt{77}}{2}$

So the exact roots of $x^3 + 9x^2 + 7x - 14 = 0$ are

$$-2, \frac{-7+\sqrt{77}}{2}, \frac{-7-\sqrt{77}}{2}$$

5 $\dfrac{(x-1)^3}{x^2} \equiv Ax + B + \dfrac{C}{x} + \dfrac{D}{x^2}$

Expanding $(x-1)^3$ gives

$$\begin{aligned}(x-1)^3 &= (x-1)(x-1)^2\\ &= (x-1)(x^2-2x+1)\\ &= x^3 - 2x^2 + x - x^2 + 2x - 1\\ &= x^3 - 3x^2 + 3x - 1\end{aligned}$$

Now divide this polynomial by x^2.

$$\begin{array}{r l} & x - 3 \\ x^2 \big) & x^3 - 3x^2 + 3x - 1 \\ & x^3 \\ \hline & 0 - 3x^2 \\ & \quad -3x^2 \\ \hline & \qquad 0 + \ 3x - 1 \end{array}$$

This means that

$$\frac{(x-1)^3}{x^2} = (x-3) + \frac{3x-1}{x^2}$$

So the quotient is $(x-3)$ and the remainder is $(3x-1)$.

Now express the proper fraction in partial fractions, with repeated root x^2.

So $$\frac{3x-1}{x^2} \equiv \frac{C}{x} + \frac{D}{x^2} \equiv \frac{Cx+D}{x^2}$$

Equating the numerators,

$$3x - 1 \equiv Cx + D$$

Comparing coefficients, $C = 3$ and $D = -1$.

So $$\frac{(x-1)^3}{x^2} = x - 3 + \frac{3}{x} - \frac{1}{x^2}$$

This gives $A = 1$, $B = -3$, $C = 3$ and $D = -1$.

6 Writing $y = \dfrac{2x+1}{x(2x-1)^2}$ in partial fractions,

$$\frac{2x+1}{x(2x-1)^2} \equiv \frac{A}{x} + \frac{B}{2x-1} + \frac{C}{(2x-1)^2}$$

$$= \frac{A(2x+1)^2 + Bx(2x-1) + Cx}{x(2x-1)^2}$$

Equating the numerators,

$$2x + 1 \equiv A(2x-1)^2 + Bx(2x-1) + Cx$$

Find A, B and C by substitution.

Let $x = 0$. Then

$$1 = A \quad \Rightarrow \quad A = 1$$

Let $x = \frac{1}{2}$. Then

$$2 = \tfrac{1}{2}C \quad \Rightarrow \quad C = 4$$

Let $x = 1$. Then

$$3 = 1 + B + 4 \quad \Rightarrow \quad B = -2$$

So $$y = \frac{1}{x} - \frac{2}{2x-1} + \frac{4}{(2x-1)^2}$$

7 For $x - 3$ to be a factor of $f(x) = x^3 - 2x^2 - x - 6$, then $f(3)$ must equal zero.

$$f(3) = 27 - 18 - 3 - 6 = 0$$

So $x^3 - 2x^2 - x - 6 \equiv (x-3)(x^2 + kx + 2)$

Find k by comparing coefficients of x^2.

$$-2 = -3 + k$$

$$\Rightarrow \quad k = 1$$

This means that $x^3 - 2x^2 - x - 6 = (x-3)(x^2 + x + 2)$

Now try and factorise $(x^2 + x + 2)$.

There isn't a pair of integers whose product is 2 and whose sum is 1. This means that there are no real roots to the quadratic equation $x^2 + x + 2 = 0$.

So the cubic equation $f(x) = 0$ has only one real root; $x = 3$.

At the points of intersection of the graphs of $y = x^2 - 2x - 1$ and $y = \dfrac{6}{x}$, we know that

$$x^2 - 2x - 1 = \frac{6}{x}$$

$$\Rightarrow \quad x^3 - 2x^2 - x = 6$$

$$\Rightarrow \quad x^3 - 2x^2 - x - 6 = 0$$

We already know that this cubic equation has only one real root. Therefore the graphs of $y = x^2 - 2x - 1$ and $y = \dfrac{6}{x}$ intersect only once, at $x = 3$.

8 **a** $p(x) = 2x^3 - 5x^2 - 28x + 15$

Using the factor theorem and substituting factors of 15 into $p(x)$,

$$p(1) = 2 - 5 - 28 + 15 \neq 0$$

$$p(3) = 54 - 45 - 84 + 15 \neq 0$$

$$p(5) = 250 - 125 - 140 + 15 = 0$$

So $(x-5)$ is a factor of $p(x)$.

Now $2x^3 - 5x^2 - 28x + 15 \equiv (x-5)(2x^2 + kx - 3)$

Find k by comparing coefficients of x^2.

$$-5 = -10 + k$$

$$\Rightarrow \quad k = 5$$

This means $\quad p(x) = (x-5)(2x^2 + 5x - 3)$

Now factorise the quadratic.

$$2x^2 + 5x - 3 = (x+3)(2x-1)$$

Therefore $\quad p(x) = (x+3)(2x-1)(x-5)$

b $p(x) = 0$ when $(x+3)(2x-1)(x-5) = 0$

$\Rightarrow \quad x = -3, x = \frac{1}{2}$ or $x = 5$.

c The graph of $y = p(x)$ crosses the x-axis when $p(x) = 0$ at $(-3, 0)$, $(\frac{1}{2}, 0)$ and $(5, 0)$.

The graph crosses the y-axis at $y = 15$ when $x = 0$.

The graph is a cubic curve, which can now be sketched.

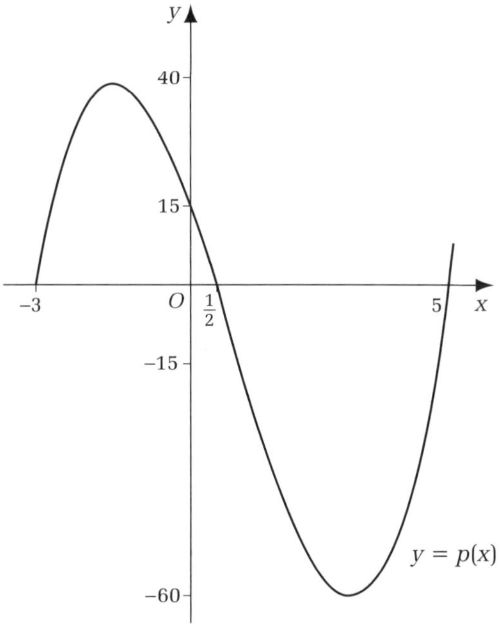

d The inequality $p(x) \geq 0$ is satisfied by those values of x for which the graph of $y = p(x)$ is on or above the x-axis; that is $-3 \leq x \leq \frac{1}{2}$ and $x \geq 5$.

9 $y = \dfrac{x+3}{x-5}$

This is an improper algebraic fraction which can be rewritten as $y = 1 + \dfrac{8}{x-5}$.

Step 1 The curve crosses the axes at $(0, -\frac{3}{5})$ and $(-3, 0)$.

Step 2 $\dfrac{dy}{dx} = \dfrac{-8}{(x-5)^2}$. Since $\dfrac{dy}{dx} < 0$ for all values of x, there are no stationary points on the curve.

Step 3 $y \to 1^+$ as $x \to +\infty$
$y \to 1^-$ as $x \to -\infty$
So $y = 1$ is a horizontal asymptote.

Step 4 The function is undefined when $x = 5$, so there is a vertical asymptote at $x = 5$.

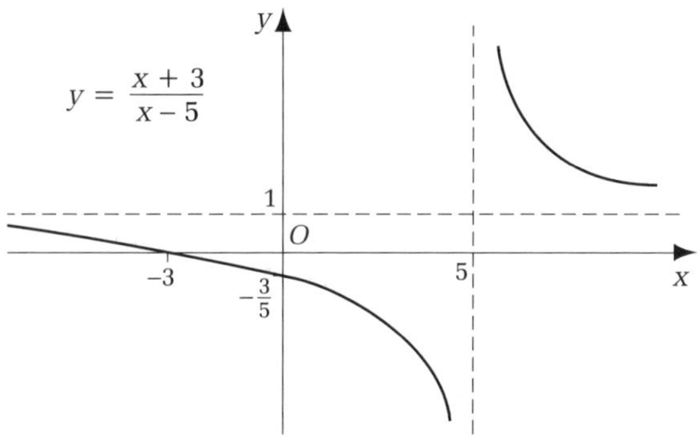

10 **a** $f(x) = \dfrac{2x+1}{x+2}$

This is an improper algebraic fraction which can be rewritten as $f(x) = 2 - \dfrac{3}{x+2}$.

Step 1 The curve crosses the axes at $(0, \frac{1}{2})$ and $(-\frac{1}{2}, 0)$.

Step 2 $f'(x) = \dfrac{3}{(x+2)^2}$. Since $f'(x) > 0$ for all values of x, there are no stationary points on the curve.

Step 3 $f(x) \to 2^-$ as $x \to +\infty$
$f(x) \to 2^+$ as $x \to -\infty$
So $y = 2$ is a horizontal asymptote.

Step 4 The function is undefined for $x = -2$, so there is vertical asymptote $x = -2$.

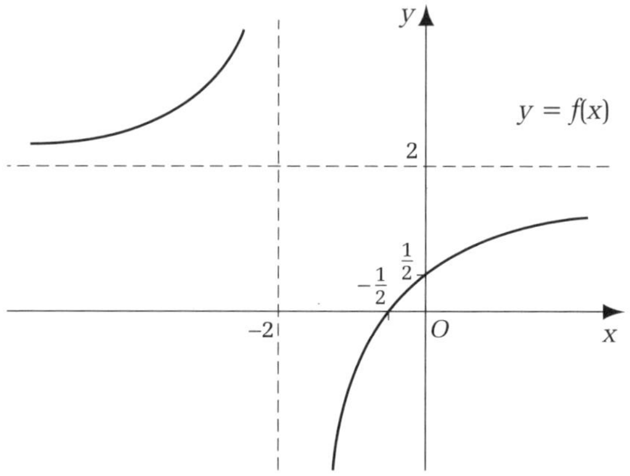

b To obtain the graph of $y = |f(x)|$, reflect in the x-axis those parts of the graph of $y = f(x)$ that lie below the x-axis.

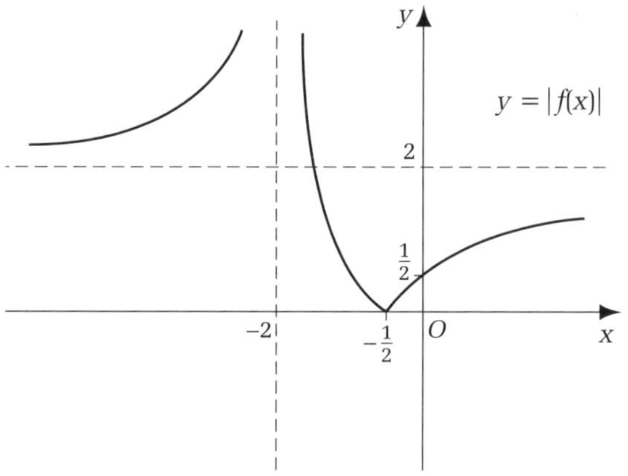

7 Exponentials and Logarithms (p. 316)

1
$$\frac{\log 9\sqrt{3} - \log 4\sqrt{2}}{\log 3 - \log 2} = \frac{\log\left(\frac{9\sqrt{3}}{4\sqrt{2}}\right)}{\log\left(\frac{3}{2}\right)}$$
$$= \frac{\log\left[\left(\frac{3}{2}\right)^2\left(\frac{\sqrt{3}}{\sqrt{2}}\right)\right]}{\log\left(\frac{3}{2}\right)}$$
$$= \frac{\log\left(\frac{3}{2}\right)^{\frac{5}{2}}}{\log\left(\frac{3}{2}\right)}$$
$$= \frac{\frac{5}{2}\log\left(\frac{3}{2}\right)}{\log\left(\frac{3}{2}\right)}$$
$$= \tfrac{5}{2}$$

2 **a**
$$\ln(x+2) < 3$$
$$\Rightarrow \quad (x+2) < e^3$$
$$\Rightarrow \quad x < e^3 - 2$$
$$\Rightarrow \quad x < 18.1 \text{ (3 s.f.)}$$

b
$$(0.7)^x > 3$$
$$\Rightarrow \quad \log(0.7)^x > \log 3$$
$$\Rightarrow \quad x\log(0.7) > \log 3$$

Note that log (0.7) is negative, so reverse the inequality when dividing both sides by log (0.7).

$$\Rightarrow \quad x < \frac{\log 3}{\log(0.7)}$$
$$\Rightarrow \quad x < -3.08 \text{ (3 s.f.)}$$

3 **a** When car is new, $t = 0$.

$$\Rightarrow \quad V = 8500 \times e^{-\lambda\times 0}$$
$$\Rightarrow \quad = 8500 \times 1 = 8500$$

So value when new is £8500.

b When $t = 2$, $V = 6580$

$$\Rightarrow \quad 6580 = 8500\ e^{-2\lambda}$$
$$\Rightarrow \quad e^{-2\lambda} = \frac{6580}{8500} = \frac{329}{425}$$
$$\Rightarrow \quad e^{2\lambda} = \frac{425}{329}$$

Taking logarithms of both sides,

$$2\lambda = \ln\left(\frac{425}{329}\right)$$
$$= 0.256\,0314$$
$$\Rightarrow \quad \lambda = 0.128 \text{ (3 s.f.)}$$

c After three years, $t = 3$

$$V = 8500\ e^{-0.128\times 3}$$
$$= 8500\ e^{-0.384}$$
$$= 5788.5 \text{ (1 d.p.)}$$
$$= 5800 \text{ (2 s.f.)}$$

After three years the estimated value is £5800.

d Half of the original value is £4250.

So we need to solve

$$8500\ e^{-0.128\times t} = 4250$$
$$\Rightarrow \quad e^{-0.128\times t} = \tfrac{1}{2}$$

Taking logarithms of both sides,

$$-0.128 \times t = \ln(\tfrac{1}{2})$$
$$= -0.693$$
$$\Rightarrow \quad t = 5.41$$

So the car halves in value after approximately $5\frac{1}{2}$ years.

4 **a** $A_{n+1} = 1.07 \times A_n$ and $A_0 = 100$.

Using this algebraic rule,

$$A_0 = 100$$
$$A_1 = 1.07 \times A_0 = 1.07 \times 100$$
$$A_2 = 1.07 \times A_1 = 1.07 \times (1.07 \times 100)$$
$$= 1.07^2 \times 100$$
$$A_3 = 1.07 \times A_2 = 1.07 \times (1.07^2 \times 100)$$
$$= 1.07^3 \times 100$$

Comparing the subscript of A with the power of 1.07, we see that

$$A_n = 1.07^n \times 100$$

b 1.07 represents the annual growth rate of the investment; that is 7% increase per annum.

c We require $A_n = 200$ (double the initial investment).

$$200 = 1.07^n \times 100$$
$$\Rightarrow \quad 2 = 1.07^n$$

Take logarithms of both sides,

$$\log 2 = \log 1.07^n$$
$$= n \log 1.07$$

So
$$n = \frac{\log 2}{\log 1.07}$$
$$= \frac{0.301}{0.0294}$$
$$= 10.2 \text{(3 s.f.)}$$

So the investment would double within 11 years.

5 **a**
$$y = 10^x \quad \Rightarrow \quad y^2 = (10^x)^2$$
$$= 10^{2x}$$
$$= (10^2)^x$$
$$= 100^x$$

b
$$y = 10^x \quad \Rightarrow \quad \frac{y}{10} = \frac{10^x}{10}$$
$$= 10^x \times 10^{-1}$$
$$= 10^{x-1}$$

c
$$100^x - 10001 \times (10^{x-1}) + 100 = 0$$
$$\Rightarrow \quad y^2 - (10001 \times \frac{y}{10}) + 100 = 0$$
$$\Rightarrow \quad 10y^2 - 10001y + 1000 = 0$$

d
$$10y^2 - 10001y + 1000 = 0$$
$$\Rightarrow \quad (10y - 1)(y - 1000) = 0$$

So $\quad y = \frac{1}{10}$ or $y = 1000$

That is $\quad 10^x = \frac{1}{10}$ or $10^x = 1000$

or $\quad 10^x = 10^{-1}$ or $10^x = 10^3$

So $\quad x = -1$ or $x = 3$

6 **a** In 1990, $t = 0$, so

$$e^{kt} = e^0$$
$$= 1$$

So using $\quad P = P_0 e^{kt}$

$\Rightarrow \quad 10.5$ million $= P_0 \times 1$ in 1990

That is $\quad P_0 = 10\,500\,000$

$$= 1.05 \times 10^7$$

b Growth rate is 0.5% per annum. To increase the population by 0.5% we need a multiplier of 1.005. Now, 1991 is one year after 1990 so the population projection is 10.5 million (the population in 1990) multiplied by 1.005 (the annual increase multiplier). That is,

population projection $= 1.05 \times 10^7 \times (1.005)$

c We know that $P_0 = 1.05 \times 10^7$. From **b** we know that when $t = 1$, $P = 1.05 \times 10^7 \times (1.005)$. Substituting these values into $P = P_0 e^{kt}$

$$1.05 \times 10^7 \times 1.005 = 1.05 \times 10^7 \times e^{k\times 1}$$

$$\Rightarrow \quad 1.005 = e^k$$

$$\Rightarrow \quad k = \ln(1.005)$$

$$\Rightarrow \quad k = 4.99 \times 10^{-3} \text{ (3 s.f.)}$$

d In 2000, $t = 10$, since it is 10 years after 1990. Using $P = P_0 e^{kt}$, the predicted population for 2000 is

$$P = 1.05 \times 10^7 \times e^{0.00499\times 10}$$

$$= 1.05 \times 10^7 \times e^{0.0499}$$

$$= 1.10 \times 10^7 \text{ (3 s.f.)}$$

$$= 11.0 \text{ million}$$

7 $y = \log_b 45 + \log_b 25 - 2\log_b 75$

$$= \log_b(45 \times 25) - \log_b 75^2$$

$$= \log_b\left(\frac{45 \times 25}{75^2}\right)$$

$$= \log_b\left(\frac{45}{9 \times 25}\right)$$

$$= \log_b\left(\tfrac{1}{5}\right)$$

When $b = 5$, $y = \log_5\left(\tfrac{1}{5}\right) = -1$

8 **a** $T = Ax^n$

Taking the natural logarithm of both sides,

$$\ln T = \ln(Ax^n)$$

$$= \ln A + \ln(x^n)$$

$$= \ln A + n\ln x$$

Plotting $\ln x$ horizontally against $\ln T$ vertically gives a straight line. This is because $\ln T = n\ln x + \ln A$, in the form $y = mX + C$, where the gradient will be n and the vertical intercept $\ln A$. From the table in the question we can construct a new table of values.

ln x	4.06	4.68	5.43	6.66
ln T	−1.43	−0.48	0.63	2.47

From these values we can draw the graph.

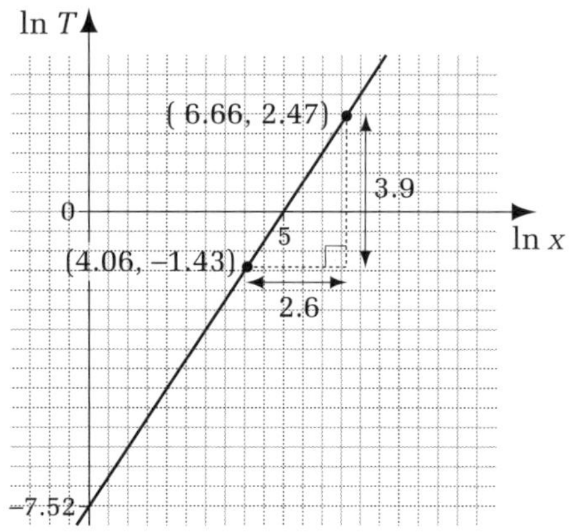

From the graph, gradient $n \approx \dfrac{3.9}{2.6}$

$$\approx 1.5$$

Substituting this value of n with the first values of $\ln T$ and $\ln x$ from the table into the equation gives,

$$-1.43 = 1.5 \times 4.06 + \ln A$$

$$\Rightarrow \quad \ln A = -7.52$$

$$\Rightarrow \quad A = e^{-7.52}$$

$$\Rightarrow \quad = 5.4 \times 10^{-4} \text{(2 s.f.)}$$

For the Earth, the period of the orbit $T = 1$ year. Putting this into the model with the calculated values of n and A,

$$T = Ax^n$$

$$1 = 5.4 \times 10^{-4} \times x^{1.5}$$

Dividing by 5.4×10^{-4}.

$$x^{1.5} = 1851.851\,852$$

and taking the logarithm of both sides

$$\Rightarrow \quad 1.5 \times \ln x = \ln(1.852 \times 10^3)$$

$$\Rightarrow \quad \ln x = 5.02 \text{ (3 s.f.)}$$

$$\Rightarrow \quad x = e^{5.02}$$

$$= 150 \text{ (2 s.f.)}$$

9

$$y = e^{-\frac{x^2}{2}}$$

$$\Rightarrow \quad \ln y = -\frac{x^2}{2}$$

$$\Rightarrow \quad -2\ln y = x^2$$

$$\Rightarrow \quad \ln\left(y^{-2}\right) = x^2$$

That is, $x^2 = \ln\left(\dfrac{1}{y^2}\right)$

$$x = \sqrt{\ln\left(\frac{1}{y^2}\right)}$$

Therefore $f^{-1}(x) = \sqrt{\ln\left(\dfrac{1}{x^2}\right)}$

10 **a** At A, $x = 0$ and at C, $x = 2.5$.

Substitute these values into equation to get the height of the supports.

At A: $\quad y = 3 \times 10^0$

$$= 3 \times 1 = 3$$

At C: $\quad y = 3 \times 10^{-2.5/2}$

$$= 3 \times 0.05623 = 0.169 \text{(3 s.f.)}$$

So the supports at A and C are 3 m and 16.9 cm high respectively.

b The support at A is 3 m high, so the support at B is 1.5 m high. Substituting the value at B into the equation.

$$1.5 = 3 \times 10^{-x/2}$$

$$\Rightarrow \quad 0.5 = 10^{-x/2}$$

$$\Rightarrow \quad \log(0.5) = -\frac{x}{2}$$

So $x = -2\log(0.5)$

$= 0.602 (3 \text{ s.f.})$

So support B should be 60.2 cm from support A.

8 Sequences and Series (p. 354)

1 We first need to find out how many terms there are in the series. Use the formula for the nth term of an AP with $a = 2$ and $d = 3$.

$$u_n = a + (n-1)d$$

For the last term,

$$398 = 2 + (n-1) \times 3$$

$$\Rightarrow \quad \frac{398-2}{3} = n - 1$$

$$\Rightarrow \quad n = 133$$

Now use $S_n = \frac{n}{2}[2a + (n-1)d]$ to find the sum of the 133 terms.

$$S_{133} = \frac{133}{2}[4 + (133-1) \times 3]$$

$$= \frac{133}{2} \times 400$$

$$= 26\,600$$

2 **a** Tenth term is zero

$\Rightarrow \quad u_{10} = a + 9d = 0$ [1]

Sum of first ten terms is 15

$\Rightarrow \quad S_{10} = 5(2a + 9d) = 15$

$\Rightarrow \quad 10a + 45d = 15$ [2]

These are simultaneous equations.

Dividing equation [2] by 5,

$2a + 9d = 3$ [3]

Subtracting equation [1] from equation [3],

$a = 3$

Substituting this value of a into equation [1]

$$3 + 9d = 0$$

$$\Rightarrow \quad d = -\tfrac{1}{3}$$

b If the sum of n terms, S_n, is zero,

$$\frac{n}{2}[6 + (n-1) \times -\tfrac{1}{3}] = 0$$

Multiplying throughout by 6,

$$\Rightarrow \quad 3n[6 - \tfrac{1}{3}(n-1)] = 0$$

$$\Rightarrow \quad -n^2 + 18n + n = 0$$

$$\Rightarrow \quad n(19 - n) = 0$$

So $n = 0$ or $n = 19$. Since there already ten terms, $n = 0$ can be discarded. Therefore an additional nine terms must be added.

3 Using $u_n = ar^{n-1}$ the third term, $u_3 = ar^2 = 27$ and the sixth term, $u_6 = ar^5 = 8$.

a The common ratio is r.

$$\frac{u_6}{u_3} = \frac{ar^5}{ar^2} = r^3 = \frac{8}{27}$$

$$\Rightarrow \quad r = \sqrt[3]{\frac{8}{27}} = \frac{2}{3}$$

b Since the third term $ar^2 = 27$,

$$a \times \left(\frac{2}{3}\right)^2 = 27$$

$$\Rightarrow \quad a = 27 \times \tfrac{9}{4}$$

$$= 60\tfrac{3}{4}$$

c Using $S_\infty = \dfrac{a}{1-r}$, with $a = 60\frac{3}{4}$ and $r = \frac{2}{3}$,

$$S_\infty = \frac{60\frac{3}{4}}{1 - \frac{2}{3}}$$

$$= \frac{\frac{243}{4}}{\frac{1}{3}}$$

$$= \frac{729}{4}$$

$$= 182\tfrac{1}{4}$$

d The difference between the sum of the first ten terms and the sum to infinity is

$$S_\infty - S_{10} = \frac{a}{1-r} - \frac{a\left(1 - r^{10}\right)}{1-r}$$

$$= \frac{a}{1-r}\left[1 - \left(1 - r^{10}\right)\right]$$

$$= 182\tfrac{1}{4} \times r^{10}$$

$$= 182\tfrac{1}{4} \times (\tfrac{2}{3})^{10}$$

$$= 3.1605$$

$$= 3.16 \text{ (3 s.f.)}$$

4 The series is arithmetic with $a = 11$, $d = \frac{1}{2}$ and $u_n = 40$.

Using $u_n = a + (n-1)d$ for the last term

$$\Rightarrow \quad 40 = 11 + (n-1) \times \tfrac{1}{2}$$

$$\Rightarrow \quad \tfrac{1}{2}(n-1) = 40 - 11$$

$$\Rightarrow \quad n - 1 = 58$$

$$\Rightarrow \quad n = 59$$

Now find the sum of the first 59 terms, using

$$S_n = \frac{n}{2}[2a + (n-1)d].$$

$$S_{59} = \tfrac{59}{2}[(2 \times 11) + (58 \times \tfrac{1}{2})]$$

$$= \tfrac{59}{2}(22 + 29)$$

$$= 1504\tfrac{1}{2}$$

So the sum of the series is $1504\frac{1}{2}$.

5 **a** $u_n = 95(\frac{4}{5})^n$

So $u_1 = 95(\frac{4}{5})^1$

$= 76$

$u_2 = 95(\frac{4}{5})^2$

$= 60.8$

b $u_{21} = 95(\frac{4}{5})^{21}$

$= 0.876$

c This is a geometric series with $a = 76$ and $r = \frac{4}{5}$, so the sum of the first 15 terms is

$$\sum_{n=1}^{15} u_n = \frac{a(1-r^n)}{1-r} = \frac{76\left[1-\left(\frac{4}{5}\right)^{15}\right]}{\frac{1}{5}}$$

$$= 5 \times 76[1-(\tfrac{4}{5})^{15}]$$

$$= 5 \times 76 \times 0.964\,815\,6$$

$$= 367 \text{ (3 s.f.)}$$

d $S_\infty = \dfrac{a}{1-r}$

$$= \frac{76}{1-\frac{4}{5}}$$

$$= 380$$

6 To expand $(1+3t)^{\frac{2}{3}}$, use the binomial theorem for $(1+x)^n$ with $x = 3t$ and $n = \frac{2}{3}$.

$$(1+x)^n = 1 + nx + \frac{n(n-1)}{2!}x^2 + \ldots$$

$$(1+3t)^{\frac{2}{3}} = 1 + \left(\tfrac{2}{3} \times 3t\right) + \frac{\left(\frac{2}{3}\right) \times \left(-\frac{1}{3}\right) \times (3t)^2}{2} + \ldots$$

$$= 1 + 2t - t^2 + \ldots$$

The series is valid for $|x| < 1$. That is, $|3t| < 1$ or $|t| < \frac{1}{3}$.

7 **a** Sequence A is given by $u_1 = 2$, $u_{n+1} = 3 - u_n$

So $u_2 = 3 - u_1$
$= 3 - 2 = 1$

$u_3 = 3 - u_2$
$= 3 - 1 = 2$

$u_4 = 3 - u_3$
$= 3 - 2 = 1$

$u_5 = 3 - u_4$
$= 3 - 1 = 2$

So the first five terms of Sequence A are 2, 1, 2, 1, 2.

Sequence B is given by $u_1 = 2, u_{n+1} = u_n + \dfrac{1}{2^n}$

So $u_2 = u_1 + \dfrac{1}{2^1}$
$= 2 + \frac{1}{2} = 2\frac{1}{2}$

$u_3 = u_2 + \dfrac{1}{2^2}$
$= 2\frac{1}{2} + \frac{1}{4} = 2\frac{3}{4}$

$u_4 = u_3 + \dfrac{1}{2^3}$
$= 2\frac{3}{4} + \frac{1}{8} = 2\frac{7}{8}$

$u_5 = u_4 + \dfrac{1}{2^4}$
$= 2\frac{7}{8} + \frac{1}{16} = 2\frac{15}{16}$

So the first five terms of Sequence B are $2, 2\frac{1}{2}, 2\frac{3}{4}, 2\frac{7}{8}, 2\frac{15}{16}$.

b Sequence A oscillates between 2 and 1.

Sequence B converges to a limit of 3.

8 $\sqrt{1+2x} = (1+2x)^{\frac{1}{2}}$ in index form.

Expand using the binomial theorem for $(1+x)^n$.

$$(1+x)^n = 1 + nx + \frac{n(n-1)}{2!}x^2 + \ldots$$

$$(1+2x)^{\frac{1}{2}} = 1 + \left(\tfrac{1}{2} \times 2x\right) + \frac{\frac{1}{2} \times \left(-\frac{1}{2}\right) \times (2x)^2}{2!} + \frac{\frac{1}{2} \times \left(-\frac{1}{2}\right) \times \left(-\frac{3}{2}\right) \times (2x)^3}{3!} + \ldots$$

$$= 1 + x - \frac{x^2}{2} + \frac{x^3}{2} + \ldots$$

9 Extracting a common factor x from inside the bracket

$$(x+2x^2)^{10} = [x(1+2x)]^{10}$$

$$= x^{10}(1+2x)^{10}$$

Recall the general result for expanding $(a+b)^n$.

$$(a+b)^n = a^n + \ldots + \binom{n}{r}a^{n-r}b^r + \ldots + b^n$$

Replacing a with 1, b with $2x$, and n with 10 gives

$$(x+2x^2)^{10} = x^{10}\left[1 + \binom{10}{1}2x + \binom{10}{2}(2x)^2 + \ldots\right]$$

$$= x^{10}[1 + (10 \times 2x) + (45 \times 4x^2) + \ldots]$$

$$= x^{10} + 20x^{11} + 180x^{12} + \ldots$$

So the coefficient of x^{12} is 180.

10 **a** Using the binomial theorem,

$$(x+y)^4 = x^4 + \binom{4}{1}x^3y + \binom{4}{2}x^2y^2 + \binom{4}{3}xy^3 + y^4$$

$$= x^4 + 4x^3y + 6x^2y^2 + 4xy^3 + y^4$$

b Substitute $y = \dfrac{2}{x}$ into the expansion in **a**.

$$\left(x + \frac{2}{x}\right)^4 = x^4 + 4x^3 \times \left(\frac{2}{x}\right) + 6x^2 \times \left(\frac{4}{x^2}\right) + 4x \times \left(\frac{8}{x^3}\right) + \left(\frac{16}{x^4}\right)$$

$$= x^4 + 8x^2 + 24 + \frac{32}{x^2} + \frac{16}{x^4}$$

c $(1+x^2)\left(x + \dfrac{2}{x}\right)^4 =$

$$(1+x^2)\left(x^4 + 8x^2 + 24 + \frac{32}{x^2} + \frac{16}{x^4}\right)$$

The x^2 term is given by

$(1 \times 8x^2) + (x^2 \times 24) = 8x^2 + 24x^2$

$= 32x^2$

So the coefficient of x^2 is 32.

9 Integration I (p. 397)

1 **a** $\int(2x-3)^2\mathrm{d}x = \int(4x^2-12x+9)\mathrm{d}x$

$= \frac{4}{3}x^3 - 6x^2 + 9x + c$

b $\frac{\mathrm{d}y}{\mathrm{d}x} = 6x - 1$

So $y = \int(6x-1)\mathrm{d}x$

$= 3x^2 - x + c$

We know that when $x = 1$, $y = 9$. Substituting these values into the equation

$9 = 3 \times 1^2 - 1 + c$

$9 = 3 - 1 + c$

$\Rightarrow \quad c = 7$

So, the equation of the curve is $y = 3x^2 - x + 7$

2 **a** $\int x(3-x)\mathrm{d}x = \int(3x - x^2)\mathrm{d}x$

$= \frac{3}{2}x^2 - \frac{1}{3}x^3 + c$

b The graph of $y = x(3-x)$ crosses the x-axis when $y = 0$

$\Rightarrow \quad x(3-x) = 0$

That is, it crosses when $x = 0$ and $x = 3$.

$$\text{area} = \int_0^3 x(3-x)\,\mathrm{d}x$$

$= \left[\frac{3}{2}x^2 - \frac{1}{3}x^3\right]_0^3$

$= \left(\frac{3}{2}\times 3^2 - \frac{1}{3}\times 3^3\right) - (0-0)$

$= \frac{27}{2} - \frac{27}{3}$

$= \frac{27}{6}$

$= \frac{9}{2}$ square units

3 **a** Differentiating $y = 5x^4 - x^5$,

$\frac{\mathrm{d}y}{\mathrm{d}x} = 20x^3 - 5x^4$

At stationary points on the curve, $\frac{\mathrm{d}y}{\mathrm{d}x} = 0$

$\Rightarrow \quad 20x^3 - 5x^4 = 0$

$\Rightarrow \quad 5x^3(4-x) = 0$

$\Rightarrow \quad x = 0$ or $x = 4$

When $x = 0$, $y = 5 \times 0^4 - 0^5$
$= 0$

When $x = 4$, $y = 5 \times 4^4 - 4^5$
$= 256$

So the coordinates of the stationary points are (0, 0) and (4, 256).

b The graph of $y = 5x^4 - x^5$ crosses the x-axis when $y = 0$

$\Rightarrow \quad 5x^4 - x^5 = 0$

$\Rightarrow \quad x^4(5-x) = 0$

That is, it crosses when $x = 0$ and $x = 5$.

$$\text{area} = \int_0^5 \left(5x^4 - x^5\right)\,\mathrm{d}x$$

$= \left[x^5 - \frac{1}{6}x^6\right]_0^5$

$= \left(5^5 - \frac{1}{6}\times 5^6\right) - (0-0)$

$= 5^5\left(1 - \frac{5}{6}\right) = 5^5\left(\frac{1}{6}\right)$

$= 520.83$ square units (2 d.p.)

c $$\text{area} = \int_0^6 \left(5x^4 - x^5\right)\,\mathrm{d}x$$

$= \left[x^5 - \frac{1}{6}x^6\right]_0^6$

$= \left(6^5 - \frac{1}{6}\times 6^6\right) - (0-0)$

$= 6^5(1-1)$

$= 0$ square units

This result means that the area enclosed by the curve *above* the x-axis between $x = 0$ and $x = 5$ is equal to the area enclosed by the curve *below* the x-axis between $x = 5$ and $x = 6$. The negative value of the latter cancels out the positive value of the former.

4 **a** $y = 7 + 6x - x^2$

P and Q lie on the x-axis, where $y = 0$.

So $\quad 7 + 6x - x^2 = 0$

$\Rightarrow \quad x^2 - 6x - 7 = 0$

$\Rightarrow \quad (x+1)(x-7) = 0$

$\Rightarrow \quad x = -1$ or $x = 7$

So the coordinates of P and Q are (7, 0) and (−1, 0), respectively.

R lies on the y-axis, where $x = 0$

When $x = 0$, $y = 7$, so the coordinates of R are (0, 7).

We now know that $y = 7$ at point S.

Therefore $\quad 7 + 6x - x^2 = 7$

$\Rightarrow \quad 6x - x^2 = 0$

$\Rightarrow \quad x(6-x) = 0$

$\Rightarrow \quad x = 0 \quad$ or $\quad x = 6$

We already know that the coordinates of R are (0, 7), so the coordinates of S are (6, 7).

b Starting from the left, label the areas A, B and C as shown.

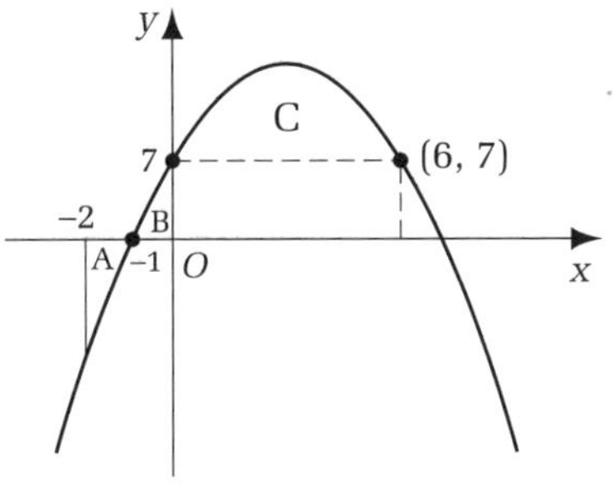

$$\text{Area A} = \int_{-2}^{-1} \left(7 + 6x - x^2\right) dx$$
$$= \left[7x + 3x^2 - \tfrac{1}{3}x^3\right]_{-2}^{-1}$$
$$= \left(-7 + 3 + \tfrac{1}{3}\right) - \left(-14 + 12 + \tfrac{8}{3}\right)$$
$$= -4 + \tfrac{1}{3} + 2 - \tfrac{8}{3}$$
$$= -2 - \tfrac{7}{3}$$
$$= -\tfrac{13}{3}$$

So area A $= 4\frac{1}{3}$ square units with the negative sign indicating that the area is *below* the x-axis.

$$\text{Area B} = \int_{-1}^{0} \left(7 + 6x - x^2\right) dx$$
$$= \left[7x + 3x^2 - \tfrac{1}{3}x^3\right]_{-1}^{0}$$
$$= (0) - \left(-7 + 3 + \tfrac{1}{3}\right)$$
$$= 4 - \tfrac{1}{3}$$
$$= \tfrac{11}{3}$$

So area B $= 3\frac{2}{3}$ square units

$$\text{Area C} = \int_{0}^{6} \left(7 + 6x - x^2\right) dx - (OR \times RS)$$
$$= \left[7x + 3x^2 - \tfrac{1}{3}x^3\right]_{0}^{6} - (7 \times 6)$$
$$= (42 + 108 - 72) - (0 + 0 - 0) - 42$$
$$= 78 - 42$$
$$= 36$$

So area C $=$ 36 square units

5 **a** The equation of the tangent at point A (1, 0) is given by $y = mx + c$, where m is the gradient of the tangent, and c is the y-axis intercept.

$$y = x^3 - x$$
$$\Rightarrow \quad \frac{dy}{dx} = 3x^2 - 1$$

Substituting $x = 1$, the gradient of the curve, and therefore the tangent, at point A is
$m = 3 \times 1^2 - 1 = 2.$

So the equation of the tangent is $y = 2x + c$.

The tangent passes through (1, 0), so substituting these coordinates into the equation for the tangent,

$$0 = 2 + c$$
$$\Rightarrow \quad c = -2$$

So the equation of the tangent at A is $y = 2x - 2$.

If the point B lies on the curve and the tangent, its coordinates must satisfy both equations.

Substituting $x = -2$ in $y = x^3 - x$,

$$y = (-2)^3 - (-2) = -8 + 2 = -6$$

Substituting $x = -2$ in $y = 2x - 2$,

$$y = (2 \times -2) - 2 = -4 - 2 = -6$$

So the tangent at A meets the curve again at B $(-2, -6)$.

b The area between graphs $y = f(x)$ and $y = g(x)$, where $f(x) > g(x)$, between $x = a$ and $x = b$ is given by

$$\text{area} = \int_a^b [f(x) - g(x)]\, dx$$

From the diagram, it follows that $f(x) = x^3 - x$ and $g(x) = 2x - 2$, since the curve is above the line.

$$\text{shaded area} = \int_{-2}^{1} \left[\left(x^3 - x\right) - (2x - 2)\right] dx$$
$$= \int_{-2}^{1} \left(x^3 - 3x + 2\right) dx$$
$$= \left[\tfrac{1}{4}x^4 - \tfrac{3}{2}x^2 + 2x\right]_{-2}^{1}$$
$$= \left(\tfrac{1}{4} - \tfrac{3}{2} + 2\right) - (4 - 6 - 4)$$
$$= \tfrac{3}{4} + 6$$
$$= \tfrac{27}{4}$$

area of shaded region $= 6\frac{3}{4}$ square units

6 **a**

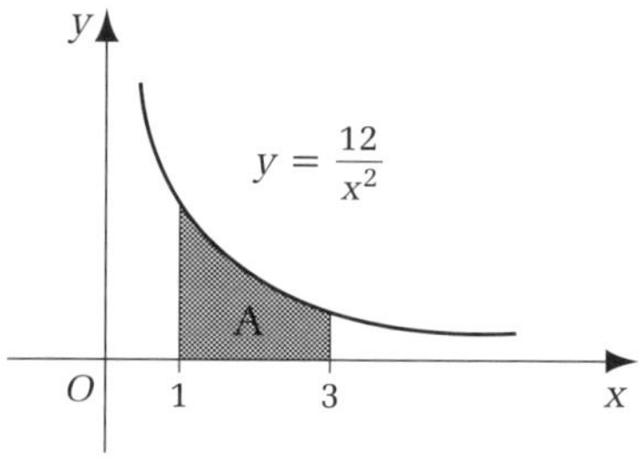

$$\text{area} = \int_1^3 \frac{12}{x^2}\, dx$$
$$= \int_1^3 12x^{-2}\, dx$$
$$= \left[\frac{12}{-1}x^{-1}\right]_1^3$$
$$= \left[-\frac{12}{x}\right]_1^3$$
$$= \left(-\tfrac{12}{3}\right) - \left(-\tfrac{12}{1}\right)$$
$$= -4 + 12 = 8$$

area $= 8$ square units

b

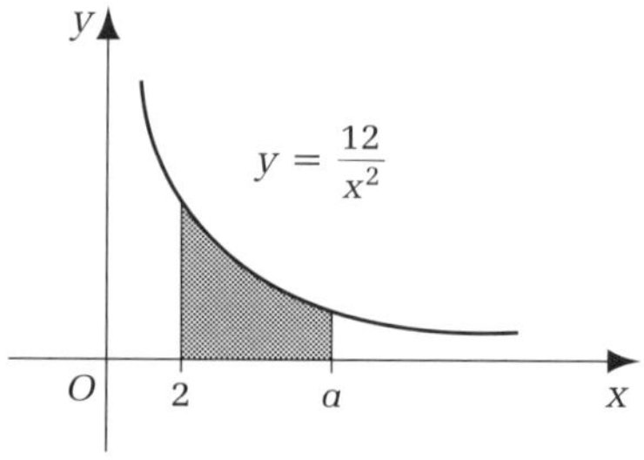

Since the shaded area is 3.6 square units, it follows that

$$3.6 = \int_2^a \frac{12}{x^2}\,dx$$

$$\Rightarrow \quad 3.6 = \left[-\frac{12}{x}\right]_2^a$$

$$\Rightarrow \quad 3.6 = \left(-\frac{12}{a}\right) - \left(-\frac{12}{2}\right)$$

$$\Rightarrow \quad 3.6 = -\frac{12}{a} + 6$$

$$\Rightarrow \quad \frac{12}{a} = 6 - 3.6$$

$$\Rightarrow \quad \frac{12}{a} = 2.4$$

$$\Rightarrow \quad a = \frac{12}{2.4}$$

So $\quad a = 5$

7 To calculate volumes of revolution formed by rotating about the x-axis we use

$$\text{volume} = \int_a^b \pi y^2\,dx$$

We have $y = \sqrt{1+x^3}$, so $y^2 = 1 + x^3$. If follows that the volume formed when the shaded region is rotated 360° about the x-axis is given by

$$\text{volume} = \int_0^2 \pi\left(1+x^3\right)dx$$

$$= \pi\int_0^2 \left(1+x^3\right)dx$$

$$= \pi\left[x + \tfrac{1}{4}x^4\right]_0^2$$

$$= \pi[(2+4) - (0+0)]$$

$$= 6\pi$$

So volume $= 6\pi$ cubic units

8 **a** $\int(2\cos x - 3\sin x)dx = 2\sin x + 3\cos x + c$

b $\int(1 - 6\sin 3x)dx = x + 2\cos 3x + c$

c

$$\int_0^{\frac{\pi}{2}} 2\cos 2t\ dt = \left[2 \times \tfrac{1}{2}\sin 2t\right]_0^{\frac{\pi}{2}}$$

$$= \left[\sin 2t\right]_0^{\frac{\pi}{2}}$$

$$= \sin\pi - \sin 0$$

$$= 0 - 0 = 0$$

d

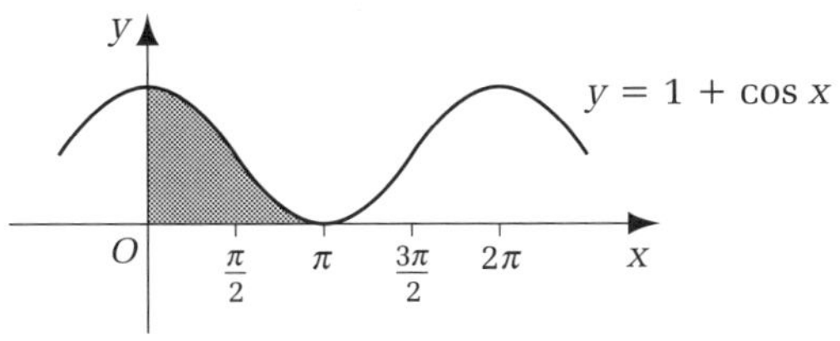

$$\text{area} = \int_0^{\pi} (1 + \cos x)\ dx$$

$$= \left[x + \sin x\right]_0^{\pi}$$

$$= (\pi + \sin\pi) - (0 + \sin 0)$$

$$= (\pi + 0) - (0 + 0)$$

$$= \pi$$

So area $= \pi$ square units

9 **a** The curve $3x^2 + 2y^2 - 12y = 32$ has its highest point (according to the graph) when $x = 0$. Substituting $x = 0$ into the equation of the curve,

$$2y^2 - 12y = 32$$

$$\Rightarrow \quad y^2 - 6y - 16 = 0$$

$$\Rightarrow \quad (y+2)(y-8) = 0$$

$$\Rightarrow \quad y = -2 \text{ or } y = 8$$

So the height of the paperweight is 8 cm.

b To calculate volumes of revolution formed by rotation about the y-axis, we use

$$\text{volume} = \int_b^a \pi x^2 dy$$

Rearranging the equation of the curve to find x^2,

$$3x^2 + 2y^2 - 12y = 32$$

$$\Rightarrow \quad 3x^2 = -2y^2 + 12y + 32$$

$$\Rightarrow \quad x^2 = \frac{-2y^2 + 12y + 32}{3}$$

The volume of the paperweight is therefore given by

$$\text{volume} = \int_0^8 \pi x^2\ dy$$

$$= \int_0^8 \pi\left(\frac{-2y^2 + 12y + 32}{3}\right) dy$$

$$= \frac{\pi}{3}\int_0^8 \left(-2y^2 + 12y + 32\right) dy$$

$$= \frac{\pi}{3}\left[-\tfrac{2}{3}y^3 + 6y^2 + 32y\right]_0^8$$

$$= \frac{\pi}{3}\left[\left(-\tfrac{1024}{3} + 384 + 256\right) - (0+0+0)\right]$$

$$= \frac{\pi}{3}\left(\frac{-1024+1152+768}{3}\right)$$

$$= \frac{896\pi}{9}$$

So volume $= \frac{896\pi}{9}$ cubic units

10 **a**

$$(\cos x - \sin x)^2 = \cos^2 x - 2\cos x\sin x + \sin^2 x$$

$$= \cos^2 x + \sin^2 x - 2\cos x \sin x$$

$$= 1 - 2\sin x\cos x$$

$$= 1 - \sin 2x$$

b The volume of revolution about the x-axis is given by

$$\text{volume} = \int_a^b \pi y^2 \, dx$$

where $y = \cos x - \sin x$, $a = 0$ and $b = \frac{\pi}{4}$

$$= \int_0^{\frac{\pi}{4}} \pi(\cos x - \sin x)^2 \, dx$$

$$= \pi \int_0^{\frac{\pi}{4}} (1 - \sin 2x) \, dx$$

$$= \pi \left[x + \tfrac{1}{2}\cos 2x\right]_0^{\frac{\pi}{4}}$$

$$= \pi\left[\left(\tfrac{\pi}{4} + \tfrac{1}{2}\cos\tfrac{\pi}{2}\right) - \left(0 + \tfrac{1}{2}\cos 0\right)\right]$$

$$= \pi\left[\left(\tfrac{\pi}{4} + 0\right) - \left(0 + \tfrac{1}{2}\right)\right]$$

$$= \pi\left(\tfrac{\pi}{4} - \tfrac{1}{2}\right)$$

$$= \tfrac{\pi}{4}(\pi - 2)$$

So volume $= \frac{\pi}{4}(\pi - 2)$ cubic units

(or 0.8966 cubic units)

10 Trigonometry II (p. 426)

1

$$\cos\theta + \cos 5\theta = 2\cos 2\theta$$

$$2\cos\left(\tfrac{\theta+5\theta}{2}\right)\cos\left(\tfrac{\theta-5\theta}{2}\right) = 2\cos 2\theta$$

Cancelling the common factor 2,

$$\cos 3\theta\cos(-2\theta) = \cos 2\theta$$

Since $\cos(-x) = \cos x, \cos(-2\theta) = \cos 2\theta$.

$$\cos 3\theta \cos 2\theta = \cos 2\theta$$

$$\cos 2\theta(\cos 3\theta - 1) = 0$$

So $\cos 2\theta = 0$ or $(\cos 3\theta - 1) = 0$

$\cos 2\theta = 0$ or $\cos 3\theta = 1$

If $\cos 2\theta = 0$,

then $2\theta = \frac{\pi}{2}, \frac{3\pi}{2}, \frac{5\pi}{2}, \frac{7\pi}{2}$

$\theta = \frac{\pi}{4}, \frac{3\pi}{4}, \frac{5\pi}{4}, \frac{7\pi}{4}$

If $\cos 3\theta = 1$,

then $3\theta = 0, 2\pi, 4\pi, 6\pi$

$\theta = 0, \frac{2\pi}{3}, \frac{4\pi}{3}, 2\pi$

So $\theta = 0, \frac{\pi}{4}, \frac{2\pi}{3}, \frac{3\pi}{4}, \frac{5\pi}{4}, \frac{4\pi}{3}, \frac{7\pi}{4}, 2\pi$

2 $63\sin x + 16\cos x \equiv R\sin(x + \alpha)$

$= R\sin x\cos\alpha + R\cos x\sin\alpha$

Equating coefficients of $\sin x$ and $\cos x$,

$\sin x$: $\quad 63 = R\cos\alpha$ [1]

$\cos x$: $\quad 16 = R\sin\alpha$ [2]

Dividing equation [2] by equation [1].

$$\tan\alpha = \tfrac{16}{63}$$

$$\alpha = \tan^{-1}\left(\tfrac{16}{63}\right)$$

$$= 0.249 \text{ (3 s.f.)}$$

Adding the squares of equations [1] and [2],

$$R^2(\sin^2\alpha + \cos^2\alpha) = 16^2 + 63^2$$

$$R^2 = 256 + 3969 = 4225$$

$$R = \sqrt{4225} = 65$$

So $\quad 63\sin x + 16\cos x = 65\sin(x + 0.249)$

a Since $\quad 63\sin x + 16\cos x = 50$

then $\quad 65\sin(x + 0.249) = 50$

$\Rightarrow \quad \sin(x + 0.249) = \frac{50}{65} = \frac{10}{13}$

$\Rightarrow \quad x + 0.249 = \sin^{-1}\left(\frac{10}{13}\right)$

$= 0.8776$

$\Rightarrow \quad x = 0.629$ (3 s.f.)

b When $\quad 63\sin x + 16\cos x = 0$

then $\quad 65\sin(x + 0.249) = 0$

$\Rightarrow \quad \sin(x + 0.249) = 0$

$\Rightarrow \quad x + 0.249 = \sin^{-1}(0)$

$= 0, \pi, 2\pi, \ldots$

For an obtuse angle, we require the solution

$$x = \pi - 0.249$$

$$= 2.89 \text{ (3 s.f.)}$$

3 $4\sin\theta - 3\cos\theta \equiv R\sin(\theta - \alpha)$

$= R\sin\theta\cos\alpha - R\cos\theta\sin\alpha$

Equating coefficients of $\sin\theta$ and $\cos\theta$,

$\sin\theta$: $\quad 4 = R\cos\alpha$ [1]

$\cos\theta$: $\quad 3 = R\sin\alpha$ [2]

Dividing equation [2] by equation [1],

$$\tan\alpha = \tfrac{3}{4}$$

$$\alpha = \tan^{-1}\left(\tfrac{3}{4}\right)$$

$$= 0.644 \text{ (3 s.f.)}$$

Adding the squares of equations [1] and [2],

$$R^2(\sin^2\alpha + \cos^2\alpha) = 4^2 + 3^2$$

$$R^2 = 25$$

$$R = \sqrt{25} = 5$$

So $\quad 4\sin\theta - 3\cos\theta = 5\sin(\theta - 0.644)$

Now examine the given expression.

$$\frac{1}{10 - 3\cos\theta + 4\sin\theta} = \frac{1}{10 + 4\sin\theta - 3\cos\theta}$$

$$= \frac{1}{10 + 5\sin(\theta - 0.644)}$$

The maximum and minimum values of the sine term are 1 and -1 only, giving the following results:

$$\frac{1}{10+5} = \frac{1}{15} \quad \text{and} \quad \frac{1}{10 + (-5)} = \frac{1}{5}$$

So the greatest value of the expression is $\frac{1}{5}$ and the least value is $\frac{1}{15}$.

4 $\cos 2\theta - \cos 4\theta \equiv 2\cos^2\theta - 2\cos^2 2\theta$

To prove this identity use could be made of the factor formulae, but this would create terms in sine that would then need to be simplified. A more elegant proof relies on noting that $\cos 4\theta = \cos 2(2\theta)$. That is, $\cos 4\theta$ can be treated as a double angle.

$$\cos 2\theta - \cos 4\theta \equiv \cos 2\theta - \cos 2(2\theta)$$
$$\equiv \cos 2\theta - (2\cos^2 2\theta - 1)$$
$$\equiv (\cos 2\theta + 1) - 2\cos^2 2\theta$$
$$\equiv 2\cos^2\theta - 2\cos^2 2\theta$$
$$\equiv \text{RHS}$$

Putting $\theta = 36°$ into the identity,

$$\cos 72° - \cos 144° = 2\cos^2 36° - 2\cos^2 72°$$

Notice that the RHS will factorise: 2 is a common factor and there is a difference of two squares.

On the LHS use the fact that $\cos 144° = -\cos 36°$

Now $\cos 72° + \cos 36° = 2(\cos 36° - \cos 72°)(\cos 36° + \cos 72°)$

Dividing by $(\cos 72° + \cos 36°)$,

$$1 = 2(\cos 36° - \cos 72°)$$

So $\cos 36° - \cos 72° = \frac{1}{2}$, as required.

Use the double angle formulae again to find $\cos 36°$ in the form $a + b\sqrt{5}$.

$$\cos 36° - \cos 72° = \tfrac{1}{2}$$
$$\Rightarrow \cos 36° - (2\cos^2 36° - 1) = \tfrac{1}{2}$$
$$\Rightarrow \cos 36° - 2\cos^2 36° + 1 = \tfrac{1}{2}$$
$$\Rightarrow 2\cos^2 36° - \cos 36° - \tfrac{1}{2} = 0$$
$$\Rightarrow \cos^2 36° - \tfrac{1}{2}\cos 36° - \tfrac{1}{4} = 0$$

This quadratic equation can be solved by completing the square.

$$\cos^2 36° - \tfrac{1}{2}\cos 36° - \tfrac{1}{4} = 0$$
$$\Rightarrow (\cos 36° - \tfrac{1}{4})^2 = \tfrac{1}{4} + \tfrac{1}{16}$$
$$\Rightarrow (\cos 36° - \tfrac{1}{4})^2 = \tfrac{5}{16}$$
$$\Rightarrow \cos 36° - \tfrac{1}{4} = \sqrt{\tfrac{5}{16}} = \tfrac{\sqrt{5}}{4}$$
$$\Rightarrow \cos 36° = \frac{1 + \sqrt{5}}{4}$$

which is in the required form with $a = \frac{1}{4}$ and $b = \frac{1}{4}$.

5 $3\sin\theta - 4\cos\theta \equiv R\sin(\theta - \alpha)$

$$= R\sin\theta\cos\alpha - R\cos\theta\sin\alpha$$

Equating coefficients of $\sin\theta$ and $\cos\theta$,

$\sin\theta$: $3 = R\cos\alpha$ [1]

$\cos\theta$: $4 = R\sin\alpha$ [2]

Dividing equation [2] by equation [1],

$$\tan\alpha = \tfrac{4}{3}$$
$$\alpha = \tan^{-1}(\tfrac{4}{3})$$
$$= 0.927 \text{ (3 d.p.)}$$

Adding the squares of equations [1] and [2],

$$R^2(\sin^2\alpha + \cos^2\alpha) = 3^2 + 4^2$$
$$R^2 = 25$$
$$R = \sqrt{25} = 5$$

So $3\sin\theta - 4\cos\theta = 5\sin(\theta - 0.927)$

a $(3\sin\theta - 4\cos\theta)^2 = [5\sin(\theta - 0.927)]^2$

$$= 25\sin^2(\theta - 0.927)$$

This has its maximum value when $\sin^2(\theta - 0.927)$ is a maximum and has its minimum value when $\sin^2(\theta - 0.927)$ is a minimum.

The maximum value of $\sin^2(\theta - 0.927)$ is 1.

The minimum value of $\sin^2(\theta - 0.927)$ is 0 (recall that square numbers are positive).

So the greatest value of $(3\sin\theta - 4\cos\theta)^2$ is 25.

The least value of $(3\sin\theta - 4\cos\theta)^2$ is 0.

b $$\frac{1}{(3\sin\theta - 4\cos\theta)^2 + 1} = \frac{1}{[5\sin(\theta - 0.927)]^2 + 1}$$

This expression has a maximum value when $\sin^2(\theta - 0.927)$ is as small as possible, and it has a minimum value when $\sin^2(\theta - 0.927)$ is as large as possible. From **a** we know that the smallest and largest values of $\sin^2(\theta - 0.927)$ are 0 and 1 respectively. So the greatest value of the above expression is 1 and the least value is $\frac{1}{26}$.

At the least value,

$$\sin(\theta - 0.927) = 1$$
$$\theta - 0.927 = \tfrac{\pi}{2}$$
$$\Rightarrow \theta = \tfrac{\pi}{2} + 0.927$$
$$= 2.498 \text{ (3 d.p.)}$$

6 **a** $3\cos x° + 4\sin x° \equiv R\sin(x + \alpha)°$

In the analysis we can remove the ° notation, but remember to find α in degrees and not radians.

$$3\cos x + 4\sin x \equiv R\sin x\cos\alpha + R\cos x\sin\alpha$$

Equating coefficients of $\cos x$ and $\sin x$,

$\cos x$: $3 = R\sin\alpha$ [1]

$\sin x$: $4 = R\cos\alpha$ [2]

Dividing equation [1] by equation [2],

$$\tan\alpha = \tfrac{3}{4}$$
$$\alpha = \tan^{-1}(\tfrac{3}{4})$$
$$= 36.9° \text{(3 s.f.)}$$

Adding the squares of equations [1] and [2],

$$R^2(\sin^2\alpha + \cos^2\alpha) = 3^2 + 4^2$$
$$R^2 = 25$$
$$R = \sqrt{25} = 5$$

So $3\cos x° + 4\sin x° = 5\sin(x + 36.9)°$

b The maximum value of $3\cos x° + 4\sin x°$ is the same as the maximum value of $5\sin(x + 36.9)°$. Since the sine function has a maximum value of 1 the expression cannot exceed the value of 5.

c $3\cos x^\circ + 4\sin x^\circ = 0$

$\Rightarrow \quad 5\sin(x + 36.9)^\circ = 0$

$\sin(x + 36.9)^\circ = 0$

$\Rightarrow \quad x + 36.9^\circ = 0, 180^\circ, 360^\circ$

$\Rightarrow \quad x = -36.9^\circ, (180^\circ - 36.9^\circ),$ and $(360^\circ - 36.9^\circ)$

So $\quad x = 143.1^\circ, 323.1^\circ$

7 $\sqrt{3}\sin\theta - \cos\theta \equiv R\sin(\theta - \alpha)$

$= R\sin\theta\cos\alpha - R\cos\theta\sin\alpha$

Equating coefficients of $\sin\theta$ and $\cos\theta$,

$\sin\theta$: $\sqrt{3} = R\cos\alpha$ [1]

$\cos\theta$: $1 = R\sin\alpha$ [2]

Dividing equation [2] by equation [1],

$$\tan\alpha = \frac{1}{\sqrt{3}}$$

$$\alpha = \tan^{-1}\left(\tfrac{1}{\sqrt{3}}\right)$$

Notice that this is a 'special' angle.

$$\alpha = 30^\circ$$

Adding the squares of equations [1] and [2],

$$R^2(\sin^2\alpha + \cos^2\alpha) = (\sqrt{3})^2 + 1^2$$

$$R^2 = 3 + 1 = 4$$

$$R = 2$$

So $\quad \sqrt{3}\sin\theta - \cos\theta = 2\sin(\theta - 30)^\circ$

Use this equality to solve the equation.

$$\sqrt{3}\sin\theta - \cos\theta = \sqrt{2}$$

$\Rightarrow \quad 2\sin(\theta - 30)^\circ = \sqrt{2}$

$\Rightarrow \quad \sin(\theta - 30)^\circ = \frac{\sqrt{2}}{2} = \frac{1}{\sqrt{2}}$

Notice that this is also a 'special' angle.

$$\theta - 30^\circ = 45^\circ, 135^\circ$$

$\Rightarrow \quad \theta = 75^\circ, 165^\circ$

8 To prove $(\cos A + \cos B)^2 + (\sin A + \sin B)^2 \equiv 2 + 2\cos(A - B)$ begin with the LHS.

$$\begin{aligned}\text{LHS} &= (\cos A + \cos B)^2 + (\sin A + \sin B)^2 \\ &= (\cos^2 A + 2\cos A\cos B + \cos^2 B) \\ &\quad + \sin^2 A + 2\sin A\sin B + \sin^2 B) \\ &= (\cos^2 A + \sin^2 A) + (\cos^2 B + \sin^2 B) \\ &\quad + 2(\cos A\cos B + \sin A\sin B) \\ &= 1 + 1 + 2\cos(A - B) \\ &= 2 + 2\cos(A - B) \\ &= \text{RHS}\end{aligned}$$

To solve the equation, $(\cos 4\theta + \cos\theta)^2 + (\sin 4\theta + \sin\theta)^2 = 2\sqrt{3}\sin 3\theta$ we can use the result that we have just proved, with $A = 4\theta$ and $B = \theta$, to give

$$2 + 2\cos(4\theta - \theta) = 2\sqrt{3}\sin 3\theta$$

or $\quad 2 + 2\cos 3\theta = 2\sqrt{3}\sin 3\theta$

Dividing by 2 and rearranging the equation,

$$\sqrt{3}\sin 3\theta - \cos 3\theta = 1$$

Now examine the trigonometric expression on the LHS.

$$\sqrt{3}\sin 3\theta - \cos 3\theta \equiv R\sin(3\theta - \alpha)$$

$$= R\sin 3\theta\cos\alpha - R\cos 3\theta\sin\alpha$$

Equating coefficients of $\sin 3\theta$ and $\cos 3\theta$,

$\sin 3\theta$: $\quad \sqrt{3} = R\cos\alpha$ [1]

$\cos 3\theta$: $\quad 1 = R\sin\alpha$ [2]

Dividing equation [2] by equation [1],

$$\tan\alpha = \tfrac{1}{\sqrt{3}}$$

$$\alpha = \tan^{-1}\left(\tfrac{1}{\sqrt{3}}\right)$$

$$= 30^\circ$$

Adding the squares of equations [1] and [2],

$$R^2(\cos^2\alpha + \sin^2\alpha) = (\sqrt{3})^2 + 1^2$$

$$R^2 = 4$$

$$R = 2$$

So $\quad \sqrt{3}\sin 3\theta - \cos 3\theta = 2\sin(3\theta - 30^\circ)$

To find the general solution of the equation $\sqrt{3}\sin\theta - \cos 3\theta = 1$, we need to solve the equivalent equation

$$2\sin(3\theta - 30^\circ) = 1$$

$\Rightarrow \quad \sin(3\theta - 30^\circ) = \tfrac{1}{2}$

So $\quad 3\theta - 30^\circ = \sin^{-1}\left(\tfrac{1}{2}\right)$

The general solution is

$$3\theta - 30^\circ = 180^\circ n + (-1)^n 30^\circ$$

$$3\theta = 180^\circ n + 30^\circ + 30^\circ(-1)^n$$

$$\theta = 60^\circ n + 10^\circ(1 + (-1)^n)$$

9 **a** $2\cos x - \sin x \equiv R\cos(x + \alpha)$

$= R\cos x\cos\alpha - R\sin x\sin\alpha$

Equating coefficients of $\cos x$ and $\sin x$,

$\cos x$: $\quad 2 = R\cos\alpha$ [1]

$\sin x$: $\quad 1 = R\sin\alpha$ [2]

Dividing equation [2] by equation [1],

$$\tan\alpha = \tfrac{1}{2}$$

$$\alpha = \tan^{-1}\left(\tfrac{1}{2}\right)$$

$$= 0.464 \text{ (3 s.f.)}$$

Adding the squares of equations [1] and [2],

$$R^2(\sin^2\alpha + \cos^2\alpha) = 2^2 + 1^2$$

$$R^2 = 5$$

$$R = \sqrt{5}$$

This means that

$$2\cos x - \sin x \equiv \sqrt{5}\cos(x + 0.464)$$

b The equation $2\cos x - \sin x = 1$ is equivalent to

$$\sqrt{5}\cos(x + 0.464) = 1$$

$$\Rightarrow \quad \cos(x + 0.464) = \tfrac{1}{\sqrt{5}}$$

$$\Rightarrow \quad x + 0.464 = \cos^{-1}\left(\tfrac{1}{\sqrt{5}}\right)$$

$$\Rightarrow \quad x + 0.464 = 2n\pi \pm 1.107$$

When $x + 0.464 = 2n\pi + 1.107$

$$x = 2n\pi + 0.643$$

When $x + 0.464 = 2n\pi - 1.107$

$$x = 2n\pi - 1.571$$

11 Differentiation II (p. 488)

1 Let $y = uv$ where $u = x$ and $v = e^{-3x}$.
Then $u' = 1$ and $v' = -3e^{-3x}$.

Using the product rule,

$$\frac{dy}{dx} = uv' + vu'$$

$$= -3xe^{-3x} + e^{-3x}$$

$$= (1 - 3x)e^{-3x}$$

At the stationary point on the curve,

$$\frac{dy}{dx} = 0$$

So $(1 - 3x)e^{-3x} = 0$

But $e^{-3x} > 0$ for all values of x, so

$$1 - 3x = 0$$

$$x = \tfrac{1}{3}$$

When $x = \frac{1}{3}$, $y = \frac{1}{3}e^{-3\left(\frac{1}{3}\right)}$

$$= \tfrac{1}{3}e^{-1}$$

$$= \frac{1}{3e}$$

So the stationary point is located at $\left(\frac{1}{3}, \frac{1}{3e}\right)$

2 **a** $x = t^2 + 3t$ and $y = 2t - t^2$.
Differentiating both equations with respect to t,

$$\frac{dx}{dt} = 2t + 3 \text{ and } \frac{dy}{dt} = 2 - 2t.$$

Using the chain rule,

$$\frac{dy}{dx} = \frac{dy}{dt} \times \frac{dt}{dx}$$

$$= \left(\frac{dy}{dt}\right) \Big/ \left(\frac{dx}{dt}\right)$$

$$= \frac{2 - 2t}{2t + 3}$$

b If the normal to the curve at P has a gradient of $\frac{3}{8}$, then the gradient of the tangent is $-\frac{8}{3}$. So the gradient of the curve at P is $-\frac{8}{3}$. That is, at P, $\frac{dy}{dx} = -\frac{8}{3}$.

$$\Rightarrow \quad \frac{2 - 2t}{2t + 3} = -\frac{8}{3}$$

$$\Rightarrow \quad 3(2 - 2t) = -8(2t + 3)$$

$$\Rightarrow \quad 6 - 6t = -16t - 24$$

$$\Rightarrow \quad 30 = -10t$$

$$\Rightarrow \quad t = -3$$

When $t = -3$,

$$x = (-3)^2 + 3(-3)$$

$$= 9 - 9 = 0$$

and $y = 2(-3) - (-3)^2$

$$= -6 - 9 = -15$$

So the coordinates of P are $(0, -15)$.

3 **a** Let $y = uv$, where $u = x - 1$ and $v = \sqrt{6 - x}$.

Then $u' = 1$ and $v' = \dfrac{-1}{2\sqrt{6 - x}}$.

Using the product rule,

$$\frac{dy}{dx} = uv' + vu'$$

$$= \left[(x - 1) \times \frac{-1}{2\sqrt{6 - x}}\right] + \left[\sqrt{6 - x} \times 1\right]$$

$$= \frac{-x + 1}{2\sqrt{6 - x}} + \sqrt{6 - x}$$

$$= \frac{-x + 1 + \sqrt{6 - x}(2\sqrt{6 - x})}{2\sqrt{6 - x}}$$

$$= \frac{-x + 1 + 2(6 - x)}{2\sqrt{6 - x}}$$

$$= \frac{-x + 1 + 12 - 2x}{2\sqrt{6 - x}}$$

$$= \frac{13 - 3x}{2\sqrt{6 - x}}$$

b When $x = 4$,

$$\frac{dy}{dx} = \frac{13 - (3 \times 4)}{2\sqrt{6 - 4}}$$

$$= \frac{13 - 12}{2\sqrt{2}}$$

$$= \frac{1}{2\sqrt{2}}$$

So the gradient is positive when $x = 4$. This means that the function is still increasing. So the x-coordinate of the maximum point on the graph $y = (x - 1)\sqrt{6 - x}$ is greater than 4.

4 **a** $x = 2\cos t + \sin 2t$ and $y = \cos t - 2\sin 2t$.
Differentiating both equations with respect to t,

$$\frac{dx}{dt} = -2\sin t + 2\cos 2t$$

$$\frac{dy}{dt} = -\sin t - 4\cos 2t$$

b Using the chain rule,

$$\frac{dy}{dx} = \frac{dy}{dt} \times \frac{dt}{dx}$$
$$= \left(\frac{dy}{dt}\right) \Big/ \left(\frac{dx}{dt}\right)$$
$$= \frac{-(\sin t + 4\cos 2t)}{-2\sin t + 2\cos 2t}$$
$$= \frac{\sin t + 4\cos 2t}{2\sin t - 2\cos 2t}$$

When $t = \frac{\pi}{4}$,

$$\frac{dy}{dx} = \frac{\sin\left(\frac{\pi}{4}\right) + 4\cos\left(\frac{\pi}{2}\right)}{2\sin\left(\frac{\pi}{4}\right) - 2\cos\left(\frac{\pi}{2}\right)}$$
$$= \frac{\sin\left(\frac{\pi}{4}\right) + 0}{2\sin\left(\frac{\pi}{4}\right) - 0}$$
$$= \frac{1}{2}$$

c At P, where $t = \frac{\pi}{4}$,

$$x = 2\cos\left(\frac{\pi}{4}\right) + \sin\left(\frac{\pi}{2}\right)$$
$$= 2\left(\frac{1}{\sqrt{2}}\right) + 1$$
$$= \sqrt{2} + 1$$

and $y = \cos\left(\frac{\pi}{4}\right) - 2\sin\left(\frac{\pi}{2}\right)$

$$= \frac{1}{\sqrt{2}} - 2$$

Since $\frac{dy}{dx} = \frac{1}{2}$ at P, the gradient of the normal to the curve at P is -2. Using $y - y_1 = m(x - x_1)$ to find the equation of the normal,

$$y - \left(\frac{1}{\sqrt{2}} - 2\right) = -2\left[x - \left(\sqrt{2} + 1\right)\right]$$
$$= -2x + 2\sqrt{2} + 2$$
$$\Rightarrow \quad y = -2x + 2\sqrt{2} + 2 + \frac{1}{\sqrt{2}} - 2$$
$$= -2x + 2\sqrt{2} + \frac{\sqrt{2}}{2}$$

So $\quad y = -2x + \frac{5}{2}\sqrt{2}$

is the equation of the normal at P.

5 The Maclaurin series expansion for some function $f(x)$ is given by

$$f(x) = f(0) + xf'(0) + \frac{1}{2!}x^2f''(0) + \frac{1}{3!}x^3f'''(0) + \frac{1}{4!}x^4f^{(4)}(0) + \frac{1}{5!}x^5f^{(5)}(0) + \dots$$

If $f(x) = \tan^{-1}x$, then we know the standard result,

$$f'(x) = \frac{1}{1+x^2} = (1+x^2)^{-1}$$

Using the chain rule,

$$f''(x) = -(1+x^2)^{-2} \times 2x = \frac{-2x}{(1+x^2)^2}$$

Now use the quotient rule to find $f'''(x)$.
Let $u = -2x$ and $v = (1+x^2)^2$.
Then $u' = -2$ and $v' = 4x(1+x^2)$.

$$f'''(x) = \frac{vu' - uv'}{v^2}$$
$$= \frac{-2(1+x^2)^2 - [-2x \times 4x(1+x^2)]}{(1+x^2)^4}$$
$$= \frac{-2(1+x^2) + 8x^2}{(1+x^2)^3}$$
$$= \frac{2(3x^2 - 1)}{(1+x^2)^3}$$

Using the quotient rule again, with $u = 2(3x^2 - 1)$ and $v = (1+x^2)^3$, we have $u' = 12x$ and $v' = 6x(1+x^2)^2$.

$$f^{(4)}(x) = \frac{vu' - uv'}{v^2}$$
$$= \frac{12x(1+x^2)^3 - 12x(3x^2-1)(1+x^2)^2}{(1+x^2)^6}$$
$$= \frac{12x[(1+x^2) - (3x^2 - 1)]}{(1+x^2)^4}$$
$$= \frac{24x(1-x^2)}{(1+x^2)^4}$$

And using the quotient rule once more with $u = 24x(1-x^2)$ and $v = (1+x^2)^4$, we have $u' = 24 - 72x^2$ and $v' = 8x(1+x^2)^3$.

$$f^{(5)}(x) = \frac{vu' - uv'}{v^2}$$
$$= \frac{(24 - 72x^2)(1+x^2)^4 - 192x^2(1-x^2)(1+x^2)^3}{(1+x^2)^8}$$
$$= \frac{24[(1-3x^2)(1+x^2) - 8x^2(1-x^2)]}{(1+x^2)^5}$$
$$= \frac{24(1 - 2x^2 - 3x^4 - 8x^2 + 8x^4)}{(1+x^2)^5}$$
$$= \frac{24(1 - 10x^2 + 5x^4)}{(1+x^2)^5}$$

Substituting $x = 0$ into each of these expressions,

$$f(0) = 0 \quad f'(0) = 1 \quad f''(0) = 0$$
$$f'''(0) = -2 \quad f^{(4)}(0) = 0 \quad f^{(5)}(0) = 24$$

So the first three non-zero terms of the Maclaurin series for $\tan^{-1}x$ are

$$\tan^{-1}x \approx xf'(0) + \frac{1}{3!}x^3f'''(0) + \frac{1}{5!}x^5f^{(5)}(0)$$
$$\approx x - \frac{2}{3!}x^3 + \frac{24}{5!}x^5$$
$$\approx x - \frac{1}{3}x^3 + \frac{1}{5}x^5$$

So for x sufficiently small for this expansion to be valid,

$$(1+x^2)\tan^{-1}x \approx (1+x^2)(x - \frac{1}{3}x^3 + \frac{1}{5}x^5)$$
$$\approx x - \frac{1}{3}x^3 + \frac{1}{5}x^5 + x^3 - \frac{1}{3}x^5 + \dots$$
$$\approx x + \frac{2}{3}x^3 - \frac{2}{15}x^5$$

as required.

6 **a** Substituting $x = 1$ and $y = 2$ into the equation for the curve,

$$x^2y + y^2 - 3x - 3 = (1^2 \times 2) + (2^2) - (3 \times 1) - 3$$
$$= 2 + 4 - 3 - 3$$
$$= 0, \text{ as required.}$$

So A (1, 2) lies on the curve.

b Since B is the point where the curve crosses the x-axis, $y = 0$ at B. Substituting $y = 0$ into the equation for the curve,

$$-3x - 3 = 0$$

$$\Rightarrow \quad x = -1$$

So the coordinates of B are $(-1, 0)$

c Differentiating each term in the equation with respect to x (using the chain rule where necessary), collecting like terms and factorising,

$$x^2\frac{dy}{dx} + 2xy + 2y\frac{dy}{dx} - 3 = 0$$

$$\Rightarrow \quad (x^2 + 2y)\frac{dy}{dx} = 3 - 2xy$$

$$\Rightarrow \quad \frac{dy}{dx} = \frac{3 - 2xy}{x^2 + 2y}$$

At A, $x = 1$ and $y = 2$, so

$$\frac{dy}{dx} = \frac{3-4}{1+4} = -\tfrac{1}{5}$$

So the gradient of the curve at A is $-\frac{1}{5}$.

d The gradient of the normal to the curve at A is 5. Using $y - y_1 = m(x - x_1)$, the equation of this normal is given by

$$y - 2 = 5(x - 1)$$

$$\Rightarrow \quad y = 5x - 3$$

7 $x = e^{2t} - 5t$ and $y = e^{2t} - 2t$.
Differentiating both equations with respect to t,

$$\frac{dx}{dt} = 2e^{2t} - 5$$

$$\frac{dy}{dt} = 2e^{2t} - 2$$

Using the chain rule,

$$\frac{dy}{dx} = \frac{dy}{dt} \times \frac{dt}{dx}$$

$$= \left(\frac{dy}{dt}\right) \Big/ \left(\frac{dx}{dt}\right)$$

$$= \frac{2e^{2t} - 2}{2e^{2t} - 5}$$

At the point on the curve where the gradient is 2, $\frac{dy}{dx} = 2$. So,

$$\frac{2e^{2t} - 2}{2e^{2t} - 5} = 2$$

$$\Rightarrow \quad 2e^{2t} - 2 = 2(2e^{2t} - 5)$$

$$\Rightarrow \quad 2e^{2t} - 2 = 4e^{2t} - 10$$

$$\Rightarrow \quad 2e^{2t} = 8$$

$$e^{2t} = 4$$

Taking logarithms of both sides,

$$2t = \ln 4$$

$$= 2\ln 2$$

So $\quad t = \ln 2$

8 **a** $y = \dfrac{3x - 9}{x^2 - x - 2} = \dfrac{3x - 9}{(x-2)(x+1)}$

Substitution of $x = -1$ or $x = 2$ into this expression would lead to division by zero, and an infinite value of y. So the equations of the vertical asymptotes are $x = -1$ and $x = 2$.

b Let $y = \frac{u}{v}$ where $u = 3x - 9$ and $v = x^2 - x - 2$. Then $u' = 3$ and $v' = 2x - 1$.

Using the quotient rule,

$$\frac{dy}{dx} = \frac{vu' - uv'}{v^2}$$

$$= \frac{3(x^2 - x - 2) - (3x - 9)(2x - 1)}{(x^2 - x - 2)^2}$$

$$= \frac{3x^2 - 3x - 6 - (6x^2 - 21x + 9)}{(x^2 - x - 2)^2}$$

$$= \frac{-3x^2 + 18x - 15}{(x^2 - x - 2)^2}$$

$$= \frac{-3(x^2 - 6x + 5)}{(x^2 - x - 2)^2}$$

At maximum and minimum points on the curve, $\frac{dy}{dx} = 0$.

$$\Rightarrow \quad \frac{-3(x^2 - 6x + 5)}{(x^2 - x - 2)^2} = 0$$

$$\Rightarrow \quad x^2 - 6x + 5 = 0$$

$$\Rightarrow \quad (x - 1)(x - 5) = 0$$

Thus $x = 1$ or $x = 5$.

When $x = 1$,

$$y = \frac{3 - 9}{1 - 1 - 2}$$

$$= \frac{-6}{-2} = 3$$

When $x = 5$,

$$y = \frac{15 - 9}{25 - 5 - 2}$$

$$= \frac{6}{18} = \frac{1}{3}$$

So the coordinates of A and B are $(5, \frac{1}{3})$ and $(1, 3)$, respectively.

9 $y = \dfrac{1}{1 + \sin 2x}$

$$= (1 + \sin 2x)^{-1}$$

Using the chain rule,

$$\frac{dy}{dx} = 2\cos 2x \times -(1 + \sin 2x)^{-2}$$

$$= \frac{-2\cos 2x}{(1 + \sin 2x)^2}$$

Use the quotient rule to find $\dfrac{d^2y}{dx^2}$.
Let $\dfrac{dy}{dx} = \dfrac{u}{v}$, where $u = -2\cos 2x$ and $v = (1 + \sin 2x)^2$.
Then $u' = 4\sin 2x$ and $v' = 4\cos 2x(1 + \sin 2x)$.

Using the quotient rule,

$$\frac{d^2y}{dx^2} = \frac{vu' - uv'}{v^2}$$

$$= \frac{(1+\sin 2x)^2 \times 4\sin 2x + 2\cos 2x \times 4\cos 2x(1+\sin 2x)}{(1+\sin 2x)^4}$$

$$= \frac{(1+\sin 2x) \times 4\sin 2x + 8\cos^2 2x}{(1+\sin 2x)^3}$$

$$= \frac{4\sin 2x + 4\sin^2 2x + 8\cos^2 2x}{(1+\sin 2x)^3}$$

$$= \frac{4\sin 2x + 4(\sin^2 2x + \cos^2 2x) + 4\cos^2 2x}{(1+\sin 2x)^3}$$

$$= \frac{4\sin 2x + 4\cos^2 2x + 4}{(1+\sin 2x)^3}$$

When $x = 0$, $\sin 2x = 0$ and $\cos^2 2x = 1$. So

$$\frac{d^2y}{dx^2} = \frac{0+4+4}{(1+0)^2} = 8$$

If $y = f(x)$, where $f(x) = \dfrac{1}{1+\sin 2x}$,

then the first three terms of the Maclaurin series for y are:

$$f(x) = f(0) + xf'(0) + \tfrac{1}{2}x^2 f''(0)$$

We have

$$f(x) = \frac{1}{1+\sin 2x}$$

$$f'(x) = \frac{dy}{dx} = \frac{-2\cos 2x}{(1+\sin 2x)^2}$$

$$f''(x) = \frac{d^2y}{dx^2} = \frac{4\sin 2x + 4\cos^2 2x + 4}{(1+\sin 2x)^3}$$

So $f(0) = 1$, $f'(0) = -2$ and $f''(0) = 8$.

So the first three terms of the Maclaurin series for y are

$$f(x) = f(0) + xf'(0) + \tfrac{1}{2}x^2 f''(0)$$
$$= 1 - 2x + 4x^2$$

a $$\int_{-0.1}^{0.1} y\,dx \approx \int_{-0.1}^{0.1} (1 - 2x + 4x^2)\,dx$$

$$= \left[x - x^2 + \tfrac{4}{3}x^3\right]_{-0.1}^{0.1}$$

$$= [0.1 - 0.1^2 + \tfrac{4}{3}(0.1)^3]$$

$$-[-0.1 - (-0.1)^2 + \tfrac{4}{3}(-0.1)^3]$$

$$= 0.2 + \tfrac{8}{3}(0.001)$$

$$= 0.2027 \text{ (4 d.p.)}$$

b Differentiating the Maclaurin series for y obtained above, the first two terms of the Maclaurin series for $\frac{dy}{dx}$ are

$$f'(0) + xf''(0) = -2 + 8x$$

10

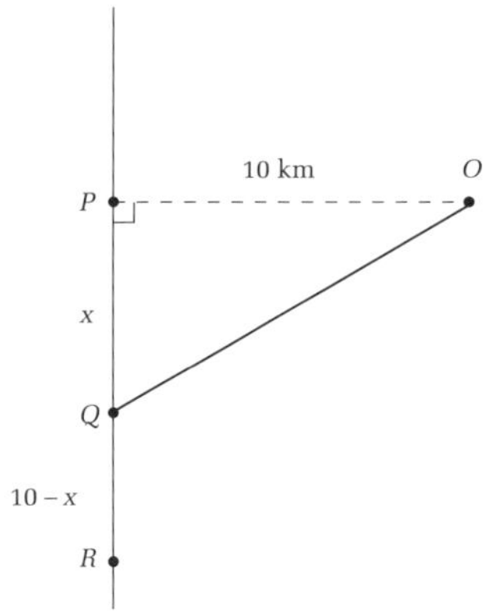

$$OQ = \sqrt{10^2 + x^2}$$
$$= \sqrt{100 + x^2} \text{ km}$$

Total cost of underwater pipeline is

£$5\sqrt{100+x^2}$ million

$QR = (10 - x)$ km

Total cost of coastline pipeline is

£$3(10 - x)$ million = £$(30 - 3x)$ million

So the total cost of the project is £C million, where

$$C = 5\sqrt{100+x^2} + 30 - 3x.$$

Using the chain rule,

$$\frac{dC}{dx} = \frac{5x}{\sqrt{100+x^2}} - 3$$

When $\frac{dC}{dx} = 0$,

$$\frac{5x}{\sqrt{100+x^2}} = 3$$

$$\Rightarrow \quad \frac{25x^2}{100+x^2} = 9$$

$$\Rightarrow \quad 25x^2 = 9(100+x^2)$$

$$\Rightarrow \quad 25x^2 = 900 + 9x^2$$

$$\Rightarrow \quad 16x^2 = 900$$

$$\Rightarrow \quad 4x = 30$$

$$\Rightarrow \quad x = 7.5 \text{ km}$$

To find $\frac{d^2C}{dx^2}$, differentiate again with respect to x, use the quotient rule.

Let $u = 5x$ and $v = \sqrt{100+x^2}$.

Then $u' = 5$ and $v' = x(100+x^2)^{-\frac{1}{2}}$.

$$\frac{d^2C}{dx^2} = \frac{vu' - uv'}{v^2}$$

$$= \frac{5\sqrt{100+x^2} - \frac{5x^2}{\sqrt{100+x^2}}}{100+x^2}$$

$$= \frac{5(100+x^2) - 5x^2}{(100+x^2)^{\frac{3}{2}}}$$

$$= \frac{500}{(100+x^2)^{\frac{3}{2}}}$$

When $x = 7.5$,

$$\frac{d^2C}{dx^2} = \frac{500}{(100+x^2)^{\frac{3}{2}}}$$
$$= 0.256$$

Since $\frac{d^2C}{dx^2} > 0$, C is a minimum when $x = 7.5$.

12 Integration II (p. 546)

1 **a**

$$\int \frac{1}{x^2}\,dx = \int x^{-2}\,dx$$
$$= \frac{1}{-1}x^{-1} + c$$
$$= -\frac{1}{x} + c$$

b

$$\int 4\cos(2x+7)\,dx = 4\int \cos(2x+7)\,dx$$
$$= 4 \times \tfrac{1}{2}\sin(2x+7) + c$$
$$= 2\sin(2x+7) + c$$

c

$$\int (\sin 2x + \cos\tfrac{1}{2}x)\,dx = -\tfrac{1}{2}\cos 2x + 2\sin\tfrac{1}{2}x + c$$

d

$$\int \sec^2 4x\,dx = \tfrac{1}{4}\tan 4x + c$$

e Using the product formula (Chapter 10),

$$\sin A \sin B \equiv \tfrac{1}{2}\cos(A-B) - \tfrac{1}{2}\cos(A+B)$$

With $A = 5x$ and $B = x$, it follows that

$$\sin 5x \sin x \equiv \tfrac{1}{2}\cos 4x - \tfrac{1}{2}\cos 6x$$

So

$$\int \sin 5x \sin x\,dx \equiv \int (\tfrac{1}{2}\cos 4x - \tfrac{1}{2}\cos 6x)\,dx$$
$$= \tfrac{1}{2}\int (\cos 4x - \cos 6x)\,dx$$
$$= \tfrac{1}{2}(\tfrac{1}{4}\sin 4x - \tfrac{1}{6}\sin 6x) + c$$
$$= \tfrac{1}{8}\sin 4x - \tfrac{1}{12}\sin 6x + c$$

f

$$\int (1 + e^{3x})\,dx = x + \tfrac{1}{3}e^{3x} + c$$

g

$$\int \frac{1}{e^{5x}}\,dx = \int e^{-5x}\,dx$$
$$= -\tfrac{1}{5}e^{-5x} + c$$
$$= -\frac{1}{5e^{5x}} + c$$

h

$$\int \frac{4x+6}{x^2+3x}\,dx = 2\int \frac{2x+3}{x^2+3x}\,dx$$
$$= 2\ln|x^2+3x| + c$$

i

$$\int \cot 3x\,dx = \int \frac{\cos 3x}{\sin 3x}\,dx$$
$$= \tfrac{1}{3}\int \frac{3\cos 3x}{\sin 3x}\,dx$$
$$= \tfrac{1}{3}\ln|\sin 3x| + c$$

j

$$\int \frac{1}{25+x^2}\,dx = \frac{1}{5^2+x^2}\,dx$$
$$= \tfrac{1}{5}\tan^{-1}\left(\frac{x}{5}\right) + c$$

k

$$\int \frac{1}{\sqrt{25-x^2}}\,dx = \int \frac{1}{\sqrt{5^2-x^2}}\,dx$$
$$= \sin^{-1}\left(\frac{x}{5}\right) + c$$

l First write $\dfrac{x}{x+3}$ as a proper fraction.

$$\frac{x}{x+3} = \frac{(x+3)-3}{x+3}$$
$$= 1 - \frac{3}{x+3}$$

Then

$$\int \frac{x}{x+3}\,dx = \int \left(1 - \frac{3}{x+3}\right)dx$$
$$= x - 3\ln|x+3| + c$$

2 Use integration by parts,

$$\int u\frac{dv}{dx}\,dx = uv - \int v\frac{du}{dx}\,dx,$$

with $u = \ln x$ and $\frac{dv}{dx} = x^4$. Then $\frac{du}{dx} = \frac{1}{x}$ and $v = \frac{1}{5}x^5$.

$$\int x^4 \ln x\,dx = \ln x \times \tfrac{1}{5}x^5 - \int \tfrac{1}{5}x^4\,dx$$
$$= \tfrac{1}{5}x^5\ln x - \tfrac{1}{25}x^5 + c$$

3 Use integration by parts,

$$\int u\frac{dv}{dx}\,dx = uv - \int v\frac{du}{dx}\,dx,$$

with $u = x$ and $\frac{dv}{dx} = e^{3x}$. Then $\frac{du}{dx} = 1$ and $v = \frac{1}{3}e^{3x}$.

$$\int_0^1 xe^{3x}\,dx = \left[x \times \tfrac{1}{3}e^{3x}\right]_0^1 - \int_0^1 (\tfrac{1}{3}e^{3x} \times 1)\,dx$$
$$= \left[\tfrac{1}{3}xe^{3x}\right]_0^1 - \tfrac{1}{3}\int_0^1 e^{3x}\,dx$$
$$= \left[\tfrac{1}{3}xe^{3x} - \tfrac{1}{3} \times \tfrac{1}{3}e^{3x}\right]_0^1$$
$$= \left[\tfrac{1}{3}xe^{3x} - \tfrac{1}{9}e^{3x}\right]_0^1$$
$$= (\tfrac{1}{3}e^3 - \tfrac{1}{9}e^3) - (0 - \tfrac{1}{9}e^0)$$
$$= \tfrac{1}{3}e^3 - \tfrac{1}{9}e^3 + \tfrac{1}{9}$$
$$= \tfrac{2}{9}e^3 + \tfrac{1}{9}$$
$$= \tfrac{1}{9}(1 + 2e^3)$$

4 First express the integrand as partial fractions.

$$\frac{1}{(x+1)(2x+1)} \equiv \frac{A}{x+1} + \frac{B}{2x+1}$$
$$= \frac{A(2x+1) + B(x+1)}{(x+1)(2x+1)}$$

Equating the numerators,

$$1 \equiv A(2x+1) + B(x+1)$$

Let $x = -1$. Then $1 = -A \Rightarrow A = -1$

Let $x = -\tfrac{1}{2}$. Then $1 = \tfrac{1}{2}B \Rightarrow B = 2$

So $\dfrac{1}{(x+1)(2x+1)} = \dfrac{-1}{x+1} + \dfrac{2}{2x+1}$

$$\int_0^1 \frac{1}{(x+1)(2x+1)}\,dx = \int_0^1 \left(\frac{-1}{x+1} + \frac{2}{2x+1}\right) dx$$
$$= -\int_0^1 \frac{1}{x+1}\,dx + \int_0^1 \frac{2}{2x+1}\,dx$$
$$= -\Big[\ln|x+1|\Big]_0^1 + \Big[\ln|2x+1|\Big]_0^1$$
$$= -(\ln 2 - \ln 1) + (\ln 3 - \ln 1)$$
$$= -\ln 2 + \ln 3$$
$$= \ln(\tfrac{3}{2})$$

This is in the required form with $a = \frac{3}{2}$

5 Notice that the denominator is the difference of two squares and so can be factorised easily.

$$\frac{3}{x^2-9} = \frac{3}{(x+3)(x-3)}$$

Writing this in partial fractions,

$$\frac{3}{(x+3)(x-3)} \equiv \frac{A}{x+3} + \frac{B}{x-3}$$
$$= \frac{A(x-3)+B(x+3)}{(x+3)(x-3)}$$

Equating the numerators,

$$3 \equiv A(x-3) + B(x+3)$$

Let $x = 3$. Then $3 = 6B \Rightarrow B = \frac{1}{2}$

Let $x = -3$. Then $3 = -6A \Rightarrow A = -\frac{1}{2}$

So $\dfrac{3}{x^2-9} = \dfrac{-1}{2(x+3)} + \dfrac{1}{2(x-3)}$.

Now $$\int_0^2 \frac{3}{x^2-9}\,dx = \int_0^2 \left[\frac{-1}{2(x+3)} + \frac{1}{2(x-3)}\right] dx$$
$$= -\tfrac{1}{2}\int_0^2 \frac{1}{x+3}\,dx + \tfrac{1}{2}\int_0^2 \frac{1}{x-3}\,dx$$
$$= -\tfrac{1}{2}\Big[\ln|x+3|\Big]_0^2 + \tfrac{1}{2}\Big[\ln|x-3|\Big]_0^2$$
$$= -\tfrac{1}{2}(\ln 5 - \ln 3) + \tfrac{1}{2}(\ln|-1| - \ln|-3|)$$
$$= -\tfrac{1}{2}\ln(\tfrac{5}{3}) + \tfrac{1}{2}(\ln 1 - \ln 3)$$
$$= -\tfrac{1}{2}\ln(\tfrac{5}{3}) - \tfrac{1}{2}\ln 3$$
$$= -\tfrac{1}{2}[\ln(\tfrac{5}{3}) + \ln 3]$$
$$= -\tfrac{1}{2}[\ln(\tfrac{5\times 3}{3})]$$
$$= -\tfrac{1}{2}\ln 5$$

This is in the required form, with $a = 2$ and $b = 5$.

6 Using the substitution $u = 1 + 3\sin\theta$,

we have $\frac{du}{d\theta} = 3\cos\theta$

When $\theta = 0$, $u = 1$

When $\theta = \frac{\pi}{2}$, $u = 1 + 3\sin(\frac{\pi}{2})$
$= 1 + 3 = 4$

As a result of this substitution, the integral becomes

$$\int_0^{\frac{\pi}{2}} \frac{\cos\theta}{\sqrt{1+3\sin\theta}}\,d\theta \equiv \int_{u=1}^{u=4} \frac{\cos\theta}{\sqrt{u}}\frac{d\theta}{du}\,du$$
$$= \int_1^4 \frac{\cos\theta}{\sqrt{u}} \times \frac{1}{3\cos\theta}\,du$$
$$= \tfrac{1}{3}\int_1^4 \frac{1}{\sqrt{u}}\,du$$
$$= \tfrac{1}{3}\int_1^4 u^{-\frac{1}{2}}\,du$$
$$= \tfrac{1}{3}\Big[2u^{\frac{1}{2}}\Big]_1^4$$
$$= \tfrac{1}{3}(2\sqrt{4} - 2\sqrt{1})$$
$$= \tfrac{1}{3}(4-2)$$
$$= \tfrac{2}{3}$$

7 Using the substitution $u = x - 3$

we have $\frac{du}{dx} = 1$

When $x = 3$, $u = 0$

When $x = 4$, $u = 1$

As a result of this substitution, the integral becomes

$$\int_3^4 x(x-3)^{17}dx \equiv \int_{u=0}^{u=1} x \times u^{17}\frac{dx}{du}\,du$$
$$= \int_0^1 (u+3) \times u^{17} \times 1\,du$$
$$= \int_0^1 (u^{18} + 3u^{17})du$$
$$= \left[\frac{u^{19}}{19} + \frac{u^{18}}{6}\right]_0^1$$
$$= (\tfrac{1}{19} + \tfrac{1}{6}) - (0+0)$$
$$= \tfrac{25}{114}$$

8 $\dfrac{dy}{dx} = ye^x$

Separating the variables and integrating both sides of this differential equation,

$$\int \frac{1}{y}\,dy = \int e^x\,dx$$
$\Rightarrow$ $\ln y = e^x + c$

When $x = 0$, $y = 1$. Substituting these values of x and y,

$\ln 1 = e^0 + c$

$\Rightarrow$ $0 = 1 + c$

$\Rightarrow$ $c = -1$

So $\ln|y| = e^x - 1$

$|y| = e^{e^x - 1}$

$y = \pm e^{e^x - 1}$

9 **a** The rate of change of surface area with time

$$\frac{ds}{dt} \propto \frac{1}{\sqrt{s}}$$

So $\dfrac{ds}{dt} = \dfrac{k}{\sqrt{s}}$

where k is a constant of proportionality.

When $s = 900$, $\frac{ds}{dt} = 60$. Substituting these values into the differential equation

$$60 = \frac{k}{\sqrt{900}}$$

$\Rightarrow \quad 60 = \dfrac{k}{30}$

$\Rightarrow \quad k = 1800$

So $\dfrac{ds}{dt} = \dfrac{1800}{\sqrt{s}}$

b Separating the variables and integrating both sides of the differential equation

$$\int \sqrt{s}\, ds = \int 1800\, dt$$

$\Rightarrow \quad \displaystyle\int s^{\frac{1}{2}}\, ds = \int 1800\, dt$

$\Rightarrow \quad \frac{2}{3}s^{\frac{3}{2}} = 1800t + c$

$s^{\frac{3}{2}} = 2700t + A$

where A is some constant.

When $t = 0$, $s = 400$. Substituting these values,

$400^{\frac{3}{2}} = A$

$\Rightarrow \quad A = 8000$

So $s^{\frac{3}{2}} = 2700t + 8000$

c When $s = 900$,

$900^{\frac{3}{2}} = 2700t + 8000$

$\Rightarrow \quad 27\,000 = 2700t + 8000$

$\Rightarrow \quad 19\,000 = 2700t$

$\Rightarrow \quad t = 7.03$

So to the nearest second, $t = 7$.

10 **a** The rate of change of mass with time,

$$\frac{dx}{dt} \propto -x$$

So $\dfrac{dx}{dt} = -kx$

where k is a constant of proportionality.
Separating the variables and integrating,

$$\int \frac{1}{x}\, dx = \int -k\, dt$$

$\Rightarrow \quad \ln x = -kt + c$

$\Rightarrow \quad x = e^{-kt+c}$

$= e^{-kt} \times e^c$

$\Rightarrow \quad x = Be^{-kt}$

where B is some constant.

When $t = 0$, $x = A$, so

$A = Be^0$

$A = B$

$x = Ae^{-kt}$

b When $t = 10$, $x = \frac{1}{3}A$. Substituting these values into the equation for x,

$\frac{1}{3}A = Ae^{-10k}$

$\Rightarrow \quad \frac{1}{3} = e^{-10k}$

$\Rightarrow \quad 3 = e^{10k}$

$\Rightarrow \quad \ln 3 = 10k$

$\Rightarrow \quad k = \dfrac{\ln 3}{10}$

$= 0.1099$ (4 d.p.)

Now, when $x = \frac{1}{2}A$,

$\frac{1}{2}A = Ae^{-0.1099t}$

$\Rightarrow \quad \frac{1}{2} = e^{-0.1099t}$

$\Rightarrow \quad 2 = e^{0.1099t}$

$\Rightarrow \quad \ln 2 = 0.1099t$

$\Rightarrow \quad t = \dfrac{\ln 2}{0.1099}$

$= 6.31$ (2 d.p.)

13 Numerical Methods (p. 583)

1 Let $f(x) = \sin x - 2\cos x - 1$

Then $f'(x) = \cos x + 2\sin x$

Taking $x_0 = 1.5$ as a first approximation to the root, use Newton's method to find better approximations.

$$x_1 = x_0 - \frac{f(x_0)}{f'(x_0)}$$

$$= 1.5 - \frac{\sin 1.5 - 2\cos 1.5 - 1}{\cos 1.5 + 2\sin 1.5}$$

$= 1.5697$ (4 d.p.)

$$x_2 = x_1 - \frac{f(x_1)}{f'(x_1)}$$

$$= 1.5697 - \frac{\sin 1.5697 - 2\cos 1.5697 - 1}{\cos 1.5697 - 2\sin 1.5697}$$

$= 1.5708$

So after two iterations of the Newton–Raphson method, the root of the equation $\sin x - 2\cos x - 1 = 0$ is $x = 1.57$ (2 d.p.).

2 Substitute the four ordinates $x_0 = 0, x_1 = 1, x_2 = 2$ and $x_3 = 3$ into $y = e^{-\frac{1}{2}x^2}$.

Working to five decimal places this gives

$y_0 = 1 \qquad y_1 = 0.606\,53$

$y_2 = 0.135\,34 \quad y_3 = 0.011\,11$

Now use the trapezium rule,

$$\int_0^3 e^{-\frac{1}{2}x^2}\,dx \approx \frac{h}{2}[y_0 + 2(y_1 + y_2) + y_3]$$

where the strip width $h = 1$,

$$\approx \tfrac{1}{2}[1 + 2(0.606\,53 + 0.135\,34) + 0.011\,11]$$

$$\approx 1.247\,43 \text{ (5 d.p.)}$$

This means that the area under the curve $y = e^{-\frac{1}{2}x^2}$ between $x = 0$ and $x = 3$ is 1.247 square units, correct to three decimal places.

3

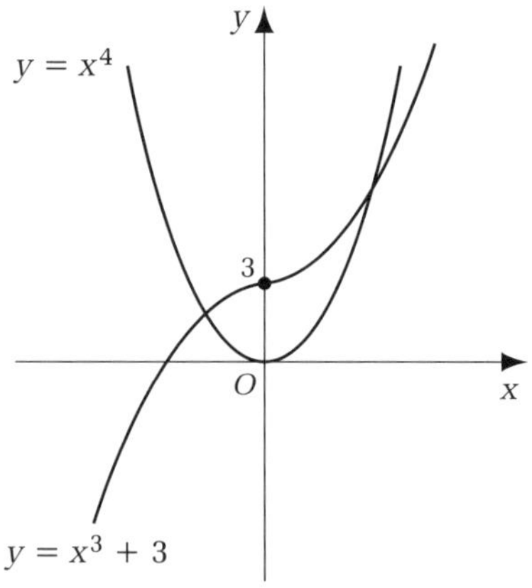

The equation $x^4 = x^3 + 3$ has two roots because the graphs of $y = x^4$ and $y = x^3 + 3$ intersect at two distinct points.

Using the iterative formula

$$x_{n+1} = \tfrac{1}{10}\left(9x_n + \frac{3}{x_n^3} + 1\right)$$

and taking the initial value, $x_0 = -1$, it follows that

$$x_1 = \tfrac{1}{10}\left[9(-1) + \frac{3}{(-1)^3} + 1\right]$$

$$= \tfrac{1}{10}(-9 - 3 + 1)$$

$$= -1.1$$

Repeating the process,

$$x_2 = \tfrac{1}{10}\left[9(-1.1) + \frac{3}{(-1.1)^3} + 1\right] = -1.1154$$

$$x_3 = \tfrac{1}{10}\left[9(-1.1154) + \frac{3}{(-1.1154)^3} + 1\right] = -1.1200$$

$$x_4 = \tfrac{1}{10}\left[9(-1.1200) + \frac{3}{(-1.1200)^3} + 1\right] = -1.1215$$

$$x_5 = \tfrac{1}{10}\left[9(-1.1215) + \frac{3}{(-1.1215)^3} + 1\right] = -1.1220$$

$$x_6 = \tfrac{1}{10}\left[9(-1.1220) + \frac{3}{(-1.1220)^3} + 1\right] = -1.1222$$

$$x_7 = \tfrac{1}{10}\left[9(-1.1222) + \frac{3}{(-1.1222)^3} + 1\right] = -1.1223$$

$$x_8 = \tfrac{1}{10}\left[9(-1.1223) + \frac{3}{(-1.1223)^3} + 1\right] = -1.1223$$

So the negative root of the equation $x^4 = x^3 + 3$ is -1.12 to two decimal places.

4 Substituting the ordinates $x_0 = 0, x_1 = 0.25, x_2 = 0.5, x_3 = 0.75$ and $x_4 = 1$ into $y = x \sec x$ gives the following results.

$y_0 = 0$ $\quad y_1 = 0.25802 \quad y_2 = 0.56975$

$y_3 = 1.02503 \quad y_4 = 1.85082$

Using Simpson's rule

$$\int_0^1 x\sec x\,dx \approx \frac{h}{3}[y_0 + y_4 + 4(y_1 + y_3) + 2y_2]$$

where the strip width $h = 0.25$

$$\approx \frac{0.25}{3}[0 + 1.85082 + 4(0.25802 + 1.02503) + 2(0.56975)]$$

$$\approx 0.677 \text{ (3 d.p.)}$$

The volume of revolution generated by rotating the area bounded by the curve $y = \sqrt{x\sec x}$, the x-axis from $x = 0$ to $x = 1$, and the line $x = 1$, completely about the x-axis, is given by

$$V \approx \int_0^1 \pi y^2\,dx$$

$$\approx \int_0^1 \pi x\sec x\,dx$$

$$\approx \pi\int_0^1 x\sec x\,dx$$

$$\approx \pi \times 0.677$$

$$\approx 2.1 \text{ cubic units (2 s.f.)}$$

5 Draw $y = e^x$. Now draw $y = 3 - x$ on the same axes.

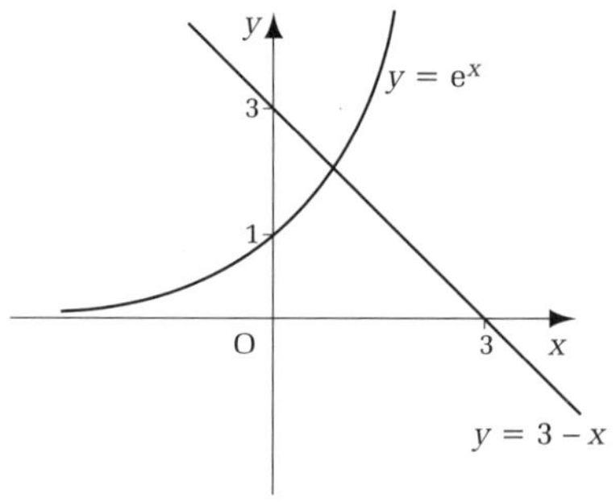

The equation $e^x + x - 3 = 0$ has exactly one root because the curve $y = e^x$ and the line $y = 3 - x$ intersect at only one place.

a If $\quad f(x) = e^x + x - 3$

then $\quad f(0.7) = e^{0.7} + 0.7 - 3$

≈ -0.29

and $\quad f(0.8) = e^{0.8} + 0.8 - 3$

≈ 0.03

Notice that there is a change of sign in the value of $e^x + x - 3$ between $x = 0.7$ and $x = 0.8$. This means that the root α lies in the interval $0.7 < \alpha < 0.8$.

b With $\quad f(x) = e^x + x - 3$

then $\quad f'(x) = e^x + 1$

Taking $x_0 = 0.8$ as a first approximation to α and using the Newton–Raphson method, a second approximation is given by

$$x_1 = x_0 - \frac{f(x_0)}{f'(x_0)}$$
$$= 0.8 - \frac{e^{0.8} + 0.8 - 3}{e^{0.8} + 1}$$
$$= 0.792 \text{ (3 d.p.)}$$

c Rearranging $e^x + x - 3 = 0$,
gives $e^x = 3 - x$
Taking the natural logarithm of both sides,

$$x = \ln(3 - x)$$

Using the iterative formula

$$x_{n+1} = \ln(3 - x_n)$$

with $x_0 = 0.8$

$$x_1 = \ln(3 - x_0)$$
$$= \ln(3 - 0.8)$$
$$= \ln(2.2)$$
$$= 0.788 \quad \text{(3 d.p.)}$$

and $x_2 = \ln(3 - x_1)$
$$= \ln(3 - 0.788\,46)$$
$$= \ln(2.211\,54)$$
$$= 0.794 \quad \text{(3 d.p.)}$$

6 **a**

x	0	3	6	9	12	15	18
d	0	1.708	2.309	2.598	2.582	2.141	0

These results can be obtained directly from a graphical calculator.

b Using the trapezium rule, with these seven ordinates, the area of the cross-section of the river can be estimated by

$$\text{area} \approx \tfrac{1}{2}h[d_0 + 2(d_1 + d_2 + d_3 + d_4 + d_5) + d_6]$$

where the strip width $h = 3$

$$\approx \tfrac{1}{2} \times 3[0 + 2(1.708 + 2.309 + 2.598 + 2.582 + 2.141) + 0]$$
$$\approx 34.0 \text{ m}^2 \text{ (1 d.p.)}$$

c volume of water passing per minute
= area of cross-section × speed
≈ 34.0 × 100 metres per minute
≈ 3400 m^3

7 If $f(x) = x^5 - 5x - 6$
then $f(1) = -10$
and $f(2) = 16$
Since there is a change of sign in the value of $x^5 - 5x - 6$ between $x = 1$ and $x = 2$, the equation $x^5 - 5x - 6 = 0$ must have a root in the interval $1 < x < 2$.

Rearranging $x^5 - 5x - 6 = 0$
into $x^5 = 5x + 6$
it follows that $x = \sqrt[5]{5x + 6}$

So the iterative formula

$$x_{n+1} = (5x_n + 6)^{\frac{1}{5}}$$

can be used to find this root. This is in the required form with $p = 5$, $q = 6$ and $r = 5$.

Taking an initial value $x_0 = 1$,

$$x_1 = (5x_0 + 6)^{1/5}$$
$$= (5 + 6)^{1/5} = 1.61539 \text{ (5 d.p.)}$$
$$x_2 = (5x_1 + 6)^{1/5} = (5 \times 1.61539 + 6)^{1/5} = 1.69708$$
$$x_3 = (5x_2 + 6)^{1/5} = (5 \times 1.69708 + 6)^{1/5} = 1.70681$$
$$x_4 = (5x_3 + 6)^{1/5} = (5 \times 1.70681 + 6)^{1/5} = 1.70796$$
$$x_5 = (5x_4 + 6)^{1/5} = (5 \times 1.70796 + 6)^{1/5} = 1.70809$$
$$x_6 = (5x_5 + 6)^{1/5} = (5 \times 1.70809 + 6)^{1/5} = 1.70811$$
$$x_7 = (5x_6 + 6)^{1/5} = (5 \times 1.70811 + 6)^{1/5} = 1.70811$$

This means that the root to the equation $x^5 - 5x - 6 = 0$ in the interval $1 < x < 2$ is 1.708, correct to three decimal places.

8 If $f(x) = x^3 - 9x^2 + 5x - 7$
then $f(8) = -31$
and $f(9) = 38$
Since there is a change of sign in the value of $f(x)$ between $x = 8$ and $x = 9$, the equation $x^3 - 9x^2 + 5x - 7 = 0$ must have a root in the interval $8 < x < 9$.

When $f(x) = x^3 - 9x^2 + 5x - 7$
$f'(x) = 3x^2 - 18x + 5$

Taking $x_0 = 9$ as a first approximation to the root, and using the Newton–Raphson method, a second approximation is then given by

$$x_1 = x_0 - \frac{f(x_0)}{f'(x_0)}$$
$$= 9 - \frac{38}{86}$$
$$= 8.6 \text{ (1 d.p.)}$$

9

a $$A = \int_0^{\frac{\pi}{2}} \sin x \, dx$$
$$= \Big[-\cos x\Big]_0^{\frac{\pi}{2}}$$
$$= (0) - (-1)$$
$$= 1 \text{ square unit}$$

b $$B = \int_0^{\frac{\pi}{2}} x \sin x \, dx$$

Use integration by parts,

$$\int u\frac{dv}{dx}dx = uv - \int v\frac{du}{dx}dx$$

with $u = x$ and $\frac{dv}{dx} = \sin x$.

Then $\frac{du}{dx} = 1$ and $v = \int \sin x \, dx = -\cos x$.

$$\int_0^{\frac{\pi}{2}} x\sin x\,dx = \left[-x\cos x\right]_0^{\frac{\pi}{2}} - \int_0^{\frac{\pi}{2}} -\cos x\,dx$$

$$= \left[-x\cos x\right]_0^{\frac{\pi}{2}} + \int_0^{\frac{\pi}{2}} \cos x\,dx$$

$$= \left[-x\cos x\right]_0^{\frac{\pi}{2}} + \left[\sin x\right]_0^{\frac{\pi}{2}}$$

$$= (0-0) + (1-0)$$

$$= 1 \text{ square unit}$$

c

$$C = \int_0^{\frac{\pi}{2}} \sqrt{x}\sin x\,dx$$

i Replacing $\sin x$ with its Taylor approximation $\sin x \approx x - \frac{1}{6}x^3$

$$C \approx \int_0^{\frac{\pi}{2}} \sqrt{x}\left(x - \tfrac{1}{6}x^3\right)dx$$

$$\approx \int_0^{\frac{\pi}{2}} \left(x^{\frac{3}{2}} - \tfrac{1}{6}x^{\frac{7}{2}}\right)dx$$

$$\approx \left[\tfrac{2}{5}x^{\frac{5}{2}} - \tfrac{1}{6}\times\tfrac{2}{9}x^{\frac{9}{2}}\right]_0^{\frac{\pi}{2}}$$

$$\approx \left[\tfrac{2}{5}x^{\frac{5}{2}} - \tfrac{1}{27}x^{\frac{9}{2}}\right]_0^{\frac{\pi}{2}}$$

$$\approx \left[\tfrac{2}{5}\left(\tfrac{\pi}{2}\right)^{\frac{5}{2}} - \tfrac{1}{27}\left(\tfrac{\pi}{2}\right)^{\frac{9}{2}}\right] - (0-0)$$

$$\approx 0.954 \text{ (3 d.p.)}$$

ii Substitute the ordinates $x_0 = 0$, $x_1 = \frac{\pi}{4}$ and $x_2 = \frac{\pi}{2}$ into $y = \sqrt{x}\sin x$. This gives

$$y_0 = 0$$

$$y_1 = 0.626\,66$$

$$y_2 = 1.253\,31$$

Using Simpson's rule with two strips,

$$C \approx \int_0^{\frac{\pi}{2}} \sqrt{x}\sin x\,dx$$

$$\approx \tfrac{1}{3}h(y_0 + 4y_1 + y_2)$$

where strip width $h = \frac{\pi}{4}$

$$\approx \tfrac{\pi}{12}(0 + 4\times 0.626\,66 + 1.253\,31)$$

$$\approx 0.984 \text{ (3 d.p.)}$$

The accuracy of answer **i** could be improved by using more than the first two terms of the Taylor approximation for $\sin x$. The accuracy of answer **ii** could be improved by using Simpson's rule with more than two intervals.

10 **a** **i** At the maximum point A on the graph $\frac{dy}{dx} = 0$. Using the product rule to differentiate $y = (4-x)\ln x$ with respect to x

$$\frac{dy}{dx} = (4-x)\times\frac{1}{x} - \ln x$$

$$= \frac{4}{x} - 1 - \ln x$$

Therefore m, the x-coordinate at A, must satisfy the equation

$$\frac{4}{x} - 1 - \ln x = 0$$

or $\quad 1 + \ln x - \dfrac{4}{x} = 0$

ii If $\quad f(x) = 1 + \ln x - \dfrac{4}{x}$

then $\quad f(2.2) = -0.03$

and $\quad f(2.3) = 0.09$

Since there is a change of sign in the value of $f(x)$, m must lie between 2.2 and 2.3.

iii Using the iterative formula

$$x_{n+1} = \frac{4}{1+\ln x_n}$$

and taking $x_0 = 2.2$ as a first approximation

$$x_1 = \frac{4}{1+\ln x_0} = \frac{4}{1+\ln 2.2} = 2.236\,564\,4$$

$$x_2 = \frac{4}{1+\ln x_1} = \frac{4}{1+\ln 2.236\,564\,4} = 2.216\,139\,0$$

$$x_3 = \frac{4}{1+\ln x_2} = \frac{4}{1+\ln 2.216\,139\,0} = 2.227\,461\,1$$

$$x_4 = \frac{4}{1+\ln x_3} = \frac{4}{1+\ln 2.227\,461\,1} = 2.221\,158\,0$$

$$x_5 = \frac{4}{1+\ln x_4} = \frac{4}{1+\ln 2.221\,158\,0} = 2.224\,658\,6$$

$$x_6 = \frac{4}{1+\ln x_5} = \frac{4}{1+\ln 2.224\,658\,6} = 2.222\,711\,9$$

$$x_7 = \frac{4}{1+\ln x_6} = \frac{4}{1+\ln 2.222\,711\,9} = 2.223\,793\,7$$

$$x_8 = \frac{4}{1+\ln x_7} = \frac{4}{1+\ln 2.223\,793\,7} = 2.223\,192\,3$$

So $\quad m = 2.22$ (2 d.p.)

b

$$R = \int_1^4 (4-x)\ln x\,dx$$

Use integration by parts,

$$\int u\frac{dv}{dx}\,dx = uv - \int u\frac{du}{dx}dx$$

with $u = \ln x$ and $\dfrac{dv}{dx} = 4 - x$.

Then $\dfrac{du}{dx} = \dfrac{1}{x}$ and $v = \int(4-x)dx = 4x - \frac{1}{2}x^2$.

$$\int_1^4 (4-x)\ln x\,dx = \left[\left(4x - \tfrac{1}{2}x^2\right)\ln x\right]_1^4 - \int_1^4 \tfrac{1}{x}\left(4x - \tfrac{1}{2}x^2\right)dx$$

$$= \left(8\ln 4 - \tfrac{7}{2}\ln 1\right) - \int_1^4 \left(4 - \tfrac{1}{2}x\right)dx$$

$$= 8\ln 4 - \left[4x - \tfrac{1}{4}x^2\right]_1^4$$

$$= 8\ln 4 - \left(12 - \tfrac{15}{4}\right)$$

$$= 8\ln 4 - \tfrac{33}{4}$$

14 Vectors (p. 657)

1 **a**

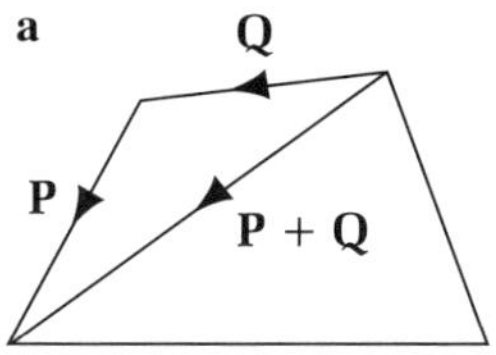

b

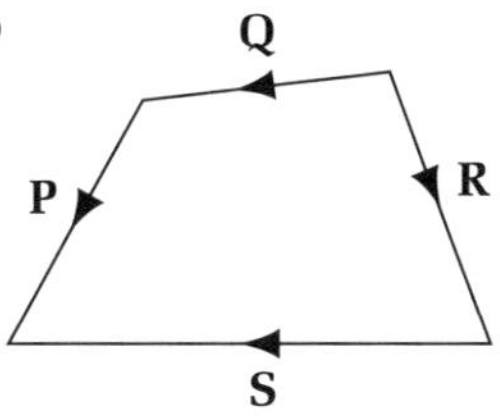

c

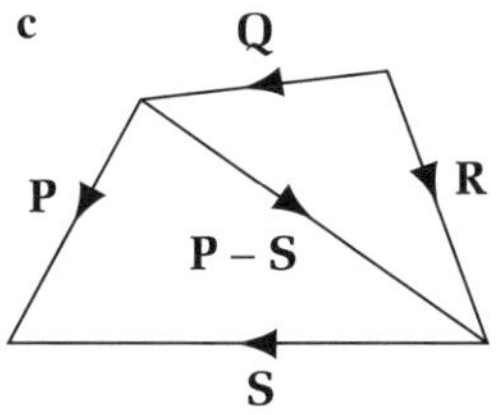

2 **a**

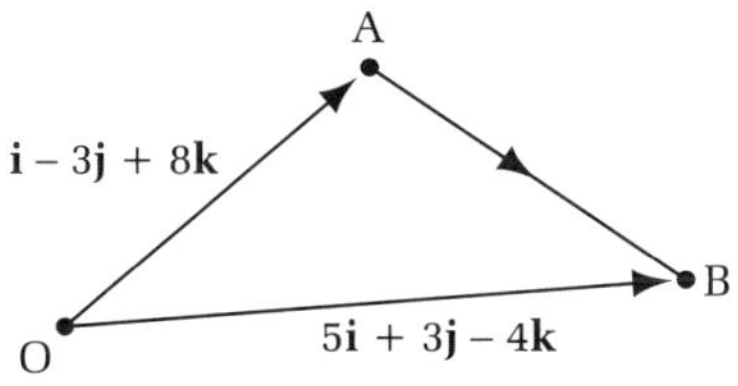

$$\overrightarrow{AB} = \overrightarrow{AO} + \overrightarrow{OB}$$
$$= -(\mathbf{i} - 3\mathbf{j} + 8\mathbf{k}) + (5\mathbf{i} + 3\mathbf{j} - 4\mathbf{k})$$
$$= 4\mathbf{i} + 6\mathbf{j} - 12\mathbf{k}$$

b
$$\left|\overrightarrow{AB}\right| = \sqrt{4^2 + 6^2 + (-12)^2}$$
$$= \sqrt{16 + 36 + 144}$$
$$= \sqrt{196}$$
$$= 14$$

So a unit vector in the direction of $\overrightarrow{AB}$ is

$$\tfrac{1}{14}(4\mathbf{i} + 6\mathbf{j} - 12\mathbf{k}) = \tfrac{2}{7}\mathbf{i} + \tfrac{3}{7}\mathbf{j} - \tfrac{6}{7}\mathbf{k}$$

c

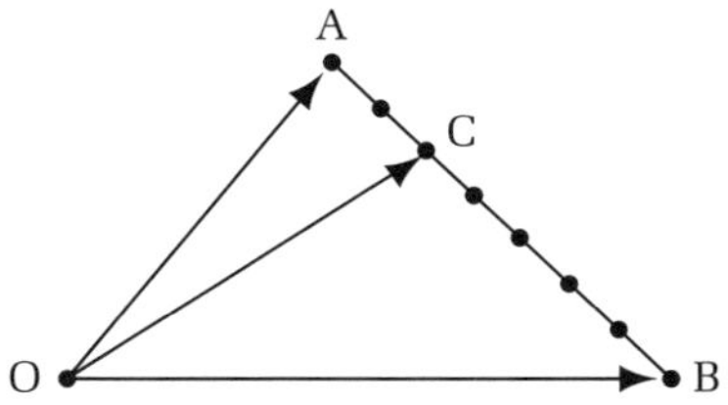

Since C divides AB in the ratio 2 : 5

$$\overrightarrow{AC} = \tfrac{2}{7}\overrightarrow{AB}$$

Now the position vector of C is $\overrightarrow{OC}$, where

$$\overrightarrow{OC} = \overrightarrow{OA} + \overrightarrow{AC}$$
$$= \overrightarrow{OA} + \tfrac{2}{7}\overrightarrow{AB}$$
$$= (\mathbf{i} - 3\mathbf{j} + 8\mathbf{k}) + \tfrac{2}{7}(4\mathbf{i} + 6\mathbf{j} - 12\mathbf{k})$$
$$= \tfrac{15}{7}\mathbf{i} - \tfrac{9}{7}\mathbf{j} + \tfrac{32}{7}\mathbf{k}$$

3 We know that the scalar product,

$$\mathbf{a}.\mathbf{b} = ab\cos\theta = a_1b_1 + a_2b_2 + a_3b_3 \begin{cases} > 0 \text{ means } \theta \text{ is acute} \\ < 0 \text{ means } \theta \text{ is obtuse} \\ = 0 \text{ means } \theta \text{ is a right angle} \end{cases}$$

Substituting the known vectors $\mathbf{a} = \mathbf{i} + \mathbf{j}$, $\mathbf{b} = \mathbf{i} + \mathbf{j} + p\mathbf{k}$ and the known angle $\theta = \frac{\pi}{4}$, we have

$$(\mathbf{i} + \mathbf{j}).(\mathbf{i} + \mathbf{j} + p\mathbf{k}) = |\mathbf{i} + \mathbf{j}| \times |\mathbf{i} + \mathbf{j} + p\mathbf{k}| \times \cos\tfrac{\pi}{4}$$
$$\Rightarrow \quad 1 + 1 + 0 = \sqrt{2} \times \sqrt{2 + p^2} \times \tfrac{1}{\sqrt{2}}$$
$$\Rightarrow \quad 2 = \sqrt{2 + p^2}$$
$$\Rightarrow \quad 4 = 2 + p^2$$
$$\text{So} \quad p = \pm\sqrt{2}$$

4 **a** Using the dot product we know that $\overrightarrow{OA}$ and $\overrightarrow{OB}$ are perpendicular if $\overrightarrow{OA}.\overrightarrow{OB} = 0$.

$$\text{Now } \overrightarrow{OA}.\overrightarrow{OB} = \begin{pmatrix} 4 \\ -3 \\ 2 \end{pmatrix}.\begin{pmatrix} 2 \\ -2 \\ -7 \end{pmatrix}$$
$$= 8 + 6 - 14$$
$$= 0$$

So $\overrightarrow{OA}$ and $\overrightarrow{OB}$ are perpendicular.

b

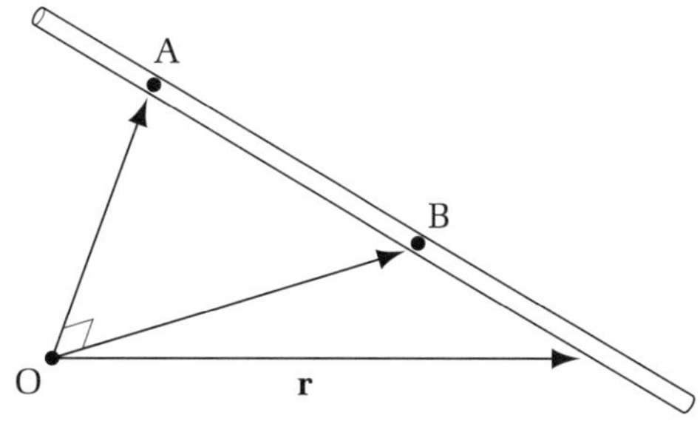

The vector parallel to the line AB is $\overrightarrow{AB}$, which can be found using vector addition.

$$\overrightarrow{AB} = \overrightarrow{AO} + \overrightarrow{OB}$$
$$= -\begin{pmatrix} 4 \\ -3 \\ 2 \end{pmatrix} + \begin{pmatrix} 2 \\ -2 \\ -7 \end{pmatrix}$$
$$= \begin{pmatrix} -2 \\ 1 \\ -9 \end{pmatrix}$$

Now use the vector form of the equation of a straight line to write line AB as

$$\mathbf{r} = \overrightarrow{OA} + t\,\overrightarrow{AB}$$
$$= \begin{pmatrix} 4 \\ -3 \\ 2 \end{pmatrix} + t\begin{pmatrix} -2 \\ 1 \\ -9 \end{pmatrix}$$

The two lines $\mathbf{r}(s)$ and $\mathbf{r}(t)$ intersect when

$$\mathbf{r}(t) = \mathbf{r}(s)$$
$$\Rightarrow \quad \begin{pmatrix} 4 \\ -3 \\ 2 \end{pmatrix} + t\begin{pmatrix} -2 \\ 1 \\ -9 \end{pmatrix} = \begin{pmatrix} 3 \\ -1 \\ 2 \end{pmatrix} + s\begin{pmatrix} -1 \\ 1 \\ -3 \end{pmatrix}$$
$$\Rightarrow \quad \begin{pmatrix} 4 - 2t \\ -3 + t \\ 2 - 9t \end{pmatrix} = \begin{pmatrix} 3 - s \\ -1 + s \\ 2 - 3s \end{pmatrix}$$

Equating components of the vectors,

$$4 - 2t = 3 - s \qquad [1]$$
$$-3 + t = -1 + s \qquad [2]$$
$$2 - 9t = 2 - 3s \qquad [3]$$

Adding equation [1] and equation [2],

$$1 - t = 2$$
$$\Rightarrow \quad t = -1$$

Substituting $t = -1$ in equation [1],

$$4 - 2(-1) = 3 - s$$
$$\Rightarrow \quad 4 + 2 = 3 - s$$
$$\Rightarrow \quad s = -3$$

Substituting $s = -3$ and $t = -1$ in equation [3],

$$2 - 9(-1) = 2 - 3(-3)$$
$$\Rightarrow \quad 11 = 11$$

So since these values of s and t satisfy all three equations the lines must intersect.

Substituting $s = -3$ and $t = -1$ into the vector equations for the two lines,

$$\mathbf{r}(t) = \begin{pmatrix} 4 \\ -3 \\ 2 \end{pmatrix} + t\begin{pmatrix} -2 \\ 1 \\ -9 \end{pmatrix}$$

becomes $$\mathbf{r}(-1) = \begin{pmatrix} 4 \\ -3 \\ 2 \end{pmatrix} - \begin{pmatrix} -2 \\ 1 \\ -9 \end{pmatrix}$$

$$= \begin{pmatrix} 6 \\ -4 \\ 11 \end{pmatrix}$$

and $$\mathbf{r}(s) = \begin{pmatrix} 3 \\ -1 \\ 2 \end{pmatrix} + s\begin{pmatrix} -1 \\ 1 \\ -3 \end{pmatrix}$$

becomes $$\mathbf{r}(-3) = \begin{pmatrix} 3 \\ -1 \\ 2 \end{pmatrix} - 3\begin{pmatrix} -1 \\ 1 \\ -3 \end{pmatrix}$$

$$= \begin{pmatrix} 6 \\ -4 \\ 11 \end{pmatrix}$$

So the lines intersect at the point (6, –4, 11).

5

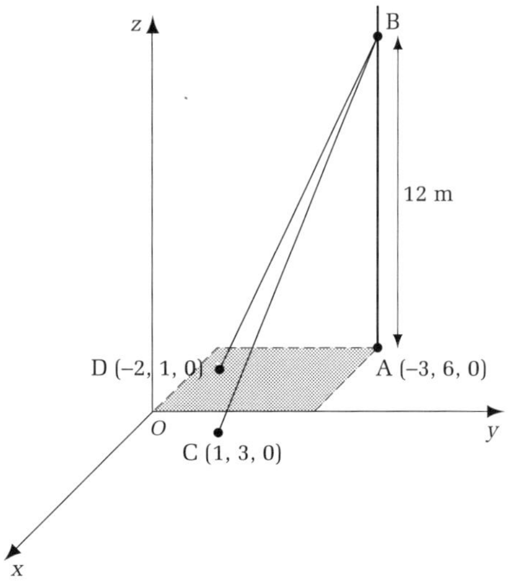

a B is 12 units higher than A, with the same x- and y-coordinates. So the coordinates of B are $(-3, 6, 12)$.

b The angle of elevation of B from O is the angle between the vectors $\overrightarrow{OB}$ and $\overrightarrow{OA}$, because $\overrightarrow{OA}$ lies in the horizontal plane.

$$|\overrightarrow{OA}| = \sqrt{(-3)^2 + 6^2 + 0^2}$$
$$= \sqrt{45}$$
$$= 3\sqrt{5}$$

and we already know that $|\overrightarrow{AB}| = 12$.
Let θ be the angle of elevation. Using trigonometry, it follows that

$$\tan\theta = \frac{12}{3\sqrt{5}} = \frac{4}{\sqrt{5}}$$
$$\theta = \tan^{-1}\left(\frac{4}{\sqrt{5}}\right)$$
$$= 60.79^\circ \text{ (2 d.p.)}$$
$$\approx 61^\circ$$

c

$$\overrightarrow{BC} = \overrightarrow{BO} + \overrightarrow{OC}$$
$$= -\begin{pmatrix} -3 \\ 6 \\ 12 \end{pmatrix} + \begin{pmatrix} 1 \\ 3 \\ 0 \end{pmatrix}$$
$$= \begin{pmatrix} 4 \\ -3 \\ -12 \end{pmatrix}$$

$$\overrightarrow{BD} = \overrightarrow{BO} + \overrightarrow{OD}$$
$$= -\begin{pmatrix} -3 \\ 6 \\ 12 \end{pmatrix} + \begin{pmatrix} -2 \\ 1 \\ 0 \end{pmatrix}$$
$$= \begin{pmatrix} 1 \\ -5 \\ -12 \end{pmatrix}$$

Using the scalar product,

$$\overrightarrow{BC}.\overrightarrow{BD} = \begin{pmatrix} 4 \\ -3 \\ -12 \end{pmatrix}.\begin{pmatrix} 1 \\ -5 \\ -12 \end{pmatrix}$$
$$= 4 + 15 + 144$$
$$= 163$$

d

$$|\overrightarrow{BC}| = \sqrt{4^2 + (-3)^2 + (-12)^2}$$
$$= \sqrt{16 + 9 + 144}$$
$$= \sqrt{169}$$
$$= 13$$

$$|\overrightarrow{BC}| = \sqrt{1^2 + (-5)^2 + (-12)^2}$$
$$= \sqrt{1 + 25 + 144}$$
$$= \sqrt{170}$$

Using the scalar product

$$\overrightarrow{BC} \cdot \overrightarrow{BD} = |\overrightarrow{BC}| \times |\overrightarrow{BD}| \times \cos\alpha$$

where α is the angle between the two cables

So $163 = 13 \times \sqrt{170} \times \cos\alpha$

$$\Rightarrow \quad \cos\alpha = \frac{163}{13\sqrt{170}}$$

$$\Rightarrow \quad \alpha = \cos^{-1}\left(\frac{163}{13\sqrt{170}}\right)$$

$$= 15.9^\circ \text{ (1 d.p.)}$$

So, to the nearest degree, the angle between the cables is 16°.

6

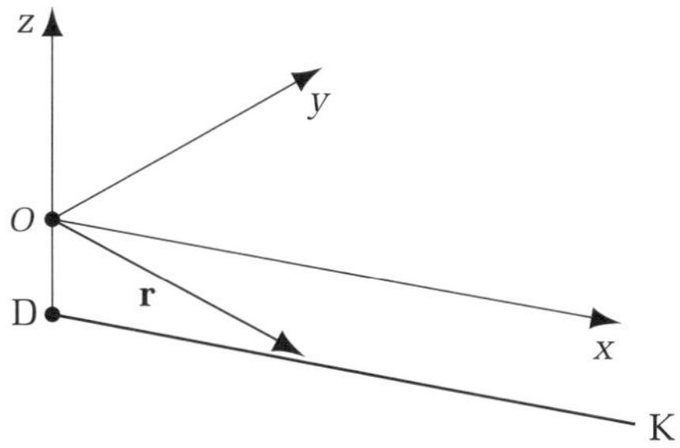

The equation of line K, which runs parallel to the x-axis but 1 unit below it is given by

$$\mathbf{r} = \overrightarrow{OD} + s\begin{pmatrix}1\\0\\0\end{pmatrix}$$

$$= \begin{pmatrix}0\\0\\-1\end{pmatrix} + s\begin{pmatrix}1\\0\\0\end{pmatrix}$$

$$= \begin{pmatrix}s\\0\\-1\end{pmatrix}$$

Lines K and L will intersect when

$$\begin{pmatrix}s\\0\\-1\end{pmatrix} = \begin{pmatrix}5\\5\\0\end{pmatrix} + t\begin{pmatrix}5\\-5\\-1\end{pmatrix}$$

$$\Rightarrow \quad \begin{pmatrix}s\\0\\-1\end{pmatrix} = \begin{pmatrix}5+5t\\5-5t\\-t\end{pmatrix}$$

Equating components to find s and t,

$s = 5 + 5t$ [1]

$0 = 5 - 5t$ [2]

$-1 = -t$ [3]

From equation [3], $t = 1$. So from equation [1], $s = 10$. Substituting these values into the vector equations of the two lines,

$$\mathbf{r}(t) = \begin{pmatrix}5\\5\\0\end{pmatrix} + t\begin{pmatrix}5\\-5\\-1\end{pmatrix}$$

becomes $$\mathbf{r}(1) = \begin{pmatrix}5\\5\\0\end{pmatrix} + \begin{pmatrix}5\\-5\\-1\end{pmatrix}$$

$$= \begin{pmatrix}10\\0\\-1\end{pmatrix}$$

and $$\mathbf{r}(s) = \begin{pmatrix}s\\0\\-1\end{pmatrix}$$

becomes $$\mathbf{r}(10) = \begin{pmatrix}10\\0\\-1\end{pmatrix}$$

So the coordinates of the point of intersection, C, are (10, 0, −1).

To find the vector equation of the line BD, start by drawing a diagram.

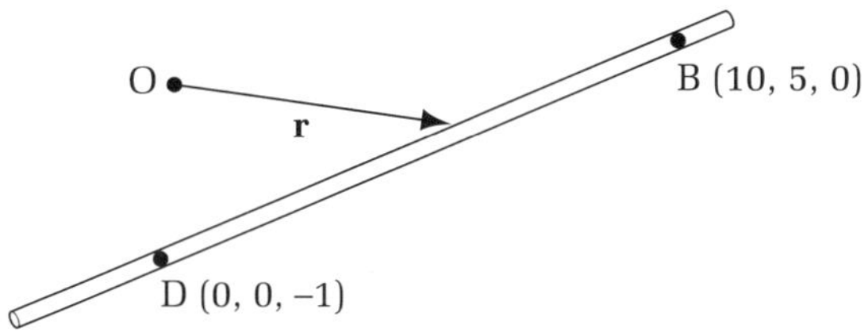

$$\overrightarrow{BD} = \overrightarrow{BO} + \overrightarrow{OD}$$

$$= \begin{pmatrix}-10\\-5\\0\end{pmatrix} + \begin{pmatrix}0\\0\\-1\end{pmatrix}$$

$$= \begin{pmatrix}-10\\-5\\-1\end{pmatrix}$$

The vector equation of line BD is given by

$$\mathbf{r} = \overrightarrow{OB} + \lambda\overrightarrow{BD}$$

$$= \begin{pmatrix}10\\5\\0\end{pmatrix} + \lambda\begin{pmatrix}-10\\-5\\-1\end{pmatrix}$$

$$= \begin{pmatrix}10-10\lambda\\5-5\lambda\\-\lambda\end{pmatrix}$$

The cables M and L will not intersect if there are no values of t and λ for which

$$\mathbf{r}(t) = \mathbf{r}(\lambda)$$

$$\begin{pmatrix}5+5t\\5-5t\\-t\end{pmatrix} = \begin{pmatrix}10-10\lambda\\5-5\lambda\\-\lambda\end{pmatrix}$$

Equating components,

$5 + 5t = 10 - 10\lambda$ [1]

$5 - 5t = 5 - 5\lambda$ [2]

$-t = -\lambda$ [3]

From equation [3] and equation [2], $t = \lambda$. Substituting $t = \lambda$ in equation [1],

$$5 + 5\lambda = 10 - 10\lambda$$

$$15\lambda = 5$$

$$\lambda = \tfrac{1}{3}$$

So $t = \lambda = \frac{1}{3}$. Substituting these values into the vector equations of the two lines,

$$\mathbf{r}(t) = \begin{pmatrix}5+5t\\5-5t\\-t\end{pmatrix}$$

becomes $$\mathbf{r}\left(\tfrac{1}{3}\right) = \begin{pmatrix}5+\frac{5}{3}\\5-\frac{5}{3}\\-\frac{1}{3}\end{pmatrix} = \begin{pmatrix}\frac{20}{3}\\\frac{10}{3}\\-\frac{1}{3}\end{pmatrix}$$

and $\quad \mathbf{r}(\lambda) = \begin{pmatrix} 10 - 10\lambda \\ 5 - 5\lambda \\ -\lambda \end{pmatrix}$

becomes $\quad \mathbf{r}\left(\frac{1}{3}\right) = \begin{pmatrix} 10 - \frac{10}{3} \\ 5 - \frac{5}{3} \\ -\frac{1}{3} \end{pmatrix} = \begin{pmatrix} \frac{20}{3} \\ \frac{10}{3} \\ -\frac{1}{3} \end{pmatrix}$

So the cables would intersect. Cable M cannot be laid along the line BD without meeting cable L.

The vector equation of the line AC is

$$\mathbf{r} = \begin{pmatrix} 5 \\ 5 \\ 0 \end{pmatrix} + t\begin{pmatrix} 5 \\ -5 \\ -1 \end{pmatrix}$$

and the vector equation of line BD is

$$\mathbf{r} = \begin{pmatrix} 10 \\ 5 \\ 0 \end{pmatrix} + \lambda\begin{pmatrix} -10 \\ -5 \\ -1 \end{pmatrix}$$

The angle between AC and BD is the same as the angle between

$$\begin{pmatrix} 5 \\ -5 \\ -1 \end{pmatrix} \text{ and } \begin{pmatrix} -10 \\ -5 \\ -1 \end{pmatrix}.$$

Using Pythagoras' theorem, the lengths of these two vectors are $\sqrt{51}$ and $\sqrt{126}$ respectively. If θ is the angle between them, then using the scalar product

$$\begin{pmatrix} 5 \\ -5 \\ -1 \end{pmatrix} \cdot \begin{pmatrix} -10 \\ -5 \\ -1 \end{pmatrix} = \sqrt{51}\sqrt{126}\cos\theta$$

$$\Rightarrow \quad -24 = \sqrt{51}\sqrt{126}\cos\theta$$

$$\Rightarrow \quad \cos\theta = \frac{-24}{\sqrt{51}\sqrt{126}} \approx -0.2994$$

$$\Rightarrow \quad \theta = \cos^{-1}(-0.2994)$$

$$= 107.4^\circ \text{ (1 d.p.)}$$

But we require the acute angle, which is given by

$$180^\circ - 107.4^\circ = 72.6^\circ \text{ (1 d.p.)}$$

So to the nearest degree, the angle is 73°.

7

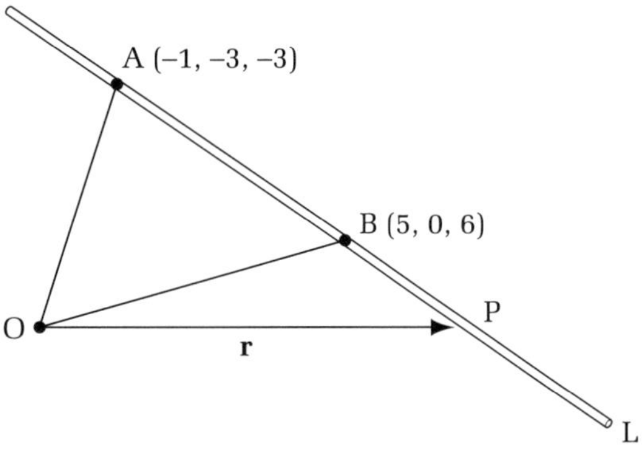

The vector equation of line L gives the position vector of a general point P on L.

Vector $\overrightarrow{AB}$ is parallel to L.

Using vector addition,

$$\overrightarrow{AB} = \overrightarrow{AO} + \overrightarrow{OB}$$

$$= -\begin{pmatrix} -1 \\ -3 \\ -3 \end{pmatrix} + \begin{pmatrix} 5 \\ 0 \\ 6 \end{pmatrix}$$

$$= \begin{pmatrix} 6 \\ 3 \\ 9 \end{pmatrix} = 3\begin{pmatrix} 2 \\ 1 \\ 3 \end{pmatrix}$$

This means that the vector $\begin{pmatrix} 2 \\ 1 \\ 3 \end{pmatrix}$ is parallel to L.

The vector equation of L is given by

$$\mathbf{r} = \overrightarrow{OA} + \lambda\begin{pmatrix} 2 \\ 1 \\ 3 \end{pmatrix}$$

$$\mathbf{r} = \begin{pmatrix} -1 \\ -3 \\ -3 \end{pmatrix} + \lambda\begin{pmatrix} 2 \\ 1 \\ 3 \end{pmatrix} = \begin{pmatrix} -1 + 2\lambda \\ -3 + \lambda \\ -3 + 3\lambda \end{pmatrix}$$

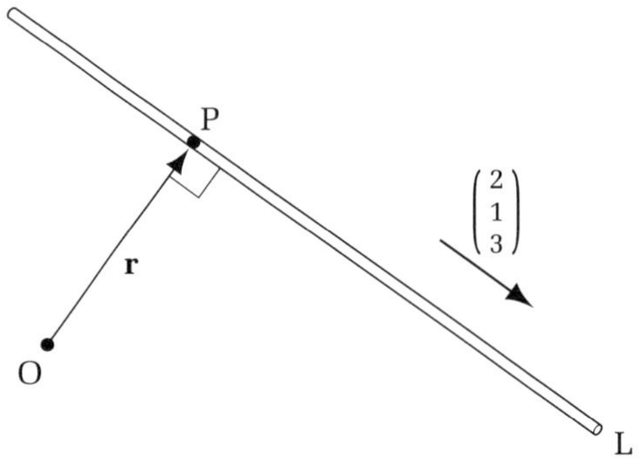

There will be a value of λ for which OP is perpendicular to L.

$$\mathbf{r} = \overrightarrow{OP} = \begin{pmatrix} -1 + 2\lambda \\ -3 + \lambda \\ -3 + 3\lambda \end{pmatrix}$$

Since vector $\begin{pmatrix} 2 \\ 1 \\ 3 \end{pmatrix}$ is parallel to L, OP is perpendicular to L if

$$\begin{pmatrix} -1 + 2\lambda \\ -3 + \lambda \\ -3 + 3\lambda \end{pmatrix} \cdot \begin{pmatrix} 2 \\ 1 \\ 3 \end{pmatrix} = 0$$

$$\Rightarrow \quad 2(-1 + 2\lambda) + (-3 + \lambda) + 3(-3 + 3\lambda) = 0$$

$$\Rightarrow \quad -2 + 4\lambda - 3 + \lambda - 9 + 9\lambda = 0$$

$$14\lambda - 14 = 0$$

$$\lambda = 1$$

So when $\lambda = 1$, OP is perpendicular to L, and the position vector of P is given by

$$\mathbf{r} = \overrightarrow{OP} = \begin{pmatrix} -1 + 2\lambda \\ -3 + \lambda \\ -3 + 3\lambda \end{pmatrix}$$

$$= \begin{pmatrix} -1 + 2 \\ -3 + 1 \\ -3 + 3 \end{pmatrix}$$

$$= \begin{pmatrix} 1 \\ -2 \\ 0 \end{pmatrix}$$

So the coordinates of P for which OP is perpendicular to L are (1, –2, 0).

The minimum distance from the origin to the line is the length of OP.

$$OP = \sqrt{1^2 + (-2)^2 + 0^2}$$
$$= \sqrt{5}$$

8 The equation of the plane $2x - 3y + z = 5$ can be written in vector form as

$$\mathbf{r}.\begin{pmatrix}2\\-3\\1\end{pmatrix} = 5$$

Now we know that the vector $\begin{pmatrix}2\\-3\\1\end{pmatrix}$ is perpendicular to the plane.

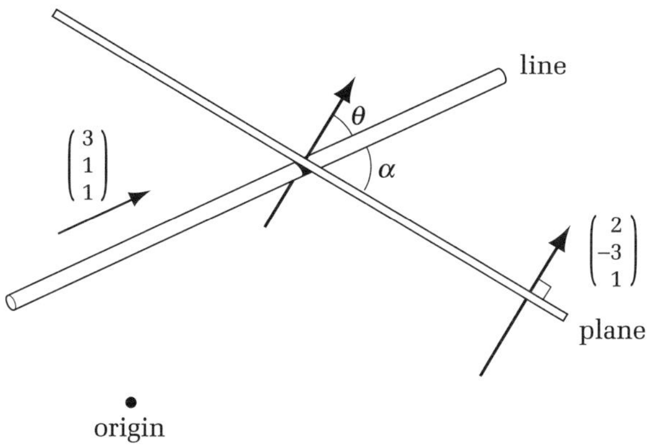

To find the angle between the line and the plane, first find the angle θ between the vector parallel to the line and the vector perpendicular to the plane; that is the angle between $\begin{pmatrix}3\\1\\1\end{pmatrix}$ and $\begin{pmatrix}2\\-3\\1\end{pmatrix}$.

Using Pythagoras' theorem, the lengths of these two vectors are $\sqrt{11}$ and $\sqrt{14}$ respectively. Therefore, using the scalar product,

$$\begin{pmatrix}3\\1\\1\end{pmatrix}\cdot\begin{pmatrix}2\\-3\\1\end{pmatrix} = \sqrt{11}\sqrt{14}\cos\theta$$
$$\Rightarrow \quad 4 = \sqrt{154}\cos\theta$$
$$\Rightarrow \quad \cos\theta = \frac{4}{\sqrt{154}}$$
$$\Rightarrow \quad \theta = \cos^{-1}\left(\frac{4}{\sqrt{154}}\right)$$
$$= 71.2^\circ \text{ (1 d.p.)}$$

So the equation between the line and the plane, α, is given by

$$\alpha = 90^\circ - \theta$$
$$= 90^\circ - 71.2^\circ$$
$$= 18.8^\circ \text{ (1 d.p.)}$$

9 First find $\overrightarrow{BA}$ and $\overrightarrow{BC}$.

$$\overrightarrow{BA} = \overrightarrow{BO} + \overrightarrow{OA}$$
$$= -\begin{pmatrix}4\\3\\4\end{pmatrix} + \begin{pmatrix}2\\3\\5\end{pmatrix}$$
$$= \begin{pmatrix}-2\\0\\1\end{pmatrix}$$

$$\overrightarrow{BC} = \overrightarrow{BO} + \overrightarrow{OC}$$
$$= -\begin{pmatrix}4\\3\\4\end{pmatrix} + \begin{pmatrix}6\\-3\\1\end{pmatrix}$$
$$= \begin{pmatrix}2\\-6\\-3\end{pmatrix}$$

Using Pythagoras' theorem, the lengths of these two vectors are $\sqrt{5}$ and 7 respectively. Since θ is the angle between $\overrightarrow{BA}$ and $\overrightarrow{BC}$, their scalar product gives

$$\begin{pmatrix}-2\\0\\1\end{pmatrix}\cdot\begin{pmatrix}2\\-6\\-3\end{pmatrix} = \sqrt{5}\times 7\times\cos\theta$$
$$\Rightarrow \quad -7 = 7\sqrt{5}\cos\theta$$
$$\Rightarrow \quad \cos\theta = \frac{-1}{\sqrt{5}}$$

Since the cosine ratio is negative, θ must be an obtuse angle.

Using the identity $\quad \sin^2\theta + \cos^2\theta \equiv 1$

gives $\quad \sin^2\theta + \left(\frac{-1}{\sqrt{5}}\right)^2 = 1$

$$\Rightarrow \quad \sin^2\theta + \tfrac{1}{5} = 1$$
$$\Rightarrow \quad \sin^2\theta = \tfrac{4}{5}$$
$$\Rightarrow \quad \sin\theta = \pm\tfrac{2}{\sqrt{5}}$$

But we know that θ is obtuse, which means that the sine ratio must be positive.

So $\quad \sin\theta = \frac{2}{\sqrt{5}}$

Area of cross-section is the area of the triangle ABC

$$= \tfrac{1}{2}\times AB\times BC\times\sin\theta$$
$$= \tfrac{1}{2}\times\sqrt{5}\times 7\times\tfrac{2}{\sqrt{5}}$$
$$= 7 \text{ square units}$$

Volume of prism = area of cross-section × length

$$= 7\times 14$$
$$= 98 \text{ cubic units}$$

a

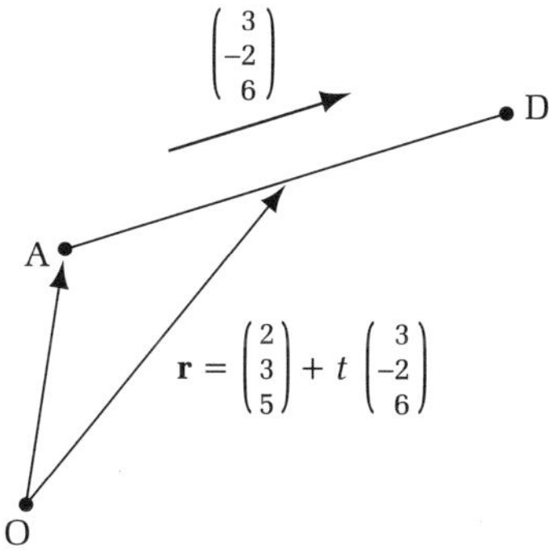

We know that the vector equation of AD is

$$\mathbf{r} = \begin{pmatrix}2\\3\\5\end{pmatrix} + t\begin{pmatrix}3\\-2\\6\end{pmatrix}.$$

This means that $\begin{pmatrix}3\\-2\\6\end{pmatrix}$ is parallel to $\overrightarrow{AD}$. Using Pythagoras' theorem, the length of this vector is 7.

However, we know that $AD = BE = CF = 14$, and so to reach D from A we need to travel twice this distance: using $t = 2$.

So the position vector of D is

$$\mathbf{r}(2) = \begin{pmatrix} 2 \\ 3 \\ 5 \end{pmatrix} + 2\begin{pmatrix} 3 \\ -2 \\ 6 \end{pmatrix}$$

$$= \begin{pmatrix} 8 \\ -1 \\ 17 \end{pmatrix}$$

b We have

$$\overrightarrow{AB} = \begin{pmatrix} 2 \\ 0 \\ -1 \end{pmatrix}$$

$$\text{and } \overrightarrow{AD} = 2 \times \begin{pmatrix} 3 \\ -2 \\ 6 \end{pmatrix} = \begin{pmatrix} 6 \\ -4 \\ 12 \end{pmatrix}$$

from **a**. Find the scalar product of both $\overrightarrow{AB}$ and $\overrightarrow{AD}$ with vector **n**.

$$\mathbf{n}.\overrightarrow{AB} = \begin{pmatrix} 2 \\ 15 \\ 4 \end{pmatrix}.\begin{pmatrix} 2 \\ 0 \\ -1 \end{pmatrix}$$

$$= 4 + 0 - 4$$

$$= 0$$

$$\mathbf{n}.\overrightarrow{AD} = \begin{pmatrix} 2 \\ 15 \\ 4 \end{pmatrix}.\begin{pmatrix} 6 \\ -4 \\ 12 \end{pmatrix}$$

$$= 12 - 60 + 48$$

$$= 0$$

So **n** is perpendicular to both $\overrightarrow{AB}$ and $\overrightarrow{AD}$, and is therefore perpendicular to the lines AB and AD. This also means that **n** is perpendicular to the plane ABED. The vector equation of this plane can be written in the form

$$\mathbf{r}.\mathbf{n} = d, \quad \text{where } \mathbf{n} = \begin{pmatrix} 2 \\ 15 \\ 4 \end{pmatrix}$$

$$\Rightarrow \quad \mathbf{r}.\begin{pmatrix} 2 \\ 15 \\ 4 \end{pmatrix} = d$$

Since this is true for all points on the plane, it must be true for point A, which lies on the plane and has position vector $\begin{pmatrix} 2 \\ 3 \\ 5 \end{pmatrix}$.

$$\text{So} \quad \begin{pmatrix} 2 \\ 3 \\ 5 \end{pmatrix}.\begin{pmatrix} 2 \\ 15 \\ 4 \end{pmatrix} = d$$

$$\Rightarrow \quad 4 + 45 + 20 = d$$

$$\Rightarrow \quad d = 69$$

So the vector equation of the plane ABED is

$$\mathbf{r}.\begin{pmatrix} 2 \\ 15 \\ 4 \end{pmatrix} = 69$$

10 The lines have a common point if there are values of s and t such that

$$\begin{pmatrix} 0 \\ 2 \\ -3 \end{pmatrix} + s\begin{pmatrix} 1 \\ -1 \\ -1 \end{pmatrix} = \begin{pmatrix} -1 \\ 6 \\ -1 \end{pmatrix} + t\begin{pmatrix} 2 \\ 1 \\ -1 \end{pmatrix}$$

$$\text{or} \quad \begin{pmatrix} s \\ 2 - s \\ -3 - s \end{pmatrix} = \begin{pmatrix} -1 + 2t \\ 6 + t \\ -1 - t \end{pmatrix}$$

Equating components,

$$s = -1 + 2t \qquad [1]$$

$$2 - s = 6 + t \qquad [2]$$

$$-3 - s = -1 - t \qquad [3]$$

Adding equation [1] and equation [2],

$$2 = 5 + 3t$$

$$\Rightarrow \quad t = -1$$

Substituting $t = -1$ into equation [1],

$$s = -1 + 2t$$

$$= -1 - 2$$

$$s = -3$$

Check these values by substituting them into equation [3]. Substituting $s = -3$ and $t = -1$ into the vector equations for the two lines,

$$\mathbf{r}(s) = \begin{pmatrix} 0 \\ 2 \\ -3 \end{pmatrix} + s\begin{pmatrix} 1 \\ -1 \\ -1 \end{pmatrix}$$

$$\text{becomes} \quad \mathbf{r}(-3) = \begin{pmatrix} 0 \\ 2 \\ -3 \end{pmatrix} - 3\begin{pmatrix} 1 \\ -1 \\ -1 \end{pmatrix}$$

$$= \begin{pmatrix} -3 \\ 5 \\ 0 \end{pmatrix}$$

$$\text{and} \quad \mathbf{r}(t) = \begin{pmatrix} -1 \\ 6 \\ -1 \end{pmatrix} + t\begin{pmatrix} 2 \\ 1 \\ -1 \end{pmatrix}$$

$$\text{becomes} \quad \mathbf{r}(-1) = \begin{pmatrix} -1 \\ 6 \\ -1 \end{pmatrix} - \begin{pmatrix} 2 \\ 1 \\ -1 \end{pmatrix}$$

$$= \begin{pmatrix} -3 \\ 5 \\ 0 \end{pmatrix}$$

So the lines have a common point at (–3, 5, 0).

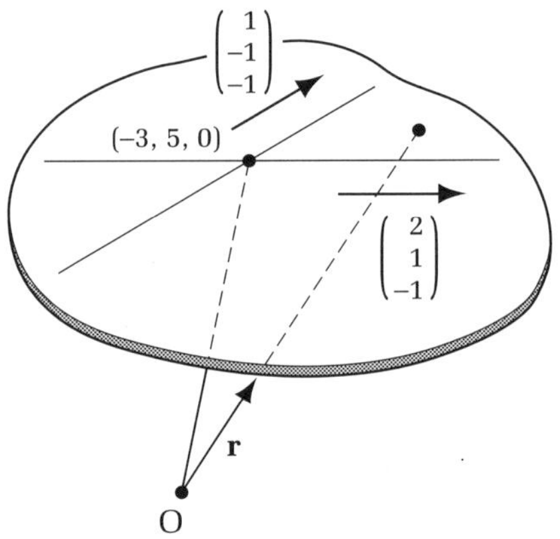

We know that the point $(-3, 5, 0)$ and vectors parallel to

both $\begin{pmatrix} 1 \\ -1 \\ -1 \end{pmatrix}$ and $\begin{pmatrix} 2 \\ 1 \\ -1 \end{pmatrix}$ lie in the plane containing the two lines. So the parametric equation of the plane containing the lines is

$$\mathbf{r} = \begin{pmatrix} -3 \\ 5 \\ 0 \end{pmatrix} + \lambda \begin{pmatrix} 1 \\ -1 \\ -1 \end{pmatrix} + \mu \begin{pmatrix} 2 \\ 1 \\ -1 \end{pmatrix}$$

15 Proof and Mathematical Argument (p. 684)

1 This argument proceeds as follows.

$1 = 1$		[1]
$x = x$	Multiply both sides by x	[2]
$x^2 = x^2$	Square both sides	[3]
$x^2 - x^2 = x^2 - x^2$	Subtract x^2 from both sides	[4]
$x(x - x) = (x + x)(x - x)$	Correct factorisation	[5]
$x = (x + x)$	Dividing by $(x - x)$, which is zero	[6]
$1 = 2$	Dividing by x	[7]

Statement [6] is invalid, since we cannot divide by zero.

2 We can write all integers using one of the following forms:

$6n \qquad 6n \pm 1 \qquad 6n \pm 2$

$6n \pm 3 \qquad 6n \pm 4 \qquad 6n \pm 5$

Now consider which of these are prime numbers

$6n$	Clearly not prime since all integers of this form have a factor 6
$6n \pm 1$	
$6n \pm 2$	Clearly not prime since all integers of this form are even and so have a factor 2
$6n \pm 3$	Clearly not prime since all integers of this form have a factor 3
$6n \pm 4$	Clearly not prime since all integers of this form are even and so have a factor 2
$6n \pm 5$	By rewriting 5 as $6 - 1$, we can write these numbers as $6n \pm (6 - 1) = 6m \pm 1$ (where $m = n \pm 1$).

Since these forms cover all possible integers, $6n$, $6n \pm 2$, $6n \pm 3$ and $6n \pm 4$ cannot be prime and numbers of the form $6n \pm 5$ can also be written in the form $6m \pm 1$, prime numbers can all be written in the form $6n \pm 1$.

3 The conjecture is that $2n^2 + 5$ is prime if n is an integer.

Look for a counter-example.

$n = 1$	$2n^2 + 5 = 2 + 5 = 7$	prime
$n = 2$	$2n^2 + 5 = 8 + 5 = 13$	prime
$n = 3$	$2n^2 + 5 = 18 + 5 = 23$	prime
$n = 4$	$2n^2 + 5 = 32 + 5 = 37$	prime
$n = 5$	$2n^2 + 5 = 50 + 5 = 55$	not prime

So the conjecture is not true for all n.

4 When $n = 1$ the AP has one term, a. Substituting $n = 1$

$$\text{sum} = \tfrac{1}{2} \times 1 \times 2a = a$$

So the formula is correct for $n = 1$.

Assume that it is true for $n = k$. The next term in the AP will be $(a + kd)$, and the sum then becomes

$$\begin{aligned} &\tfrac{1}{2}k[2a + (k-1)d] + (a + kd) \\ &= ak + \tfrac{1}{2}k(k-1)d + a + kd \\ &= a(k+1) + d\left[\tfrac{1}{2}k(k-1) + k\right] \\ &= a(k+1) + \tfrac{1}{2}kd[k - 1 + 2] \\ &= a(k+1) + \tfrac{1}{2}kd(k+1) \\ &= \tfrac{1}{2}(k+1)[2a + kd] \\ &\equiv \tfrac{1}{2}n[2a + (n-1)d] \text{ with } n = k + 1 \end{aligned}$$

So if it is true for $n = k$, then it is also true for $n = k + 1$. The statement has been proved true by induction.

5 Proving the statement is going to be very difficult, so instead search for a counter-example to prove that it is not true.

Recall that integration can be used to find the area under a curve. Choose a function $f(x)$ that crosses the x-axis (so producing positive and negative areas), and choose p and q to make the areas have the same magnitude.

So, for example, consider $f(x) = \cos x$ with $p = \frac{\pi}{2}$ and $q = \pi$.

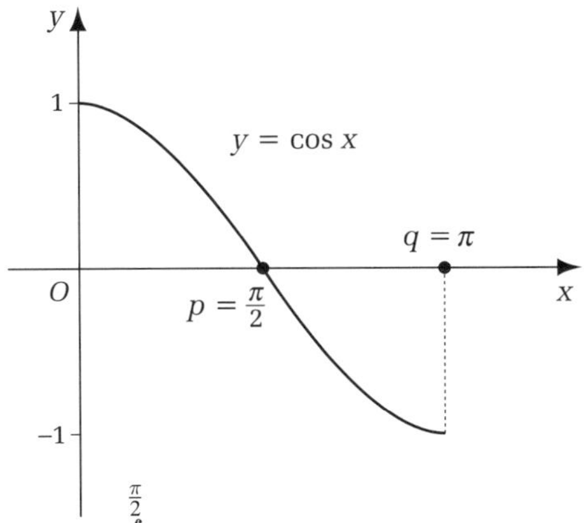

We know that $\int_0^{\frac{\pi}{2}} \cos x \, dx$ is positive, and we also know that

$$\int_0^{\frac{\pi}{2}} \cos x \, dx + \int_{\frac{\pi}{2}}^{\pi} \cos x \, dx = 0.$$

So

$$\int_0^{\frac{\pi}{2}} \cos x \, dx > \int_0^{\pi} \cos x \, dx$$

This clearly contradicts the statement in the question.

APPLICATIONS AND ACTIVITIES

ANSWERS AND SOLUTIONS

(refer to *Complete Advanced Level Mathematics: Pure Mathematics*)

1 Algebra I (p. 47)

1 **a** $x^2 - 6x - 16 = (x-3)^2 - 25$
b $x^2 + 2x - 8 = (x+1)^2 - 9$
c $x^2 - 2x - 3 = (x-1)^2 - 4$
d $2x^2 - 2x - 1 = 2(x - \frac{1}{2})^2 - \frac{3}{2}$
e $x^2 - 6x + 1 = (x-3)^2 - 8$
f $-3x^2 + 8x + 7 = -3(x - \frac{4}{3})^2 + \frac{37}{3}$

2 **a** $(3,\ -25)$
b $(-1,\ -9)$
c $(1,\ -4)$
d $(\frac{1}{2},\ -\frac{3}{2})$
e $(3,\ -8)$
f $(\frac{4}{3}, \frac{37}{3})$

You should have noticed that once the quadratic has been expressed in the form $a(x+b)^2 + c$, it is easy to locate the top or bottom of the curve. The coordinates will always be $(-b, c)$.

2 Coordinate Geometry (p. 85)

Constructing a circle through three known points

1 This task can be very frustrating but can initially be solved by paper-folding. Call the three points A, B and C. Fold the paper so that A is on top of B and crease the fold. Repeat for A and C and for B and C. The folds are the perpendicular bisectors of AB, AC and BC respectively. The point of intersection of the three folds is the centre of the required circle.

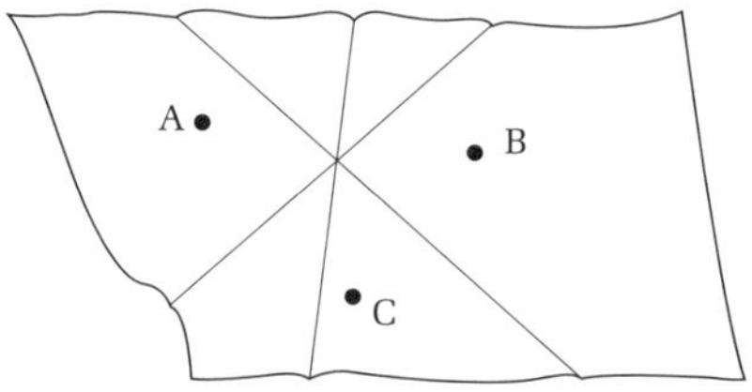

It cannot always be done. If A, B and C lie in a straight line it is impossible (excluding circles of infinite radius so that the straight line ABC is part of an arc of a great circle).

2 **a** When the activity is repeated on graph paper with coordinate points:

- Find the gradient of AB and then find the mid-point of AB. You should then be able to determine the equation of the perpendicular bisector of AB.
- Repeat for AC and BC.
- The equations of these three perpendicular bisectors can then be solved simultaneously (using the third equation for checking) to find the coordinates of the centre of the circle.

b Use Pythagoras' theorem to find the distance from the centre of the circle to each of A, B and C. They should be equal to each other, as these three distances are radii of the circle.

c The coordinates of the centre of the circle and the radius can be substituted into

$$(x-a)^2 + (y-b)^2 = r^2$$

Responses can always be checked on a graphical calculator.

3 Trigonometry I (p. 136)

2 GRAD means grade. It is a system of 'new degrees', where the circle has been divided into 400 equal parts and the right-angle has 100 parts. A grade can then be subdivided into centi-grades (equivalent to 100 'new minutes') and each of these into 100 'new seconds'.

Notation for this system uses g, c and cc.

$$1^{g} = 100^{c} = 10\,000^{cc}$$

So if we think of the right angle

$$90^\circ = 5400' \quad = 324\,000''$$

$$100^{g} = 10\,000^{c} \quad = 1\,000\,000^{cc}$$

In surveying it is still customary to use the degree, but it is subdivided decimally, not into the navigation systems of minutes and seconds.

3 The grad system has obvious advantages if angles are to be represented as percentages (see the Highway Code).

4 Functions (p. 188)

1 **a** The possible functions mapping elements in the set $\{a, b, c\}$ across to elements in the set $\{p, q\}$ are shown below.

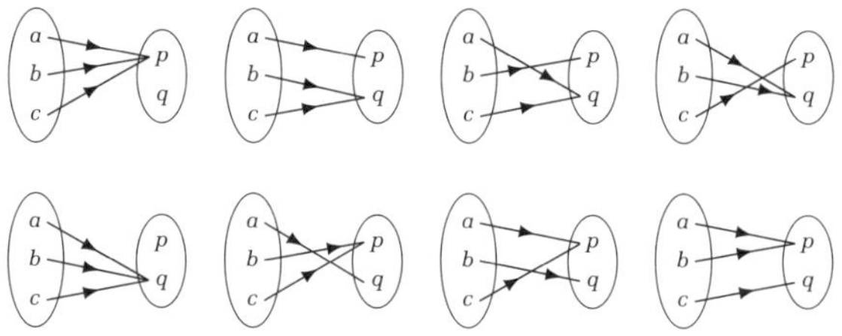

Notice that there are two 'three-to-one' functions and six 'two-to-one' functions. (These six 'two-to-one' functions are made up of three possible pairs in $\{a, b, c\} \times$ two possible singles in $\{p, q\}$.) So six of the possible functions have $\{p, q\}$ as their range, and two of them have $\{p, q\}$ as their co-domain.

b With three elements in the domain and two in the co-domain, there are eight possible functions altogether.

c Consider mapping four elements in the domain to two elements in the co-domain. There are 16 possible functions mapping elements in the set $\{a, b, c, d\}$ across to elements in the set $\{p, q\}$, as shown below.

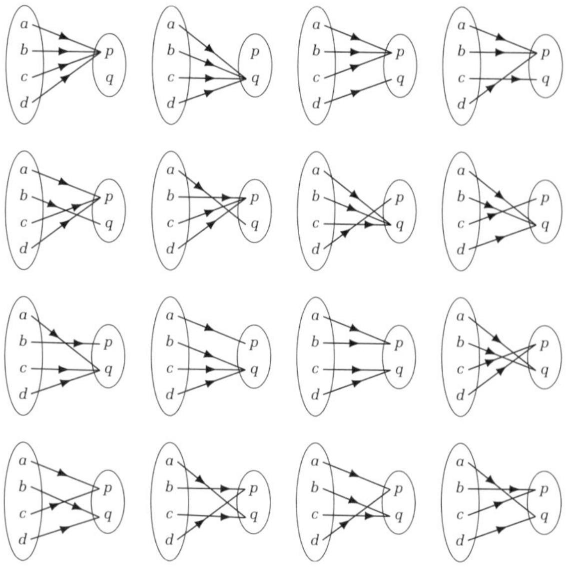

There are two 'four-to-one' functions, eight 'three-to-one' functions (made up of four possible 'triples' in $\{a,b,c,d\}\times$ two possible 'singles' in $\{p,q\}$) and six 'two-to-one' functions (made up of three possible pairs in $\{a,b,c,d\}\times$ two possible 'singles' in $\{p,q\}$). So altogether there are 16 possible functions mapping elements in the set $\{a,b,c,d\}$ across to elements in the set $\{p,q\}$.

By the same process we find that there are 32 possible functions mapping elements in the set $\{a,b,c,d,e\}$ across to elements in the set $\{p,q\}$ (not shown here). This, and the earlier results, suggest that, with m elements in the domain mapping to just two elements in the co-domain, there are 2^m possible functions.

Further investigation reveals that, with careful counting, there are n^m possible functions with m elements in the domain and n elements in the co-domain.

5 Differentiation I (p. 253)

1 **a**

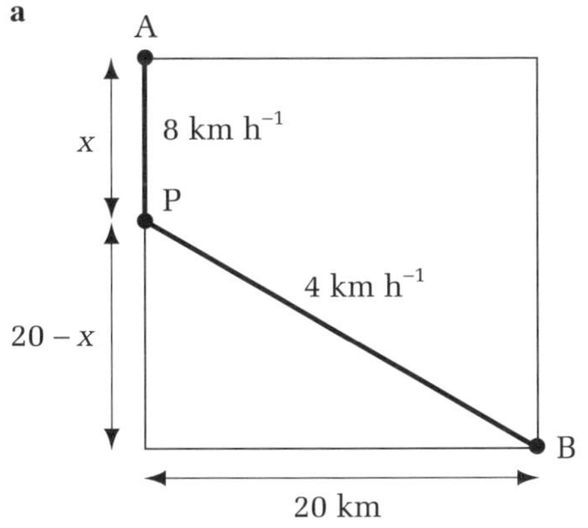

Let x be the distance between A and P. Using Pythagoras' theorem, the distance between P and B is

$$\sqrt{(20-x)^2+20^2} = \sqrt{800-40x+x^2}$$

Using time taken $= \dfrac{\text{distance travelled}}{\text{speed}}$

$$\text{time A}\to\text{P} = \frac{\text{distance A}\to\text{P}}{\text{speed A}\to\text{P}} = \frac{x}{8}$$

$$\text{time P}\to\text{B} = \frac{\text{distance P}\to\text{B}}{\text{speed P}\to\text{B}} = \frac{\sqrt{800-40x+x^2}}{4}$$

So the total time taken in travelling from A to B is

$$T = \frac{x}{8} + \frac{\sqrt{800-40x+x^2}}{4}$$

This is dependent only on the value of x; the distance between A and P.

b Use the chain rule to differentiate T with respect to x.

Since $\quad T = \dfrac{x}{8} + \frac{1}{4}(800-40x+x^2)^{\frac{1}{2}}$

then $\quad \dfrac{dT}{dx} = \frac{1}{8} + \frac{1}{8}(2x-40)(800-40x+x^2)^{-\frac{1}{2}}$

$$= \frac{1}{8} + \frac{x-20}{4\sqrt{800-40x+x^2}}$$

At maximum and minimum values of the total time T,

$$\frac{dT}{dx} = 0$$

So $\quad \dfrac{1}{8} + \dfrac{x-20}{4\sqrt{800-40x+x^2}} = 0$

Multiplying throughout by 8,

$$1 + \frac{2x-40}{\sqrt{800-40x+x^2}} = 0$$

$\Rightarrow \quad 40-2x = \sqrt{800-40x+x^2}$

$\Rightarrow \quad (40-2x)^2 = 800-40x+x^2$

$\Rightarrow \quad 1600-160x+4x^2 = 800-40x+x^2$

$\Rightarrow \quad 3x^2-120x+800 = 0$

Using the quadratic formula,

$$x = \frac{120 \pm \sqrt{120^2-9600}}{6}$$

$$= 20 \pm \frac{\sqrt{4800}}{6}$$

$$= 20 \pm \frac{20\sqrt{3}}{3}$$

$= 8.453$ km or 31.457 km (to the nearest metre)

(Ignore the second solution because the square is only 20 km $\times$ 20 km.)

Substituting $x = 8.453$ km into the expression for total time T,

$$T = 6.83\text{h}$$

$= 6\,\text{h}\ 50\,\text{min}$ (to the nearest minute)

6 Algebra II (p. 292)

Parametric equations

1
a circle, centre (0, 0), radius 1
b circle, centre (3, 4), radius 1
c circle, centre (–3, 4), radius 1
d circle, centre (–3, –4), radius 1
e circle, centre (3, –4), radius 1

2 Consider the generalised result given below.

If $\quad x = a + \cos\theta$ and $y = b + \sin\theta$

then $\quad \cos\theta = (x-a)$ and $\sin\theta = (y-b)$

Using $\quad \sin^2\theta + \cos^2\theta = 1$ gives

$$(x-a)^2 + (y-b)^2 = 1$$

which is the Cartesian equation of a circle, centre (a, b), radius 1.

So we could predict that $x = 2\cos\theta$, $y = 2\sin\theta$ is the equation of a circle, centre (0, 0), radius 2.

The general parametric form of this type is

$$x = a + r\cos\theta \text{ and } y = b + r\sin\theta$$

which reduces to $(x-a)^2 + (y-b)^2 = r^2$

This is the Cartesian equation of a circle, centre (a, b), radius r.

So $x = 5 + 2\cos\theta$ and $y = 1 + 2\sin\theta$ are the parametric equations of a circle, centre (5, 1), radius 2.

3 Consider the following families of parametric equations:

a $x = a + b\cos\theta,\ y = c + d\sin\theta$

These generate an ellipse, centre (a, c).

b $x = t - \frac{1}{t},\ y = t + \frac{1}{t}$

These generate hyperbolae.

c $x = \theta - \sin\theta,\ y = 1 - \cos\theta$

These generate cycloids.

d $x = \cos^3 t,\ y = \sin^3 t$

These generate astroids.

Parametric equations can be sketched quickly on most graphical calculators.

7 Logarithms and Exponentials (p. 319)

Fibonacci numbers

1 The model suggested in the text, $y = ab^x$, only works approximately for large values of x.

Students should be able to calculate values of a and b for the data given.

For significantly larger Fibonacci numbers the model can be refined to $y = ab^x + cb^{-x}$. (This provides opportunities for investigation using a spreadsheet as necessary.)

The Logarithmic Spiral

2 The logarithmic spiral is one of a whole family of spirals that can be investigated and examples of which can be found in the real world. Consider also Archimedean spirals (after Archimedes 287–212BC), Reciprocal spirals and Pythagorean spirals.

Possible starting points are:

- a sunflower head
- snail shells
- passion flowers and the spirals they use to attach themselves as they grow.

8 Sequences and Series (p. 357)

Loan repayments

Loan repayment problems can be solved by using the formula for the sum to n terms of a GP and the laws of logarithms. APR is the *annual percentage rate*, which gives a yearly multiplier for interest charges. The monthly interest rate can be found by raising this figure to the power $\frac{1}{12}$.

1 Borrow £1000 at 14% APR, and annually repay £300.

The interest multiplier is 1.14 annually.

Balance at 0 years = 1000

Balance at 1 year = (1000 × 1.14) − 300

Balance at 2 years = [(1000 × 1.14 − 300) × 1.14 − 300]

. . .

$$\begin{aligned}\text{Balance at } n \text{ years} &= 1000 \times 1.14^n - 300 \times 1.14^{n-1} - 300 \times 1.14^{n-2} - \dots - 300\\ &= 1000 \times 1.14^n - 300(1.14^{n-1} + 1.14^{n-2} + \dots + 1)\\ &= 1000 \times 1.14^n - 300(1 + \dots + 1.14^{n-2} + 1.14^{n-1})\end{aligned}$$

The bracketed terms form a geometric series with n terms, first term 1 and common ratio 1.14. So the sum of this GP is

$$\frac{1 - 1.14^n}{1 - 1.14} \quad \text{or} \quad \frac{1.14^n - 1}{1.14 - 1} = \frac{1.14^n - 1}{0.14}$$

The balance after n years will be zero if the loan is repaid.

$$\begin{aligned}\text{So} \qquad 0 &= 1000 \times 1.14^n - 300\left(\frac{1.14^n - 1}{0.14}\right)\\ &= 1000 \times 1.14^n - \frac{300 \times 1.14^n}{0.14} + \frac{300}{0.14}\\ &= 1.14^n\left(1000 - \frac{300}{0.14}\right) + \frac{300}{0.14}\end{aligned}$$

$$\begin{aligned}\text{So} \qquad 1.14^n &= \frac{\left(\frac{300}{0.14}\right)}{\left(\frac{300}{0.14} - 1000\right)}\\ &= 1.875\end{aligned}$$

Using the laws of logarithms to find n,

$$\begin{aligned}\log 1.14^n &= \log 1.875\\ \Rightarrow \quad n \log 1.14 &= \log 1.875\\ \Rightarrow \quad n &= \frac{\log 1.875}{\log 1.14}\\ &= 4.798 \text{ (3 d.p.)}\end{aligned}$$

So the loan is repaid after 4.8 years, or 5 years on an annual repayment basis.

2 Borrow £1000 at 14% APR and repay debt in three years.

Calculate this problem over 36 months to allow comparisons with financial institutions' illustrations.

APR multiplier is 1.14.

Monthly interest multiplier is $(1.14)^{\frac{1}{12}} = 1.010\,98$ (5 d.p.)

Let each payment by £P.

Balance at 0 months = 1000

Balance at 1 month = (1000 × 1.010 98) − P

Balance at 2 months

= [(1000 × 1.010 98) − P] × 1.010 98 − P

. . .

Balance at 36 months

$= 1000 \times (1.010\,98)^{36} - P(1.010\,98^{35} + \dots + 1)$

The bracketed term is a geometric series with 36 terms, with a first term of 1 and a common ratio of 1.01098. The sum of this GP is

$$\frac{1 - 1.010\,98^{36}}{1 - 1.010\,98} \quad \text{or} \quad \frac{1.010\,98^{36} - 1}{1.010\,98 - 1} = \frac{1.010\,98^{36} - 1}{0.010\,98}$$

The balance after 36 months will be zero since the loan will have been repaid. So

$$0 = 1000 \times 1.010\,98^{36} - P \times \left(\frac{1.010\,98^{36} - 1}{0.010\,98}\right)$$

This equation can be rearranged to give

$$\begin{aligned}P &= \frac{1000 \times 1.010\,98^{36} \times 0.010\,98}{1.010\,98^{36} - 1}\\ &= 33.779 \text{ (3 d.p.)}\end{aligned}$$

This means you should repay £33.78 each month.

Check that £33.78 × 36 = £1216.08.

The total interest on the loan over three years at 14% APR is £216.08.

3 In many cases the quoted figures in repayment tables will not be equal to the figures derived using the techniques in 1 and 2. Often this is due to rounding errors and interest charges being calculated on a timescale other than a monthly one – often daily.

Raise students' awareness to the fact that in general the lower the APR the better the deal.

9 Integration I (p. 401)

Volume of a pyramid

1 **a** Let w be the width of the horizontal cross-section z cm from the top.

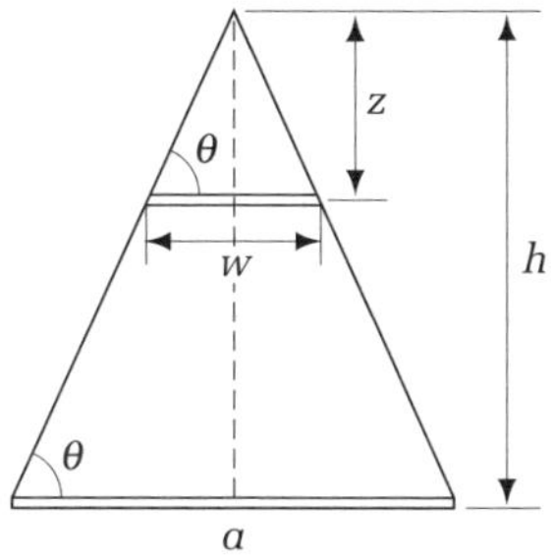

Using similar triangles,

$$\tan\theta = \frac{z}{\left(\frac{w}{2}\right)} = \frac{h}{\left(\frac{a}{2}\right)}$$

$$\Rightarrow \quad w = \frac{za}{h}$$

The area of the horizontal cross-section is dependent on the value of z, with

$$A(z) = w^2 = \frac{z^2a^2}{h^2}$$

b The volume of a thin cross-section of thickness δz is $\frac{z^2a^2}{h^2}\delta z$. The sum of all such thin cross-sections from $z = 0$ to $z = h$ is given by

$$\sum_{z=0}^{h} A(z)\delta z$$

This can be evaluated by integration.

c
$$\sum_{z=0}^{h} A(z)\delta z = \int_0^h A(z)\ \mathrm{d}z = \int_0^h \frac{z^2a^2}{h^2}\ \mathrm{d}z = \frac{a^2}{h^2}\int_0^h z^2\ \mathrm{d}z = \frac{a^2}{h^2}\left[\tfrac{1}{3}z^3\right]_0^h = \frac{a^2}{h^2}\left(\tfrac{1}{3}h^3 - 0\right) = \tfrac{1}{3}a^2h$$

So the volume of a pyramid

$$= \tfrac{1}{3} \times (\text{area of base}) \times (\text{height})$$

This is true for all pyramids, and not just right pyramids.

Area under a curve

2 **a** If $h = 3$,

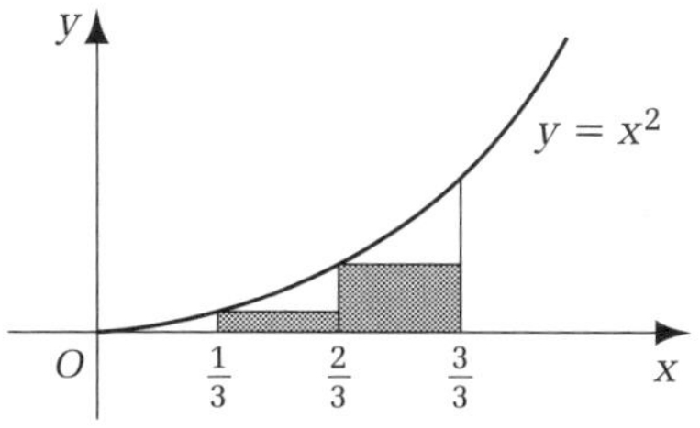

If $h = 4$,

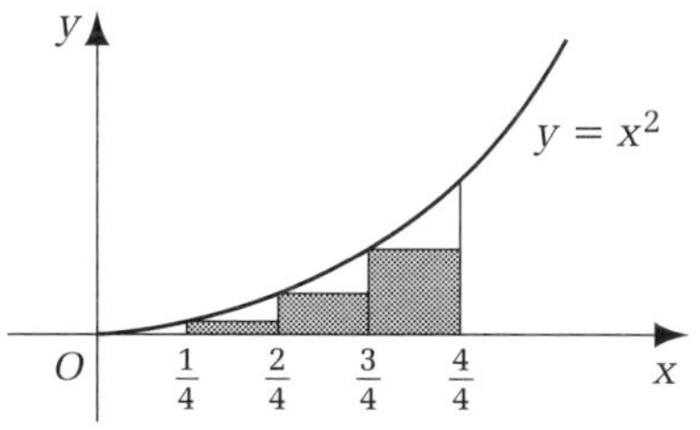

If there are h rectangles, the first one always has a height of zero. The sum of the areas of all the rectangles is

$$\frac{1}{h}(0)^2 + \frac{1}{h}\left(\frac{1}{h}\right)^2 + \frac{1}{h}\left(\frac{2}{h}\right)^2 + \frac{1}{h}\left(\frac{3}{h}\right)^2 + \ldots + \frac{1}{h}\left(\frac{h-1}{h}\right)^2$$

$$= \frac{1}{h} \times \frac{1}{h^2}\left(1^2 + 2^2 + 3^2 + \ldots + (h-1)^2\right)$$

$$= \frac{1}{h^3}\sum_{r=1}^{h-1} r^2$$

$$= \frac{1}{h^3} \times \frac{1}{6} \times (h-1) \times h \times (2h-1)$$

$$= \frac{1}{6h^2}(h-1)(2h-1) \text{ as required}$$

b If h is large, more rectangles would yield a more accurate answer.

$h = 10$ sum $= \frac{1}{600} \times 9 \times 19 = 0.285$

$h = 20$ sum $= \frac{1}{6\times 20^2} \times 19 \times 39 = 0.308\,75$

$h = 30$ sum $= \frac{1}{6\times 30^2} \times 29 \times 59 = 0.316\,85$

$h = 40$ sum $= \frac{1}{6\times 40^2} \times 39 \times 79 = 0.320\,937\,5$

$h = 50$ sum $= \frac{1}{6\times 50^2} \times 49 \times 99 = 0.3234$

$h = 60$ sum $= \frac{1}{6\times 60^2} \times 59 \times 119 = 0.325\,05$

It appears that the sum is approaching a limiting value of 0.33.

c The top of the rectangle is always below the curve $y = x^2$. Therefore the sum of the rectangles will always give an underestimate of the area, and so

$$\int_0^1 x^2\ \mathrm{d}x > \frac{1}{6h^2}(h-1)(2h-1)$$

d Extracting a factor of h from each bracket.

$$\frac{1}{6h^2}(h-1)(2h-1) = \frac{1}{6h^2}h\left(1-\tfrac{1}{h}\right)h\left(2-\tfrac{1}{h}\right)$$
$$= \tfrac{1}{6}\left(1-\tfrac{1}{h}\right)\left(2-\tfrac{1}{h}\right)$$

Now as $h \to \infty, \frac{1}{h} \to 0$, and so

$$\frac{1}{6h^2}(h-1)(2h-1) = \frac{1}{6}\left(1-\tfrac{1}{h}\right)\left(2-\tfrac{1}{h}\right)$$
$$\to \frac{1}{6}(1-0)(2-0)$$
$$\to \frac{1}{3}$$

So as the number of rectangles becomes very large, the sum of their areas converges towards a value of $\frac{1}{3}$. This is what we would expect, since using integration the exact value for the area under the curve is

$$\int_0^1 x^2 \, dx = \left[\frac{1}{3}x^3\right]_0^1 = \frac{1}{3}$$

10 Trigonometry II (p. 429)

Mathematical representation of a wave

1 The factor formula for $\sin X + \sin Y$ can be used provided the sound waves have the same amplitude $(A_1 = A_2)$.

Let $A = A_1 = A_2$.

Then $y_1 + y_2 = A\sin(k_1x - \omega_1 t) + A\sin(k_2 x - \omega_2 t)$

$$= 2A\sin\left[\tfrac{(k_1+k_2)x-(\omega_1+\omega_2)t}{2}\right]\cos\left[\tfrac{(k_1-k_2)x-(\omega_1-\omega_2)t}{2}\right]$$

This equation may look quite complicated but can, by a change of variable, (since k_1, k_2, ω_1 and ω_2 are constants anyway), be reduced to

$$y_1 + y_2 = 2A\sin(Kx - Wt)\cos(Lx - Mt)$$

where K, W, L and M depend on k_1, k_2, ω_1 and ω_2.

If $Lx - Mt = \frac{\pi}{2}$ or $Kx - Wt = \pi$ then $y_1 + y_2 = 0$. This means that the two sound waves will cancel out giving something called 'destructive interference'. The two original waves are completely out of phase with each other.

If $\sin(Kx - Wt)\cos(Lx - Mt) = 1$ then $y_1 + y_2 = 2A$. This means that the two sound waves are in phase with each other. The result is called 'constructive interference'.

This phenomenon has obvious implications for the acoustics in concert halls and could be experienced by students with a stereo system.

Simple harmonic motion (SHM)

2 If displacement x is written in the form $x = A\cos\omega t + B\sin\omega t$ then it can easily be converted into the form $R\sin(\omega t \pm \alpha)$ or $R\cos(\omega t \pm \alpha)$. This allows us to analyse SHM in various ways. We can allow the motion to have either $x = 0$ or $x = \pm\alpha$ when the timing begins $(t = 0)$. This is achieved by choosing an appropriate value for α. In problems where the choice of position at $t = 0$ is left to the problem solver this is a very useful technique. The displacement equation can be differentiated to find an expression for velocity, and differentiated again to find acceleration.

11 Differentiation II (p. 491)

Making a cone

1 **a**

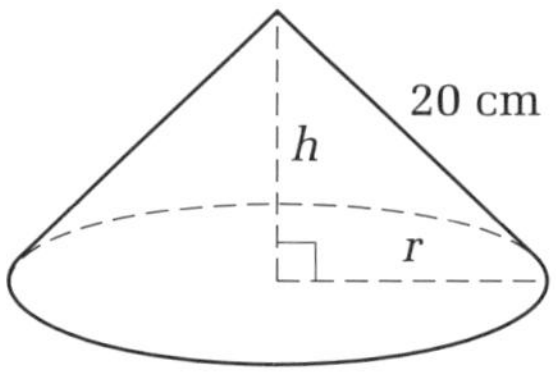

Using Pythagoras' theorem,

$$h^2 + r^2 = 20^2$$
$$\Rightarrow \quad h = \sqrt{400 - r^2}$$

b The volume V of a right circular cone of perpendicular height h and base radius r is given by

$$V = \tfrac{1}{3}\pi r^2 h$$

Substituting $h = \sqrt{400 - r^2}$ gives

$$V = \tfrac{1}{3}\pi r^2\sqrt{400 - r^2}$$

c Using the product rule to differentiate V with respect to r,

$$\frac{dV}{dr} = \tfrac{2}{3}\pi r\sqrt{400 - r^2} - \frac{\pi r^3}{3\sqrt{400 - r^2}}$$
$$= \frac{\pi r}{3\sqrt{400 - r^2}}\left[2\left(400 - r^2\right) - r^2\right]$$
$$= \frac{\pi r}{3\sqrt{400 - r^2}}\left(800 - 3r^2\right)$$

When $\frac{dV}{dr} = 0, r = 0$ or $800 - 3r^2 = 0$.

Ignoring the first (trivial) solution, we have

$$800 - 3r^2 = 0$$
$$r = \sqrt{\tfrac{800}{3}}$$
$$\approx 16.3 \text{ cm}$$

Substituting $r = \sqrt{\frac{800}{3}}$ into $V = \frac{1}{3}\pi r^2\sqrt{400 - r^2}$,

$$V = \tfrac{1}{3}\pi\left(\tfrac{800}{3}\right)\sqrt{400 - \tfrac{800}{3}}$$
$$= \frac{800\pi}{9}\sqrt{\frac{400}{3}}$$
$$\approx 3224.5 \text{ cm}^3$$

To show that this corresponds to the maximum possible volume of the cone, look at the value of $\frac{dV}{dr}$ on either side of $r = \sqrt{\frac{800}{3}}$.

When $r = 16$ cm, $\frac{dV}{dr} \approx 44.7$

When $r = 17$ cm, $\frac{dV}{dr} \approx -113.2$

Since the gradient changes from positive to negative, the volume of the cone has a maximum value when $r = \sqrt{\frac{800}{3}}$.

d

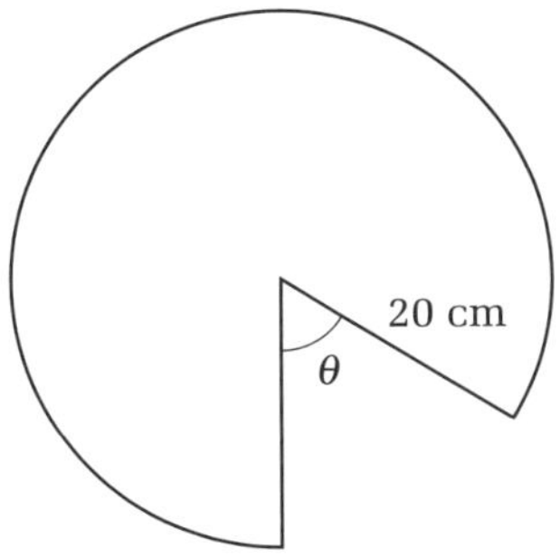

If θ is measured in radians, the length of the circular edge of the piece of card from which the cone is made is given by

$$2\pi \times \text{radius} - \text{radius} \times \theta = 20(2\pi - \theta)$$

Equating this to the circumference of the base of the cone gives

$$20(2\pi - \theta) = 2\pi r$$

$$\Rightarrow \quad \theta = 2\pi - \frac{\pi r}{10}$$

Substituting $r = \sqrt{\frac{800}{3}}$,

$$\theta = 2\pi - \frac{\pi}{10}\sqrt{\frac{800}{3}}$$
$$= 1.153 \text{ (3 d.p.)}$$
$$\equiv 66.06^\circ \text{ (2 d.p.)}$$

So the cone with the maximum possible volume is obtained by cutting a sector of 66.06° from the circular piece of card.

e Using a circular piece of card of radius R cm, the height and volume of the resulting cone would be

$$h = \sqrt{R^2 - r^2}$$

and $$V = \tfrac{1}{3}\pi r^2\sqrt{R^2 - r^2}$$

Using the product rule to differentiate V with respect to r,

$$\frac{dV}{dr} = \frac{\pi r}{3\sqrt{R^2 - r^2}}(2R^2 - 3r^2)$$

It follows that $\frac{dV}{dr} = 0$ when $r = 0$ or $r = \sqrt{\frac{2}{3}}R$ cm. Ignoring the (trivial) solution $r = 0$, the value of V corresponding to $r = \sqrt{\frac{2}{3}}R$ cm is $V = \frac{2\pi R^3}{9\sqrt{3}}$ cm^3. Equating the length of the circular edge of the piece of card from which the cone is made with the circumference of the base of the cone itself,

$$R(2\pi - \theta) = 2\pi r$$
$$= 2\pi\sqrt{\tfrac{2}{3}}R$$
$$\Rightarrow \quad \theta = 2\pi\left(1 - \sqrt{\tfrac{2}{3}}\right)$$
$$= 1.153$$
$$\equiv 66.06^\circ$$

So the cone with the maximum possible volume is obtained by cutting a sector of 66.06° from a circular piece of card, whatever its radius R.

12 Integration II (p. 549)

Cooling

1 **a** A body cools down at a rate which is proportional to the temperature difference between the body and its surroundings.

b Let θ be the temperature of the body and θ_0 be the room temperature. From Newton's law of cooling,

$$\frac{d\theta}{dt} \propto -(\theta - \theta_0)$$

giving $$\frac{d\theta}{dt} = -k(\theta - \theta_0), \text{ where } k > 0$$

c Separating the variables and integrating both sides of the differential equation,

$$\int \frac{d\theta}{\theta - \theta_0} = \int -k\,dt$$
$$\Rightarrow \quad \ln(\theta - \theta_0) = -kt + c$$
$$\Rightarrow \quad \theta - \theta_0 = e^{-kt+c}$$
$$= e^{-kt}e^c$$
$$\equiv Ae^{-kt}, \text{ where } A \text{ is some constant}$$
$$\Rightarrow \quad \theta = \theta_0 + Ae^{-kt}$$

The constants A and k can be found using the data from the table. The text suggests using the first and third entries.

d Use your formula for θ to calculate the temperature of the water at various times. Compare your results with the experimental values.

Integration spotting

2 Some of the possible answers are:

$\int \frac{1}{x} \times x^2\,dx = \frac{1}{2}x^2 + c$ $\quad$ $\int \frac{1}{x} \times x^3\,dx = \frac{1}{3}x^3 + c$

$\int \frac{1}{x} \times \ln x\,dx = \frac{1}{2}(\ln x)^2 + c$ $\quad$ $\int \frac{1}{x} \times e^{\ln x}\,dx = e^{\ln x} + c$

$\int \frac{1}{x} \times x\,dx = x + c$ $\quad$ $\int xe^{x^2}\,dx = \frac{1}{2}e^{x^2} + c$

$\int x^2e^{x^3}\,dx = \frac{1}{3}e^{x^3} + c$ $\quad$ $\int \cos x e^{\sin x}\,dx = e^{\sin x} + c$

$\int \sin x \cos x\,dx = \frac{1}{2}\sin^2 x + c$ $\quad$ $\int \sin x \cos x\,dx = -\frac{1}{2}\cos^2 x + c$

Encourage students to be systematic in their search for solutions. Remind them that differentiation can be used to check their answers.

13 Numerical Methods (p. 587)

1 **a** At point P, $x = -\frac{3h}{2}$. Substituting this into $y = ax^3 + bx^2 + cx + d$ gives

$$y_0 = a\left(-\tfrac{3h}{2}\right)^3 + b\left(-\tfrac{3h}{2}\right)^2 + c\left(-\tfrac{3h}{2}\right) + d$$
$$= -\tfrac{27}{8}ah^3 + \tfrac{9}{4}bh^2 - \tfrac{3}{2}ch + d \qquad [1]$$

In a similar way,

$$y_1 = -\tfrac{1}{8}ah^3 + \tfrac{1}{4}bh^2 - \tfrac{1}{2}ch + d \qquad [2]$$
$$y_2 = \tfrac{1}{8}ah^3 + \tfrac{1}{4}bh^2 + \tfrac{1}{2}ch + d \qquad [3]$$
$$y_3 = \tfrac{27}{8}ah^3 + \tfrac{9}{4}bh^2 + \tfrac{3}{2}ch + d \qquad [4]$$

b Adding equation [1] and equation [4] together gives

$$y_0 + y_3 = \tfrac{9}{2}bh^2 + 2d \qquad [5]$$

and adding equation [2] and equation [3] together gives

$$y_1 + y_2 = \tfrac{1}{2}bh^2 + 2d \qquad [6]$$

c The area under the curve $y = ax^3 + bx^2 + cx + d$ between $x = -\frac{3h}{2}$ and $x = \frac{3h}{2}$ is given by

$$A = \int_{-\frac{3h}{2}}^{\frac{3h}{2}} \left(ax^3 + bx^2 + cx + d\right) dx$$

$$= \left[\tfrac{1}{4}ax^4 + \tfrac{1}{3}bx^3 + \tfrac{1}{2}cx^2 + dx\right]_{-\frac{3h}{2}}^{\frac{3h}{2}}$$

$$= \left(\tfrac{81}{64}ah^4 + \tfrac{9}{8}bh^3 + \tfrac{9}{8}ch^2 + \tfrac{3}{2}dh\right)$$

$$- \left(\tfrac{81}{64}ah^4 - \tfrac{9}{8}bh^3 + \tfrac{9}{8}ch^2 - \tfrac{3}{2}dh\right)$$

$$= \tfrac{9}{4}bh^3 + 3dh$$

$$A = \frac{3h}{4}\left(3bh^2 + 4d\right) \qquad [7]$$

d Subtracting equation [6] from equation [5] gives

$$y_0 + y_3 - y_1 - y_2 = 4bh^2$$

so that $\quad 3bh^2 = \tfrac{3}{4}(y_0 + y_3 - y_1 - y_2) \qquad [8]$

Multiplying equation [6] by 9, and then subtracting equation [5] gives

$$9(y_1 + y_2) - (y_0 + y_3) = 16d$$

so that $\quad 4d = \tfrac{1}{4}(9y_1 + 9y_2 - y_0 - y_3) \qquad [9]$

Substituting for $3bh^2$ and $4d$ into equation [7], the area under the curve $y = ax^3 + bx^2 + cx + d$ between $x = -\frac{3h}{2}$ and $x = \frac{3h}{2}$ is given by

$$A = \tfrac{3h}{4}\left[\tfrac{3}{4}(y_0 + y_3 - y_1 - y_2) + \tfrac{1}{4}(9y_1 + 9y_2 - y_0 - y_3)\right]$$

$$= \tfrac{3h}{4}\left(\tfrac{1}{2}y_0 + \tfrac{3}{2}y_1 + \tfrac{3}{2}y_2 + \tfrac{1}{2}y_3\right)$$

$$= \tfrac{3h}{8}(y_0 + 3y_1 + 3y_2 + y_3)$$

Using this result to find $\int_a^b f(x)\,dx$, the area under the curve $y = f(x)$ between $x = a$ and $x = b$, the area of the first three vertical strips is approximately

$$\tfrac{3h}{8}(y_0 + 3y_1 + 3y_2 + y_3)$$

The areas of the second and third sets of three vertical strips are approximately

$$\tfrac{3h}{8}(y_3 + 3y_4 + 3y_5 + y_6) \text{ and}$$

$$\tfrac{3h}{8}(y_6 + 3y_7 + 3y_8 + y_9)$$

and eventually the area of the last set of three vertical strips is approximately

$$\tfrac{3h}{8}(y_{n-3} + 3y_{n-2} + 3y_{n-1} + y_n)$$

Adding all the separate areas together, the approximate value of $\int_a^b f(x)\,dx$, found by approximating the function $y = f(x)$ by a series of cubic curves, is given by

$$\int f(x)\,dx = \tfrac{3h}{8}\left[y_0 + y_n + 3(y_1 + y_2 + y_4 + y_5 + \ldots + y_{n-2} + y_{n-1}) + 2(y_3 + y_6 + \ldots + y_{n-3})\right]$$

where $h = \frac{b-a}{n}$, and n is a multiple of 3.

14 Vectors (p. 665)

1 **d** The three lines meet at the same point, which is the intersection of the medians of the triangle. This point is the centre of gravity of the triangle.

2 **a**

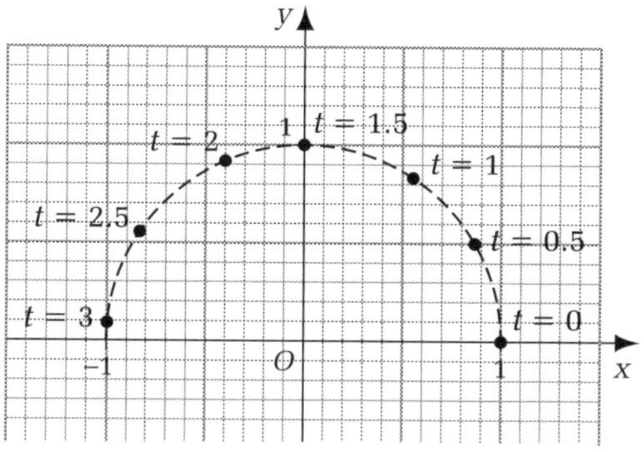

b

$$\dot{\mathbf{r}} = \frac{d}{dt}(\mathbf{r})$$

$$= \begin{pmatrix} \frac{d}{dt}\cos t \\ \frac{d}{dt}\sin t \end{pmatrix}$$

$$= \begin{pmatrix} -\sin t \\ \cos t \end{pmatrix}$$

c

$$\mathbf{r}.\dot{\mathbf{r}} = \begin{pmatrix} \cos t \\ \sin t \end{pmatrix}.\begin{pmatrix} -\sin t \\ \cos t \end{pmatrix}$$

$$= -\cos t \sin t + \sin t \cos t$$

$$= 0$$

This result indicates that at any time t, the position vector $\mathbf{r}$ and the velocity vector $\dot{\mathbf{r}}$ are at right angles to each other. This is a characteristic feature of circular motion, where the velocity is tangential to the path of the particle, and is therefore perpendicular to the radius at all times.

15 Proof and Mathematical Argument (p. 686)

1 This conjecture has been thoroughly checked by computer similar to Goldbachs Conjecture. At present no proof has been found, or counter-example produced.

2 This activity requires investigation into both Mersenne primes and perfect numbers.

Tabulate the information.

Mersenne prime	Product	Perfect number	Sum of divisors
$2^2 - 1 = 3$	$3 \times 4 = 12$	6	$1 + 2 + 3$
$2^3 - 1 = 7$	$7 \times 8 = 56$	28	$1 + 2 + 4 + 7 + 14$
$2^5 - 1 = 31$	$31 \times 32 = 992$	496	$1 + 2 + 4 + 8 + 16 + 31$ $+ 62 + 124 + 248$

The next perfect number is 8128. This also fits the pattern.

$2^7 - 1 = 127$	$127 \times 128 =$ 16256	8128	$1 + 2 + 4 + 8 + 16 + 32 + 64$ $+ 127 + 254 + 508 + 1016$ $+ 2032 + 4064$

The fifth perfect number is 33 550 336. The Mersenne prime that generates it is $2^{13} - 1 = 8191$. Check that $8191 \times 8192 \div 2 = 33550336$

The sixth to 32nd known perfect numbers are much larger. These could be investigated using a computer.

GLOSSARY

PURE MATHEMATICS

Absolute Value: The difference of a number from zero. Denoted by $|x|$ or abs(x) on some calculators. It is always positive. i.e. $|-4| = 4$ and $|4| = 4$. (See Modulus Function.)

Algebraic Fractions: Fractions that have a polynomial in the numerator and in the denominator. (See also Proper Fractions.)

Amplitude: Half the distance of the height of a wave from the top of its peak to the base of its trough.

Arc Notation: An alternative to traditional inverse trigonometric notation: for example, $\arcsin x \equiv \sin^{-1} x$.

Arc: This is part of a curve. Extensively used in questions involving sectors of circles.

Area of a Triangle: For triangle ABC,

$$\text{area} = \tfrac{1}{2}ab \sin C$$
$$= \tfrac{1}{2}\text{ base} \times \text{perpendicular height}$$

Arithmetic Progression: A sequence where the next term is found by adding a constant to the previous term. This means that there is a **common difference** between each term denoted by the letter d. The common difference can always be calculated by selecting a term and subtracting the previous term from it. For example, 4, 7, 10, 13, ... is an arithmetic progression with a common difference of 3.

Associative: The associative law of addition is

$$a + (b + c) = (a + b) + c$$

Other operators are also associative. For example $a \times (b \times c) = (a \times b) \times c$ shows that multiplication is associative.

Asymptote: A straight line that a curve can get closer and closer to but never cross, irrespective of the point chosen on the curve.

Base Vectors: The **unit vectors** that are parallel to the x-, y- and z-axes are denoted by **i, j** and **k** respectively. All vectors in two or three dimensions can be expressed in terms of these vectors.

Base: A number that is being raised to a power. For example, in 2^3, 3 is the power and 2 is the base.

Binomial Coefficients: See Pascal's Triangle

Cartesian Coordinates: These describe the position of a point by using a set of mutually perpendicular axes. In two dimensions, two axes, usually labelled x and y, are used. In three dimensions, three axes labelled x, y and z are used.

Chain Rule: The relationship $\dfrac{dy}{dx} = \dfrac{dy}{du} \times \dfrac{du}{dx}$ is called the chain rule. It is used to differentiate composite functions.

Chord: A straight line joining two points on a curve. Also called a secant line.

Co-domain: The set of possible outputs from a mapping.

Coefficient: The number in front of a letter or collection of letters in an algebraic expression. It indicates 'how many' of each term there are in the expression.

Common Difference: See Arithmetic Progression

Common Logarithm: Logarithms in base 10. The common logarithm of x is denoted by $\log x$ or $\log_{10} x$.

Common Ratio: See Geometric Progression

Commutative Law: The commutative law for addition is $a + b = b + a$. Other operators in mathematics are also commutative. For example, since $a \times b = b \times a$ and $\mathbf{a}.\mathbf{b} = \mathbf{b}.\mathbf{a}$, both multiplication and the **scalar product** are commutative.

Completing the Square: A technique that can be used for solving quadratic equations that cannot easily be solved by factorising. The technique involves writing the quadratic expression in the form $a(x + b)^2 + c$, where a, b and c are real numbers.

Components of a Vector: The components of a vector indicate how must of the vector is acting in particular directions. In two dimensions, the vector $\mathbf{a} = a_1\mathbf{i} + a_2\mathbf{j} = \begin{pmatrix} a_1 \\ a_2 \end{pmatrix}$ has two components, a_1 and a_2, parallel to the x- and y-axes. In three dimensions, the vector

$\mathbf{a} = a_1\mathbf{i} + a_2\mathbf{j} + a_3\mathbf{k} = \begin{pmatrix} a_1 \\ a_2 \\ a_3 \end{pmatrix}$ has three components, a_1, a_2 and a_3, parallel to the x-, y- and z-axes.

Composite Functions: When the output values for one function are used as input values for another function, this creates a composite function or 'function of a function'. So a composite function is any function of the form $f(g(x))$, written $fg(x)$ or $f \circ g(x)$.

Composite Number: A number that will factorise, having factors other than 1 and itself.

Compound Angle Formulas:

$$\sin(A + B) \equiv \sin A \cos B + \cos A \sin B$$
$$\sin(A - B) \equiv \sin A \cos B - \cos A \sin B$$
$$\cos(A + B) \equiv \cos A \cos B - \sin A \sin B$$
$$\cos(A - B) \equiv \cos A \cos B + \sin A \sin B$$
$$\tan(A + B) \equiv \frac{\tan A + \tan B}{1 - \tan A \tan B}$$
$$\tan(A - B) \equiv \frac{\tan A - \tan B}{1 + \tan A \tan B}$$

Conjugate Numbers: Used to rationalise the denominator of a surd. Keep the terms the same, but change the sign of the square root. For example, the conjugate of $3+\sqrt{2}$ is $3-\sqrt{2}$.

Constants: These represent a number that does not change in a given situation. Can be represented by letters, such as a_0, a_1, b, c, k etc.

Cosine Rule (see Sine Rule): In triangle ABC,

$$a^2 = b^2 + c^2 - 2bc \cos A$$
$$b^2 = a^2 + c^2 - 2ac \cos B$$
$$c^2 = a^2 + b^2 - 2ab \cos C$$

Alternatively

$$\cos A = \frac{b^2 + c^2 - a^2}{2bc}$$
$$\cos B = \frac{a^2 + c^2 - b^2}{2ac}$$
$$\cos C = \frac{a^2 + b^2 - c^2}{2ab}$$

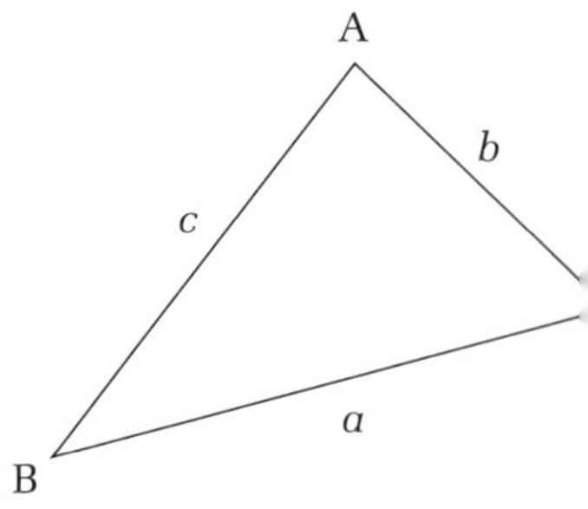

Cover-up Rule: A quick method of obtaining partial fractions when the denominator only contains distinct linear factors.

Decreasing Function: See Increasing Function

Deductive Reasoning (or Deduction): The use of initial assumptions called premises, axioms or postulates to build a proof.

Definite Integral: An integral with limits, which can be evaluated to give a numerical answer.

Derivative: The result of differentiating a function is called the derivative or first derivative of the function. If the function is then differentiated again the result is called the second derivative of the function. For example, the first derivative of $y = x^3$ is $\frac{dy}{dx} = 3x^2$ and the second derivative is $\frac{d^2y}{dx^2} = 6x$.

Difference of Two Squares: A useful technique for factorising quadratics that appear in the form:

$a^2 - b^2 = (a - b)(a + b)$

Differential equations: An equation involving a $\frac{dy}{dx}$ term is called a first order differential equation. If there is a $\frac{d^2y}{dx^2}$ term in the equation then this represents a second order differential equation. Note the variables need not be x and y.

Differentiation: The algebraic process by which the 'gradient function' of a function is found. This gradient function, or **derivative**, can be used to calculate the gradient of a curve at any point.

Discriminant: Defined as $b^2 - 4ac$ from the formula for solving a quadratic equation. This can be used to determine whether or not a quadratic equation has any roots, and if so whether they are distinct or equal.

Distributive Law: The rule $a(b + c) = ab + ac$, which is true for all real numbers. The **scalar product** also follows the distributive law because $\mathbf{a}.(\mathbf{b} + \mathbf{c}) = \mathbf{a}.\mathbf{b} + \mathbf{a}.\mathbf{c}$.

Domain: The set of inputs for a mapping.

Dot Product: See Scalar Product

Elimination: An algebraic technique used to solve simultaneous equations. One of the variables is eliminated by adding or subtracting the equations.

Equation: A statement, usually written in algebra, where the LHS and RHS are stated as being equal. For example, $x + 3 = 7$.

Even Function: A function whose graph is symmetrical about the y-axis. This means that $f(-x) = f(x)$ for all values of x in the domain of f. The functions $f(x) = x^2$ and $f(x) = \cos x$ are examples of even functions.

Exponential Function: The function e^x. This function is unique because its gradient function is equal to function itself for all real values of x.

Expression: A mathematical statement, usually written in algebra that is not written as an equation or as an identity. For example, $x^2 + 3y - 4$.

Factor Formulae (or Sum and Product Formulae): A set of trigonometric identities that convert expressions like $\sin A + \sin B$ into a product, in a way factorising the expression.

$$\sin C + \sin D \equiv 2\sin\left(\frac{C+D}{2}\right)\cos\left(\frac{C-D}{2}\right)$$

$$\sin C - \sin D \equiv 2\sin\left(\frac{C-D}{2}\right)\cos\left(\frac{C+D}{2}\right)$$

$$\cos C + \cos D \equiv 2\cos\left(\frac{C+D}{2}\right)\cos\left(\frac{C-D}{2}\right)$$

$$\cos C - \cos D \equiv -2\sin\left(\frac{C+D}{2}\right)\sin\left(\frac{C-D}{2}\right)$$

Factor: A number that divides exactly into another number.

Factorisation: The process of expressing a number or an algebraic expression as the product of its factors.

Factor Theorem: If $P(x)$ is a polynomial and $P(a) = 0$ then $(x - a)$ is a factor of $P(x)$. This theorem is used to factorise polynomials of degree 3 or higher.

Fermat's Last Theorem: Pierre de Fermat (1601–1665) is believed to have proved that $a^n + b^n = c^n$ has no integer solutions, greater than two. There is no record of the proof of this conjecture and it was not until 1994 when British born, American-based mathematician Andrew Wiles produced a proof, that one was developed. Fermat's Last Theorem had puzzled mathematicians for over three hundred years.

Fixed Point Iterative Method: An iterative process used to solve equations, which must be written in the form $x = g(x)$. The iterative formula is $x_{n+1} = g(x_n)$.

Function: Any mathematical mapping that takes an element (or number) in the domain and maps it to one and only one element in the co-domain.

Fundamental Theorem of Calculus: Discovered independently by Newton and Leibniz. The theorem shows that integration and differentiation are opposite processes.

Geometric Progression: A sequence where a term is generated by multiplying the previous term by a constant. This mean that consecutive terms have a **common ratio** denoted by the letter r. The common ratio can always be calculated by choosing any term and dividing it by the previous term. For example, $4, 2, 1, \frac{1}{2}, \frac{1}{4}, \ldots$ is a geometric progression with a common ratio of $\frac{1}{2}$.

Gradient: Measures the steepness of a line.

Hero's Formula: Area of triangle ABC is $\sqrt{s(s-a)(s-b)(s-c)}$, where $s = \frac{a+b+c}{2}$ is the semi-perimeter.

Identity: A statement of equality that is true for all values of the variable.

Implicit Differentiation: This is a technique used to differentiate expressions that are difficult or impossible to express in the form $y = f(x)$.

Improper Fractions: See Proper Fractions

Increasing Function: A function is said to be an increasing function between $x = a$ and $x = b$ if its gradient $\frac{dy}{dx}$ is always positive in this interval. If the gradient is always negative between $x = a$ and $x = b$, the function is a **decreasing function** in this interval.

Indefinite Integral: An integral with no limits. A constant term should be added at the end of the resulting algebraic expression.

Index (plural, indices): See Powers

Indices, Rules of:

$a^p \times a^q = a^{p+q}$	Property 1
$(a^p)^q = a^{p \times q} = a^{pq}$	Property 2
$a^p \div a^q = a^{p-q}$	Property 3
$a^0 = 1$	Property 4
$a^{-p} = \frac{1}{a^p}$	Property 5
$a^{\frac{1}{p}} = \sqrt[p]{a}$	Property 6
$a^{\frac{p}{q}} = \sqrt[q]{a^p} = (\sqrt[q]{a})^p$	Property 7

Induction: The principle of mathematical induction is a three-step proof:

Step 1: Show the result is true for a natural number, say $n = 1$.

Step 2: Assume the result is also true for some other natural number $n = k$, where $k > 1$.

Step 3: Show that the result is also true for the next natural number $n = (k + 1)$, given that it is true for $n = k$.

Integers: The set of positive and negative whole numbers ... $-3, -2, -1, 0, 1, 2, 3, \ldots$, denoted by the symbol $\mathbb{Z}$.

Integrand: The term that is being integrated.

Integration: The algebraic process by which an expression for the function $f(x)$ can be found from its gradient function. It is the reverse process to **differentiation**. Integration is also used to calculate the area under a curve.

Integration by Parts: A technique used to integrate the product of two functions. The formula is

$\int u\frac{dv}{dx}\,dx = uv - \int v\frac{du}{dx}\,dx$, which can also be written

as $\int uv'\,dx = uv - \int vu'\,dx$.

Interval Bisection: An iterative process, involving bisecting an interval, used to solve an equation in the form $f(x) = 0$.

Inverse Function: Reverses the effect of a function f. The notation used is f^{-1}. Note: this is not the reciprocal of the function f.

Irrational Numbers: Numbers that cannot be expressed as fractions. Examples of irrational numbers are π, e, and $\sqrt{2}$, which can only be described by a decimal with an infinite number of non-recurring decimal places.

Linear Equation: Equations containing only x and constant terms but no terms containing x^2, x^3 or higher powers. Examples are $3x + 7 = x - 5$ and $3(x - 10) = 2$.

Linear Functions: Functions that can be written in the form $y = ax + b$ or $f(x) = ax + b$ where a and b are constants and x is the variable

Linear Interpolation: An iterative method used to solve equations when they are written in the form $f(x) = 0$. It converges faster than the **interval bisection** method.

Local Maximum/Minimum Points: See Turning Points

Maclaurin's Series: A polynomial that approximates a function $f(x)$ when x is close to zero. The series is:

$$f(x) = f(0) + f'(0)x + \frac{f''(0)}{2!}x^2 + \frac{f^{(3)}(0)}{3!}x^3 + \ldots\ldots + \frac{f^{(n)}(0)}{n!}x^n,$$

where $f^{(n)}(0)$ is the value of the nth derivative of $f(x)$ at $x = 0$.

Magnitude of a Vector: The length of the directed line segment representing a vector, denoted by $|\mathbf{a}|$ or a.

For a vector $\mathbf{a} = \begin{pmatrix} a_1 \\ a_2 \end{pmatrix}$ in two dimensions, $|\mathbf{a}| = \sqrt{a_1^2 + a_2^2}$. In three dimensions, if $\mathbf{a} = \begin{pmatrix} a_1 \\ a_2 \\ a_3 \end{pmatrix}$, then $|\mathbf{a}| = \sqrt{a_1^2 + a_2^2 + a_3^2}$.

Mapping: A rule that relates one set of elements or numbers to another.

Mathematical Modelling: A system of representing a real-world situation mathematically, providing a mathematical solution and then applying this mathematical solution back to the real-world problem.

Mid-ordinate Rule: A numerical method of estimating the area under a curve, which involves adding up the areas of a series of rectangles. Area under the curve is given by

$\int_a^b f(x)\,dx \approx h(f_1 + f_2 + \ldots. + f_n)$, where h is the width of each rectangle, f_1 is the height of the ith rectangle and n is the number of rectangles.

Modulus Function: This is denoted by $y = |x|$. When $x \geq 0$ the graph is the same as $y = x$, and when $x < 0$ it is the same as $y = -x$. Overall a V-shape graph, symmetrical about the y-axis, is formed. This function reflects the negative value into their positive equivalents. It also indicates the distance of a number from zero: $|-3| = 3$ and $|+3| = 3$.

Natural Logarithms: Logarithms in base e. The natural logarithm of x is denoted by $\ln x$ or $\log_e x$

Natural Numbers: The whole numbers used for counting. They are 1, 2, 3, ... denoted by $\mathbb{N}$.

Newton–Raphson Method: An iterative process used to solve equations that can be written in the form $f(x) = 0$. Provided the function can be differentiated, the iteration formula is

$$x_{n+1} = x_n - \frac{f(x_n)}{f'(x_n)}.$$

Normal: The straight line that is perpendicular to the tangent to a curve. The vector normal to a plane is the vector that is perpendicular to the surface of the plane.

Odd Function: A function that has rotational symmetry of order 2 about the origin. It also has the property $f(-x) = f(x)$ for all values x in the domain of f. The functions $f(x) = x^3$ and $f(x) = \sin x$ are examples of odd functions.

Orthogonal Vectors (see Base Vectors): Vectors that are mutually perpendicular such as **i** and **j** in two dimensions, and **i**, **j** and **k** in three dimensions.

Paradox: A seemingly absurd statement that appears to be self-contradictory but, in reality, is based on truth.

Parameter: See Parametric Equation

Parametric Equation: These are usually denoted by letters such as $s, t, \theta, \phi, \lambda, \mu, \ldots$. In two dimensions a parameter forms a link between the x- and y-coordinates of a point on a curve. For example, the parametric equation of a curve could be $x = t^2$ and $y = 2t$. Each value of the parameter t defines a unique point on the curve. Curves that are difficult to describe using a Cartesian equation can sometimes be defined using parametric equations. Some common examples are shown in this table:

Curve	Cartesian equation	Parametric equation
Circle	$x^2 + y^2 = r^2$	$x = r\cos\theta, y = r\sin\theta$
Ellipse	$\frac{x^2}{a^2} + \frac{y^2}{b^2} = 1$	$x = a\cos\theta, y = b\cos\theta$
Parabola	$y^2 = 4ax$	$x = at^2, y = 2at$
Rectangular hyperbola	$xy = c^2$	$x = ct, y = \frac{c}{t}$

Partial Fractions: Proper fractions whose denominators factorise can be expressed as the sum or difference of two or more simpler fractions, known as partial fractions. There are several types of partial fractions:

- linear factors in the denominator: e.g. $\frac{3+2x}{(x+2)(x-1)} \equiv \frac{A}{(x+2)} + \frac{B}{(x-1)}$
- repeated factors in the denominator: e.g. $\frac{1}{x(x+4)^2} \equiv \frac{A}{x} + \frac{B}{(x+4)} + \frac{C}{(x+4)^2}$
- quadratic factors in the denominator: e.g. $\frac{x}{(1+x^2)(x-7)} \equiv \frac{Ax+B}{(1+x^2)} + \frac{C}{(x-7)}$

Particular Solution: The solution of a differential equation in which the numerical values of any constants have been determined.

Pascal's Triangle: A triangular pattern of numbers.

$$\begin{array}{ccccccccc} & & & & 1 & & & & \\ & & & 1 & & 1 & & & \\ & & 1 & & 2 & & 1 & & \\ & 1 & & 3 & & 3 & & 1 & \\ 1 & & 4 & & 6 & & 4 & & 1 \end{array}$$

The numbers represent the coefficients in the expansion of $(a+b)^n$. The first row corresponds to $(a+b)^0$. The fifth row represents the coefficients of $(a+b)^4$. These coefficients are called the binomial coefficients. They can be generated using the formula

$\binom{n}{r} = \frac{n!}{r!(n-r)!}$, which represents the coefficient of $a^{n-r}b^r$ in the expansion of $(a+b)^n$.

Perfect Number: A number that is equal to the sum of its proper divisors, but excluding the number itself.

Periodic Function: A function that repeats itself. It has the property $f(x+a) \equiv f(x)$, where a is the period. The function $f(x) = \sin x$ is an example of periodic function, with period 2π.

Perpendicular: At right angles to a line or a surface.

Polynomial: The general form of an algebraic expression with many terms, it comes from the Greek 'poly' meaning many and 'nomial' meaning terms or names.

Position Vector: The position vector of a point describes how to get to the point from the origin.

Powers: Eight cubed is written as 8^3. This is also known as 'eight to the power three'. The small number, 3, at the top is called the power or the index. The number being raised to the power, 8, is called the base.

Product Rule: A rule used to differentiate the product of two functions. If $y = u(x)v(x)$ then

$$\begin{aligned} \frac{dy}{dx} &= u\frac{dv}{du} + v\frac{du}{dx} \\ &= uv' + vu' \end{aligned}$$

Proof by Contradiction (or Reductio ad Absurdum): The idea is to first assume the opposite of what is being attempted to be proven. Then, via a mathematical argument, it is shown that a false (absurd) conclusion is reached. If all reasoning and deduction is correct, then the only incorrect statement must be the initial one. Since the initial statement was opposite to what was being attempted to be proven, this means that the original statement must be true.

Proof by Exhaustion: A method of proving a statement where every possibility is systematically considered and the results are checked.

Proper Fractions: Fractions that have a numerator that is less than the denominator. In algebra, algebraic fractions of the form $\frac{f(x)}{g(x)}$ are proper if the degree of the polynomial $f(x)$ is less than the degree of the polynomial $g(x)$. If this is not the case it is an improper fraction. Some examples of proper algebraic fractions are

$\frac{1}{x+4}$, $\frac{x+5}{x(x-2)}$ and $\frac{2}{(x-3)(x+1)}$;

whilst $\frac{x^2}{(x+5)(x-7)}$ is an improper algebraic fraction.

Property: A possession of quality; for example a square has the property of possessing four equal sides.

Pythagorean Identities: In the unit circle, Pythagoras' theorem can be used to show that for all values of θ,

$\sin^2\theta + \cos^2\theta \equiv 1.$

Two further identities can be established by dividing both sides by $\cos^2\theta$ and $\sin^2\theta$ respectively:

$\tan^2\theta + 1 \equiv \sec^2\theta$

$1 + \cot^2\theta \equiv \operatorname{cosec}^2\theta$

QED (Quod Erat Demonstrandum): Placed after a proof means, 'which was demonstrated to an incontrovertible conclusion'.

Quadrant, First: The quadrant between the positive x- and y-axes. In this quadrant, θ is acute. (Remember that θ is measured anticlockwise from the positive x-axis.) The trigonometric ratios $\sin\theta$, $\cos\theta$ and $\tan\theta$ are all positive in this quadrant.

Quadrant, Second: The quadrant between the positive y-axis and negative x-axis, in which θ is obtuse. In this quadrant, $\sin\theta$ is positive and both $\cos\theta$ and $\tan\theta$ are negative.

Quadrant, Third: The quadrant between the negative x- and y-axes. Only $\tan\theta$ is positive in this quadrant; both $\sin\theta$ and $\cos\theta$ are negative.

Quadrant, Fourth: The quadrant between the positive x-axis and negative y-axis, in which only $\cos\theta$ is positive; both $\sin\theta$ and $\tan\theta$ are negative.

Quadratic Equation: Any equation that can be written in the form $ax^2 + bx + c = 0$, where $a \neq 0$.

Quotient: Functions in the form $\frac{g(x)}{h(x)}$ are often referred to as quotients (see Quotient Rule). When a polynomial is divided by another polynomial the result is a polynomial plus a remainder. The polynomial in the answer to the division is also referred to as the quotient. Algebraically, if

$\frac{g(x)}{h(x)} = Q(x) + R$, then $Q(x)$ is the quotient.

Quotient Rule: A rule used to differentiate a quotient. If

$y = \frac{u(x)}{v(x)}$ then

$$\begin{aligned} \frac{dy}{dx} &= \frac{v\left(\frac{du}{dx}\right) - u\left(\frac{dv}{dx}\right)}{v^2} \\ &= \frac{vu' - uv'}{v^2} \end{aligned}$$

Radian: The angle θ at the centre of a circle is said to be one radian when the length of the arc corresponding to the angle is equal to the radius of the circle.

Range of a Function: The set of images or output values to which objects in the domain map. This can be the entire co-domain or a subset of the co-domain. In the first case, the domain is said to map onto the co-domain, in the second case, it is said to map into the co-domain.

Rational Numbers: All the numbers that can be written as a fraction, where the numerator and denominator are integers. These numbers are denoted by $\mathbb{Q}$.

Rationalising the denominator: Writing numbers without a surd in the denominator is known as rationalising the denominator. For example

$$\frac{1}{\sqrt{2}} = \frac{\sqrt{2}}{2} \text{ and } \frac{3}{1+\sqrt{5}} = \frac{-3(1-\sqrt{5})}{4}$$

Real Numbers: All the numbers which make up the number line.

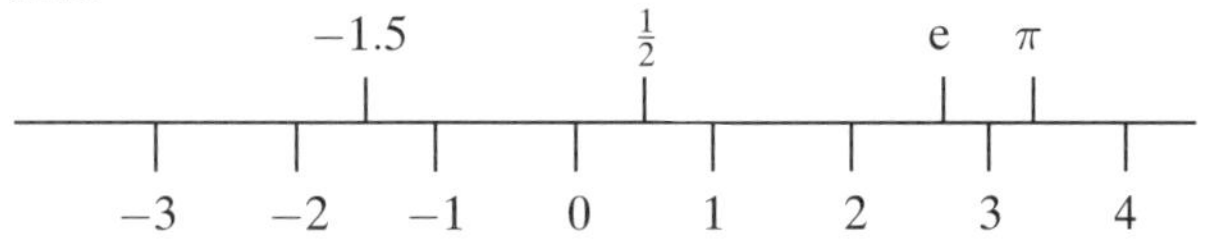

They include all the natural numbers, integers, rational and irrational numbers. The set of real numbers is denoted by $\mathbb{R}$.

Reciprocal Ratios: $\operatorname{cosec}\theta = \dfrac{1}{\sin\theta}$

$$\sec\theta = \frac{1}{\cos\theta}$$

$$\cot\theta = \frac{1}{\tan\theta}$$

Recurrence Relationship: The rule that must be applied to a particular term (or terms) in order to find the next term in the sequence.

Reduction to Linear Form: A technique used to establish the relationship between two quantities. The horizontal and/or vertical axes are re-labelled using logarithms in order to obtain a straight line relationship between the two quantities.

Remainder Theorem: If the polynomial $P(x)$ is divided by $(x - a)$ then the remainder is $P(a)$. This is a quick way of calculating the remainder when a polynomial is divided by a linear factor.

Roots: The values of x that satisfy an equation.

Scalar Product: Also called the **dot product**. It is based on the relationship between two vectors and the angle between them. It is extensively used to find the angle between two vectors. The scalar product is defined as

$$\mathbf{a}.\mathbf{b} = ab\cos\theta = a_1b_1 + a_2b_2 + a_3b_3,$$

where $\mathbf{a} = \begin{pmatrix} a_1 \\ a_2 \\ a_3 \end{pmatrix}$ and $\mathbf{b} = \begin{pmatrix} b_1 \\ b_2 \\ b_3 \end{pmatrix}$.

Note that **a.b** is a **scalar** quantity; that is, a number.

Scalars: Quantities that can be described by using just a number to represent their magnitude. Numbers themselves are regarded as scalars.

Second Derivative: See Derivative

Sector: The region enclosed inside a circle between two radii.

Self-inverse Functions: If the graphs of a function $f(x)$ and its inverse $f^{-1}(x)$ are symmetrical in the line $y = x$, then $f(x) = f^{-1}(x)$ and the function is said to be self-inverse. For example, the function

$f(x) = \dfrac{1}{x}$ is a self-inverse function since $f^{-1}(x) = \dfrac{1}{x}$.

Sigma Notation: Σ, which means 'the sum of'.

Simpson's Rule: A numerical method used to estimate the area under a curve by fitting a quadratic curve over a pair of strips. The area of each pair of strips is given by the formula $\frac{1}{3}h(y_0 + 4y_1 + y_2)$, and this is used repeatedly to estimate the area under a curve divided into an even number of strips.

$$\int_a^b f(x)\mathrm{d}x \approx \tfrac{1}{3}h[y_0 + 4(y_1 + y_3 + \ldots + y_{n-1}) + 2(y_2 + y_4 + \ldots + y_{n-2}) + y_n]$$

where h is the width of equal strips, n is the number of strips, and y_i is the height of the curve at the ith value along the x-axis.

Sine Ratio: See Trigonometry

Sine Rule: (see Cosine Rule)

$\dfrac{a}{\sin A} = \dfrac{b}{\sin B} = \dfrac{c}{\sin C}$ for finding missing lengths

$\dfrac{\sin A}{a} = \dfrac{\sin B}{b} = \dfrac{\sin C}{c}$ for finding missing angles

Solid of Revolution: See Volume of Revolution

Stationary Points: These are points on a graph where its gradient $\dfrac{\mathrm{d}y}{\mathrm{d}x} = 0$. They can be maximum or minimum points, or points of inflexion.

Sum and Product Formulae: See Factor Formulae.

Surd: A number that cannot be determined exactly since it is irrational and contains a square root. Examples include $\sqrt{2}$, $3\sqrt{5}$, and $\dfrac{1}{1+\sqrt{3}}$.

Tangent Ratio: See Trigonometry

Tangent: A straight line that just touches a curve at just one point.

Trace and Zoom: The 'trace' facility on a graphical calculator allows a point to move along the graph drawn whilst, simultaneously, showing the x- and y-coordinates of the point. The 'zoom' facility enables you to look more closely at a region or section of a plotted graph.

Transcendental Numbers: Any number that cannot satisfy an algebraic equation whose coefficients are integers.

Translation: A transformation that moves an object, usually in the x–y plane. It can be described by a vector.

Trapezium Rule: A numerical method of estimating the area under a curve, which involves adding up the areas of a series of trapeziums of equal width:

$$\int_a^b f(x)\ \mathrm{d}x \approx \tfrac{1}{2}h[y_0 + 2(y_1 + y_2 + \ldots. + y_{n-1}) + y_n],$$

where h is the width of each strip, n is the number of trapezia, and y_i is the height of the curve at the ith value along the x-axis.

Trial-and-improvement: The technique of successive attempts to solve an equation where each attempt is an improvement on the previous attempt.

Trigonometric Equation: An equation that contains a trigonometric function such as sine, cosine or tangent.

Trigonometric Functions: Sine, cosine and tangent and their reciprocals. They are periodic and have many-to-one mappings for the domain of real numbers.

Trigonometry: The study of angular measure, in particular, the study and calculation of triangle measure. In right-angled triangles:

- Cosine ratio (abbreviated to cos):

$$\cos\theta = \frac{\text{side adjacent to } \theta}{\text{hypotenuse}}$$

- Sine ratio (abbreviated to sin):

$$\sin\theta = \frac{\text{side opposite to } \theta}{\text{hypotenuse}}$$

- Tangent ratio (abbreviated to tan):

$$\tan\theta = \frac{\text{side opposite } \theta}{\text{side adjacent to } \theta}$$

Turning Points: The points on a graph where the curve has a local maximum or a local minimum point.

Unit Circle: A circle of radius one unit that is used to define trigonometric ratios for any angle.

Unit Vector: Any vector whose magnitude is unity (one). The unit vector in the direction of **a** is denoted by **â**.

Variables: These are denoted by letters (often x and y), which can take any value in a given range. If a variable is used to form an equation then only certain values of the variable will satisfy the equation.

Variables Separable: A method of rearranging certain first-order differential equations. If a differential equation can be written in the form

$$\frac{dy}{dx} = \frac{f(x)}{g(y)}$$

or $$\frac{dy}{dx} = f(x)g(y)$$

or $$\frac{dy}{dx} = \frac{g(y)}{f(x)}$$

then the variables are said to be separable and the equation can be solved by integrating both sides.

Vector Equation of a Straight Line: This is defined by the position vector **r** of a general point on a line. The equation can always be written in the form $\mathbf{r} = \mathbf{a} + \lambda\mathbf{b}$, where **a** is the position vector of a point on the line, **b** is a vector parallel to the line and λ is a parameter. Each value of λ corresponds to a unique point on the line.

Vector Equation of a Plane: This is defined by the position vector of a general point on the plane. There are several ways of describing a plane:

- if the vector normal to the plane, denoted by **n**, is known, the vector equation is $\mathbf{r.n} = d$, where d is a constant.
- performing this scalar product gives the Cartesian equation $ax + by + cz = d$, where a, b, c and d are constants.
- if three points are known, the equation of the plane passing through these is given by $\mathbf{r} = \mathbf{a} + \lambda\mathbf{b} + \mu\mathbf{c}$, where **a** is the position vector of one of the points on the plane, and **b** and **c** are any two vectors that lie in the plane. The letters λ and μ are parameters. Each combination of values of λ and μ defines a unique point on the plane.

Vector: A directed line segment used to model quantities that possess both direction and magnitude.

Volume of Revolution: The volume of the solid formed when a plane area is rotated around the x-axis or the y-axis or some other straight line. The **surface of revolution** is the surface of the volume of revolution.

SYNOPTIC ASSESSMENT

Introduction

Advanced level specifications include a minimum of 20% synoptic assessment, which aims to assess candidates' understanding of the connections between different elements of the subject. This means that a question in an examination paper may require use of methods from the core content or from earlier stages of the course.

Examiners may set problems that require candidates to use and apply methods and techniques developed at different stages of the advanced level programme.

Throughout the core book, examples and questions have linked new methods to those introduced in previous chapters. This section of the Teacher Resource File provides additional worked examples to those in the Question Bank, and a range of questions, with answers, that can be used with students to draw together knowledge, skills and understanding from several chapters of the core book. Students who have worked through this course thoroughly should already be well prepared to tackle problems that require knowledge from more than one area of pure mathematics. This section of the Teacher Resource File contains more examples and questions of the synoptic assessment type.

Worked Examples

1 A curve has equation $y = \dfrac{3x+4}{(x-2)(2x+1)}$

a Express $\dfrac{3x+4}{(x-2)(2x+1)}$ in partial fractions.

See Algebra II, section 6.3, p.270.

b Show that $\dfrac{dy}{dx} = \dfrac{2}{(2x+1)^2} - \dfrac{2}{(x-2)^2}$ and hence, or otherwise, show that the curve has a turning point when $x = -3$.

Determine the value of x at the other stationary point on the curve.

See Differentiation I, section 5.4, p.226 and Algebra I, section 1.5, p.26.

c Find $\dfrac{d^2y}{dx^2}$ and hence determine the nature of the turning point when $x = -3$.

See Differentiation I, section 5.4, p.226 and Differentiation II, section 11.4, p.455.

d Find $\displaystyle\int \frac{3x+4}{(x-2)(2x+1)}\,dx$. Hence show that the area of the region bounded by the curve, the x-axis and the lines $x = 4$ and $x = 12$ is equal to ln 15.

See Integration II, section 12.2, p. 507 and Exponentials and Logarithms, section 7.3, p.304.

(*AEB*)

Solution

a Notice that the denominator contains only linear factors. In Algebra II, section 6.3, there are two techniques that can be used to find the partial fractions, the quickest being the cover-up rule.

Let $\dfrac{3x+4}{(x-2)(2x+1)} \equiv \dfrac{A}{(x-2)} + \dfrac{B}{(2x+1)}$

Cover up the factor $(x - 2)$ and substitute $x = 2$.

$$\frac{(3 \times 2) + 4}{(\quad)(2 \times 2 + 1)} = \frac{6+4}{5} \quad \Rightarrow \quad A = 2$$

Now cover up the factor $(2x + 1)$ and substitute $x = -\frac{1}{2}$

$$\frac{(3 \times -\frac{1}{2}) + 4}{(-\frac{1}{2} - 2)(\quad)} = \frac{2\frac{1}{2}}{-2\frac{1}{2}} \quad \Rightarrow \quad B = -1$$

This means that

$$\frac{3x+4}{(x-2)(2x+1)} \equiv \frac{2}{(x-2)} - \frac{1}{(2x+1)}$$

Check that you get the same result using the other technique.

b To show the stated result for $\frac{dy}{dx}$ we should differentiate the partial fractions we found in **a**. If we rewrite them first using the rules of indices, we can get the partial fractions into a form where the chain rule can be used directly.

An alternative approach is to use the quotient rule, Differentiation II, section 11.1, p.433.

So $\quad y = 2(x-2)^{-1} - (2x+1)^{-1}$

Then $\quad \frac{dy}{dx} = 2(-1)(x-2)^{-2} - (-1)(2x+1)^{-2}(2)$

$$= \frac{-2}{(x-2)^2} + \frac{2}{(2x+1)^2} = \frac{2}{(2x+1)^2} - \frac{2}{(x-2)^2} \text{ as required.}$$

At stationary points $\frac{dy}{dx} = 0$. This means that

$$(x-2)^2 = (2x+1)^2$$

$$\Rightarrow \quad x^2 - 4x + 4 = 4x^2 + 4x + 1$$

$$\Rightarrow \quad 3x^2 + 8x - 3 = 0$$

Notice that the quadratic equation can be solved by factorisation (recall the PAFF technique outlined in Algebra I, section 1.5).

$$3x^2 + 8x - 3 = (x+3)(3x-1)$$

$$\Rightarrow \quad x = -3 \text{ or } x = \tfrac{1}{3}$$

c To find the second derivative we can use the rules of indices and then the chain rule again on our result in **b**.

$$\frac{dy}{dx} = -2(x-2)^{-2} + 2(2x+1)^{-2}$$

$$\Rightarrow \quad \frac{d^2y}{dx^2} = -2(-2)(x-2)^{-3} + 2(-2)(2x+1)^{-3}(2)$$

$$= \frac{4}{(x-2)^3} - \frac{8}{(2x+1)^3}$$

Now substitute $x = -3$. This gives

$$\frac{d^2y}{dx^2} = 0.032$$

See Differentiation II, section 11.4, p.455 on how to use the second derivative test.

Since $\frac{d^2y}{dx^2} > 0$, we conclude that the stationary point is a minimum point.

As in Differentiation I, section 5.2, p.204 we can check the sign of the first derivative on either side of $x = -3$ to confirm that we have a minimum point.

d Instead of trying to integrate this algebraic fraction directly we can use the partial fractions found in **a**. Since the denominators are linear and the fractions are 'proper', they will give natural logarithms when they are integrated.

So $\quad \int \frac{3x+4}{(x-2)(2x+1)}\,dx = \int \left(\frac{2}{x-2} - \frac{1}{2x+1}\right)dx$

$$= 2\ln|x-2| - \tfrac{1}{2}\ln|2x+1| + c$$

Substituting the limits gives

$$\text{area} = \left[2\ln|x-2| - \tfrac{1}{2}\ln|2x+1|\right]_4^{12}$$

$$= (2\ln 10 - \tfrac{1}{2}\ln 25) - (2\ln 2 - \tfrac{1}{2}\ln 9)$$

$$= \ln 100 - \ln 5 - \ln 4 + \ln 3$$

$$= \ln\frac{100 \times 3}{5 \times 4}$$

$$= \ln 15$$

Remember the rule to find the area under a curve developed in Integration I, section 9.2, p.367.

Use the logarithm laws in Exponentials and Logarithms, section 7.3, p.304 and remember that power $\frac{1}{2}$ means the same as square root.

2 The parametric equations of a curve are $x = 2\cos^3 t$ and $y = 2\sin^3 t$, where the parameter t takes all values such that $0 \le t \le \frac{\pi}{2}$.

a Show that the equation of the tangent at the point P $(2\cos^3 p, 2\sin^3 p)$ is $x\sin p + y\cos p = 2\sin p\cos p$.
See Differentiation II, section 11.6, p.472, Coordinate Geometry, section 2.3, p.64 and Trigonometry I, section 3.2, p.99.

b This tangent crosses the axes at A and B. Show that the length of AB is independent of p.
See Coordinate Geometry, section 2.1, p.50 and Trigonometry I, section 3.2, p.99.

(*NEAB*)

Solution

a Given $x = 2\cos^3 t$ and $y = 2\sin^3 t$, we can differentiate with respect to t.

So $\dfrac{dx}{dt} = -6\cos^2 t\sin t$

and $\dfrac{dy}{dt} = 6\sin^2 t\cos t$

Now use the chain rule to find $\dfrac{dy}{dx}$.

$$\frac{dy}{dx} = \left(\frac{dy}{dt}\right) \Big/ \left(\frac{dx}{dt}\right)$$
$$= \frac{6\sin^2 t\cos t}{-6\cos^2 t\sin t}$$
$$= -\frac{\sin t}{\cos t}$$

The equation of the tangent is given by $y - y_1 = m(x - x_1)$.

See Coordinate Geometry, section 2.3, p.64.

Use this result with the coordinates of P. This means that $(x_1, y_1) = (2\cos^3 p, y_1 = 2\sin^3 p)$ and $m = -\dfrac{\sin p}{\cos p}$. Notice that at P the value of the parameter t is p. Putting these two results together gives

$$y - 2\sin^3 p = \frac{-\sin p}{\cos p}\left(x - 2\cos^3 p\right)$$

$$\Rightarrow \quad y\cos p - 2\sin^3 p\cos p = -x\sin p + 2\sin p\cos^3 p$$

$$\Rightarrow \quad y\cos p + x\sin p = 2\sin^3 p\cos p + 2\sin p\cos^3 p$$

Now factorise the RHS.

$$y\cos p + x\sin p = 2\sin p\cos p(\sin^2 p + \cos^2 p)$$

We can now use the Pythagorean identity $\sin^2\theta + \cos^2\theta \equiv 1$ to simplify this result.
See Trigonometry I, section 3.2, p.99.

$$y\cos p + x\sin p = 2\sin p\cos p$$

This is the equation we were asked to show.

b Let A be the point on the y-axis and B be the point on the x-axis.
So at A, $x = 0$ and $y = 2\sin p$. Similarly at B, $x = 2\cos p$ and $y = 0$.
The length of the line AB is given by

$$|AB| = \sqrt{(x_1 - x_2)^2 + (y_1 - y_2)^2}$$

See Coordinate Geometry, section 2.1, p.50.

$$= \sqrt{(2\cos p - 0)^2 + (0 - 2\sin p)^2}$$
$$= \sqrt{4\cos^2 p + 4\sin^2 p}$$
$$= \sqrt{4(\sin^2 p + \cos^2 p)}$$
$$= \sqrt{4} = 2$$

So the length of AB is independent of p.

Question Bank

1 $f(x) \equiv \dfrac{5x^2 - 8x + 1}{2x(x-1)^2} \equiv \dfrac{A}{x} + \dfrac{B}{x-1} + \dfrac{C}{(x-1)^2}$

a Find the values of the constants A, B and C.
See Algebra II, section 6.4, p.281.

b Hence find $\int f(x)\,dx$.
See Integration II, section 12.2, p.507.

c Hence show that $\int_4^9 f(x)\,dx = \ln\left(\frac{32}{3}\right) - \frac{5}{24}$

See Integration I, section 9.2, p.367 and Exponentials and Logarithms, section 7.3, p.304.

(ULEAC)

2 It is given that $y = x^{\frac{3}{2}} + \frac{48}{x}, \quad x > 0.$

a Find the value of x and the value of y when $\frac{dy}{dx} = 0$.
See Algebra I, section 1.2, p.8 and Differentiation I, section 5.1, p.191.

b Show that the value of y that you found in **a** is a minimum.
See Differentiation I, section 5.2, p.204 and Differentiation II section 11.4, p.455.

c The finite region R is bound by the curve with equation $y = x^{\frac{3}{2}} + \frac{48}{x}$, the lines $x = 1, x = 4$ and the x-axis. Find, by integration, the area of R giving your answer in the form $p + q\ln r$, where the numbers p, q and r are to be found.
See Integration I, section 9.2, p.367 and Exponentials and Logarithms, section 7.3, p.304.

(ULEAC)

3 **a** Express $\dfrac{3}{x^2 - 3x + 2}$ in the form $\dfrac{A}{x-1} + \dfrac{B}{x-2}$.

Hence find $\displaystyle\int \frac{3}{x^2 - 3x + 2}\,dx$.

See Algebra II, section 6.3, p.270 and Integration II, section 12.2, p.507.

b **i** Show that $\int x^2 e^{2x}\,dx = \left(\frac{1}{2}x^2 - \frac{1}{2}x + \frac{1}{4}\right)e^{2x} + c$.
See Integration II, section 12.3, p.515.

ii Find the exact value of the volume of the solid formed when the region between the curve with equation $y = xe^x$ and the lines $y = 0, x = 0$ and $x = 1$ is rotated completely about the x-axis.
See Integration I, section 9.5, p.384.

(NEAB)

4

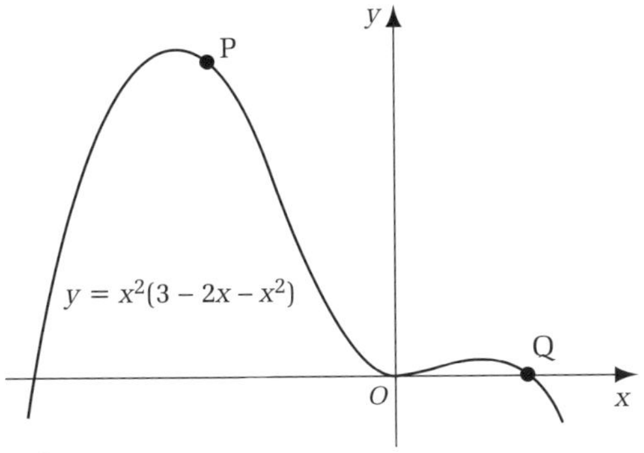

Figure 1

Figure 1 shows the curve with equation $y = x^2(3 - 2x - x^2)$. P and Q are points on the curve with coordinates $(-2, 12)$and $(1, 0)$ respectively.

a Find $\frac{dy}{dx}$.
See Differentiation II, section 11.1, p.433.

b Find the equation of the line PQ.
See Coordinate Geometry, section 2.3, p.64.

c Prove that the line PQ is a tangent to the curve at both P and Q.
See Differentiation I, section 5.1, p.191 and section 5.3, p.217.

d Find the area of the region bounded by the line PQ and that part of the curve for which $-2 \le x \le 1$.
See Integration I, section 9.2, p.367.

(MEI)

5 A circle has centre O and radius r. The chord AB is such that angle AOB $= \theta$ radians and the minor segment cut off by the chord is shaded in the diagram.

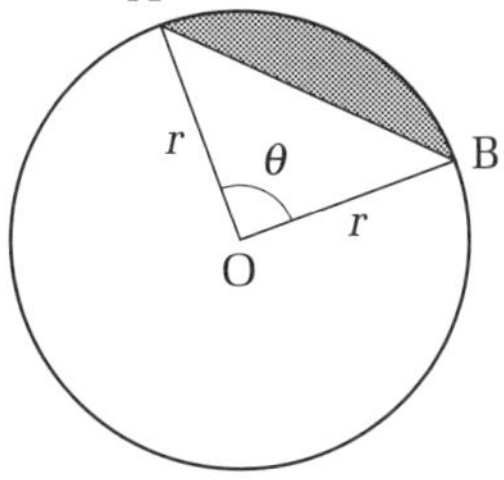

a Determine an expression for the area of the shaded region in terms of r and θ.
See Trigonometry I, section 3.5, p.123 and section 3.6, p.127.

b Given that the shaded region has area equal to one quarter of the area of the circle, show that θ satisfies the equation $\theta = \sin\theta + \frac{\pi}{2}$. Show that this equation has a root between 2.3 and 2.4.
See Numerical Methods, section 13.1, p.552.

c By writing $\theta_{n+1} = \sin\theta_n + \frac{\pi}{2}$ and using $\theta_1 = 2.3$ find θ_2, θ_3, θ_4 and θ_5, giving your answers to four decimal places. Convert the value of θ_5 into degrees, giving your answer to the nearest degree.
See Sequences and Series, section 8.1, p.322 and Trigonometry I, section 3.6, p.127.

(AEB)

6 The curve C has parametric equations $x = t^3$ and $y = t^2$, $t > 0$.

a Find an equation of the tangent to C at A (1, 1).
See Differentiation II, section 11.6, p.472.

Given that the line l with equation $3y - 2x + 4 = 0$ cuts the curve C at point B,

b find the coordinates of B,
See Differentiation II, section 11.6, p.472 and Coordinate Geometry, section 2.3, p.64.

c prove that the line l only cuts C at the point B.
See Differentiation II, section 11.6, p.472, Proof and Mathematical Argument, section 15.1, p.668 and Algebra I, section 1.5, p.26.

(ULEAC)

7 The curve shown in Figure 2 has parametric equations

$x = \dfrac{1}{1+t}$ and $y = \dfrac{1}{(1+t)(1-t)}$, $t \neq \pm 1$.

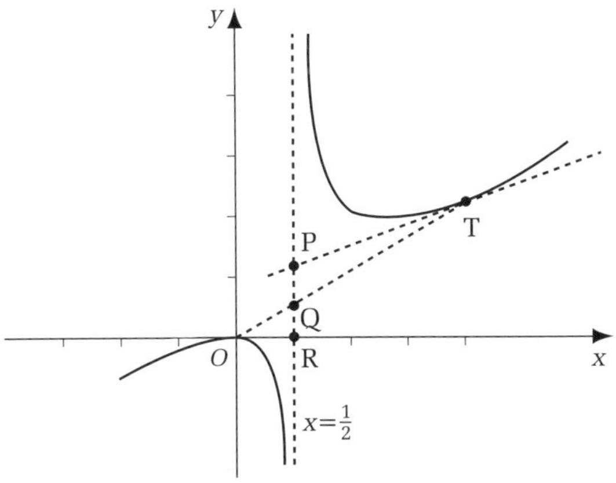

Figure 2

a Express t in terms of x. Hence show that the Cartesian equation of the curve is $y = \frac{x^2}{2x-1}$.
See Differentiation II, section 11.6, p.472.

b Find $\frac{dy}{dt}$ and $\frac{dx}{dt}$ and hence show that $\frac{dy}{dx} = -\frac{2t}{(1-t)^2}$.
See Differentiation I, section 5.4, p.226.

c You are given that the equation of the tangent at the point T having parameter t is $2tx + (1-t)^2y = 1$. Find the y-coordinate of the point P where this tangent cuts the line $x = \frac{1}{2}$.
See Differentiation II, section 11.6, p.472 and Coordinate Geometry, section 2.3, p.64.

d The point Q is the intersection of the line $x = \frac{1}{2}$ and the straight line joining the origin to the point T. Point R has coordinates $(\frac{1}{2}, 0)$. Show that $RQ = QP$.
See Coordinate Geometry, section 2.1, p.50.

(MEI)

8 **a i** Express $12\cos x + 9\sin x$ in the form $R\cos(x - \theta)$ where $R > 0$ and $0 < \theta < \frac{\pi}{2}$.
See Trigonometry II, section 10.2, p.411.

ii Use the method of part **a(i)** to find the smallest possible root α of the equation
$12\cos x + 9\sin x = 14$, giving your answer in radians correct to three decimal places.
See Trigonometry II, section 10.2, p.411 and Trigonometry I, section 3.2, p.99.

b i Show that, when x is a small angle measured in radians,

$$12\cos x + 9\sin x \approx 12 + 9x - 6x^2$$

See Differentiation I, section 5.5, p.236.

ii Use the quadratic approximation in part **b(i)** above to find an approximation to the same root α of the equation $12\cos x + 9\sin x = 14$.
See Algebra I, section 1.5, p.26.

(OCSEB)

9 The curve with equation $y = (x+2)e^{-x}$ is sketched below.

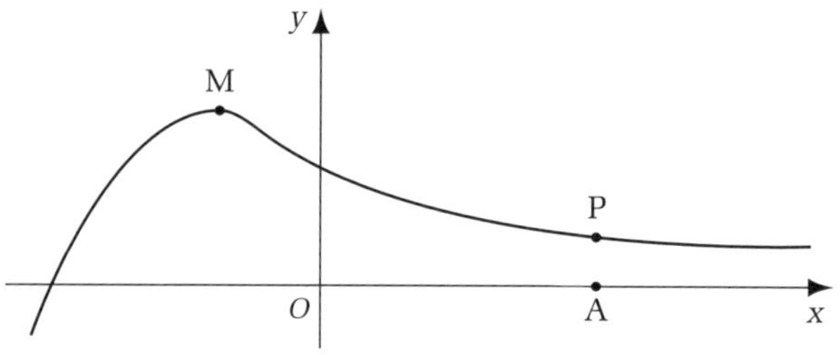

The maximum point M and the point P lie on the curve and A lies on the x-axis. The points A and P each have x-coordinate equal to 3.

a Find the value of $\frac{dy}{dx}$ at the point P, leaving your answer in terms of e.
See Exponentials and Logarithms, section 7.1, p.295 and Differentiation II, section 11.1, p.433.

b Determine the equation of the tangent to the curve at P and find the x-coordinate of the point B where this tangent crosses the x-axis.
See Differentiation I, section 5.3, p.217.

c Calculate the coordinates of the maximum point M.
See Differentiation I, section 5.2, p.204.
No credit will be given for simply reading values from the display of a graphics calculator. All working must be shown.

d Determine the coordinates of the point of inflection of the curve.
See Differentiation II, section 11.4, p.455.

e Use integration by parts to find $\int_0^3 (x+2)e^{-x}\,dx$, leaving your answer in terms of e.
See Integration II, section 12.3, p.515.
No credit will be given for a numerical approximation or for a numerical answer from a calculator without supporting working.

f The finite region bounded by the coordinate axes, the curve and the line with equation $x = 3$ is R. Show that the line OP divides R into two regions with areas in the ratio $(2e^3 - 9) : 5$.
See Trigonometry I, section 3.5, p.123 and Integration I, section 9.2, p.367.

(AEB)

10

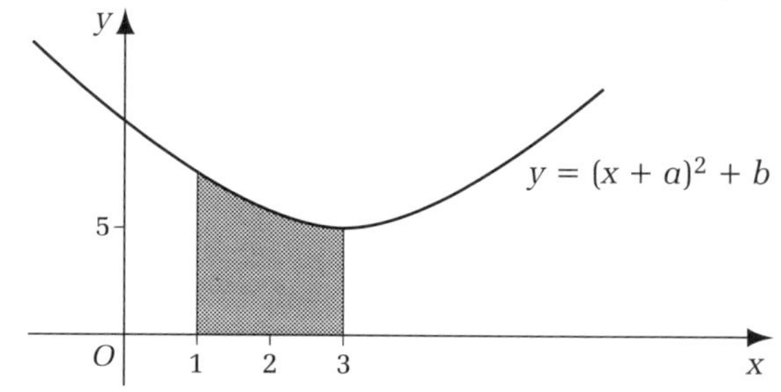

The graph of $y = (x+a)^2 + b$ is sketched above.

a Write down the values of a and b.
See Functions, section 4.4, p.166.

b Use algebraic integration to find the area of the shaded region shown above.
See Integration I, section 9.2, p.367.

(NEAB)

11 The function f is defined by $f: x \to e^x + k, x \in \mathbb{R}$, and k is a positive constant.

a State the range of f.
See Functions, section 4.1, p.140.

b Find $f(\ln k)$, simplifying your answer.
See Exponentials and Logarithms, section 7.2, p.300 and Functions, section 4.3, p.159.

c Find f^{-1}, the inverse function of f, in the form $f^{-1}: x \to \ldots$, stating its domain.
See Functions, section 4.2, p.147.

d On the same axes, sketch the curves with equations $y = f(x)$, and $y = f^{-1}(x)$ giving the coordinates of all points where the graphs cut the axes.
See Functions, section 4.2, p.147.

(ULEAC)

12 The line $3x + 4y = 24$ cuts the x-axis at point A and the y-axis at point B. The point C is the mid-point of AB.

a Find the coordinates of A, B and C. Find also the length of AB.
See Coordinate Geometry, section 2.1, p.50 and section 2.2, p.60.

b Find the equation of the circle with centre C which passes through A and B. Show that this circle also passes through the origin.
See Coordinate Geometry, section 2.5, p.77.

c The same points A and B lie on an ellipse with equation $\frac{x^2}{a^2} + \frac{y^2}{b^2} = 1$. Write down the values of a and b.
See Algebra I, section 1.6, p.39.

d The tangent to the circle at B meets the straight line through O and C at the point D. Find the coordinates of D.
See Algebra I, section 1.6, p.39.

e State, with a reason, whether D is inside, outside or on the ellipse.
See Coordinate Geometry, section 2.4, p.70 and Proof and Mathematical Argument, section 15.1, p.668.

(UCLES)

13

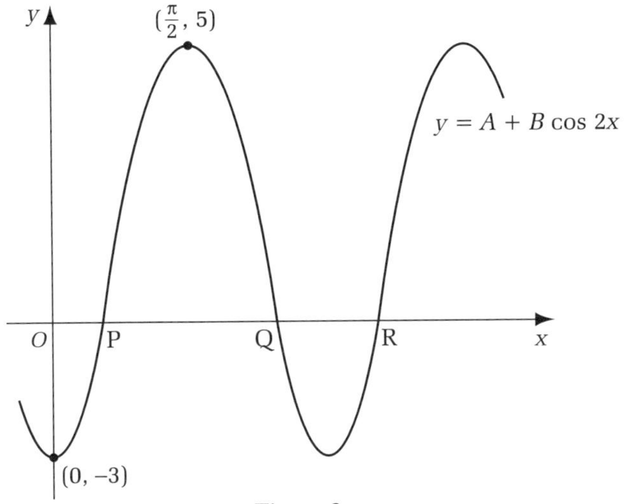

Figure 3

Figure 3 shows a sketch of part of a curve with equation $y = A + B\cos 2x$, where A and B are constants and x is in radians. The curve passes through the points $(0, -3)$ and $(\frac{\pi}{2}, 5)$ and meets the x-axis at the points P, Q and R, as shown.

a Find the values of A and B and hence show that $\frac{A}{B} = -\frac{1}{4}$.
See Functions, section 4.4, p.166 and Trigonometry I, section 3.6, p.127.

b Hence, or otherwise, determine the x-coordinates of P, Q and R, giving your answers to two decimal places.
See Trigonometry I, section 3.1, p.89.

(ULEAC)

14

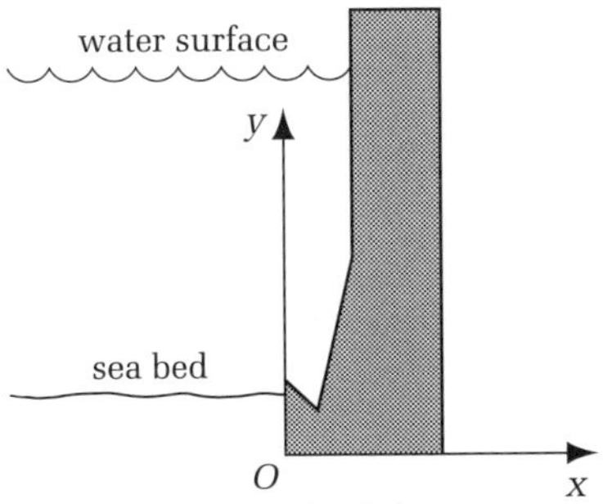

Figure 4

Figure 4 is a sketch of the cross-section of a proposed sea defence wall for a harbour. The detailed drawings of the structure give the following values for x and y, referred to the axes shown. The units are metres.

x	0	1	2	3	4	5	6	7	8
y	4	3	2	7	24	24	24	24	24

a An engineer uses these values to estimate the cross-sectional area, using the trapezium rule. What answer does the engineer obtain?
See Numerical Methods, section 13.2, p.571.

b A mathematician points out that these values are consistent with equations

$$y = x^3 - 3x^2 + x + 4 \quad \text{for} \quad 0 \le x \le 4$$

$$y = 24 \quad \text{for} \quad 4 \le x \le 8$$

and claims that consequently the area can be worked out exactly. What answer does the mathematician obtain?
See Integration I, section 9.2, p.367.

c What percentage error does the mathematician believe that the engineer has made?
GCSE Pre-requisite.

d When the wall is constructed, measurements confirm the values given in the table above. Further measurements give the following extra values.

x	0.5	1.5	2.5	3.5
y	3.5	2.5	4.5	15.5

Giving a reason, state whether the extra values support the engineer or the mathematician.
See Numerical Methods, section 13.2, p.571 and Proof and Mathematical Argument, section 15.1, p.668.

(MEI)

15 **a** Show that $2x^2 - 12x + 13$ may be written in the form $a(x - b)^2 + c$, where the values of a, b and c are to be found.
See Algebra I, section 1.4, p.20 and section 1.5, p.26.

b The curve K has equation $y = 2x^2 - 12x + 13$.

i Sketch the curve K, indicating the coordinates of its turning point A.
See Functions, section 4.4, p.166.

ii Find the equation of the tangent to K which is parallel to the line $y = 8x + 11$ and determine the coordinates of the point where the tangent intersects the tangent to K at A.
See Coordinate Geometry, section 2.2, p.60, section 2.3, p.64 and Differentiation I, section 5.3, p.217.

(WJEC)

16 The function f is defined for all real values of x by $f(x) = (\cos x - \sin x)(17\cos x - 7\sin x)$.

a By first multiplying out the brackets, show that $f(x)$ may be expressed in the form

$$5\cos 2x - 12\sin 2x + k$$

where k is a constant, and state the value of k.
See Algebra I, section 1.3, p.14 and Trigonometry I, section 3.3, p.107.

b Given that $5\cos 2x - 12\sin 2x \equiv R\cos(2x + \alpha)$ where $R > 0$ and $0 < \alpha < \frac{\pi}{2}$, state the value of R and find the value of α in radians to three decimal places.
See Trigonometry I, section 3.6, p.127 and Trigonometry II, section 10.2, p.411.

c Determine the greatest and least values of $\dfrac{39}{f(x) + 14}$ and state a value of x at which the greatest value occurs.
See Trigonometry I, section 3.1, p.89 and Trigonometry II, section 10.2, p.411.

(AEB)

17 **a** Given that $f(x) = 2x^3 - 7x^2 + x + 1$, show that $(2x - 1)$ is a factor of $f(x)$ and find the remaining quadratic factor. Hence find the exact values of the three roots of the equation $f(x) = 0$.
See Algebra I, section 1.5, p.26 and Algebra II, section 6.1, p.256.

b Solve the equation $2\cos^3\theta - 7\cos^2\theta + \cos\theta + 1 = 0$, for $0° \le \theta \le 360°$, giving your answers to the nearest degree.
See Trigonometry I, section 3.1, p.89 and section 3.2, p.99.

(AEB)

18

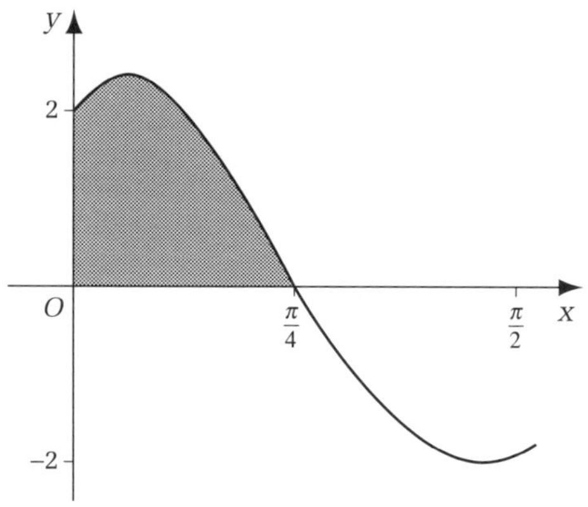

The equation of the curve shown in the diagram is $y = \dfrac{4\cos 2x}{2 - \sin 2x}$, $0 \le x \le \frac{\pi}{2}$.

a **i** Show that $\frac{dy}{dx} = -8$ at the point $\left(\frac{\pi}{4}, 0\right)$.
See Differentiation I, section 5.5, p.236 and Differentiation II, section 11.1, p.433.
ii Find the equation of the normal at the point $\left(\frac{\pi}{4}, 0\right)$.
See Trigonometry I, section 3.1, p.89, section 3.6, p.127 and Differentiation I, section 5.3, p.217.

b Find the area of the shaded region in the diagram in the form $p\ln 2$, where p is an integer to be determined.
See Integration II, section 12.2, p.507.

(NEAB)

19 The function f is defined by $f(x) = \ln(2x + 3), x > -\frac{3}{2}$.

a **i** Find an expression for $f^{-1}(x)$ in the form $ae^x + b$, where a and b are constants to be determined.
See Functions, section 4.2, p.147 and Exponentials and Logarithms, section 7.2, p.300.
ii State the domain of f^{-1}.
See Functions, section 4.1, p.140.
iii Describe geometrically a series of transformations by which the graph of $y = f^{-1}(x)$ can be obtained from the graph of $y = e^x$.
See Functions, section 4.4, p.166.

b **i** Find $f'(x)$.
See Differentiation II, section 11.2, p.438.
ii The gradient of the curve $y = f(x)$ at the point P is equal to the gradient of the line $y = 2x + 3$. Find the coordinates of P.
See Differentiation I, section 5.3, p.217.

c The line through P parallel to the line $y = 2x + 3$ meets the curve $y = f^{-1}(x)$ at two points. Show that the x-coordinate of these points must satisfy the equation $e^x - 4x - 7 = 0$.
See Coordinate Geometry, section 2.3, p.64.

d Show that there is a root of $e^x - 4x - 7$ between $x = 2$ and $x = 3$.
See Numerical Methods, section 13.1, p.552.

(NEAB)

20 **a** Show that $x - 2$ is a factor of $p(x) = x^3 - 4x^2 + x + 6$ and factorise $p(x)$ completely. Hence, or otherwise, solve the inequality $x^3 - 4x^2 + x + 6 < 0$.
See Algebra II, section 6.1, p.256 and Coordinate Geometry, section 2.4, p.70.

b The graph of $y = x^3 - 4x^2 + x + 6$ intersects the line $y = 10 + x$ at the point where $x = \alpha$. Show that α satisfies the equation $x = 4 + \frac{4}{x^2}$ and that α lies between 4 and 5. Use the iterative formula $x_{n+1} = 4 + \frac{4}{x_n^2}$ with $x_1 = 4$ to find x_5, giving your answer to four significant figures.
See Numerical Methods, section 13.1, p.552.

(AEB)

21 **a** **i** Express $2\sin 2\theta \sin\theta$ as a difference of two cosines.
See Trigonometry II, section 10.1, p.404.
ii Express $2\cos 2\theta \sin\theta$ as a difference of two sines.
See Trigonometry II, section 10.1, p.404.

b The parametric equations of a curve are given by

$$x = 3\cos\theta - \cos 3\theta \quad \text{and} \quad y = 3\sin\theta - \sin 3\theta.$$

i Show that $\dfrac{dy}{dx} = \tan 2\theta$.
See Differentiation II, section 11.6, p.472.
ii Find the equation of the tangent to the curve at the point where $\theta = \frac{\pi}{6}$ in the form $y = x\sqrt{a} + b$, where a and b are integers to be determined.
See Differentiation II, section 11.6, p.472.

(NEAB)

22 **a** Use the Maclaurin series to find a quadratic approximation to $f(x) = \ln(1 + x)$.
See Differentiation II, section 11.7, p.481.

b Use your answer to **a** to estimate the value of $\int_0^{0.5} \ln(1 + x)\, dx$.
See Integration I, section 9.2, p.367.

c Sketch the function $f(x) = \ln(1 + x)$ and its quadratic approximation on the same set of axes. Comment on the accuracy of the approximation.
See Algebra II, section 6.4, p.281.

d Use integration by parts to obtain the exact value of $\int_0^{0.5} \ln(1 + x)\, dx$.
See Integration II, section 12.3, p.515.

23 Show that $x + 3$ is a factor of $p(x) = 2x^3 - 9x^2 - 38x + 21$ and hence express $p(x)$ as a product of three linear factors with integer cofficients.
See Algebra I, section 1.4, p.20 and Algebra II, section 6.1, p.256.

a Solve the equation $2e^{3y} - 9e^{2y} - 38e^y + 21 = 0$, leaving your answers as natural logarithms.
See Exponentials and Logarithms, section 7.4, p.309.

b Determine all the solutions in radians in the interval $0 < \theta < \pi$ of the equation $2\sec^3\theta - 9\sec^2\theta - 38\sec\theta + 21 = 0$, giving your answers correct to three significant figures.
See Trigonometry I, section 3.1, p.89 and section 3.6, p.127.

(AEB)

24 The line l_1 has vector equation $\begin{pmatrix} x \\ y \\ z \end{pmatrix} = \begin{pmatrix} 1 \\ 1 \\ 7 \end{pmatrix} + \lambda \begin{pmatrix} 2 \\ -1 \\ 2 \end{pmatrix}$.

A second line l_2 passes through the points (0, 3, 8) and (−3, 0 −1).

a Find the vector equation of the line l_2.
See Vectors, section 14.2, p.600 and section 14.4, p.626.

b Show that these two lines intersect and find the coordinates of the point of intersection.
See Vectors, section 14.4, p.626.

c Find the angle between the two lines.
See Vectors, section 14.3, p.614 and Trigonometry I, section 3.1, p.89, section 3.6, p.127.

25 **a** Show that the first three terms in the expansion of $\sqrt{1 - x}$ are $1 - \frac{1}{2}x - \frac{1}{8}x^2$ when $|x| < 1$.
See Sequences and Series, section 8.5, p.346.

b Use the result in **a** to estimate the value of $\int_0^{\frac{1}{2}} \sqrt{1 - x}\, dx$.
See Integration I, section 9.2, p.367.

c Use the substitution $u = 1 - x$ to find the exact value of $\int_0^{\frac{1}{2}} \sqrt{1 - x}\, dx$.
See Integration II, section 12.4, p.521.

Question Bank Answers

1
a $A = \frac{1}{2}, B = 2, C = -1$
b $\frac{1}{x-1} + 2\ln|x-1| + \frac{1}{2}\ln|x| + c$

2
a $x = 4, y = 20$
c $\frac{62}{5} + 96\ln 2$

3
a $A = -3, B = 3;\ 3\ln|x-2| - 3\ln|x-1| + c$
b ii $\frac{\pi}{4}(e^2 - 1)$ cubic units

4
a $\frac{dy}{dx} = -2x(2x^2 + 3x - 3)$
b $y = -4x + 4$
d 8.1 square units

5
a $A = \frac{1}{2}r^2(\theta - \sin\theta)$
c $\theta_2 = 2.3165, \theta_3 = 2.3054, \theta_4 = 2.3129,$
$\theta_5 = 2.3079 \equiv 132^\circ$ (nearest degree)

6
a $y = \frac{2}{3}x + \frac{1}{3}$
b (8, 4)

7
a $t = \frac{1}{x} - 1$
b $\frac{dy}{dt} = \frac{2t}{(1+t)^2(1-t)^2}, \quad \frac{dx}{dt} = \frac{-1}{(1+t)^2}$
c $y = \frac{1}{1-t}$

8
a i $15\cos(x - 0.644)$
ii 0.277
b ii 0.271

9
a $\frac{dy}{dx} = -(x+1)e^{-x}$; at P, $\frac{dy}{dx} = -4e^{-3}$
b $y = -4e^{-3}x + 17e^{-3}; x = \frac{17}{4}$
c $(-1, e)$
d (0, 2)
e $3 - \frac{6}{e^3}$

10
a $a = -3, b = 5$
b $12\frac{2}{3}$ square units

11
a $f(x) > k$
b $f(\ln k) = 2k$
c $f^{-1}: x \to \ln(x - k)$, domain is $x \in \mathbb{R}, x > k$
d

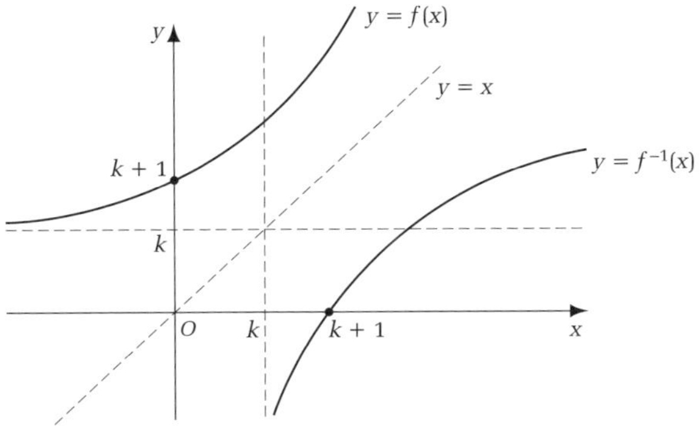

12
a A (8, 0), B (0, 6), C (4, 3); AB = 10 units
b $(x-4)^2 + (y-3)^2 = 5^2$, or $x^2 + y^2 - 8x - 6y = 0$
c $a = 8, b = 6$
d $\left(-10\frac{2}{7}, -7\frac{5}{7}\right)$
e outside

13
a $A = 1, B = -4$
b at P, $x = 0.66$; at Q, $x = 2.48$, at R, $x = 3.80$

14
a 122 m^2
b 120 m^2
c 1.67%
d The engineer – extra measurements are exactly in the middle of the original measurements, consistent with the use of trapezia. Values do not work in the mathematician's formula for $0 \le x \le 4$.

15
a $a = 2, b = 3, c = -5$
b i

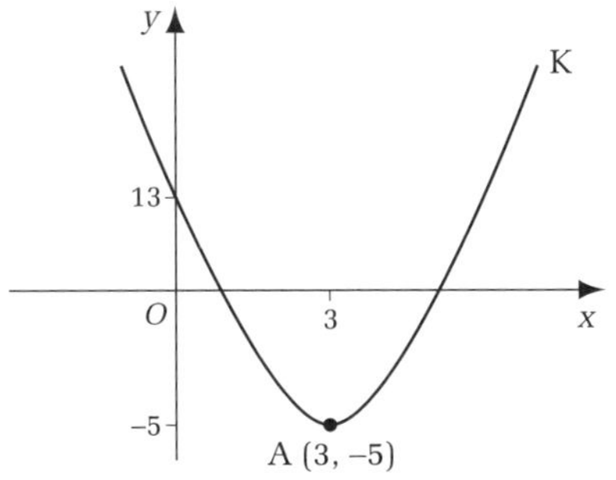

ii $y = 8x - 37; (4, -5)$

16
a $k = 12$
b $R = 13, \alpha = 1.176$
c greatest = 3, least = 1, $x = 0.983$

17
a $x^2 - 3x - 1$; roots are $\frac{1}{2}, \frac{3}{2} \pm \frac{\sqrt{13}}{2}$
b $\theta = 60^\circ, 108^\circ, 252^\circ, 300^\circ$

18
a ii $4x - 32y - \pi = 0$
b 2 ln 2

19
a i $f^{-1}(x) = \frac{1}{2}e^x - \frac{3}{2}$
ii domain of f^{-1} is $x \in \mathbb{R}$
iii translation of -3 units parallel to the y-axis followed by a stretch of factor $\frac{1}{2}$ parallel to the y-axis. Or, a stretch of factor $\frac{1}{2}$ parallel to the y-axis followed by a translation of $-\frac{3}{2}$ units parallel to the y-axis.
b i $f'(x) = \frac{2}{2x+3}$
ii $(-1, 0)$

20
a $(x-2)(x-3)(x+1); x < -1$ and $2 < x < 3$
b $x_5 = 4.224$

21
a i $\cos\theta - \cos 3\theta$
ii $\sin 3\theta - \sin\theta$
b ii $y = \sqrt{3}x - 4$

22
a $\ln(1 + x) \approx x - \frac{1}{2}x^2$
b $\frac{5}{48}$
c good approximation
d $\frac{3}{2}\ln\left(\frac{3}{2}\right) - \frac{1}{2}$

23
$p(x) = (x+3)(2x-1)(x-7)$
a $y = \ln 7, y = \ln\left(\frac{1}{2}\right)$
b $\theta = 1.43, \theta = 1.91$

24
a $\begin{pmatrix} x \\ y \\ z \end{pmatrix} = \begin{pmatrix} 0 \\ 3 \\ 8 \end{pmatrix} + t\begin{pmatrix} 1 \\ 1 \\ 3 \end{pmatrix}$
b $(-1, 2, 5)$
c 0.790 rad or 45.3°

25
b $\frac{83}{192} \approx 0.432$
c $\frac{2}{3} - \frac{1}{6}\sqrt{2}$